I0885913

Climate Urbanism and Vulnerability

This book delves into the intricate relationship between climate change, urbanisation, political attitudes, and spatial inequities, examining how these elements collectively shape vulnerability in cities. This comprehensive book, structured across ten chapters and three core themes (public perception, urban exposure, and methodological innovation) presents an integrated framework that unifies social science, spatial analysis, and data science. Unlike narrower approaches, it bridges the gap between technical metrics and social responses to urban climate challenges.

The book explores how political worldviews influence climate risk perception and how peri-urbanisation intensifies vulnerability, particularly focusing on the disproportionate impact of heat in rapidly expanding cities. It reveals that while public awareness of climate change is present, a significant gap exists between awareness and individual action, with calls for more effective government responses and integrated climate strategies. It also highlights how advanced technologies like AI, machine learning, and remote sensing are transforming the assessment and visualisation of urban climate risks. Specifically, it demonstrates how remote sensing quantifies the intensification of climate change due to urban greenfield expansion, revealing ecosystem loss and increased emissions, especially in peri-urban areas. This underscores the need for policies prioritising ecosystem protection for climate mitigation. Furthermore, the book provides a bibliometric review of urban heat vulnerability research, showcasing its rapid growth and evolution across diverse fields such as urban planning, environmental dynamics, public health, climatology, and remote sensing. It notes a crucial shift towards human-centred approaches and advocates for integrated policies to effectively manage escalating urban heat risks.

This book is an indispensable resource for practitioners in the field, including city government officials, policymakers, urban planners, and consultants. It can also be used as a reference for researchers and university students in urban studies.

Tan Yigitcanlar is a leading Australian scholar recognised internationally for contributions to urban studies, sustainability, technology, and planning. He is a Professor at Queensland University of Technology in Australia and serves as a Distinguished Visiting Professor at the University of Johannesburg in South Africa.

Climate Urbanism and Vulnerability advances an integrated systems framework linking public climate perceptions with urban growth, peri-urban transformation, and spatially explicit vulnerability mapping. By embedding social theory within geospatial analytics, remote sensing, and AI-enabled methods, the book provides a rigorous foundation for analysing and managing climate risk in a rapidly urbanising world.
Professor Xinyue Ye, *University of Alabama, US*

Addressing rapidly escalating global climate risks, this book offers a clear lens on urban vulnerability, a gateway to interdisciplinary understanding of complex urban systems, and a roadmap for intelligent innovations in future urban governance. Tan Yigitcanlar delivers a compelling vision for next-generation cities aligned with the UN Sustainable Development Goals.
Professor Zhen Wang, *Chief Technical Advisor, UNDP*

An outstanding and deeply integrative work, *Climate Urbanism and Vulnerability* redefines how urban climate vulnerability is understood by uniting public perception, political context, and cutting-edge AI-based spatial analysis. It offers rare conceptual clarity and practical insight, making it an essential reference for advancing equitable, evidence-driven responses to urban heat in a warming world.
Professor Zhaohui Lin, *Chinese Academy of Sciences, China*

Climate Urbanism and Vulnerability

Public Perceptions, Urban Risks, and Mapping Inequities in a Changing Climate

Tan Yigitcanlar

CRC Press
Taylor & Francis Group
Boca Raton London New York

CRC Press is an imprint of the
Taylor & Francis Group, an **informa** business

Designed cover image: Shutterstock

First edition published 2027
by CRC Press
2385 NW Executive Center Drive, Suite 320, Boca Raton FL 33431

and by CRC Press
4 Park Square, Milton Park, Abingdon, Oxon, OX14 4RN

CRC Press is an imprint of Taylor & Francis Group, LLC

ISBN: 978-1-041-23427-2 (hbk)
ISBN: 978-1-041-23234-6 (pbk)
ISBN: 978-1-003-73704-9 (ebk)

DOI: 10.1201/9781003737049

Typeset in Palatino
by KnowledgeWorks Global Ltd.

This book is dedicated to my daughters, Ela and Selin, whose future will be shaped by a changing climate and the cities we design today. Your presence is a constant reminder that climate knowledge must translate into action, equity, and care for generations yet to come.

Contents

Author

Tan Yigitcanlar is a leading Australian scholar internationally recognised for his contributions to urban studies, sustainability, technology, and planning. He is a Professor in the School of Architecture and Built Environment at Queensland University of Technology (QUT), Brisbane, Australia, Distinguished Visiting Professor in the Department of Urban and Regional Planning at the University of Johannesburg, Johannesburg, South Africa, and an Adjunct Professor in the Department of Urban and Regional Planning at the Sepuluh Nopember Institute of Technology, Surabaya, Indonesia, while also holding several senior leadership roles. At QUT, he leads the Urban AI Hub, and the City 4.0 Lab, and he also serves as a distinguished member of the Australian Research Council College of Experts.

With a career spanning over three decades, Tan's work has encompassed research, teaching, training, and capacity building at prestigious universities in Australia, Brazil, Finland, Indonesia, Japan, South Africa, and Turkey. His research is dedicated to addressing contemporary challenges in urban planning and development, covering a broad spectrum of economic, societal, spatial, governance, and technology-related issues. At the heart of his work lies a strong commitment to promoting smart and sustainable urbanisation.

Tan has provided research consultancy services to all levels of government, international corporations, and non-governmental organisations both in Australia and internationally. His expertise has been instrumental in helping these entities formulate forward-looking strategies, build resilience, and respond effectively to emerging disruptive challenges. He serves as the Lead Editor-in-Chief of the Elsevier Smart Cities Book Series and holds senior editorial appointments with 12 high-impact academic journals. Tan has held senior leadership roles in more than 40 major international conferences, including serving as chair for 12 of them. He has delivered more than 90 keynote and invited addresses at prominent academic and industry events.

Tan's research findings have been extensively disseminated, with over 360 articles published in high-impact journals and 36 key reference books published by esteemed international publishing houses. His work has significantly influenced urban policy, practice, and research internationally, with an h-index of over 110. He is ranked #1 in Australia and top-10 worldwide in urban and regional planning according to the 2025 Science-wide Author Databases of Standardised Citation Indicators. He was also recognised as an 'Australian Research Superstar' in the Social Sciences Category by The Australian's 2020 Research Special Report and named Australia's Social Sciences 'Research Field Leader' for Urban Studies and Planning in the 2024 edition of The Australian Research Magazine. He is acknowledged as one of the top 1% of scientists worldwide by Clarivate in 2023.

Preface

Cities have always been shaped by the forces of climate, technology, and human decision-making, yet the unprecedented pace of environmental change in the twenty-first century has placed urban areas at the forefront of global climate vulnerability. Today, more than half of the world's population lives in cities, and this proportion continues to rise. Urbanisation (especially when rapid, unregulated, or spatially uneven) creates new forms of exposure to climate risks while amplifying long-standing social and spatial inequities. Extreme heat, flooding, declining ecosystem services, and environmental degradation increasingly define the lived reality of urban dwellers. In many places, the people most affected are those who have the least capacity to respond. Understanding how these patterns emerge, how they are experienced, and how they may be mitigated is central to the ambition of *Climate Urbanism and Vulnerability: Public Perceptions, Urban Risks, and Mapping Inequities in a Changing Climate.*

This book, *Climate Urbanism and Vulnerability: Public Perceptions, Urban Risks, and Mapping Inequities in a Changing Climate,* is motivated by both scholarly gaps and pressing real-world challenges. On the one hand, urban climate vulnerability has been studied extensively, but often in disciplinary silos—some focusing on technical assessments, others on behavioural responses, and yet others on land use and planning frameworks. On the other hand, communities, planners, and policymakers are struggling to integrate these disparate perspectives into coherent strategies that are scientifically rigorous, socially inclusive, and contextually meaningful. The aim of this book is to bridge these divides by offering a multidimensional approach that connects public perceptions, urbanisation processes, and advanced analytical tools for assessing and visualising climate risk.

I first became interested in the intersection of climate change and urban development in the early stages of my academic career, while working with local governments seeking to manage growth pressures in environmentally sensitive regions. The recurring questions raised in policy workshops and community forums: Why do some groups perceive climate risk differently? Why does vulnerability cluster in certain neighbourhoods? Why do cities struggle to implement equitable adaptation strategies? These questions highlighted the importance of integrating social, political, and spatial perspectives into climate research. Over the years, as I collaborated with colleagues in urban planning, data science, and environmental studies, it became clear that a comprehensive and interdisciplinary lens was needed to understand the drivers of urban climate vulnerability. This book reflects that intellectual journey and represents the synthesis of more than a decade of research, fieldwork, community engagement, and methodological experimentation.

Climate urbanism, the conceptual anchor of this book, refers to the ways in which climate change reshapes urban spaces, governance, and lived experiences, while also exposing the tensions inherent in competing visions of urban futures. It highlights that climate risk is not merely a matter of biophysical exposure but is fundamentally embedded in social structures, political orientations, planning decisions, and historical inequities. To fully understand vulnerability in cities, therefore, we must examine not only where risk occurs, but why, and also for whom. This requires a nuanced understanding of public attitudes, urban form, spatial dynamics, and technological tools. It also demands recognition of the socio-political forces that shape climate governance and influence community responses.

Against this backdrop, the chapters in this book collectively address three critical dimensions of climate urbanism: **Public perception**, **urban exposure**, and **methodological innovation**. These themes are not isolated strands but interconnected domains that together shape how vulnerability is constructed, understood, measured, and ultimately acted upon.

Positioning the Book within Existing Literature

While numerous works have addressed urban climate adaptation, vulnerability assessment, or public perceptions of climate change individually, relatively few examine how these domains converge. Existing studies on climate perceptions often focus on national-level trends or political ideology but rarely situate these attitudes within the spatial and developmental context of rapidly urbanising regions. Similarly, many technical assessments of climate risk utilise advanced modelling or geospatial analytics yet lack integration with social science insights or behavioural responses.

This book seeks to fill these gaps by explicitly connecting public attitudes to physical patterns of urban growth and translating these insights into spatially explicit models of vulnerability. It provides a holistic framework that moves beyond dichotomies between social versus technical approaches, advocating instead for a systems-oriented understanding of climate risk in cities.

Another distinctive contribution of this book lies in its emphasis on **peri-urbanisation**, an often-overlooked frontier where climate risk, land use change, and governance challenges intersect most acutely. Peri-urban areas, those transitional zones between urban cores and rural hinterlands, are frequently characterised by fragmented governance, weak planning controls, and rapid demographic change. Their transformation has profound implications for localised climate impacts, yet public understanding of these dynamics remains limited. Several chapters in this book address this gap directly.

Finally, the volume introduces new tools and frameworks that harness the power of **artificial intelligence, machine learning, remote sensing, and advanced spatial analysis** to map climate vulnerability with far greater precision and equity-awareness than previously possible. These analytic innovations are critical in a world where climate risks are evolving faster than traditional planning approaches can accommodate.

Theoretical Foundations and Intellectual Orientation

The theoretical underpinnings of this book draw from multiple traditions: Climate psychology, urban political ecology, land use planning, geospatial science, and environmental governance. Together, these perspectives illuminate different facets of climate vulnerability.

From **climate psychology**, the book draws on research examining how political identity, values, and worldviews shape risk perception. While climate change is a scientific reality, the interpretation of that reality is mediated by social and political filters. Chapters in Part I explore these nuances in depth, showing how political orientation influences not only climate belief but also attitudes towards urban development and adaptation strategies.

From **urban political ecology**, the book adopts the view that climate vulnerability is inherently shaped by power relations, resource distribution, and historical patterns of inequity. Vulnerability is not simply a biophysical condition but a socio-spatial construction. The spatial analyses in Part II illustrate how governance failures, land use pressures, and social marginalisation converge to intensify climate risks.

From **planning theory and practice**, the book draws on frameworks related to growth management, resilience, and sustainable urban form. The chapters emphasise the importance of coordinated land use policies, peri-urban governance, and the protection of ecosystem services as essential components of climate mitigation and adaptation.

From **data science and geospatial analytics**, the book incorporates cutting-edge approaches (such as artificial intelligence and machine learning–based vulnerability modelling, remote sensing for land cover change detection, and fine-grained spatial inequality metrics) that enhance both the conceptual and practical understanding of urban climate risk. Part III offers methodological guidance that can assist researchers, practitioners, and local governments in adopting these tools.

By weaving these threads together, the book situates climate vulnerability as both a social and spatial phenomenon, one that requires integrated frameworks and cross-disciplinary collaboration to address effectively.

Organisation of the Book

The structure of the book reflects the three central themes and is designed to guide readers from foundational concepts to applied methods and practical insights.

Part I: Understanding Climate Perceptions and Urbanisation

Part I examines how political orientation, demographic factors, environmental behaviours, and local contexts shape perceptions of climate risk, particularly in peri-urban regions experiencing rapid change.

- **Chapter 1** explores how political views intersect with perceptions of climate risks associated with urbanisation. Drawing on survey data, it highlights the nuanced but important role of political attitudes in shaping public concern.
- **Chapter 2** expands on this by analysing political bias in climate change belief, identifying attitudinal clusters and mapping the challenges such polarisation poses for equitable climate governance.
- **Chapter 3** examines public understanding of peri-urban climate risks, revealing gaps between awareness and action and highlighting the need for more effective community engagement.

Part II: Localised Climate Risk and Urban Exposure

Part II focuses on the spatial dimensions of climate vulnerability, showing how urban development patterns, land use changes, and governance failures shape exposure and risk.

- **Chapter 4** analyses peri-urbanisation and its contribution to intensified vulnerability, illustrating how informal or unregulated expansion increases exposure to climate hazards.
- **Chapter 5** examines how land use change, particularly the loss of green spaces, contributes to localised anthropogenic climate impacts and the erosion of ecosystem services.

Part III: Methods and Mapping Urban Heat Vulnerability

Part III advances the methodological frontier of urban climate vulnerability assessment.

- **Chapter 6** provides a systematic review of existing heat vulnerability assessment methods, identifying strengths, limitations, and gaps.
- **Chapter 7** contextualises heat vulnerability research through a bibliometric analysis spanning five decades, highlighting emerging trends and interdisciplinary shifts.
- **Chapter 8** introduces a new conceptual framework that integrates machine learning and remote sensing to improve the accuracy and responsiveness of vulnerability assessments.
- **Chapter 9** presents a validated, equity-sensitive heat vulnerability index applied across Australian capital cities, demonstrating spatial inequalities in exposure and adaptive capacity.
- **Chapter 10** builds on this by showcasing how machine learning models, especially random forest, can predict vulnerability at fine spatial scales, revealing demographic and health factors as primary drivers.

Together, these chapters provide a comprehensive, theoretically grounded, and methodologically innovative understanding of urban climate vulnerability.

Rationale and Scope

The scope of this book is intentionally broad yet analytically focused. It draws on case studies from different urban contexts, with particular emphasis on rapidly growing cities and peri-urban regions. The decision to centre the analysis on heat vulnerability reflects its growing importance as one of the most lethal yet under-recognised climate threats. Heatwaves are silent disasters (causing more fatalities globally than any other climate-related hazard) but they often lack the visual and political immediacy of floods, fires, or storms. By foregrounding heat vulnerability, this book contributes to a more comprehensive and inclusive understanding of climate risks.

The chapters emphasise **inequity** as a central theme. Vulnerability is not evenly distributed; it follows social gradients and is shaped by gender, age, income, ethnicity, and geographic location. Understanding these inequities is fundamental for designing inclusive adaptation strategies. The book also underscores the importance of **methodological rigour** in vulnerability assessment. Poorly designed indices, inconsistent indicator selection, and inadequate validation have often led to misleading or incomplete insights. By presenting new frameworks grounded in evidence-based methods, this book contributes to improving both scientific and practical approaches.

Finally, the book highlights the role of **public attitudes and governance** in shaping climate resilience. Even the most sophisticated models and analyses cannot lead to meaningful change without political will, community participation, and institutional capacity. By exploring how people interpret climate risks and how political orientation influences these interpretations, the book offers insights that can support more inclusive, democratic, and effective climate governance.

Intended Audience

This book is written for a diverse readership.

- **Urban planners and policymakers** will find practical insights into land use management, heat adaptation strategies, and equity-oriented climate planning.
- **Researchers** in climate science, geography, urban studies, and environmental psychology will benefit from the integrated conceptual frameworks and novel methodological contributions.
- **Students and educators** will find the book useful as a comprehensive introduction to interdisciplinary approaches to climate vulnerability.
- **Data scientists and geographic information system specialists** will gain exposure to new ways of applying machine learning and remote sensing to climate risk assessment.
- **Community organisations and practitioners** may draw on the findings to support more participatory and socially aware climate adaptation initiatives.

Acknowledgements

The ideas and analyses presented in this book were made possible through the collaboration of colleagues, students, and research partners over many years. Numerous local governments generously shared data, insights, and policy experiences that enriched the empirical chapters. I am also deeply grateful to the interdisciplinary networks that fostered many of the conversations that informed this work; from urban planners and climate scientists to geospatial analysts and artificial intelligence and machine learning experts. Their expertise and perspectives helped shape the book's direction and strengthen its contributions. I also acknowledge the students and early-career researchers (particularly from the QUT Urban AI Hub) whose energy, creativity, and intellectual curiosity inspired many of the methodological innovations presented in the latter chapters. Their commitment to exploring complex urban problems through new tools and technologies is a reminder of the transformative potential of the next generation of scholars.

Final Reflections

Climate change is reshaping cities in profound and uneven ways. The risks are growing, but so too are the opportunities to rethink how we plan, govern, and inhabit urban spaces. This book approaches climate vulnerability not as an inevitable outcome but as a challenge that can be addressed through thoughtful policy, innovative technology, and inclusive public engagement. By integrating public perceptions, spatial analysis, and advanced geospatial methods, the book offers a comprehensive and actionable framework for understanding and addressing climate risk in a world where urbanisation and environmental change are deeply intertwined.

It is my hope that the insights presented in *Climate Urbanism and Vulnerability: Public Perceptions, Urban Risks, and Mapping Inequities in a Changing Climate* will empower cities, communities, and researchers to navigate climate risks with greater clarity and purpose. By illuminating the social, spatial, and technological dimensions of vulnerability, this book seeks to inspire more deliberate and justice-oriented approaches to urban adaptation. Finally, I hope it contributes, however modestly, to the collective effort of shaping urban futures that are more resilient, inclusive, and responsive to the profound climate challenges of our time.

Foreword

We stand at a critical juncture in human history, where abstract climate forecasts have crystallised into lived, everyday realities. The climate crisis has migrated from distant ice caps and future projections to the very streets we walk, the homes we inhabit, and the communities to which we belong. It is a crisis defined not only by rising temperatures and sea levels but by a profound and accelerating inequality in who bears its burdens and who is afforded protection. The question is no longer whether our cities will be transformed, but how and, crucially, for whom. Within this urgent context, *Climate Urbanism and Vulnerability: Public Perceptions, Urban Risks, and Mapping Inequities in a Changing Climate* emerges not merely as a timely academic contribution but as an essential guide and a compelling manifesto for action.

The discourse on cities and climate change has undergone a fundamental transformation over recent decades. Once dominated by biophysical, hazard-centric models, it has evolved towards an understanding of urban areas as complex social-ecological systems. This paradigm is inherently political and justice-oriented, recognising that vulnerability is neither natural nor accidental. Rather, it is produced through historical and ongoing socio-political and economic processes, including entrenched inequality, governance failures, discriminatory planning practices, and colonial legacies. Addressing urban climate risk, therefore, demands far more than technical fixes or predictive models. It requires a deep understanding of how communities themselves perceive risk and vulnerability, and how those perceptions are entangled with urban development, power, and justice. Yet, these crucial conversations too often remain siloed across disciplines and sectors. What is urgently needed is a cohesive, interdisciplinary approach that treats the city not as a passive recipient of climate impacts, but as an active, political organism in which risk is systematically produced and distributed.

In this book, Professor Tan Yigitcanlar moves decisively beyond viewing vulnerability as a simple geographic coincidence of hazard and exposure. Instead, he frames it as a constructed condition, the cumulative outcome of social, political, and spatial decisions made over time: Where infrastructure investment is prioritised, where industrial activities are permitted, and where green space is preserved or sacrificed. Vulnerability, as this book powerfully demonstrates, is shaped by power relations, resource allocation, and the perceptual filters through which different communities experience and respond to threat. Through careful empirical analysis and conceptual clarity, the book reveals that today's maps of climate risk are, in many places, palimpsests of yesterday's injustices.

What makes this book both distinctive and necessary is its explicitly multidimensional approach. It brings together public perception, urbanisation processes, and advanced analytical tools for assessing climate risk, arguing persuasively that only such an integrated perspective can capture the full complexity of climate change as it is experienced in urban and peri-urban contexts.

Firstly, the book places public perception at the centre of climate enquiry, recognising that how people understand and experience risk is as consequential as any engineered metric or model output. In an era increasingly dominated by big data and sophisticated simulations, lived experience is too often dismissed as anecdotal or subjective. This work corrects that oversight with scholarly rigour. It demonstrates that the ways in which communities interpret, prioritise, and emotionally process climate risk fundamentally shape resilience, trust, and political will. A flood model, no matter how technically refined, is ineffective if the community it is meant to protect distrusts the institutions behind it. By examining how political identity, place-based context, and worldviews influence perceptions of climate threats, the book grounds the climate crisis in everyday lived reality—positioning perception not as a secondary concern, but as an indispensable form of knowledge and the starting point for meaningful engagement and governance.

Secondly, the book offers rigorous spatial analyses of urban exposure and risk, unpacking how the very processes of urban development actively shape landscapes of vulnerability. It directs particular attention to urbanisation itself as a driver of climate risk, moving beyond iconic global city centres to examine rapidly transforming peri-urban frontiers. In these zones, where governance is fragmented, planning regimes are contested, and ecosystems are under intense pressure, the analysis demonstrates with clarity how unregulated expansion, the erosion of natural buffers, and inadequate infrastructure generate new and compounding hazards. This leads to a sobering realisation: Much contemporary urban development is not merely failing to mitigate climate change but is actively amplifying its risks. The work presented in Part II thus shifts the analytical lens from identifying hazardous places to exposing the hazardous processes embedded within prevailing models of urban growth.

Thirdly, and perhaps most innovatively, the book pushes the frontier of mapping inequity. It harnesses artificial intelligence, machine learning, and advanced geospatial analytics not simply to see vulnerability more precisely but to see it more justly. Here, the book's transformative potential becomes most tangible. Cutting-edge technologies are mobilised to ask fundamentally ethical questions, turning maps from static representations of risk into dynamic instruments for advocacy, accountability, and targeted intervention. In doing so, the book reclaims technical tools as instruments of social justice rather than neutral artefacts of analysis.

I was particularly struck by the book's courageous and sustained focus on heat vulnerability. In an era dominated by the dramatic imagery of floods, fires, and storms, the silent and pervasive disaster of extreme heat remains dangerously underestimated. Yet, as Professor Yigitcanlar makes clear, heat is among the deadliest climate-related hazards we face. It is a stealthy and deeply inequitable risk, concentrated in substandard housing, disproportionately affecting older adults and socially isolated populations, and intensified by the urban heat island effect, most severe in neighbourhoods systematically deprived of green space. By centring this often-overlooked hazard and dissecting its social and spatial anatomy with methodological precision, the author performs a vital scholarly and moral service. The equity-sensitive heat vulnerability index developed for Australian cities is more than an academic contribution; it is a tool of moral accounting. It exposes the hidden contours of risk, revealing which communities, and, by extension, which lives, are rendered most vulnerable through the interaction of climate dynamics and policy choices. Importantly, this framework is both rigorous and replicable, offering valuable lessons for cities worldwide confronting this under-prioritised threat.

This book is the product of deep, sustained engagement with communities, data, and the complex realities of governance and planning. Professor Yigitcanlar writes as a scholar-practitioner who has listened attentively in community forums, policy workshops, and collaborative research settings. The volume represents a synthesis of that journey, grounded in empirical evidence yet attentive to lived realities. It offers not only a diagnosis of urban climate vulnerability but a coherent and actionable pathway for response.

As you begin this book, your understanding of urban climate risk will likely be transformed. You will be invited to see the city anew, not as a neutral container for climate impacts, but as a dynamic political space in which risk is allocated, often with striking precision. Streets, neighbourhoods, and skylines emerge as living archives of past and present decisions that will shape survival and suffering in the decades ahead. Throughout this exploration, the author guides the reader with clarity, intellectual integrity, and an unwavering commitment to justice.

This book is far more than a significant academic contribution to urban studies, geography, and climate science. It is a practical and ethical guide for the urgent work of building urban futures that are resilient precisely because they are just. Professor Yigitcanlar offers a formidable gift: The conceptual clarity to see the problem in its full, interconnected complexity, and the innovative analytical tools to begin addressing it. In doing so, he provides not only critique but grounded hope, a conviction that through better knowledge, inclusive engagement, and deliberate action, we can steer our urban trajectories towards greater equity, resilience, and care. For planners, policymakers, scholars, advocates, and all those invested in the future of cities, this book is an indispensable companion and a powerful call to action.

Beyond its immediate scholarly contributions, this book also speaks to a broader moment of reckoning in urban governance and knowledge production. As cities across the globe experiment with digital technologies, smart city agendas, and data-driven decision-making, questions of power, transparency, and accountability have become increasingly urgent. Professor Yigitcanlar's work offers a timely corrective to techno-solutionist narratives that promise efficiency without confronting inequity. By embedding advanced analytics within a justice-oriented framework, the book demonstrates how data and algorithms can either reinforce existing disparities or be consciously designed to expose and challenge them. This insistence on ethical intentionality is one of the book's most important interventions, reminding readers that tools are never neutral, and that the values guiding their use ultimately determine their social consequences.

Equally important is the book's contribution to rethinking the role of local and metropolitan governance in an era of accelerating climate uncertainty. It highlights the limits of top-down, one-size-fits-all policies and underscores the necessity of place-sensitive, participatory approaches that recognise local knowledge as a form of expertise. In doing so, it bridges a critical gap between academic research and policy practice, offering insights that are both analytically robust and operationally relevant. The frameworks and methods presented here are not abstract ideals; they are designed to be taken up, adapted, and implemented by those working within planning departments, local councils, and community organisations.

Ultimately, this book challenges us to confront an uncomfortable truth: Climate vulnerability in cities is rarely accidental. It is planned, normalised, and reproduced through everyday decisions about land use, infrastructure, housing, and service provision. By making these processes visible, *Climate Urbanism and Vulnerability* equips readers with the intellectual tools to question prevailing urban trajectories and to imagine alternatives. It asks us not only to map risk, but to take responsibility for it, and, in doing so, to reaffirm the ethical obligation at the heart of urban planning and climate action: To protect the most vulnerable, to distribute risk more fairly, and to design cities that enable all residents to live with dignity in a rapidly changing climate.

Professor Yan Liu
The Chinese University of Hong Kong
Hong Kong SAR, P. R. China

Part I

Understanding Climate Perceptions and Urbanisation

This part of the book examines how climate change is understood, interpreted, and contested within rapidly urbanising contexts. It explores the social and political foundations of climate perception, showing how worldviews, ideology, and governance shape public understanding of climate risks. Through a focus on political perspectives and bias, the chapters reveal why climate change is experienced unevenly not only in physical terms but also cognitively and culturally. Special attention is given to peri-urban areas, where urban expansion intersects with limited infrastructure and governance gaps, amplifying uncertainty and risk. By unpacking how people perceive and respond to climate threats in these transitional spaces, this part establishes a critical foundation for understanding vulnerability as both a social construct and an urban process.

DOI: 10.1201/9781003737049-1

1

Climate Change, Urbanisation, and Political Perspectives

1.1 Introduction and Background

The impacts of climate change have now become evident across the globe (Case et al., 2021; Katelaris, 2021; Ortiz-Bobea et al., 2021; Reyer et al., 2017). In tackling climate change impacts (e.g., flood, drought, abrupt rainfall, extreme heat, bushfire, sea-level rise, salinity intrusion) and their associated externalities (e.g., loss of shelter and livelihood, increased environmental degradation, rapid population increase, and subsequent urbanisation), the pressure of constant urban growth is one of the accompanying challenges that major global cities have been predominantly confronting (Cobbinah, 2021; Egerer et al., 2021; Galderisi, 2014; Kahn, 2009).

In response to the urban growth pressures, cities have been encroaching on nearby peri-urban landscapes (Mortoja et al., 2020; Yigitcanlar, 2010). Consequently, urbanisation is rapidly taking place worldwide, predominantly in the form of peri-urbanisation. Nevertheless, these peri-urban lands are highly significant for providing biodiversity habitats, preserving ecosystem services, and maintaining environmental quality, including the sources of carbon reserves (Houston, 2005; Mahbub et al., 2011). Hence, they are also critical for mitigating climate change risks (Dizdaroglu & Yigitcanlar, 2014; Sotto et al., 2019).

Thus, the strategy of rapid urbanisation to peri-urban landscapes at the cost of overall environmental decline seems to put the ongoing initiatives of climate risk mitigation in jeopardy (Yigitcanlar & Kamruzzaman, 2014). Besides, such continuous peri-urban growth undermines the socio-political standing of peri-urban dwellers (Xu et al., 2018). For instance, peri-urban dwellers might be misguided by the instant tangible benefits of increased land values, while the long-term consequences of such ongoing peri-urbanisation either are overlooked or not properly comprehended (Foelske et al., 2019; Friedberger, 2000; Merson et al., 2010).

Besides, such an ongoing drive of peri-urbanisation sometimes does not need to face strong opposition from the public, as most of the peri-urban dwellers are not politically so united to uphold their long-term benefits (Zabik & Prytherch, 2013). While planners and policymakers are highly aware of such peri-urbanisation-driven long-term consequences, in most

DOI: 10.1201/9781003737049-2

cases, they take a pragmatic stance due to the market-led speculative forces or political short-termism (Al-Mamun & Kim, 2020). Hence, policymakers and planners, instead, provide a tacit approval to this ongoing drive of peri-urbanisation (Yue et al., 2013). In this way, along with causing continuous deterioration to the long-term scenarios of climate risk mitigation circumstances, justice to the existing peri-urban dwellers is also compromised (Liu et al., 2019).

Hence, to ensure justice as well as to promote balanced development in urban growth areas, the global ongoing drive of peri-urbanisation also should be viewed from the socio-political standpoint. Nonetheless, few studies so far have investigated the association between climate risk perception and political views (Hu et al., 2017; Unsworth & Fielding, 2014), whereas probing people's political preference against their views on climate change risk and urban growth is scarce. Besides, researchers predominantly investigate the phenomenon of climate change risk objectively.

Notwithstanding, subjective perception of climate risk is equally important to an individual's way of cognition and understanding a certain phenomenon, and thereby perceiving forthcoming consequences might be subjective (Van der Linden, 2014, 2015; Weber, 2010; Yang et al., 2014). Thus, an individual's political orientation might exert significant influences on climate risk perception (McCright & Dunlap, 2011a). Besides, someone's socioeconomic status is also linked to climate risk belief (Van der Linden, 2017).

Against this backdrop, investigating the socio-political perspective of peri-urbanisation-driven climate change risks through the lens of subjective interpretation is necessary. As a part of this investigation, the study probes the association between urbanisation, climate change risk, and the political views of the public. The study elicited the perceptions of South East Queensland's (SEQ's) peri-urban dwellers through an online survey. The perceptions are assessed through the respondents' political preferences against their views on the association between climate change and urban growth. The findings generated provide valuable insights to policymakers in pursuing appropriate measures in mitigating climate risks.

1.2 Research Design

1.2.1 Hypothesis Development

In general, liberals in Western societies are often more supportive of ensuring social and climate justice (Janoff-Bulman et al., 2009; Morrison & Miller, 2008), whereas conservatives prefer to promote market-led approaches, defend the prevailing conditions, and are more sceptical of climate change impacts (Antonio & Brulle, 2011; Costa & Kahn, 2013; McCright & Dunlap,

2011b). Accordingly, polarisation on the issue of climate risk based on political standing is already evident in most developed countries. For instance, in the US, it is reported that the liberals (e.g., Democrats) are prone to perceiving more climate risk than those of the conservatives (e.g., Republicans) (Carlton & Jacobson, 2013; Dunlap & McCright, 2008; Leiserowitz, 2005; Marquart-Pyatt et al., 2014). Similar findings are also observed in the context of European Union countries (Gregersen et al., 2020).

Nevertheless, such political polarisation for the context of other developed countries appears to be less explored. For instance, Australia is one of the major contributors to global anthropogenic climate change (Biggs, 2016) due to this country's high dependency on coal-based power plants for domestic electricity production and for being a major global coal exporter as well. Albeit political polarisation is perceived with regard to greenhouse gas emissions, in particular, other aspects of climate risk are hardly prioritised with due attention in the overall political agendas.

For instance, some major political parties in Australia comprise Liberal National Party (LNP), Australian Labour Party (ALP), and Greens. ALP is known to be left or progressive, and LNP is known to be right or conservative. Other political parties include the Katter's Australian Party (right or conservative), One Nation (right or conservative), Independent Party (right or conservative), and few other small ones.

In response to the global urge for greenhouse gas emission reduction, LNP has no intention to phase out coal, whereas ALP acknowledges that coal plants are to be withdrawn but has no plans to phase them out. Contrariwise, Greens want a complete phase-out of coal by 2030. Unlike LNP, Greens are the strident supporters of abolishing the existing Emission Reduction Fund (ERF) and want to rejuvenate the Carbon Farming Initiative (CFI). Through this CFI, along with maintaining agricultural production, farmers will be supported to bring down carbon emissions.

Thus, although Australia's progress on climate policies is evident to some extent that are predominantly centred around reducing greenhouse gas emissions, there is a clear divide between the left-wing and right-wing supporters on the issue of greenhouse gas emission reduction strategies. In this way, this country's efforts to respond to climate change risks are being continuously interrupted.

For instance, Australians predominantly supported climate actions and voted for the Labour Government in 2007, in order to empower policymakers in taking proper climate risk mitigation and adaptation measures (Beeson & McDonald, 2013). However, during the last five years, the Liberal-National Government's accomplishment on the issue of climate risk mitigation seems to be an evident leadership failure of the previous decade (The Climate Council, 2019). On that very point, Australia has been recently ranked as the last out of 54 countries in terms of its ability to cope with the impending climate risk (Nalau & Melville-Rea, 2021). Contrariwise, Australia was one of the few countries to invest millions in climate risk adaptation and established the National Climate Change Adaptation Research Facility in the year 2008. Accordingly, such an increasing political polarisation on the issue of climate risk has significantly hindered the country's burgeoning fund-raising attempts at the federal government level and put ongoing climate risk adaptation in jeopardy.

Besides, Australia is one of the most urbanised nations globally where urban growth predominantly is occurring in peri-urban locations backed up by the resource-based economy (Fisher, 2003; Head et al., 2014; Kijas, 2002). Being an arid continent, Australia has long been exposed to temporarily and spatially variable climatic conditions too (Head et al., 2014). Thus, increased vulnerabilities to climate change risk led by constant urbanisation to climate risk mitigation reservoirs (i.e., peri-urban areas) appear to pose dire consequences (Perveen et al., 2017). Consequently, Australia has increasingly become more vulnerable to climate change impacts; the total cost accrued with climate change-induced disaster events in Australia is USD 18.2 billion/year (Merzian et al., 2019).

While the country's major political parties are torn to gain a consensus on the issue of climate risk, investigating people's perception on the link between climate risk and peri-urbanisation against their political orientation might provide valuable insights to policymakers in finding the most effective ways to scale down the level of existing disagreements. In particular, such an investigation is critical to the countries that are undergoing constant peri-urbanisation, but are biologically diverse, and accommodate most of their climate risk-mitigating elements within peri-urban growth pockets. Based on the discussions as stated above, the study reported in this chapter is going to test the following hypothesis.

Hypothesis: An individual's political orientation towards the left or progressive has a positive role in perceiving increased climate risk. To be more specific, it is more likely that a respondent is believed to be a left-wing supporter (i.e., progressive) than a right-wing (i.e., conservative), if s/he agrees rather than disagrees with the issues concerning increased climate risk perception in general.

This above-mentioned hypothesis was tested through a set of multinomial logistic regression (MLR1–MLR5) (Appendix 1.A1). Firstly, the study tested an individual's political orientation (i.e., left, centre, right, other) against his/her personal traits (i.e., educational, political, moral, and religious standings in mitigating climate risk) (MLR1 of Appendix 1.A1). Nevertheless, other elements characterising an individual's personal traits such as gender, age, and income were not considered as they did not yield a significant model fit for this particular regression analysis.

Secondly, an individual's political orientation against his/her perceptions on the causes and consequences of climate risk were tested (MLR2 of Appendix 1.A1). As anthropogenic greenhouse gas emissions are widely believed to be one of the prime causes of climate risk, human-induced greenhouse gas emissions have been placed as the sole cause of climate risk perception in this particular regression analysis. In the case of the consequences of climate risk, respondents' exposures to climate risk-induced extreme events were considered.

Thirdly, as the study predominantly focused on investigating the link between political orientation and peri-urbanisation-driven climate risk impacts, respondents' political orientation was tested against the prime causes and consequences of peri-urbanisation-driven climate risk impacts (MLR3 of Appendix 1.A1).

Fourthly, respondents' political orientation was tested against a set of mitigation measures that require self-sacrifice at the individual level in mitigating peri-urbanisation-driven climate risk impacts (MLR4 of Appendix 1.A1). In this way, respondents' level of willingness to mitigate peri-urbanisation-led climate risk at the expense of self-sacrifice was tested.

Fifthly, while MLR4, as stated above, was employed to test respondents' level of willingness to self-sacrifice in mitigating peri-urbanisation-driven climate risk impacts, MLR5 (Appendix 1.A1) was used to test respondents' political orientation against their perception on peri-urbanisation-driven climate risk mitigation measures that should be initiated by the government. In this way, the differences in respondents' perceptions on self-sacrificed approaches versus government-initiated approaches in mitigating climate risk were tested based on their political orientation.

Lastly, the study tested climate risk perceptions between urban, peri-urban, and rural dwellers against their political orientation. In doing so, this study carried out an ordinal logistic regression.

1.2.2 Case Study

In order to investigate the question of how the public perceptions differ on the relationship between urban growth and climate change based on their political views, this chapter embraces a case study approach. For disclosing public perception, this study chose the SEQ region of Australia. The SEQ region comprises a total of 12 local government areas, including Brisbane, Gold Coast, Ipswich, Lockyer Valley, Logan, Moreton Bay, Noosa, Redland, Scenic Rim, Toowoomba, Somerset, and Sunshine Coast.

The SEQ region is home to nearly 3.8 million people and is expected to accommodate additional 1.9 million people by 2041 (Perveen et al., 2020). The majority of the South East Queenslanders are the proponents of two major political parties, including ALP and LNP. The SEQ electorates map is shown in Figure 1.1.

The following considerations were taken into account in selecting SEQ as a case study: (a) The politics of climate change in Australia is more global and predominantly centred around the activities of the federal government (Beeson & McDonald, 2013), and addressing climate change impacts does not seem to be a strong political agenda for either the state government or the local government; (b) SEQ generally comprises people with different political

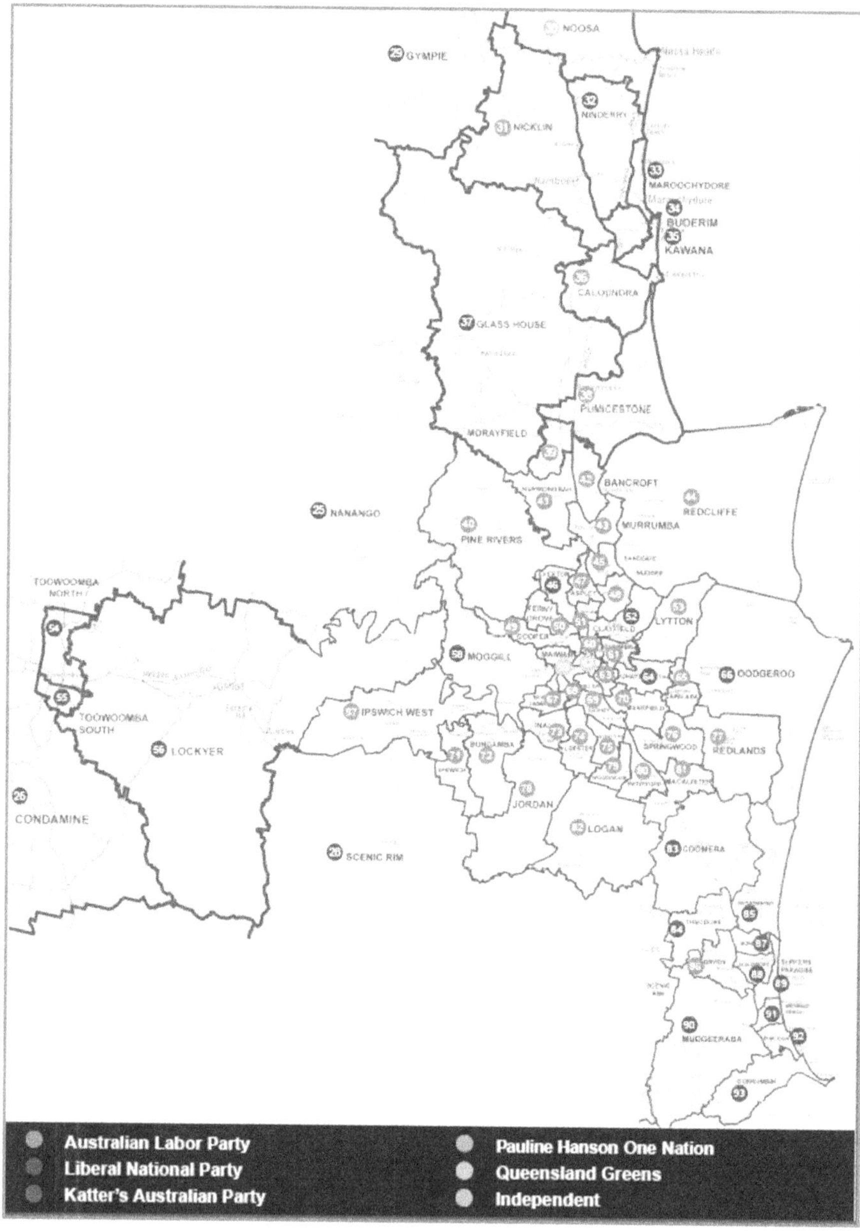

FIGURE 1.1
South East Queensland electorates. (From Queensland Parliament, 2017, p. 2.)

standings (Queensland Parliament, 2017); (c) in SEQ, incidents like extreme heat events have become more frequent and broke all records of previous extreme heat events in the last decade (2010–2020) (Mortoja & Yigitcanlar, 2020); (d) SEQ is predicted to encounter climate change–induced disaster events like flooding, drought, heatwaves, and irregular and extreme rainfall events with an intensified degree in the near future (Kankanamge et al., 2020); (e) SEQ's perpetual urban encroachments to nearby peri-urban agrarian landscapes appear to generate more anthropogenic climate change impacts compared to other Australian cities/regions (Bajracharya & Hastings, 2018; Guhathakurta & Stimson, 2007).

1.3 Survey and Participants

Respondents older than 18 years of age were considered as eligible participants for this study. The Key Survey platform was used to collect responses from the urban, peri-urban, and rural dwellers of SEQ. In order to distribute the survey among the participants, the study first identified urban, peri-urban, and rural areas of SEQ by using Mortoja and Yigitcanlar's (2021a, 2021b) study approach (Figure 1.2). Under their study approach, peri-urbanisation is placed under the microscope of fuzzy linear urban membership function set that ranges from 0 to 1. Afterwards, the values between 0 and 0.10 are categorised as rural, values between 0.11 and 0.79 are named peri-urban, and values between 0.80 and 1.0 are classified as urban (Figure 1.2).

Later on, a simple random survey technique was used to distribute the online survey link among the urban, peri-urban, and rural dwellers. The

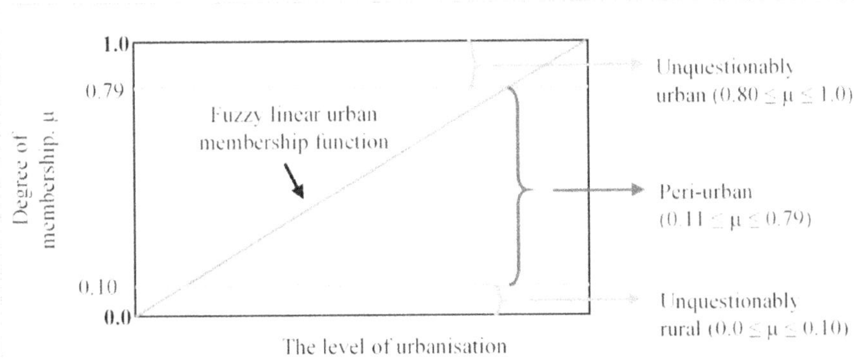

FIGURE 1.2
The mapping approach of urban, peri-urban, and rural areas. (From Mortoja & Yigitcanlar, 2020, p. 10.)

survey was conducted during January 2021. By using the Australian Bureau of Statistics (ABS) sample size calculator (https://www.abs.gov.au/websitedbs/d3310114.nsf/home/sample+size+calculator), the threshold sample size under this study was determined as 385 (estimated resident population of SEQ = 38,00,000; confidence level = 95%). Nevertheless, as most of the respondents were the registered member of a survey company, this study got more responses than expected and finally had been able to collect a total of 659 valid responses (28% response rate).

Afterwards, the surveyed sample was compared with the census data to ensure that the collected samples fairly represent the fraction of study areas' corresponding population (Table 1.1). The spatial distribution of sampling size between rural, peri-urban, and urban areas is mapped in Figure 1.3.

1.4 Analysis and Results

1.4.1 Descriptive Analysis

This study employs descriptive analysis in the SPSS platform. While performing descriptive analysis, the chi-square test was employed in order to ensure that there is a statistically significant association between the chosen categorical variables.

Participant profiles: Respondents' socioeconomic profiles are presented in Table 1.2. The study ensures nearly an equitable distribution of participation between the two most common genders i.e., male (50.8%) and female (48.6%). In terms of age composition, more than four-fifth (82.1%) of the respondents comprise the age range of 25–74. Besides, out of the respondents being surveyed, 85.1% of participants possess a minimum level of Year 12 education. With regard to political views, the majority of the participants adhere to centrism (55.1%), while political adherence to the left and right wings is nearly equal (i.e., left wing = 18.7%; right wing = 19%). This study observes a significant association to political views with gender ($\chi^2 = 13.88$, df = 6; $p = 0.031$), age ($\chi^2 = 53.92$, df = 21; $p = 0.000$), and educational level ($\chi^2 = 45.56$, df = 18; $p = 0.000$). Respondents with relatively lower age group, higher education level, and females are more favourable to the left-wing than those of the right-wing supporters.

Climate change perceptions and political views: The study observes a strong association between political ideology and climate change belief. The majority of the participants who support left wing identify the issue of climate change with the highest importance (Table 1.3). For instance, across the respondents who underestimated the issue of climate change, the left-wing supporters' share is the lowest and right-wing supporters' share is

TABLE 1.1

Distribution of Sample Size with Respect to Census Data

Variables	Category	Sample Size	Survey (%)	Census (%)
Gender	Male	335	50.8	49.40
	Female	320	48.6	50.60
Age	85 and over	8	1.2	1.8
	75–84	65	9.9	4.5
	65–74	96	14.6	8.9
	55–64	97	14.7	11.7
	45–54	104	15.8	13.5
	35–44	119	18.1	13.4
	25–34	125	19	13.7
	18–24	45	6.8	Not classified as per census
Education	Certificate	129	19.6	18.3
	Bachelor	132	20	18.3
	Advanced diploma	97	14.7	8.7
	Year 12	103	15.6	16.5
	Year 11	14	2.1	4.3
	Year 10 or below	84	12.7	20.6
Profession	Agriculture and fisheries employee	1	0.2	Not classified as per census
	Business owner	17	2.6	Not classified as per census
	Clerical or office worker	71	10.8	13.6
	Consultant	10	1.5	Not classified as per census
	Manager, executive, or director	62	9.4	12.1
	Manufacturer	11	1.7	Not classified as per census
	Professional	101	15.3	19.8
	Sales worker	35	5.3	9.7
	Semi-skilled worker	19	2.9	Not classified as per census
	Skilled tradesperson	28	4.2	Not classified as per census
	Student	53	8.0	Not classified as per census
	Technology professional	18	2.7	Not classified as per census
	Unskilled or labourer	33	5.0	Not classified as per census
	Other	200	30.3	Not classified as per census

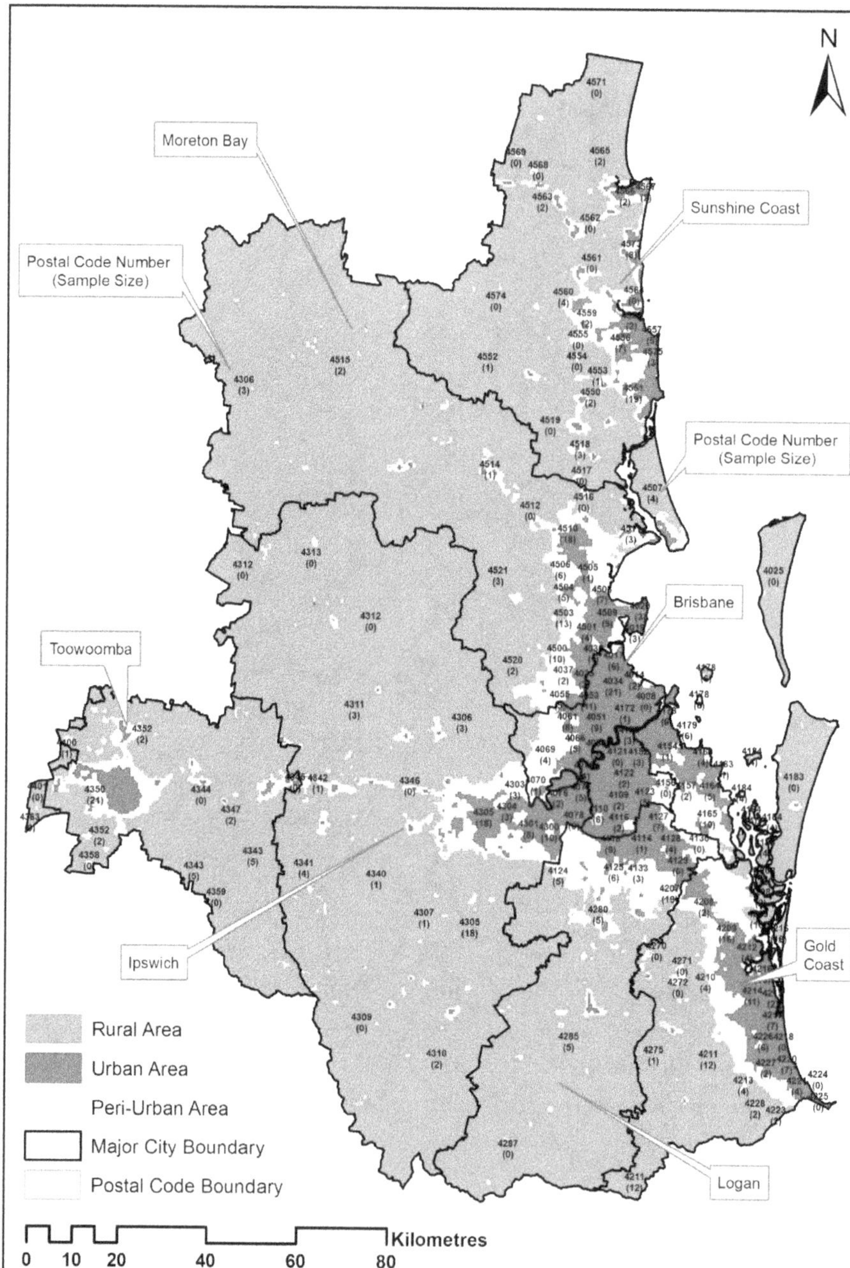

FIGURE 1.3
Spatial distribution of samples between rural, peri-urban, and urban areas of South East Queensland.

TABLE 1.2

Socioeconomic Profile of the Respondents

Socioeconomic Profile	Number	%
What is your gender?		
Female	320	48.6
Male	335	50.8
Other	4	0.6
What is your age?		
18–24	45	6.8
25–34	125	19.0
35–44	119	18.1
45–54	104	15.8
55–64	97	14.7
65–74	96	14.6
75–84	65	9.9
85 and over	8	1.2
What is your highest education qualification obtained?		
Year 10 or below	84	12.7
Year 11 or equivalent	14	2.1
Year 12 or equivalent	103	15.6
Certificate	129	19.6
Advanced diploma/diploma	97	14.7
Bachelor's degree	132	20.0
Postgraduate degree	100	15.2
Which of the following describes your political view best?		
Left	123	18.7
Centre	363	55.1
Right	125	19.0
Other (e.g., unwilling to reveal, apolitical, undecided, and do not vote)	48	7.3

the highest. Additionally, the respondents who attributed that the problem of climate change is not given due importance are predominantly the left-wing supporters and are least supported by the right-wing proponents. Nevertheless, those who advocated for centrism viewed this issue of climate change as generally correct, meaning neither overestimated nor underestimated. Besides, out of the left-wing supporters, only 0.8% do not believe in climate change, indicating that almost all of the left-wing supporters believe

TABLE 1.3

Respondents' Views About Climate Change Against Political Stances

	Thinking About Climate Change, Which One Best Describes Your Views? ($\chi^2 = 110.39$, df = 9; $p = 0.000$)			
Political Views	The Issue of Climate Change Is Generally Overestimated (%)	The Issue of Climate Change Is Generally Correct (%)	The Issue of Climate Change Is Generally Underestimated (%)	There Is No Climate Change (%)
Left	6.5	27.6	65.0	0.8
Centre	24.2	38.3	30.3	7.2
Right	43.2	24.8	16.8	15.2
Other	16.7	25.0	41.7	16.7
Total	24.0	32.8	35.1	8.2

in climate change impact. Moreover, out of the respondents who support left wing, the majority of them attributed that climate change is mainly driven by anthropogenic disturbances (Table 1.4).

Furthermore, it is found that nearly half (45.5%) of the left-wing supporters are willing to participate in climate change mitigation events, whereas around two-thirds (i.e., 71.2%) of the right-wing supporters do not have

TABLE 1.4

Respondents' Stance Between Political Views and Anthropogenic Climate Change Impacts

	Thinking About the Causes of Climate Change, Which of the Following Best Describes Your Opinion? ($\chi^2 = 137.29$, df = 18; $p = 0.000$)						
Political Views	Climate Change Is Entirely Caused by Natural Processes (%)	Climate Change Is Mainly Caused by Natural Processes (%)	Climate Change Is Partly Caused by Natural Processes and Partly by Human Activities (%)	Climate Change Is Mainly Caused by Human Activities (%)	Climate Change Is Entirely Caused by Human Activities (%)	I Think There Is No Such Thing as Climate Change (%)	Don't Know (%)
Left	1.6	4.9	20.3	52.0	19.5	0.8	0.8
Centre	5.8	7.4	35.5	29.8	8.5	4.7	8.3
Right	17.6	20.0	27.2	18.4	2.4	13.6	0.8
Other	0.0	12.5	22.9	35.4	10.4	12.5	6.3
Total	6.8	9.7	30.2	32.2	9.6	6.2	5.3

TABLE 1.5

Respondents' Willingness to Participate in Climate Change Mitigation Events

	Are You Willing to Participate in Such Events/ Activities/Protests Targeting the Mitigation of Climate Change Impacts? ($\chi^2 = 83.93$, df = 6; $p = 0.000$)		
Political Views	**Yes (%)**	**No (%)**	**Not Sure (%)**
Left	45.5	24.4	30.1
Centre	14.6	53.4	32.0
Right	15.2	71.2	13.6
Other	20.8	56.3	22.9
Total	20.9	51.6	27.5

willingness to participate (Table 1.5). Likewise, the majority of the respondents (i.e., 53.4%) who adhere to centrism are unwilling to participate in climate change mitigation–related events.

Table 1.6 lists seven ordinal variables which all were measured on a scale of 1–9 (where 1 = strongly disagree; 5 = neutral; and 9 = strongly agree) and highlights the lack of sound policies in place to ease climate change impacts at the federal, state, and local government levels of Australia.

TABLE 1.6

Respondents' Perceptions on the Government Policies Concerning Climate Change Impacts

Variables	DA (%)	Neutral (%)	A (%)	R-A	R-DA
The political party I vote for has developed sound policies to ease climate change impacts	23.4	35.7	40.9	4	7
My federal government has sound policies in place to ease climate change impacts	38.4	29.9	31.7	5	1
My federal government should declare a climate emergency, like New Zealand government did in 2020	29.4	23.1	47.5	1	5
My state government has sound policies in place to ease climate change impacts	34	37.3	28.7	6	2
My state government should declare a climate emergency like the ACT government did in 2019	29.1	27.2	43.7	2	6
My local government has sound policies in place to ease climate change impacts	32.5	40.5	27	7	3
My local government should declare a climate emergency, like the Sydney City Council did in 2019	31.1	26.4	42.5	3	4

Notes: DA = disagree, A = agree, R = rank.

1.4.2 Regression Analyses

Respondents' knowledge on climate change perceptions was tested by using a set of 20 independent variables, which are listed in Appendix 1.A1.

Except for the variable titled 'what is your highest educational qualification obtained' (which is an ordered categorical one), all other variables presented in Appendix 1.A1 are ordinal and measured on a scale of 1–9, where 1 = strongly disagree; 5 = neutral; and 9 = strongly agree. These listed variables were considered as factors (independent variables), and the respondents' political standings—which were categorised as left, centre, right, and other (Table 1.2)—are considered as the dependent variable in order to run each MLR individually.

Political views against respondents' personal traits on climate change perception: Respondents' personal traits on climate change perceptions with regard to their political standings (MLR1), the model fitting outcome proves to be significant ($\chi^2 = 197.09$, df = 90; $p < 0.000$). Out of the model outcomes, variables with statistically significant outcomes are illustrated in Table 1.7. The findings disclose that (a) it is more likely that a respondent is a left-wing prone than the right-wing, if they possess a higher educational level (higher than the category of 'Year 10 or below'); (b) it is more likely that a respondent is a left-wing supporter than the right-wing, if they strongly agreed rather than strongly disagreed with the statements: 'my concerns about climate change impacts guide my voting behaviour', and 'it is our moral responsibility to use energy-efficient and clean technologies to reduce greenhouse gas emissions'.

Political views against respondents' perception on the causes and consequences of climate change impacts: Respondents' perceptions on the causes and

TABLE 1.7

Multinomial Logistic Regression Model Parameters Underpinning Political Views Against Respondents' Personal Traits on Climate Change Perception

Which of the Following Describes Your Political View Best?[a]	Factors	Parameter Estimate		
		B	Sig.	Exp(B)
Left	[What is your highest education qualification obtained? = Year 10 or below]	−1.368	.030	.255
	[My concerns about climate change impacts guide my voting behaviour = 1.00]	−2.495	.000	.082
	[It is our moral responsibility to use energy-efficient and clean technologies to reduce greenhouse gas emissions = 1.00]	−1.739	.013	.176

Notes: [a] = The Reference Category is: Right.; N.B.: The variables with the remaining model outcomes are not shown as they were not found statistically significant.

TABLE 1.8

Multinomial Logistic Regression Model Parameters Underpinning Political Views Against Respondents' Perception on the Cause and Consequences of Climate Change Impacts

Which of the Following Describes Your Political View Best?[a]	Factors	Parameter Estimate		
		B	Sig.	Exp(B)
Left	[Climate change is occurring because of human activity–driven greenhouse gas emissions = 2.00]	−2.300	.034	.100
Centre	[Climate change is occurring because of human activity–driven greenhouse gas emissions = 1.00]	−2.088	.045	.124

Notes: [a] = The Reference Category is: Right.; N.B.: The variables with the remaining model outcomes are not shown as they were not found statistically significant.

consequences of climate change with regard to their political standings (MLR2), the model fitting information was found statistically significant (χ^2 = 204.18, df = 96; p < 0.000). The model output indicates that it is more likely that a respondent is a left-wing or central-wing supporter than the right-wing, if s/he agreed rather than disagreed with the statement: 'climate change is occurring because of human activity-driven greenhouse gases emissions' (Table 1.8).

In terms of an agreement between the factors considered under the MLR2, the statement titled 'climate change is occurring because of human activity-driven greenhouse gases emissions' was ranked as lowest (i.e., the level of agreement is 60.2%) but achieved the highest rank among the left-wing supporters (i.e., the level of agreement within the left-wing supporters = 83.7%). Contrariwise, the right-wing supporters ranked this statement with the lowest agreement level of 41.6%. Thus, it is revealed that the left- and central-wing supporters more progressively view anthropogenic disturbances as a cause of climate change impacts than those of the right-wing supporters. Nevertheless, with regard to the consequences of climate change impacts, no statistically significant information was found.

Political views against respondents' perceptions on peri-urbanisation-driven climate change impacts: Respondents' peri-urbanisation-driven climate change perceptions with regard to their political standings (MLR3), the regression analysis yields a significant model fit (χ^2 = 158.18, df = 72; p = 0.000). The model output indicates that (Table 1.9) (a) it is more likely that a respondent is a left-wing supporter than the right-wing, if s/he is neutral with the statement: 'due to rapid urban growth, we are generating more greenhouse gas emissions and exposed to climate change impacts'; (b) it is more likely that a respondent is a left-wing supporter than the right-wing, if s/he strongly agreed rather than strongly disagreed with the statement: 'urban

TABLE 1.9

Multinomial Logistic Regression Model Parameters Underpinning Political Views Against Respondents' Perceptions on Peri-Urbanisation-Driven Climate Change Impacts

Which of the Following Describes Your Political View Best?[a]	Factors	Parameter Estimate		
		B	Sig.	Exp(B)
Left	[Due to rapid urban growth, we are generating more greenhouse gas emissions and exposed to climate change impacts =5.00]	−2.436	.001	.087
	[Urban encroachment to the urban fringe causes environmental degradation and biodiversity loss = 2.00]	−3.457	0.022	0.032
Centre	[Urban encroachment to the urban fringe causes environmental degradation and biodiversity loss = 8.00]	−1.940	.005	.144

Notes: [a] = The Reference Category is: Right.; N.B.: The variables with the remaining model outcomes are not shown as they were not found statistically significant.

encroachment to the urban fringe causes environmental degradation and biodiversity loss'.

Political views against respondents' perceptions on peri-urbanisation-driven climate change impact mitigation measures that require self-sacrifice: Respondents' peri-urbanisation-driven climate change impact mitigation measures (that require self-sacrifice) with regard to their political standings (MLR4), the model fitting outcome was statistically significant ($\chi^2 = 255.87$, df = 144; p = 0.000) and unveils that (Table 1.10) (a) it is more likely that a respondent is a left-wing supporter than the right-wing, if they agreed rather than disagreed to pay more to rent/buy a home, if urbanisation predominantly is centred around the city in favour of preserving valuable ecosystems of the urban fringe; (b) it is more likely that a respondent is a left-wing supporter than the right-wing, if they strongly agreed rather than strongly disagreed with the statement: 'I am willing to pay more tax to support a more climate-resilient urban growth practice'; (c) it is more likely that a respondent is a left-wing or central-wing supporter than the right-wing, if they strongly agreed rather than strongly disagreed with the statement: 'sustainable urban development is critical for addressing and minimising climate challenge externalities'; and (d) it is more likely that a respondent is a left-wing or central-wing supporter than the right-wing, if they strongly disagreed rather than strongly agreed with the statement: 'I am in favour of imposing a more stringent control on rapid urban growth'.

Political views against respondents' perceptions on government-initiated policies concerning peri-urbanisation-driven climate change impact

TABLE 1.10

Multinomial Logistic Regression Model Parameters Underpinning Political Views Against Respondents' Perceptions on Peri-Urbanisation-Driven Climate Change Impact Mitigation Measures that Require Self-Sacrifice

Which of the Following Describes Your Political View Best?[a]	Factors	Parameter Estimate		
		B	Sig.	Exp(B)
Left	[I am willing to pay more to rent/buy a home, if urbanisation predominantly is centred around the city in favour of preserving valuable ecosystems of the urban fringe = 6.00]	1.994	.013	7.343
	[When developing policies and regulations for urban development, climate change should be an important consideration relative to other issues =5.00]	−1.904	.030	.149
	[I am willing to pay more tax to support a more climate-resilient urban growth practice = 1.00]	−2.722	.011	.066
	[Sustainable urban development is critical for addressing and minimising climate challenge externalities = 8.00]	1.402	.034	4.062
	[I am in favour of imposing a more stringent control on rapid urban growth = 1.00]	2.947	.011	19.05
Centre	[Sustainable urban development is critical for addressing and minimising climate challenge externalities − 1.00]	−2.588	.018	.075
	[I am in favour of imposing a more stringent control on rapid urban growth = 1.00]	1.753	.012	5.771
	[I have, or am keen to gain, knowledge on how to minimise my carbon footprint and contribute to sustainable development = 3.00]	−2.371	.004	.093

Notes: [a] = The Reference Category is: Right.; N.B.: The variables with the remaining model outcomes are not shown as they were not found statistically significant.

mitigation: With regard to respondents' perception on government-initiated policies concerning the prevention of urban growth to the most productive peri-urban farmlands (MLR5), this study found a good model fit ($\chi^2 = 115.752$, df = 72; p = 0.001). The model output indicates that (Table 1.11) (a) it is more likely that a respondent is a left-wing or central-wing supporter than the right-wing, if they strongly agreed rather than strongly disagreed with the

TABLE 1.11

Multinomial Logistic Regression Model Parameters Underpinning Political Views Against Respondents' Perceptions on Government-Initiated Policies Concerning Peri-Urbanisation-Driven Climate Change Impact Mitigation

Which of the Following Describes Your Political View Best?[a]	Factors	Parameter Estimate		
		B	Sig.	Exp(B)
Left	[Governments need to prevent the ongoing rapid urban encroachment to environmentally sensitive areas and agricultural farmlands = 2.00]	−3.609	.004	.027
	[Farmers should think of sub-dividing their farms to use for residential/commercial purposes in order to offset the losses from agricultural activities = 3.00]	1.831	.026	6.239
Centre	[Governments need to prevent the ongoing rapid urban encroachment to environmentally sensitive areas and agricultural farmlands = 1.00]	−1.795	.010	.166
	[Farmers should think of sub-dividing their farms to use for residential/commercial purposes in order to offset the losses from agricultural activities = 8.00]	2.831	.015	16.967

Notes: [a] = The Reference Category is: Right.; N.B.: The variables with the remaining model outcomes are not shown as they were not found statistically significant.

statements: 'governments need to prevent the ongoing rapid urban encroachment to environmentally sensitive areas and agricultural farmlands'; (b) it is more likely that a respondent is a left-wing supporter than the right-wing, if they agreed rather than disagreed with the statements: 'farmers should think of sub-dividing their farms to use for residential/commercial purposes in order to offset the losses from agricultural activities'.

Political views against climate change perceptions of urban, peri-urban, and rural workers: A cumulative odds ordinal logistic regression with proportional odds was run in order to determine the effect of the respondents' location (with respect to the city centre) on the belief of whether respondents' concerns about climate change impacts guide his/her voting behaviour. Variables used in this regression are listed in Table 1.12.

The respondents who work in the city centre or within 10 km from the city centre were considered as 'urban workers'. Whereas the respondents who work within 25–75 km and more than 75 km from the city centre were considered as 'peri-urban workers' and 'rural workers', respectively. Nevertheless, no statistically significant association was found when urban workers' climate risk perception in connection to their voting behaviour is

TABLE 1.12

Variables Underpinning Climate Change Belief of Dwellers against Their Political View

Variable Name	Variable Type	Description of Variable	Proxy Measure	Variable's Role in Regression Model	Variable Abbreviation
My concerns about climate change impacts guide my voting behaviour	Ordinal	Measured on a scale of 1–9, where 1 = strongly disagree; 5 = neutral; and 9 = strongly agree	Political stance	Dependent variable	Concern_ Climate_ Change
Where is your workplace located with respect to the city centre?[a]	Ordered categorical	Categorised as (a) work in the city centre; (b) I work within 10 km from the city centre; (c) I work within 25 km from the city centre; (d) I work within 50 km from the city centre; (e) I work within 75 km from the city centre; (f) I work within 100 km from the city centre; (g) I work within more than 100 km from the city centre; and (h) I don't work	Identifying urban, peri-urban, and rural dwellers	Factor	Workplace_ Location

Notes: [a] = The Reference Category is: (8) I don't work; Model: (Threshold)

investigated against the reference category of 'peri-urban' or 'rural' workers. Notwithstanding, a statistically significant association is observed when the respondents' such corresponding views are observed against the reference category: 'I don't work' (Table 1.12).

There were proportional odds, as assessed by a full likelihood ratio test comparing the fitted model to a model with varying location parameters, ($\chi^2 = 45.61$, df = 49; p = 0.612). Both the Pearson ($\chi^2 = 42.02$, df = 49; p = 0.858) and Deviance ($\chi^2 = 45.61$, df = 49; p = 0.931) goodness-of-fit tests indicated that the model was a good fit to the observed data, and limited cells were sparse with zero frequencies in 12.5% of cells. However, the final model statistically significantly predicted the dependent variable over and above the intercept-only model ($\chi^2 = 40.24$, df = 7; $p < 0.001$).

The odds of the respondents who work in the city centre on the belief that his/her concerns about climate change impacts guide his/her voting behaviour was 3.307 (95% CI, 2.206–4.958) times than that of those who do not work, a statistically significant effect, $\chi^2(1) = 33.52$, p < 0.001 (Table 1.13). Similarly, the odds of the respondents who work within 10 km from the city centre on the belief that his/her concerns about climate change impact guide his/her voting behaviour was 2.160 (95% CI, 1.438–3.244) times than that of those who do not work, a statistically significant effect, $\chi^2(1) = 13.77$, $p < 0.001$. Also, the odds of the respondents who work within 25 km from the city centre on the belief that his/her concerns about climate change impacts guide his/her voting behaviour was 1.486 (95% CI, 0.978–2.258) times than that of those who do not work, nearly a statistically significant effect, $\chi^2(1) = 3.44$, $p = 0.064$.

1.5 Findings and Discussion

The *first* hypothesis testing conducted under this study was probing the impacts of respondents' personal traits on climate change perception. The findings disclose that higher educational standing, a feeling of moral obligation in using green energy, and their climate change concerns on voting behaviour are linked to be reported with more likelihood of being left-wing prone. Besides, from the descriptive analysis, it was also found that penchant to be more left-wing prone than to be right-wing proponents also is linked with the participants being females, lower age groups, and higher educational attainments. Such findings are similar to recent studies (Colvin & Jotzo, 2021; Merzian et al., 2019), which investigated Australian's attitudes towards climate risk. Nevertheless, Hu et al.'s (2017) study which investigated the association between climate risk perception and political orientation in the context of the US claims that respondents' climate risk perception, in general, is not linked to their age, gender, or educational attainment at all. Nonetheless, when respondents' such perceptions are assessed based on their political views, it is not clear from their study outcomes that how age, gender, and educational attainments are associated with respondents' climate change perceptions.

Notwithstanding, while educational attainment proved to be significant, the inclusions of age and gender in the MLR1 model did not yield a statistically significant model fit and were subsequently excluded from this MLR1 regression analysis. In general, political standings are found to be statistically associated with respondents' age, gender, and educational attainments. But, in the case of investigating whether respondents' personal traits on climate change perception are further augmented by their political standings, age and gender did not play any statistically significant roles in the regression model outcomes.

TABLE 1.13

Ordinal Logistic Regression Model Parameters

Parameter estimates											
				95% Wald Confidence Interval		Hypothesis Test				95% Wald Confidence Interval for Exp(B)	
Parameter		B	Std. Error	Lower	Upper	Wald Chi-Square	df	Sig.	Exp (B)	Lower	Upper
Threshold	[Concern Climate_Chnage = 1.00]	−1.118	.1260	−1.365	−.871	78.766	1	.000	.327	.255	.418
	[Concern Climate_Chnage = 2.00]	−.751	.1196	−.985	−.516	39.403	1	.000	.472	.373	.597
	[Concern Climate_Chnage = 3.00]	−.505	.1170	−.734	−.276	18.664	1	.000	.603	.480	.759
	[Concern Climate_Chnage = 4.00]	−.337	.1158	−.564	−.110	8.493	1	.004	.714	.569	.895
	[Concern Climate_Chnage = 5.00]	.769	.1192	.535	1.003	41.630	1	.000	2.158	1.708	2.726
	[Concern Climate_Chnage = 6.00]	1.206	.1247	.962	1.451	93.629	1	.000	3.341	2.617	4.266
	[Concern Climate_Chnage = 7.00]	1.909	.1386	1.637	2.180	189.730	1	.000	6.745	5.141	8.850
	[Concern Climate_Chnage = 8.00]	2.433	.1543	2.130	2.735	248.407	1	.000	11.388	8.415	15.411
[Workplace_Location = 1.00]		1.196	.2066	.791	1.601	33.526	1	.000	3.307	2.206	4.958
[Workplace_Location = 2.00]		.770	.2075	.363	1.177	13.765	1	.000	2.160	1.438	3.244
[Workplace_Location = 3.00]		.396	.2136	−.023	.815	3.437	1	.064	1.486	.978	2.258
[Workplace_Location = 4.00]		.356	.2612	−.156	.868	1.860	1	.173	1.428	.856	2.383
[Workplace_Location = 5.00]		.135	.5424	−.928	1.198	.062	1	.803	1.145	.395	3.314
[Workplace_Location = 6.00]		.149	.4538	−.741	1.038	.108	1	.743	1.160	.477	2.824
[Workplace_Location = 7.00]		−.074	.4845	−1.023	.876	.023	1	.879	.929	.359	2.401

Moreover, among the 19 ordinal variables listed in Appendix 1.A1, the top-ranked disagreed item was the religious belief influencing respondents' climate change perceptions. Albeit, no statistically significant outcome was found against this particular variable in the MLR1 model, a statistically strong association ($\chi^2 = 49.32$, df = 24; p = 0.002) prevails between their denial of religious belief on climate change perception and the political ideology they hold (data not shown). While 'religious belief' and 'religious practices' are discernibly different, yet the impact of religious standings on climate change perception is extensively unexplored (Haluza-DeLay, 2014). However, climate change perception for the context of the US was found to be positively linked with religious practices (Hu et al., 2017). Yet, Hu et al.'s (2017) study did not explore how this perceived association—i.e., the link between respondents' climate change concerns and religious beliefs—is influenced by their political standings.

The *second* hypothesis tested under this study was investigating respondents' perceptions on the causes and consequences of climate change impacts. In this way, respondents' knowledge of climate change belief corresponding to their political views was tested. The findings unveil that both left- and central-wing proponents are more progressive in perceiving the cause of anthropogenic climate change impacts than those of the right-wing supporters. Such findings are also mirrored in the descriptive analyses (Table 1.4). For instance, the claim titled 'climate change is mainly caused by human activities' is least supported by the right-wing supporters. Inversely, believing that 'climate is entirely caused by natural process' is most supported by the right-wing and least supported by the left-wing supporters. Notwithstanding, those who advocated that climate change is caused by both natural and human processes are predominantly the central-wing supporters.

Nevertheless, regarding perceptions on the consequences of climate change impacts, the political penchants were not found to be statistically significant. Such findings can be substantiated by descriptive analysis. Appendix 1.A1 demonstrates that the variables of climate change consequences are ranked with the higher agreement and each of them possesses a higher agreement value than their corresponding disagreement value. Such a finding implies that the grave consequences of climate change impacts are broadly perceived irrespective of political penchants. Additionally, the statement titled 'my concerns about climate change impacts guide my voting behaviour' is perceived with higher disagreement (Appendix 1.A1). Such a contrasting finding implies that respondents are predominantly supportive of climate risk mitigation measures, but the issue of climate risk does not play a determinant role in deciding whom to vote for. This finding reported in this chapter conforms to Colvin and Jotzo's (2021) recent study finding for the Australian context. However, studies carried out for the US context identify climate risk perception as a determinant factor in their voting decisions (Hu et al., 2017; Marquart-Pyatt et al., 2014; McCright & Dunlap, 2011a; Unsworth & Fielding, 2014).

Notwithstanding, while the first and second hypotheses were employed to test respondents' political views influencing climate change perceptions, the *third* hypothesis tested under this exercise was fact-finding whether respondents' perceptions on the causes and consequences of peri-urbanisation-driven climate change impacts are influenced by their political standings. Nevertheless, with regard to the cause of peri-urbanisation-driven climate change impacts, the left-wing supporters demonstrate a neutral stance when their viewpoints are compared to those of the right-wing proponents. One of the grave consequences of constant peri-urbanisation is environmental degradation and biodiversity loss (McAlpine et al., 2006; Sutton et al., 2016). The analysis reveals that respondents' agreement on this particular issue is both supported and opposed by the left- and central-wing supporters, respectively, when their stances are compared with the corresponding views of the right-wing proponents.

The *fourth* hypothesis tested under this study was investigating whether respondents' perceptions on peri-urbanisation-driven climate change impact mitigation measures (that require self-sacrifice) are linked to their political standings. However, the hypothesis testings (i.e., MLR1–MLR3) as stated above are found with smaller Exp(B)—i.e., odds ratio values, some variables tested under MLR4 are observed with a very high odds ratio (Table 1.9). Such findings imply that the likelihood of the left-wing supporters embracing those agreed/disagreed mitigating measures is enormously higher than those of the right-wing proponents. For instance, the likelihood of the left-wing supporters paying more to rent or buy a home in order to compensate for the additional cost accrued with compact urban development is 7.34 times higher than those of the right-wing supporters. While the remaining statistically significant variables listed in Table 1.9 are found with more agreements, the mitigation policy of imposing a more stringent control on rapid urban growth is more strongly disagreed by the left-wing and central-wing proponents by 19.05 times and 5.77 times, respectively, than those of the right-wing proponents. Thus, adopting mitigation measures and simultaneously allowing flexibility to urban growth is the most preferred form of urban growth policy of the respondents.

The *fifth* hypothesis tested under this study was probing whether respondents' perceptions on government-initiated policies concerning peri-urbanisation-driven climate change impact mitigation are influenced by their political standings. In this regard, when the role of government comes in preventing urban growth in the environmentally sensitive areas and agricultural farmlands, a stronger agreement is observed in both the left- and central-wing supporters than those of the right-wing proponents. Besides, this particular variable is also found as the highest agreed-upon item among the variables listed in Appendix 1.A1—implying that irrespective of political standings, the urge to initiate attempts from the government part in preventing urban growth to the environmentally sensitive areas and agricultural farmlands is overwhelmingly supported. Therefore, government-initiated mitigation measures—which do not require self-sacrifice—are reported to be more welcoming irrespective of political standings. However, with regard to the mitigation measures requiring self-sacrifice, the left-wing proponents appear to be more supportive.

Still, respondents' stance on the issue of sub-dividing farming lots is more disagreed by left-wing supporters by 6.24 times than those of the right-wing proponents. In general, the policy supporting farmers in sub-dividing their farming lots is less supported at large and hence becomes the second-ranked variable in terms of disagreement (Appendix 1.A1). However, the policy concerning the government's role in preventing environmentally sensitive areas and agricultural farmlands becomes the highest-ranked variable in terms of the agreement. Besides, with regard to the perceptions on the government-initiated policies concerning climate change impacts, respondents also urged declaring climate emergency at the federal, state, and local government levels with the highest priorities (Table 1.6).

In terms of respondents' willingness to participate in climate change mitigation–related events, although the left-wing supporters demonstrated their highest interest, their intention to participate in such events appears to be less spirited being a strong proponent of the climate change issue. The findings demonstrate that more than half of the left-wing supporters are either torn or offered direct denials with regard to the question of their willingness to participate in climate change mitigation events (Table 1.5).

The *sixth* hypothesis tested under this study was identifying the variation in climate risk perceptions guiding political behaviours between urban, peri-urban, and rural workers. However, the investigation carried out under this study unveils that no statistically significant variation prevails in such climate risk perceptions between the urban, peri-urban, and rural workers. Notwithstanding, the findings from this particular analysis suggest that for those who work in the city centre or within 10–25 km from the city centre, the issue of climate risk appears to exert more influences on their voting behaviour than those who are unemployed (does not work). Furthermore, the proportional odds ratio unveils that those who work in the city centre seem to demonstrate twofold more influences on that particular issue than those who work within 25 km from the city centre (Table 1.13). However, while the probing of such perceived occurrences had not been possible to investigate under the scope of this research, such phenomena might be largely attributed to the respondents' socioeconomic and cultural standings, which require further research.

In aggregate, left-wing supporters appear to possess more risk perceptions on the general consequences of climate change impacts as well as on the consequences of peri-urbanisation-driven climate change impact in particular. Nevertheless, the likelihood of such occurrences is extremely limited. For instance, the odds ratio in most of the cases is found to be less than 0.3 (Tables 1.7–1.11)—implying that the likelihood of such occurrences in those cases is a minimum of 70% less likely to take effect. Therefore, climate risk perception is common irrespective of political standings, although peri-urbanisation-driven climate risk is not discernibly perceived.

1.6 Conclusion

This study primarily investigates the association of an individual's political ideology with climate change risk perception resulting from the constant urban encroachments into peri-urban areas. This study attributes such a particular phenomenon: 'peri-urbanisation driven climate risk perception'. The outcomes reveal that there is little to no relationship between political affiliation and climate risk perception. Nevertheless, it is found that more politically left wing inclined a person is, a slightly bit higher climate change risk perception s/he will have than those of the right-wing proponents.

Nonetheless, no significant political penchant is attached with regard to risk perception on the cause of peri-urbanisation-driven climate change impacts. Hence, respondents might be well aware of the risk of increased greenhouse gas emissions and subsequent climate change exposure in general. Notwithstanding, how these increased greenhouse gas emissions could be further intensified by rapid urban growth in particular—which this study attributed as the cause of peri-urbanisation-driven climate change impacts—is less substantiated.

Thus, while the world is at the stake of grave consequences to climate change impacts, peri-urbanisation-driven climate risk perception appears to be underestimated. Such an underestimation could be attributed to respondents' lack of knowledge. While the ongoing researches and media campaigns predominantly focus on the general causes and consequences of climate change impacts, the particular causes and consequences of climate change impacts yet are not gaining proper attention. Henceforth, climate change externalities—i.e., the increasing trend of climate migrants, changes in lifestyle and livelihoods, and the resulting consequences of rapid urbanisation in accommodating these climate externalities—yet are less understood.

The study reported under this chapter has two major limitations. Firstly, although adapting OLR would have been an ideal choice, in the case of using OLR the proportional odds assumption would be violated for the dataset collected under this study. As such, MLR is the approach this study predominantly adapted instead. Secondly, while this study identified some forms of political penchants with climate risk perception, such reported political bias might be moderated by the respondents' level of exposure to climate change events and higher-level knowledge on climate change (Akerlof et al., 2013; Ballew et al., 2020) that remains unaddressed under this study.

Despite all the limitations as stated above, given the impacts of Australia's 2019–2020 bushfire event and the ongoing global pandemic of COVID-19, this research portrays the Australian's latest perception on climate change risk with a particular focus on assessing respondents' political preference against their views on the association between climate change risks and rapid urban growth. This study bids fair hope to policymakers in pursuing climate risk mitigation measures as the majority of the respondents are supportive of this climate risk issue irrespective of political standings. However, the extent of their given supports is restrained by their political standings, educational attainments, and their willingness to incur personal costs or self-sacrifice.

Acknowledgements

This chapter, with permission from the copyright holder, is a reproduced version of the following journal article: Mortoja, M., & Yigitcanlar, T., (2022). Are climate change, urbanisation and political views correlated? Empirical evidence from South East Queensland. *Urban Climate*, *41*(1), 101061.

References

Akerlof, K., Maibach, E. W., Fitzgerald, D., Cedeno, A. Y., & Neuman, A. (2013). Do people "personally experience" global warming, and if so how, and does it matter? *Global Environmental Change*, *23*, 81–91.

Al-Mamun, M. M., & Kim, S. M. (2020). Stakeholder analysis matrix for buffer zone management in the peri-urban area of Chittagong, Bangladesh. *Environment, Development and Sustainability*, *22*, 5503–5520.

Antonio, R. J., & Brulle, R. J. (2011). The unbearable lightness of politics: Climate change denial and political polarization. *The Sociological Quarterly*, *52*, 195–202.

Bajracharya, B., & Hastings, P. (2018). A regional, strategic growth-management approach to urban and peri-urban development in South East Queensland, Australia. *Journal of Regional and City Planning*, *29*, 210–233.

Ballew, M. T., Pearson, A. R., Goldberg, M. H., Rosenthal, S. A., & Leiserowitz, A. (2020). Does socioeconomic status moderate the political divide on climate change? The roles of education, income, and individualism. *Global Environmental Change*, *60*, 102024.

Beeson, M., & McDonald, M. (2013). The politics of climate change in Australia. *Australian Journal of Politics & History*, *59*, 331–348.

Biggs, C. (2016). A resource-based view of opportunities to transform Australia's electricity sector. *Journal of Cleaner Production*, *123*, 203–217.

Carlton, S. J., & Jacobson, S. K. (2013). Climate change and coastal environmental risk perceptions in Florida. *Journal of Environmental Management*, *130*, 32–39.

Case, M. J., Johnson, B. G., Bartowitz, K. J., & Hudiburg, T. W. (2021). Forests of the future: Climate change impacts and implications for carbon storage in the pacific northwest, USA. *Forest Ecology and Management, 482*, 118886.

Cobbinah, P. B. (2021). Urban resilience in climate change hotspot. *Land Use Policy, 100*, 104948.

Colvin, R. M., & Jotzo, F. (2021). Australian voters' attitudes to climate action and their social-political determinants. *PLoS ONE, 16*, e0248268.

Costa, D. L., & Kahn, M. E. (2013). Energy conservation "nudges" and environmentalist ideology: Evidence from a randomized residential electricity field experiment. *Journal of the European Economic Association, 11*, 680–702.

Dizdaroglu, D., & Yigitcanlar, T. (2014). A parcel-scale assessment tool to measure sustainability through urban ecosystem components: The MUSIX model. *Ecological Indicators, 41*, 115–130.

Dunlap, R. E., & McCright, A. M. (2008). A widening gap: Republican and democratic views on climate change. *Environment: Science and Policy for Sustainable Development, 50*, 26–35.

Egerer, M., Haase, D., McPhearson, T., Frantzeskaki, N., Andersson, E., Nagendra, H., & Ossola, A. (2021). Urban change as an untapped opportunity for climate adaptation. *Urban Sustainability, 1*, 1–9.

Fisher, T. (2003). Differentiation of growth processes in the peri-urban region: An Australian case study. *Urban Studies, 40*, 551–565.

Foelske, L., Van Riper, C. J., Stewart, W., Ando, A., Gobster, P., & Hunt, L. (2019). Assessing preferences for growth on the rural-urban fringe using a stated choice analysis. *Landscape and Urban Planning, 189*, 396–407.

Friedberger, M. (2000). The rural-urban fringe in the late twentieth century. *Agricultural History, 74*, 502–514.

Galderisi, A. (2014). Climate change adaptation. Challenges and opportunities for a smart urban growth. *TeMA – Journal of Land Use, Mobility and Environment, 7*, 43–68.

Gregersen, T., Doran, R., Böhm, G., Tvinnereim, E., & Poortinga, W. (2020). Political orientation moderates the relationship between climate change beliefs and worry about climate change. *Frontiers in Psychology, 11*, 1573.

Guhathakurta, S., & Stimson, R. (2007). What is driving the growth of new "Sunbelt" metropolises? Quality of life and urban regimes in greater Phoenix and Brisbane-South East Queensland region. *International Planning Studies, 12*, 129–152.

Haluza-DeLay, R. (2014). Religion and climate change: Varieties in viewpoints and practices. *Wiley Interdisciplinary Reviews: Climate Change, 5*, 261–279.

Head, L., Adams, M., McGregor, H. V., & Toole, S. (2014). Climate change and Australia. *Wiley Interdisciplinary Reviews: Climate Change, 5*, 175–197.

Houston, P. (2005). Re-valuing the fringe: Some findings on the value of agricultural production in Australia's peri-urban regions. *Geographical Research, 43*(2), 209–223.

Hu, S., Jia, X., Zhang, X., Zheng, X., & Zhu, J. (2017). How political ideology affects climate perception: Moderation effects of time orientation and knowledge. *Resources, Conservation and Recycling, 127*, 124–131.

Janoff-Bulman, R., Sheikh, S., & Hepp, S. (2009). Proscriptive versus prescriptive morality: Two faces of moral regulation. *Journal of Personality and Social Psychology, 96*, 521.

Kahn, M. E. (2009). Urban growth and climate change. *Annual Review of Resource Economics, 1,* 333–350.

Kankanamge, N., Yigitcanlar, T., Goonetilleke, A., & Kamruzzaman, M. (2020). Determining disaster severity through social media analysis: Testing the methodology with South East Queensland Flood tweets. *International Journal of Disaster Risk Reduction, 42,* 101360.

Katelaris, C. H. (2021). Climate change and extreme weather events in Australia: Impact on allergic diseases. *Immunology and Allergy Clinics, 41,* 53–62.

Kijas, J. (2002). A place at the coast: Internal migration and the shift to the coastal countryside. *Transformations, 2*(1), 12.

Leiserowitz, A. (2005). American risk perceptions: Is climate change dangerous? *Risk Analysis, 25,* 1433–1442.

Liu, C., Valentine, G., Vanderbeck, R. M., Diprose, K., & McQuaid, K. (2019). Rural–urban inequality and the practice of promoting sustainability in contemporary China. *GeoJournal, 84,* 1187–1198.

Mahbub, P., Goonetilleke, A., Ayoko, G. A., Egodawatta, P., & Yigitcanlar, T. (2011). Analysis of build-up of heavy metals and volatile organics on urban roads in Gold Coast, Australia. *Water Science and Technology, 63,* 2077–2085.

Marquart-Pyatt, S. T., McCright, A. M., Dietz, T., & Dunlap, R. E. (2014). Politics eclipses climate extremes for climate change perceptions. *Global Environmental Change, 29,* 246–257.

McAlpine, C. A., Rhodes, J. R., Callaghan, J. G., Bowen, M. E., Lunney, D., Mitchell, D. L, Pullar, D. V., & Possingham, H. P. (2006). The importance of forest area and configuration relative to local habitat factors for conserving forest mammals: A case study of koalas in Queensland, Australia. *Biological Conservation, 132,* 153–165.

McCright, A. M., & Dunlap, R. E. (2011a). The politicization of climate change and polarization in the American public's views of global warming, 2001–2010. *The Sociological Quarterly, 52,* 155–194.

McCright, A. M., & Dunlap, R. E. (2011b). Cool dudes: The denial of climate change among conservative white males in the United States. *Global Environmental Change, 21,* 1163–1172.

Merson, J., Attwater, R., Ampt, P., Wildman, H., & Chapple, R. (2010). The challenges to urban agriculture in the Sydney basin and lower Blue Mountains region of Australia. *International Journal of Agricultural Sustainability, 8,* 72–85.

Merzian, R., Quicke, A., Bennett, E., Campbell, R., & Swann, T. (2019). *Climate of the Nation 2019. Tracking Australia's attitudes towards climate change and energy.* The Australia Institute.

Morrison, K. R., & Miller, D. T. (2008). Distinguishing between silent and vocal minorities: Not all deviants feel marginal. *Journal of Personality and Social Psychology, 94,* 871.

Mortoja, M., & Yigitcanlar, T. (2020). Local drivers of anthropogenic climate change: Quantifying the impact through a remote sensing approach in Brisbane. *Remote Sensing, 12,* 2270.

Mortoja, M., & Yigitcanlar, T. (2021a). Public perceptions of peri-urbanism triggered climate change: Survey evidence from South East Queensland. *Sustainable Cities and Society, 103407.*

Mortoja, M., & Yigitcanlar, T. (2021b). Why is determining peri-urban area boundaries critical for sustainable urban development? *Journal of Environmental Planning and Management.* https://doi.org/10.1080/09640568.2021.1978405.

Mortoja, M., Yigitcanlar, T., & Mayere, S. (2020). What is the most suitable methodological approach to demarcate peri-urban areas? A systematic review of the literature. *Land Use Policy, 95*, 104601.

Nalau, J., & Melville-Rea, H. (2021). *Australia ranks last out of 54 nations on its strategy to cope with climate change. The Glasgow summit is a chance to protect us all.* https://theconversation.com/australia-ranks-last-out-of-54-nations-on-its-strategy-to-cope-with-climate-change-the-glasgow-summit-is-a-chance-to-protect-us-all-169627

Ortiz-Bobea, A., Ault, T. R., Carrillo, C. M., Chambers, R. G., & Lobell, D. B. (2021). Anthropogenic climate change has slowed global agricultural productivity growth. *Nature Climate Change, 11*, 306–312.

Perveen, S., Kamruzzaman, M., & Yigitcanlar, T. (2017). Developing policy scenarios for sustainable urban growth management: A Delphi approach. *Sustainability, 9*, 1787.

Perveen, S., Yigitcanlar, T., Kamruzzaman, M., & Agdas, D. (2020). How can transport impacts of urban growth be modelled? An approach to consider spatial and temporal scales. *Sustainable Cities and Society, 55*, 102031.

Queensland Parliament (2017). *Political Representation, Fact Sheet 7.6.* https://www.parliament.qld.gov.au/documents/explore/education/factsheets/Factsheet_7.6_PoliticalRepresentation.pdf.

Reyer, C. P., Otto, I. M., Adams, S., Albrecht, T., Baarsch, F., Cartsburg, M., Coumou, D., Eden, A., Ludi, E., Marcus, R., Mengel, M., & Stagl, J. (2017). Climate change impacts in Central Asia and their implications for development. *Regional Environmental Change, 17*, 1639–1650.

Sotto, D., Philippi, A., Yigitcanlar, T., & Kamruzzaman, M. (2019). Aligning urban policy with climate action in the global south: Are Brazilian cities considering climate emergency in local planning practice? *Energies, 12*(18), 3418.

Sutton, P., Anderson, S., Costanza, R., & Kubiszewski, I. (2016). The ecological economics of land degradation: Impacts on ecosystem service values. *Ecological Economics, 129*, 182–192.

The Climate Council. (2019). *Climate policies of major Australian political parties.* https://www.climatecouncil.org.au/wp-content/uploads/2019/05/climate-policies-of-major-australian-political-parties-v2.pdf.

Unsworth, K. L., & Fielding, K. S. (2014). It's political: How the salience of one's political identity changes climate change beliefs and policy support. *Global Environmental Change, 27*, 131–137.

Van der Linden, S. (2014). On the relationship between personal experience, affect and risk perception: The case of climate change. *European Journal of Social Psychology, 44*, 430–440.

Van der Linden, S. (2015). The social-psychological determinants of climate change risk perceptions: Towards a comprehensive model. *Journal of Environmental Psychology, 41*, 112–124.

Van der Linden, S. (2017). Determinants and measurement of climate change risk perception, worry, and concern. In: *The Oxford encyclopedia of climate change communication*. Oxford University Press.

Weber, E. (2010). What shapes perceptions of climate change? *Wiley Interdisciplinary Reviews: Climate Change, 1*, 332–342.

Xu, Q., Zheng, X., & Zhang, C. (2018). Quantitative analysis of the determinants influencing urban expansion: A case study in Beijing, China. *Sustainability, 10*, 1630.

Yang, Z. J., Rickard, L. N., Harrison, T. M., & Seo, M. (2014). Applying the risk information seeking and processing model to examine support for climate change mitigation policy. *Science Communication, 36*(3), 296–324.

Yigitcanlar, T. (2010). *Sustainable urban and regional infrastructure development: Technologies, applications and management*. IGI Global.

Yigitcanlar, T., & Kamruzzaman, M. (2014). Investigating the interplay between transport, land use and the environment: A review of the literature. *International Journal of Environmental Science and Technology, 11*, 2121–2132.

Yue, W., Liu, Y., & Fan, P. (2013). Measuring urban sprawl and its drivers in large Chinese cities: The case of Hangzhou. *Land Use Policy, 31*, 358–370.

Zabik, M. J., & Prytherch, D. L. (2013). Challenges to planning for rural character: A case study from exurban Southern New England. *Cities, 31*, 186–196.

Appendix

APPENDIX 1.A1

Independent Variables Used for Hypothesis Testing by Using Multinomial Logistic Regression

Equation	Variables	Proxy Measure	Variables' Role in Hypothesis Testing	Disagree (%)	Neutral (%)	Agree (%)	R-A	R-DA
MLR1	Climate change does not matter as I have a religious view on eternal afterlife	Personal trait	Political views against respondents' personal traits on climate change perception	68.7	16.4	14.9	19	1
	My concerns about climate change impacts guide my voting behaviour	Personal trait		33.1	25.8	41.1	16	5
	It is our moral responsibility to use energy-efficient and clean technologies to reduce greenhouse gas emissions	Personal trait		15.8	16.1	68.1	2	16
	What is your highest education qualification obtained?	Personal trait		–	–	–	–	–
MLR2	Climate change is occurring because of human activity–driven greenhouse gas emissions	Cause of climate change	Political views against respondents' perception on the causes and consequences of climate change impacts	23.1	16.7	60.2	11	6
	Climate change is gradually deteriorating our environment and ecosystems	Consequence of climate change		20.8	15.2	64	6	9
	The occurrences of extreme weather events—e.g., drought, extreme or abrupt rainfall, prolonged summer, more hot events in the summer—are contributed by climate change impacts	Consequence of climate change		22.2	14.3	63.5	7	8
	Climate change is causing an increase in the frequency and severity of natural disasters—e.g., bushfires, floods	Consequence of climate change		22.5	15.2	62.3	9	7

MLR3	Due to rapid urban growth, we are generating more greenhouse gas emissions and exposed to climate change impacts	Cause of peri-urbanisation-driven climate change impacts	Political views against respondents' perceptions on peri-urbanisation-driven climate change impacts	18.1	17.6	64.3	5	12
	Urban encroachment to the urban fringe limits ecosystem services and agricultural activities in these areas	Consequences of peri-urbanisation-driven climate change impacts		15.3	20.3	64.4	4	17
	Urban encroachment to the urban fringe causes environmental degradation and biodiversity loss	Consequences of peri-urbanisation-driven climate change impacts		16.1	18.5	65.4	3	15
MLR4	I am willing to pay more to rent/buy a home, if urbanisation predominantly is centred around the city in favour of preserving valuable ecosystems of the urban fringe	Mitigation measure	Political views against respondents' perceptions on some peri-urbanisation-driven climate change impact mitigation measures	33.1	25	41.9	15	4
	When developing policies and regulations for urban development, climate change should be an important consideration relative to other issues	Mitigation measure		17.3	21.2	61.5	10	13
	I am willing to pay more tax to support a more climate-resilient urban growth practice	Mitigation measure		38.7	24.9	36.4	17	3
	Sustainable urban development is critical for addressing and minimising climate challenge externalities	Mitigation measure		16.2	20.8	63	8	14
	I am in favour of imposing a more stringent control on rapid urban growth	Mitigation measure		18.2	24.4	57.4	13	11
	I have, or am keen to gain, knowledge on how to minimise my carbon footprint and contribute to sustainable development	Mitigation measure		19.6	26.6	53.8	14	10

(Continued)

APPENDIX 1.A1 ***(Continued)***

Independent Variables Used for Hypothesis Testing by Using Multinomial Logistic Regression

Equation	Variables	Proxy Measure	Variables' Role in Hypothesis Testing	Disagree (%)	Neutral (%)	Agree (%)	R-A	R-DA
MLR5	Governments need to prevent the ongoing rapid urban encroachment to environmentally sensitive areas and agricultural farmlands	Proposed mitigation policy	Political views against respondents' perceptions on some policies concerning peri-urbanisation-driven climate change impact mitigation	10.9	18.4	70.7	1	19
	Farmers should think of sub-dividing their farms to use for residential/commercial purposes in order to offset the losses from agricultural activities	Proposed mitigation policy		48.1	29	22.9	18	2
	Governments should develop policy for farmers not to convert their farms into large industrial or residential uses	Proposed mitigation policy		14.7	26.1	59.2	12	18

Notes: DA = disagree, A = agree, R = rank.

2

Political Bias in Climate Change Perception

2.1 Introduction and Background

Climate change is here, and it is disrupting every country on every continent; hence, urgent and effective government action is needed to sustain our existence on the planet (Limaye et al., 2020; Metaxiotis et al., 2010; Sotto et al., 2019). Despite the clear scientific evidence, there are still a significant quantity of populations that deny the climate change reality (Antonio & Brulle, 2011; Häkkinen & Akrami, 2014; Wong-Parodi & Feygina, 2020). It is assumed that the concerns about climate change should be dependent upon the level of knowledge someone possesses on the issues that trigger climate risk impacts (Arvai et al., 2012; Walsh & Tsurusaki, 2014). Thus, a plethora of studies investigated public perceptions on the climate risk issue (Bord et al., 2000; Kellstedt et al., 2008; Menny et al., 2011; Stevenson et al., 2014).

The dimensions of climate risk knowledge, however, in turn, impact differently on the public's climate risk perception (Mortoja & Yigitcanlar, 2022a). For instance, it is found that a higher level of knowledge concerning the causes and consequences of climate risk appears to be linked with a higher level of public perception (Mortoja & Yigitcanlar, 2021a; Shi et al., 2016). This is because human activity-driven greenhouse gas (GHG) emission have already been established as a prime cause of climate risk impact (Mahbub et al., 2011; Tobler et al., 2012; Yigitcanlar & Kamruzzaman, 2014). Therefore, many people self-report themselves liable for the relevant activities which trigger GHG emissions—which in turn generate more climate risk concerns within them.

Similarly, the evident consequences of climate risk—e.g., floods, bushfires, drought, extreme rainfall events, and extreme heat events—may result in an elevated concern about climate risk perception (Goonetilleke et al., 2014; Shi et al., 2015; Yigitcanlar, 2010). Such events might remind people about the consequences of these events, as stated above which, in turn, could generate more public concerns about climate risk. Nevertheless, physical knowledge concerning climate risk—e.g., investigating some key questions including whether fossil fuel produces carbon dioxide (CO_2), CO_2 depletes the ozone layer, the coal power plant generates more GHG emissions, CO_2 is more ominous to climate risk than methane, and so on—has had a negative or no effects on shaping public's climate risk concern (Kamruzzaman et al.,

DOI: 10.1201/9781003737049-3

2015; Macintosh, 2013; Shi et al., 2016; Yigitcanlar et al., 2008) Therefore, not all forms of climate knowledge exert significant influence or are important in shaping the public's climate risk concern (Tranter & Lester, 2017). Thus, it is critical to investigate the public's climate risk knowledge concerning the causes and consequences of climate change impacts.

Albeit the idea that 'climate change is real' is gaining an increasing concern (Contestabile, 2014), climate risk perception, in Western societies, is polarised predominantly between two groups: 'climate risk believers' and 'climate risk deniers' (Aklin & Mildenberger, 2020; Williams et al., 2015). Nonetheless, such polarisation between the believers and deniers was evenly balanced until the mid-2000s (Boykoff & Smith, 2010), and this perceived polarisation has been later intensified in recent decades (Tangney, 2021). The state of such polarisation primarily depends on an individual's exposure to climate risk events (Myers et al., 2013), moral stance (Feinberg & Willer, 2013), socioeconomic standing (Whitmarsh, 2011), socio-political orientation (McCright & Dunlap, 2011), and the level of scientific understanding (Malka et al., 2009). However, such an existing polarisation is reported to be best explained as a socio-political conflict between these two opposing groups, as it is evident that climate risk could significantly be influenced by an individual's political orientation (Bliuc et al., 2015; McCright & Dunlap, 2011).

For instance, Liberals (or left-wing political orientation) in most Western countries are found to be more supportive of upholding climate risk perception (Farstad, 2019; Janoff-Bulman et al., 2009), whereas Conservatives (or right-wing political orientation) are more unconvinced of climate risk belief (Antonio & Brulle, 2011; Costa & Kahn, 2013; Gonzalez, 2022). Accordingly, for the context of the US, the Democrats—who are known to be left-wing prone—are reported to be more believers in climate risk issues, and the Republicans—who are known to be left-wing prone—are reported to be more deniers of climate risk issues (Gregersen et al., 2020; Marquart-Pyatt et al., 2014). While the prevailing disagreement between these two opposing groups has significant implications for climate risk governance (O'Neill & Boykoff, 2010), few studies have delved into a detailed investigation of these two opposing groups.

In addition, while the concern about climate risk issue now has become too familiar among the public (Lorenzoni et al., 2007), such an issue yet is not highly prioritised when the climate risk issue is particularly compared with other environmental problems (Poortinga & Pidgeon, 2004). Besides, a limited fraction of people usually practise climate risk mitigation measures (e.g., using of solar and wind energy, energy monitoring devices, rainwater harvesting, composting) to date (Norton & Leaman, 2004). Hence, the required level of public participation in mitigating climate risk yet is reported to be missing particularly from partisan and ideological standpoints as discussed above (Nisbet, 2009). Again, public perceptions, opinions and subsequent polarisations stemmed from the climate risk perception of the public vary between countries (Taylor et al., 2014).

Against this brief backdrop, this chapter aims to identify different groups of respondents based on their level of knowledge concerning climate risk against their political orientation. This in return helps in understanding political bias in forming a climate change belief. In doing so, through an online survey that was carried out in April 2021, this study investigated the perceptions of South East Queensland's (SEQ's) (Australia) dwellers (n = 659). The findings generated from this study provide valuable insights to overcome the knowledge gaps between climate risk believers and deniers.

2.2 Research Design

2.2.1 Political Parties in Australia

The major political parties in Australia include Australian Labour Party (ALP), Liberal National Party (LNP), and the Greens. LNP is known to be right-wing prone or conservative. Contrariwise, ALP and the Greens are known to be left-wing prone or progressive. Other notable political parties include the Katter's Australian Party, One Nation, and Independent Party that all are known to be right-wing prone or conservative (Holmes & Star, 2018).

Although, Australia has made some progress in reducing anthropogenic GHG emissions, the left-wing and right-wing political parties as listed above are highly divided in terms of GHG reduction strategies (Holmes & Star, 2018). Besides, all ruling Governments of Australia have been significantly incapable of addressing the country's burgeoning climate risk issue (Palutikof et al., 2019; The Climate Council, 2019; Tranter, 2020).

2.2.2 Case Study

The chapter adopts a case study approach to investigate the relationship between the political orientations of residents and their climate change beliefs (Yin, 1992). To identify different groups of political orientations based on their level of climate change belief, this study uses the SEQ region of Australia as a testbed.

The reasons for the case area selection include: (a) SEQ comprises a larger area coverage, including 12 local government areas, with about 22,888 km^2 (Dizdaroglu et al., 2012); (b) SEQ hosts political parties predominantly with two major ideological standings—i.e., 'progressive' or 'centre-left and left-wing' prone such as ALP and the Greens, and 'conservative' or 'right-wing or far-right' prone such as LNP, One Nation, and Independent Party (Queensland Parliament, 2017)—the electoral map and federal seats are available at https://documents.parliament.qld.gov.au/members/photosheet/MPElectorateMap.pdf; (c) Anthropogenic climate risk are reportedly higher in SEQ than those of other Australian regions (Mortoja & Yigitcanlar, 2020); (d) The residents of SEQ are already exposed to significant climate risk impacts (Bajracharya & Hastings, 2018); I SEQ, on top of its current population of about 3.8 million, is going to accommodate around 2 million additional people by 2041 (Perveen at al., 2020); (f) Newly proposed settlements are predominantly located in peri-urban areas, which are one of the most vulnerable chunk of impending climate risk impacts (Infrastructure Australia, 2015), and; (g) Despite looming climate risk impacts, no solid consensus on climate risk mitigation and adaptation is achieved within the political parties (The Climate Council, 2019).

2.2.3 Survey and Participants

One of the principles of data collection was to capture public views from different geographical subregions of the SEQ region—as political orientations across urban, peri-urban, and rural areas are highly diverse in the study area. This study, thus, investigates the public opinions of the peri-urban, urban, and rural dwellers in relation to climate risk impacts. Thus, to distribute the online survey questionnaire link, urban, peri-urban, and rural areas were identified by Mortoja and Yigitcanlar's (2021b) mapping approach first (Figure 2.1). In their study approach, urban, peri-urban, and rural areas are mapped based on the level of urbanisation through the application of fuzzy set theory (Mortoja & Yigitcanlar, 2021b, 2022b). In doing so, they utilised night-time light data and placed the level of urbanisation within the framework of fuzzy linear membership function that lies between 0 and 1. Figure 2.1

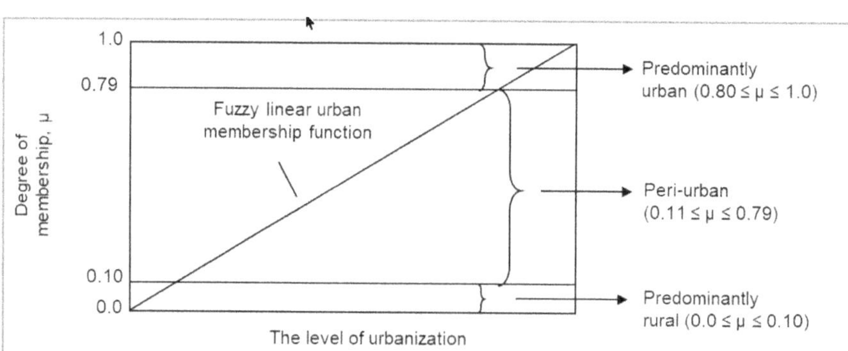

FIGURE 2.1
The value range of the fuzzy linear urban membership function set categorising rural, urban, and peri-urban areas. (Mortoja & Yigitcanlar, 2020, p. 10.)

depicts the value range of the fuzzy linear urban membership function set categorising rural, urban, and peri-urban areas.

The sample size was determined based on the SEQ's estimated resident population of 2020, which is about 3.8 million at present (Australian Bureau of Statistics (ABS), 2020). Afterwards, the Australian Bureau of Statistics' (ABS) sample calculator, which calculates sample size by using a simple random survey technique, was used to determine the minimum sample size (see https://www.abs.gov.au/websitedbs/d3310114.nsf/home/sample+size+calculator).

Under the simple random sampling survey technique, with a 95% confidence interval, the minimum sample size for this survey was identified as 385. However, due to avoid the complexity and procrastination in data collection, this study employed a private survey panel company to carry out the survey. The survey was operationalised through the Key Survey Tool—an enterprise survey platform—and 659 valid responses were obtained from the residents of the SEQ region (Figure 2.2). The collected responses exceeded the designed threshold sample size of 385. All the questions under this survey were closed-ended and hence no case was found with missing data. Therefore, all collected samples were used for analysis.

2.2.4 Statistical Analysis

The methodological approach primarily comprises two major parts: (a) Identification of major items underpinning knowledge factors for climate risk analysis, and; (b) Cluster analysis by using those identified items.

The study derived the items for knowledge factor through the light of the findings of Mortoja and Yigitcanlar's (2021a) factor analysis as their study approach listed the latest findings for the Australian context. According to their study, the items of knowledge factor are categorised into two major parts: (a) Cause of climate risk, and; (b) Consequence of climate risk. To investigate respondents' knowledge on the cause of climate risk, they identified human activity driven GHG emissions as the prime cause of climate risk. Whereas to investigate the consequences of climate risk, they identified environmental deterioration and ecosystem degradation, extreme weather events (extreme heat and rainfall events), bushfires, floods, drought, and sea-level rise as major consequences of climate risk.

Next, the selected items underpinning this knowledge factor were further gone through a two-step cluster analysis procedure. In step one, the study used hierarchical cluster analysis (Cluster method: Ward's method; Interval: Squared Euclidean Distance) to identify the number of clusters under this selected factor (see https://www.youtube.com/watch?v=3BzfOLnIY9w). While performing the hierarchical cluster analysis, dendrogram plotting was chosen to better understand the inter-relationships that lie between the selected items describing the knowledge factor. The Ward's Cluster method was chosen for this analysis due to its inherent robustness in identifying the clusters with equal size (Murtagh & Legendre, 2014). While doing the cluster

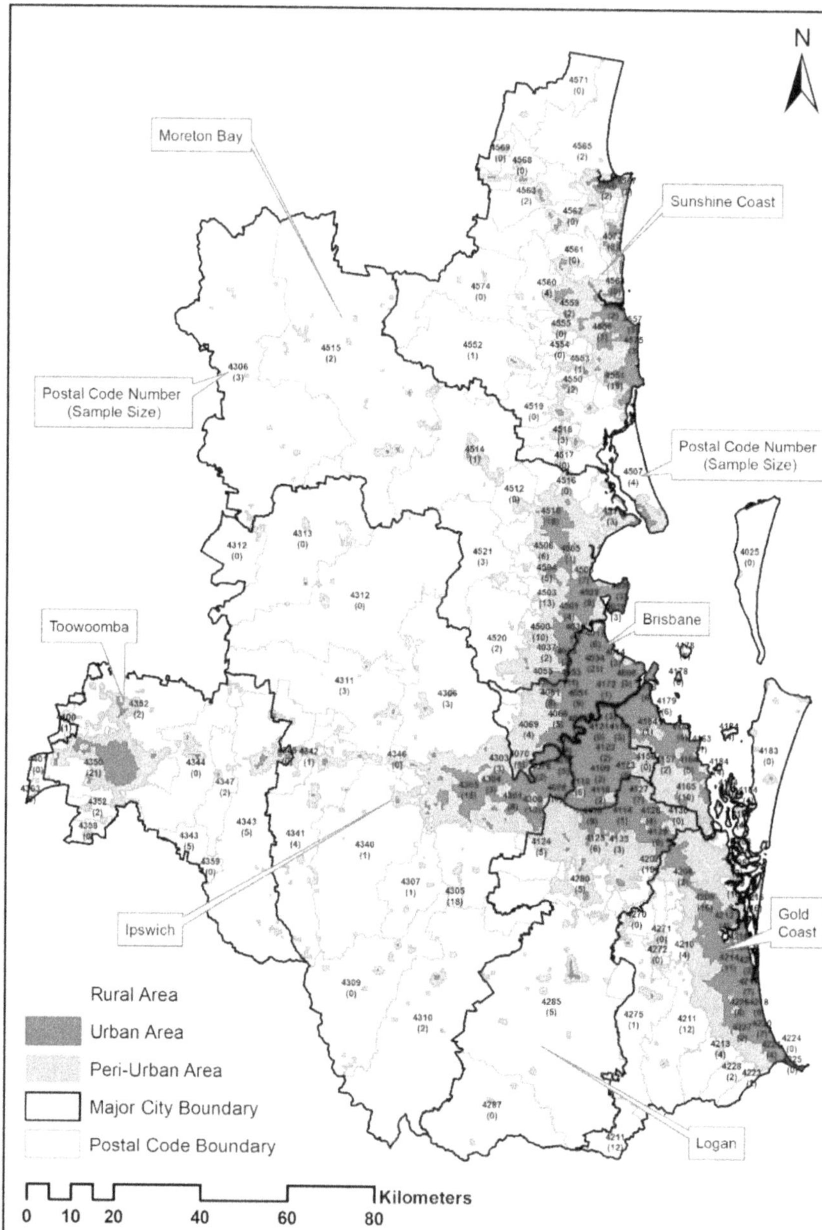

FIGURE 2.2
Sample distribution between peri-urban, urban, and rural areas of SEQ.

analysis, no value standardisation was performed as all the selected variables under this knowledge factor were measured on a scale of 1–9 (where 1 = strong disagreement; 5 = neural; and 9 = strong agreement). In this way, the required number of clusters for the selected datasets was determined.

Later, the K-means clustering technique was used to interpret the clusters—i.e., how the items of knowledge factor contribute to form each cluster (see https://www.youtube.com/watch?v=wdVtiVx_mAk&t=16s). More importantly, K-means in an iterate algorithm and as such this study used the maximum iteration of 99 to run this operation to get the most desirable clustering outcome. Afterwards, items describing each cluster were explored through the examination of ANOVA table, cluster membership, and the distance from cluster centre.

Thus, while the number of clusters was determined through the hierarchical cluster analysis, the K-means clustering technique was employed in deciding how the selected items of knowledge factor are accommodating into each cluster.

Finally, each cluster was analysed based on its political orientation.

2.3 Analysis and Results

2.3.1 Survey Participant Profiles

The socioeconomic characteristic of the respondents is presented in Table 2.1. Respondents' political orientations have been categorised into four types including left, centre, right, and other (e.g., apolitical, unwilling to reveal, do not vote, undecided, and so on). The survey outcome reveals that most of the respondents are inclined to centrism (55.1%) political doctrine, while the remaining chunk is almost equally distributed to left-wing (18.7%) and right-wing (19%) proponents.

2.3.2 General Description of Clusters

Regarding the knowledge factor, two contrasting clusters—i.e., Clusters 1 and 2—were found here (Table 2.1). From those, Cluster 1 appears to be least concerned about climate risk and Cluster 2 is highly concerned about climate risk. Besides, the average value of each item of the knowledge factor as listed in Table 2.2 is surprisingly different between these two generated clusters. All items were found highly statistically significant too—meaning the generated clusters are discernibly different from each other.

Nonetheless, albeit the reported clusters are discernibly different, they do share some similarities too. For instance, it is revealed that the anthropogenic cause of climate risk—i.e., 'climate change is taking place due to the

TABLE 2.1

Socioeconomic Profile of the Respondents

Socioeconomic Profile	Number	%
How do you identify yourself in terms of gender?		
Female	320	48.6
Male	335	50.8
Other	4	0.6
Which age group do you belong?		
18–24	45	6.8
25–34	125	19.0
35–44	119	18.1
45–54	104	15.8
55–64	97	14.7
65–74	96	14.6
75–84	65	9.9
≥85	8	1.2
Which group do you belong in terms of your highest educational attainment?		
Year 10 or below	84	12.7
Year 11 or equivalent	14	2.1
Year 12 or equivalent	103	15.6
Certificate	129	19.6
Advanced diploma/diploma	97	14.7
Bachelor's degree	132	20.0
Postgraduate degree	100	15.2
How do you identify yourself in terms of your most preferred political ideology?		
Left	123	18.7
Centre	363	55.1
Right	125	19.0
Other (e.g., apolitical, unwilling to reveal, do not vote, undecided, etc.)	48	7.3
What is your occupation/job title? (If you are unemployed due to COVID-19, please state your pre-COVID-19 occupation/job title)		
Agriculture and fisheries employee	1	0.20
Consultant	10	1.50
Manufacturer	11	1.70
Business owner	17	2.60

(Continued)

TABLE 2.1 ***(Continued)***

Socioeconomic Profile of the Respondents

Socioeconomic Profile	Number	%
Technology professional	18	2.70
Semi-skilled worker	19	2.90
Skilled tradesperson	28	4.20
Unskilled or labourer	33	5.00
Sales worker	35	5.30
Student	53	8.00
Manager, executive or director	62	9.40
Clerical or office worker	71	10.80
Professional	101	15.30
Other	200	30.30

TABLE 2.2

Clusters Based on Knowledge Factors

Knowledge Factor	Proxy Variable	Cluster 1: Least Concerned/Mostly Disagreed (Average Value of Items On a Scale of 1–9)	Cluster 2: Highly Concerned/Mostly Agreed (Average Value of Items On a Scale of 1–9)
Climate change is continuously degrading our valuable ecosystems and surrounding environment	Consequences of climate risk	3.36	7.76
Extreme or unusual weather events—e.g., abrupt, or extreme rainfall, drought, abrupt hot events in the summer, prolonged summer—are intensified by climate change impacts		3.25	7.75
Sea-level-rise is intensified by climate change impacts		3.25	7.65
The severity and frequency of natural disaster events—e.g., bushfires, floods—are intensified by climate change impacts		3.15	7.73
Climate change is taking place due to the GHG emissions produced by human activities	Cause of climate risk	3.13	7.64

GHG emissions produced by human activities' is least acknowledged by the respondents of both two clusters. Similarly, regarding the consequence of climate risk, the items—e.g., 'Climate change is continuously degrading our valuable ecosystems and surrounding environment' and 'extreme or unusual weather events—e.g., abrupt, or extreme rainfall, drought, abrupt hot events in the summer, prolonged summer—, are intensified by climate risk impacts' are most agreed by the respondents.

In general, the value below five implies disagreement and beyond five implies agreement. Thus, in sum, the findings imply that Cluster 1 mostly disagrees with the consequences and cause of climate change impacts. Whereas Cluster 2 predominantly agrees with the corresponding consequences and cause of climate risk. Therefore, higher disagreement in Cluster 1, in turn, generates more agreement in Cluster 2.

2.3.3 Socioeconomic Profile of Clusters

A cross-tabulation of respondents' gender, age, educational standing, occupation, and income status against these generated clusters was made. Such cross-tabulation nullifies any significant correlation of these generated clusters either with gender or income, but exhibits a strong correlation with age, educational standing, and occupation. It is evident that Cluster 2 dominates the presence of respondents that share the age range of 18–44, whereas Cluster 1 dominates the respondents that share the age range of 45 and above (Table 2.3).

In terms of educational standing, the percentage distribution of respondents having Year 10 to advanced diploma/diploma education is higher

TABLE 2.3

Age Distribution of Clusters 1 and 2 ($\chi^2 = 16.31$; df = 7; $p = 0.02$)

	Cluster 1: Least Concerned/Mostly Disagreed Group		Cluster 2: Highly Concerned/Mostly Agreed Group	
Age Group	**Frequency**	**%**	**Frequency**	**%**
18–24	15	6.4	30	7.1
25–34	39	16.7	86	20.2
35–44	29	12.4	90	21.2
45–54	40	17.1	64	15.1
55–64	35	15.0	62	14.6
65–74	40	17.1	56	13.2
75–84	31	13.2	34	8.0
≥85	5	2.1	3	0.7
Total	234	100.0	425	100.0

TABLE 2.4
Educational Standing of Clusters 1 and 2 ($\chi^2 = 34.82$; df = 6; $p = 0.000$)

	Cluster 1: Least Concerned/Mostly Disagreed Group		Cluster 2: Highly Concerned/Mostly Agreed Group	
Educational Level	**Frequency**	**%**	**Frequency**	**%**
Year 10 or below	44	18.8	40	9.4
Year 11 or equivalent	7s	3.0	7	1.6
Year 12 or equivalent	40	17.1	63	14.8
Certificate	54	23.1	75	17.6
Advanced diploma/diploma	38	16.2	59	13.9
Bachelor's degree	33	14.1	99	23.3
Postgraduate degree	18	7.7	82	19.3
Total	234	100.0	425	100.0

in Cluster 1, whereas the presence of respondents having Bachelor or Post Graduate degrees is higher in Cluster 2 (Table 2.4).

As for the case of occupational status, albeit the contrast in percentage distribution between clusters is not highly conspicuous, respondents' professional status is found to be significantly correlated with educational standing too ($\chi^2 = 305.50$; df = 78; $p = 0.000$). Therefore, respondents comprising unskilled or labourer, semi-skilled worker, skilled tradesperson, manager, executive or director, manufacturer, business owner, and respondents categorised as other that include pensioners, homemaker, freelancer, traffic controller, etc. are a bit more inclined to be accommodated in Cluster 1 (Table 2.5). Whereas respondents including professionals, clerical or office workers, and students are a bit more prone to be accommodated in Cluster 2.

2.3.4 Distribution of Political Orientations Within the Clusters

Throughout this chapter, respondents' political orientation is meant to refer to their most preferred political ideologies—e.g., left, centre, right, and other. Whereas respondents' political bias is meant to clarify whether respondents' concern on climate risk guides their voting behaviour or decision.

In terms of percentage distribution, the major differences between Clusters 1 and 2 are perceived in accommodating left-wing and right-wing supporters within these two clusters (Table 2.6). While Cluster 2 accommodates more than thrice left-wing supporters (in terms of percentage) than those of Cluster 1, the percentage of right-wing supporters in Cluster 1 is more than double than those of Cluster 2. Thus, it is implied that right-wing supporters mostly fall within Cluster 1 and the left-wing supporters mostly fall within Cluster 2, while supporters adhering to the 'centrism' and 'other' political

TABLE 2.5

Occupational Status of Clusters 1 and 2 ($\chi^2 = 46.64$; df = 13; $p = 0.000$)

	Cluster 1: Least Concerned/Mostly Disagreed Group		Cluster 2: Highly Concerned/Mostly Agreed Group	
Occupational Status	**Frequency**	**%**	**Frequency**	**%**
Agriculture and fisheries employee	0	0	1	0.20
Technology professional	3	1.30	15	3.50
Consultant	4	1.70	6	1.40
Manufacturer	6	2.60	5	1.20
Student	7	3.00	46	10.80
Business owner	8	3.40	9	2.10
Sales worker	10	4.30	25	5.90
Semi-skilled worker	12	5.10	7	1.60
Unskilled or labourer	15	6.40	18	4.20
Skilled tradesperson	17	7.30	11	2.60
Clerical or office worker	22	9.40	49	11.50
Professional	22	9.40	79	18.60
Manager, executive or director	26	11.10	36	8.50
Other (e.g., pensioners, homemaker, freelancer, traffic controller, etc.)	82	35.00	118	27.80
Total	234	100.0	425	100.0

doctrines are almost equally distributed (in terms of percentage) between these two generated clusters.

Nevertheless, both generated clusters are dominated by the central-wing supporters in general, and out of the total respondents that adhere to the centrism political ideology: nearly two-third of them (i.e., $(229/363) \times 100 = 63.1\%$)

TABLE 2.6

Political Orientations of Clusters 1 and 2 ($\chi^2 = 42.90$; df = 3; $p = 0.000$)

	Cluster 1: Least Concerned/Mostly Disagreed Group		Cluster 2: Highly Concerned/Mostly Agreed Group	
Political Orientations	**Frequency**	**%**	**Frequency**	**%**
Left	19	8.1	104	24.5
Centre	134	57.3	229	53.9
Right	68	29.1	57	13.4
Other	13	5.6	35	8.2
Total	234	100.0	425	100.0

are accommodated in Cluster 2. Besides, out of the total respondents (n = 659): only 35.5% of them are accommodated in Cluster 1—implying that nearly two-third of the respondents are accommodated in Cluster 2.

Later, this study investigated whether respondents' perceptions of climate risk guide their voting behaviour. In doing so, the respondents were asked to rate their agreement on the statement that: 'my concerns about climate risk guide my voting behaviour'. Out of the total respondents (n = 659): those who did not agree (218; 33.1%) are described as 'politically non-biased', and those who agreed (271; 41.1%) are described as 'politically biased', and the remaining fraction is considered as 'politically neutral' (170; 25.8%) (Table 2.7). Those who are politically biased are not significantly correlated with age, gender, or income but has a strong correlation with educational standing ($\chi^2 = 50.93$; df = 12; p = 0.000). Within this politically biased group, nearly half of the respondents (i.e., 48.3%) hold a bachelor or post graduate degree.

Besides, while investigating respondents' such reported political bias as stated above against their respective political orientations, there is a strong correlation found (Table 2.7). It is also revealed that out of the respondents those are the followers of 'centrism' and 'other' political doctrines: their stance on concerns about climate risk guiding voting behaviour is nearly equally distributed between politically biased, politically neutral, and politically non-biased groups with slightly a higher portion is reported to be 'politically biased'. Whereas, out of the total left-wing supporters, 70.7% of them are politically biased on that issue. Contrariwise, the highest percentage share (52%) that belongs to the politically non-biased group, are the right-wing supporters. Besides, the politically neutral group is seen to be least contributed by the left-wing supporters (i.e., 11.4%), and is highest contributed with almost equal share by the central-wing (31.4%) supporters and proponents of 'other' political doctrine (31.3%).

TABLE 2.7

Respondents' Political Orientations Against Their Recoded Stances that are Derived Based on the Statement That 'Concerns About Climate Change Impacts Guide Their Voting Behaviour' ($\chi^2 = 72.78$; df = 6; $p = 0.000$)

Political Orientation	Politically Non-Biased		Politically Neutral		Politically Biased		Total	
	Frequency	%	Frequency	%	Frequency	%	Frequency	%
Left	22	17.9	14	11.4	87	70.7	123	100
Centre	116	32.0	114	31.4	133	36.6	363	100
Right	65	52	27	21.6	33	26.4	125	100
Other	15	31.3	15	31.3	18	37.5	48	100
Total	218	33.1	170	25.8	271	41.1	659	100

TABLE 2.8

The Distribution of Political Bias in Highly Concerned/Mostly Agreed Group (χ^2 = 34.64; df = 6; p = 0.000)

Political Orientation	Politically Non-Biased		Politically Neutral		Politically Biased		Total	
	Frequency	%	Frequency	%	Frequency	%	Frequency	%
Left	9	8.7	11	10.6	84	80.8	104	100.0
Centre	42	18.3	65	28.4	122	53.3	229	100.0
Right	17	29.8	16	28.1	24	42.1	57	100.0
Other	7	20.0	12	34.3	16	45.7	35	100.0
Total	75	17.6	104	24.5	246	57.9	425	100.0

Surprisingly, when cluster-wise crosstabulation was made in identifying such reported political bias as stated above, Cluster 1 does not exhibit any significant correlation, but Cluster 2 does (Table 2.8). In Cluster 2, out of the left-wing supporters, 80.8% of them are politically biased. Surprisingly, out of the total right-wing supporters in Cluster 2, a significant portion (42.1%) of them is reported to be politically biased.

2.3.5 Political Bias Within the Knowledge Factor

To understand political bias in forming a climate change belief, first, respondents' political bias and over the knowledge factor is investigated. This investigation is carried out over three sampling considerations: (a) Overall sample (n = 659); (b) Excluding political bias—i.e., 'politically biased' group was removed from the sample (n = 659 − 271 = 388), and; (c) Only including politically biased group (n = 271). Then, respondents' perceptions on the knowledge factor of climate risk against their political bias were measured.

About the knowledge factor, the overall views of the respondents concerning the cause of climate risk—i.e., human activity-driven climate risk impacts—are presented in Table 2.9. From the overall sample, it is evident that Cluster 1 predominantly denies (62.8%) their level of agreement on the cause of human activity driven climate risk impacts, whereas Cluster 2 overwhelmingly supports (90.8%) this cause of climate risk.

Even when the politically biased group is removed from the sample, no significant changes in the percentage distribution is found in Cluster 1—implying that Cluster 1 is politically least affected to their perception on the cause of human activity driven climate risk. Contrastingly, in the absence of political bias, the level of agreement of the respondents in Cluster 2 declines to 81.6% from 90.8%. In addition, it is revealed that when political

TABLE 2.9

Views of the Respondents Concerning Climate Risk Cause

Sample Type	Level of Agreement On the Statement: 'Climate Change Is Taking Place Due to the GHG Emissions Produced by Human Activities'	Cluster 1: Least Concerned/Mostly Disagreed Group		Cluster 2: Highly Concerned/Mostly Agreed Group	
		Frequency	%	Frequency	%
Overall sample ($\chi^2 = 488.6$; df = 2; $p = 0.000$)	Disagree	147	62.8	5	1.2
	Neutral	76	32.5	34	8.0
	Agree	11	4.7	386	90.8
	Total	234	100.0	425	100.0
Excluding political bias ($\chi^2 = 211.5$; df = 2; $p = 0.000$)	Disagree	130	62.2	5	2.8
	Neutral	70	33.5	28	15.6
	Agree	9	4.3	146	81.6
	Total	209	100.0	179	100.0
Only including politically biased group ($\chi^2 = 254.03$; df = 2; $p = 0.000$)	Disagree	17	68.0	0	0
	Neutral	6	24.0	6	2.4
	Agree	2	8.0	240	97.6
	Total	25	100.0	246	100.0

bias is removed from the overall sample, the sample size of Cluster 2 is significantly reduced. It is calculated that the representation of politically biased respondents in Cluster 1 is 10.68%, whereas in Cluster 2 such percentage is 57.9%.

Cluster 1, however, exhibits higher disagreement (68%) on the cause of climate risk when the only politically biased group is included in the sample. Such reported disagreement in Cluster 1 is comparatively higher correspond to the remaining sampling considerations. Whereas the level of disagreement on the cause of climate risk in Cluster 2 is just 'nil' and such level of agreement reaches to 97.6% when only the politically biased group is included in the sample. Therefore, while perceiving human activity driven climate risk impacts, political bias is highly evident in Cluster 2, whereas the presence of political bias in Cluster 1 has been observed as extremely limited. Similar findings also are evident when respondents' stance is compared with the corresponding consequences of climate change impacts (refer to Appendices 2.A–2.C).

2.3.6 Exposure to Climate Risk and Political Bias

To understand the level of political bias in forming a climate change belief, then, the research investigated the differences in the perceptions of the

TABLE 2.10

Exposure to Climate Risk Events for Clusters 1 and 2

Exposure to Climate Risk Events	Cluster 1: Least Concerned/ Mostly Disagreed Group				Cluster 2: Highly Concerned/ Mostly Agreed Group			
	Overall Sample		Excluding Political Bias		Overall Sample		Excluding Political Bias	
	Frequency	%	Frequency	%	Frequency	%	Frequency	%
Floods	56	15.2	52	15.8	236	19.5	81	18.0
Bushfires	58	15.8	54	16.4	226	18.7	89	19.7
Drought	51	13.9	46	14.0	239	19.8	93	20.6
Extreme heat events	62	16.8	57	17.3	293	24.2	109	24.2
Extreme rainfall events	40	10.9	35	10.6	200	16.5	69	15.3
No, climate risk events still are not evident in my local area/region	101	27.4	85	25.8	15	1.2	10	2.2
Total	368		329	100.0	1209	100.0	451	100.0

Note: The total frequencies exceeded the sample size as multiple responses were allowed.

respondents who are already exposed to climate risk impacts—i.e., floods, bushfires, extreme heat events, and extreme rainfall events. The analysis reveals that Cluster 2 is more exposed to climate risk events than those of Cluster 1 (Table 2.10). Contrastingly, Cluster 1 is highly dominated by the group of respondents who perceived no climate risk events that took place in their local area/region. Besides, when the politically biased group is excluded from the sample, no significant changes in the percentage distribution between these two clusters are observed—implying that respondents' exposure to climate change events does not exert sufficient influence in weakening political bias.

2.4 Findings and Discussion

This study aims to identify different groups of respondents based on their level of knowledge concerning climate risk against their political orientations. This in return helps in understanding political bias in forming a climate change belief. For this end, the chapter identified groups of respondents based on their level of knowledge concerning climate risk, and such group's identification has been accomplished through cluster analysis. In terms of the knowledge factor—that investigates respondents' knowledge on the prime consequences and cause of anthropogenic climate risk, the cluster analysis carried out under this study identifies two major groups of respondents: (a) Cluster 1 named as 'least concerned/mostly disagreed group', and; (b) Cluster 2 named as 'highly concerned/mostly agreed group'. The analysis unveils that Cluster 1—i.e., 'least concerned/mostly disagreed group'—predominantly denies the consequences and cause of climate risk. Whereas Cluster 2—i.e., 'highly concerned/mostly agreed group'—is overwhelmingly agreed with the consequences and cause of climate risk. This finding is at par with what other studies reported (Bliuc et al., 2015; Corry & Jørgensen, 2015; Tyagi et al., 2020).

In terms of societal characteristics of the identified clusters, the 'least concerned/mostly disagreed group' predominantly comprises the older people and respondents with relatively lower educational standing. Contrariwise, younger adults and respondents with higher educational standing are mostly accommodated in the 'highly concerned/mostly agreed group'. Recent studies also have confirmed Australian's such climate risk perception with their corresponding age and educational attainment (Chiw & Ling, 2019; Leviston et al., 2011). Similar findings have also been reported in the studies concern of the US context (Weber, 2016).

On the one hand, these two generated clusters are not significantly correlated with their gender or income status. On the other, in case of other developed countries (e.g., Sweden, Germany, Norway, and the US), out of the gender, age, income, and educational status: only gender is reported to be statistically correlated with climate risk perception and females are found to be more worrying on climate risk issue (Lujala et al., 2015; Sundblad et al., 2007).

In addition, the 'highly concerned/mostly agreed group' is a bit more prone to accommodating different professionals, students, and office workers. Whereas the least concerned/mostly disagreed group' is slightly skewed to accommodating unskilled or labourer, semi-skilled worker, skilled tradesperson, manager, executive or director, manufacturer, business owner, pensioners, homemaker, freelancer, traffic controller, and so on. As the respondents' occupational status and educational standing are reported to be statistically correlated under this investigation, thus it can be implied that those who adapted higher education believe more in climate risk knowledge.

Accordingly, the 'least concerned/mostly disagreed group' is slightly skewed to accommodating unskilled or labourer, semi-skilled workers, skilled tradesperson who mostly did not adopt higher education. Besides, the denials of climate risk belief from the manager, executive or director, manufacturer, business owners are alarming in the sense that they can play the key role in practising an environmentally friendly business environment in mitigating climate risk. However, pensioners being senior citizens, may not be aware of climate risk perception, as the issue of climate risk has got momentum since 1990s (Wilby et al., 2009; Wilson, 2013). Henceforth, relatively younger adults are likely to be more aware of climate risk perception than those of elder adults.

These two generated clusters are also significantly correlated with their political orientations—i.e., right-wing, and left-wing supporters are more inclined to be with the 'least concerned/mostly disagreed group' and 'highly concerned/mostly agreed group' respectively. Hence, the right-wing and left-wing supporters are stationed in the two opposite ends of the totem pole concerning climate risk knowledge that has been tested under this study. A similar finding has also been reported for the context of the US (Boykoff, 2011; Jenkins, 2011).

Central-wing supporters being the most dominant group in the overall sample are almost equally distributed (in terms of percentage) between these two contrasting clusters. Such a finding implies that the key respondents are the left-wing and right-wing supporters for interpreting respondents political bias, and such reported political bias is inversely correlated between each group—i.e., left-wing supporters are more positively biased and right-wing supporters are more negatively biased in perceiving the consequences and cause of climate risk. In terms of the extent of such perceived bias, the left-wing supporters are found to be more positively biased than the level of negative bias that is reported within the right-wing supporters. Thus, it can also be inferred that out of the right-wing supporters, those are accommodated in the 'highly concerned/mostly agreed group': their concerns of climate risk are negatively impacted in perceiving climate risk knowledge.

In addition, while investigating whether the voting decisions of these two contrasting clusters are influenced by their climate risk concerns, it is found that climate risk concerns of the 'least concerned/mostly disagreed group' do not influence significantly in guiding their voting decisions. Therefore, those who do not believe in the consequences and cause of climate risk, do believe so irrespective of political orientations. Even respondents' such denials on climate risk perception are unaffected by their exposure to climate risk events, albeit it is claimed that exposure to climate risk events increases climate risk perception (Ballew et al., 2020), which in turn reduces political bias (Akerlof et al., 2013).

For instance, the findings unveil that a major fraction of respondents within the same geographic setting agreed with the view that climate change has already taken place in their local regions, while the remaining portion denied on occurring of such. Such findings imply that albeit all the respondents are exposed to different climate risk events, the respondents of the 'least concerned/mostly disagreed group' predominantly do not believe that such events are caused/intensified by climate change impacts—unveiling that these events are believed to be mainly natural than anthropogenic climate risk induced.

To this very point, within the consequences of climate risk, it is revealed that the highest disagreement in the 'least concerned/mostly disagreed group' is reported under the statement that climate change is causing an increase in the frequency and severity of natural disasters—e.g., bushfires, floods. Hence, the perception that events of natural disasters could be climate risk induced is least acknowledged by this 'least concerned/mostly disagreed group'. Whereas in the 'highly concerned/mostly agreed group', this item of knowledge is least disagreed. Thus, it can be inferred that the respondents of the 'least concerned/mostly disagreed group' are highly rigid on the stance that there is nothing like anthropogenic climate change. Therefore, the 'least concerned/mostly disagreed group' predominantly believe that the globally acknowledged anthropogenic climate risk events occurring around are mostly natural and are least affected by anthropogenic disturbances.

The remaining group of respondents—i.e., 'highly concerned/mostly agreed group'—those who predominantly believe in the consequences and cause of climate risk, are also politically biased on the premise that their concerns on climate risk guide their voting decisions. Nevertheless, the state of such reported bias is correlated with higher educational standing but is not linked to age, gender, or income status of the respondents—implying that education is the key determinant in deciding whom to vote for when the concern of climate risk becomes an important issue in the overall political agenda.

2.5 Conclusion

The findings reported in this study unveil a clear-cut stance of both left-wing and right-wing supporters concerning the knowledge factor underpinning climate change impacts. However, the findings disclose that their such stances about climate risk knowledge are rigid and simply distributed between the two groups—i.e., the right-wing supporters predominantly denies and the left-wing predominantly agrees on the consequences and cause of climate risk. This highlights the strong bias in climate change and risk perceptions in parallel to political views.

In addition, among the three sampling considerations, as investigated under this study, when only the politically biased group (i.e., those who agree with the view that their concerns on climate risk guide their voting decision) is sampled, the level of disagreement concerning the climate risk knowledge within the 'least concerned/mostly disagreed group' reaches to its peak. Contrariwise, the level of agreement in such a particular circumstance is found to be the highest in the 'highly concerned/mostly agreed group'. Besides, most of the respondents are accommodated in the 'highly concerned/mostly agreed group'. Therefore, it can be inferred that most of the respondents—whose perceptions were examined under this investigation—are aware of the climate risk knowledge in aggregate.

This study used anthropogenic GHG emissions as a single cause of climate risk, because such GHG emissions have already been unanimously acknowledged as one of the leading elements impacting climate risk across countries. Whereas the consequences of climate risk could be different based on the socioeconomic and geographical conditions of a country. As this study used the SEQ region of Australia as a testbed, the climate risk consequences that are endemic to this region were considered for this investigation.

In the findings reported under this study, however, the anthropogenic GHG emission as the cause of climate risk is least acknowledged irrespective of cluster-wise distribution whereas the consequences of climate risk have been reported with a higher agreement. Such findings could be explained on the premise that climate change-induced events—e.g., environmental degradation, ecosystem decline, extreme rainfall and heat events, bushfires, floods, drought—are generally evident across countries and respondents are exposed to these climate change-induced events directly or indirectly. Whereas the cause of climate risk is perceived based on the knowledge gained from different sources—e.g., newspapers, radio/TV, internet/social media. Thus, while the cause of climate change is theoretical, the consequences of climate risk are predominantly tangible and empirical. Hence, consequences are more perceived than the corresponding cause of climate risk in general.

The findings generated from this study disclose the characteristics of two opposing groups about climate risk belief and help us in understanding political bias in forming a climate change belief. The study thereby provides invaluable insights to overcome the knowledge gaps between climate risk believers and deniers. In that perspective particularly being subjected to climate change impacts (the experience) is shown to be the most effective way to curb the bias against the climate action. This is evidenced in the recent 2022 Australian Federal elections. While the polarisation of the far-left and far-right political are deepening, the centre-left kept its stance but the traditional centre-right had received the bigger lost, due to its climate change inaction while they were in power (in coalition with Australian National Party) for nine years (2013–2022).

Just before the 2022 election the centre-right party (Liberal Party) gave birth to a new political formation, so-called the Teals or Teal independents—mainly female independent candidates strongly advocating for increased climate change action (Brett, 2022). The raise of Teals and Greens in Australia, after the election now turns up the heat on the now governing party's (Australian Labor Party) climate policy (Skarbek & Malos, 2022). In sum, in this Australian case, public opinion and their experience with ongoing environmental catastrophes and natural disasters have led to a U-turn in the climate denial policy of the Federal government—where the climate change is also directly linked with urban policy and land use decisions.

The recent Federal election results gave hope for positive move towards climate action in Australia. However, the political polarisation is still a significant issue in Australia, particularly in the context or urban versus regional Australia (Melleuish, 2022). This is also the case for other countries such as the US where the political polarisation is getting deeper (Flowers, 2022; Gonzales, 202). In that instance, the study findings inform experts and policymakers in formulating more effective climate communication strategies, which in turn helps develop behavioural changes among the public in mitigating climate risk, and decision-makers taking adequate measures to address climate risks and develop appropriate land use decisions.

Lastly, while the items of knowledge factor that accommodate each cluster are found to be mutually exclusive within the clusters, the outcome is primarily derived through the Ward's method of clustering. Thus, how other clustering techniques—e.g., nearest neighbour, furthest neighbour, centroid clustering, medium clustering—differ from this study outcome remained unaddressed which in turn warrants further research.

Acknowledgements

This chapter, with permission from the copyright holder, is a reproduced version of the following journal article: Mortoja, M., & Yigitcanlar, T., (2022). Understanding political bias in climate change belief: A public perception study from South East Queensland. *Land Use Policy, 122*(1), 106350.

References

ABS (Australian Bureau of Statistics) (2020). *Regional population*. https://www.abs.gov.au/statistics/people/population/regional-population/latest-release.

Akerlof, K., Maibach, E. W., Fitzgerald, D., Cedeno, A. Y., & Neuman, A. (2013). Do people "personally experience" global warming, and if so how, and does it matter? *Global Environmental Change, 23*, 81–91.

Aklin, M., & Mildenberger, M. (2020). Prisoners of the wrong dilemma: Why distributive conflict, not collective action, characterizes the politics of climate change. *Global Environmental Politics, 20*(4), 4–27.

Antonio, R. J., & Brulle, R. J. (2011). The unbearable lightness of politics: Climate change denial and political polarization. *The Sociological Quarterly, 52*(2), 195–202.

Arvai, J., Gregory, R., Bessette, D., & Campbell-Arvai, V. (2012). Decision support for developing energy strategies. *Issues in Science and Technology, 28*(4), 43–52.

Bajracharya, B., & Hastings, P. (2018). A regional, strategic growth-management approach to urban and peri-urban development in South East Queensland, Australia. *Journal of Regional and City Planning, 29*, 210–233.

Ballew, M. T., Pearson, A. R., Goldberg, M. H., Rosenthal, S. A., & Leiserowitz, A. (2020). Does socioeconomic status moderate the political divide on climate change? The roles of education, income, and individualism. *Global Environmental Change, 60*, 102024.

Bliuc, A. M., McGarty, C., Thomas, E. F., Lala, G., Berndsen, M., & Misajon, R. (2015). Public division about climate change rooted in conflicting socio-political identities. *Nature Climate Change, 5*(3), 226–229.

Bord, R. J., O'Connor, R. E., & Fisher, A. (2000). In what sense does the public need to understand global climate change? *Public Understanding of Science, 9*(3), 205.

Boykoff, M. T. (2011). *Who speaks for the climate? Making sense of media reporting on climate change.* Cambridge University Press.

Boykoff, M. T., & Smith, J. (2010). Media presentations of climate change. In *Routledge handbook of climate change and society* (pp. 240–248). Routledge.

Brett, J. (2022). *How the Liberals lost the 'moral middle class', and now the teal independents may well cash in.* Accessed on 25 July 2022 from https://theconversation.com/how-the-liberals-lost-the-moral-middle-class-and-now-the-teal-independents-may-well-cash-in-182293.

Chiw, A., & Ling, H. (2019). *Young people of Australia and climate change: perceptions and concerns.* Millennium Kids Inc.

Contestabile, M. (2014). Americans' views. *Nature Climate Change, 4*(2), 86.

Corry, O., & Jørgensen, D. (2015). Beyond 'deniers' and 'believers': Towards a map of the politics of climate change. *Global Environmental Change, 32*, 165–174.

Costa, D. L., & Kahn, M. E. (2013). Energy conservation "nudges" and environmentalist ideology: evidence from a randomized residential electricity field experiment. *Journal of the European Economic Association, 11*, 680–702.

Dizdaroglu, D., Yigitcanlar, T., & Dawes, L. (2012). A micro-level indexing model for assessing urban ecosystem sustainability. *Smart and Sustainable Built Environment, 1*(3), 291–315.

Farstad, F. M. (2019). Does size matter? Comparing the party politics of climate change in Australia and Norway. *Environmental Politics, 28*(6), 997–1016.

Feinberg, M., & Willer, R. (2013). The moral roots of environmental attitudes. *Psychological Science, 24*(1), 56–62.

Flowers, P. (2022). *Will Roe v Wade be overturned, and what would this mean? The US abortion debate explained.* Accessed on 25 July 2022 from https://theconversation.com/will-roe-v-wade-be-overturned-and-what-would-this-mean-the-us-abortion-debate-explained-173156.

Gonzalez, Y. (2022). Political polarisation and the extension of government: A revisited concept. *Social Identities, 28*(2), 252–266.

Gonzales-Iwanciw, J., Karlsson-Vinkhuyzen, S., & Dewulf, A. (2025). Multi-level learning in climate change adaptation planning: comparing three experiences from Latin America. *Regional Environmental Change, 25*(3), 108.

Goonetilleke, A., Yigitcanlar, T., Ayoko, G. A., & Egodawatta, P. (2014). *Sustainable urban water environment: climate, pollution and adaptation.* Edward Elgar.

Gregersen, T., Doran, R., Böhm, G., Tvinnereim, E., & Poortinga, W. (2020). Political orientation moderates the relationship between climate change beliefs and worry about climate change. *Frontiers in Psychology, 11*, 1573.

Häkkinen, K., & Akrami, N. (2014). Ideology and climate change denial. *Personality and Individual Differences, 70*, 62–65.

Holmes, D., & Star, C. (2018). Climate change communication in Australia: The politics, mainstream media and fossil fuel industry nexus. In *Handbook of climate change communication: Vol. 1* (pp. 151–170). Springer.

Infrastructure Australia (2015). *Population estimates and projections. Australian Infrastructure audit background paper April 2015*. Canberra.

Janoff-Bulman, R., Sheikh, S., & Hepp, S. (2009). Proscriptive versus prescriptive morality: Two faces of moral regulation. *Journal of Personality and Social Psychology, 96*, 521.

Jenkins, J. C. (2011). Democratic politics and the long march on global warming: Comments on McCright and Dunlap. *The Sociological Quarterly, 52*(2), 211–219.

Kamruzzaman, M., Hine, J., & Yigitcanlar, T. (2015). Investigating the link between carbon dioxide emissions and transport-related social exclusion in rural Northern Ireland. *International Journal of Environmental Science and Technology, 12*(11), 3463–3478.

Kellstedt, P. M., Zahran, S., & Vedlitz, A. (2008). Personal efficacy, the information environment, and attitudes toward global warming and climate change in the United States. *Risk Analysis, 28*(1), 113–126.

Leviston, Z., Walker, I., & Morwinski, S. (2013). Your opinion on climate change might not be as common as you think. *Nature Climate Change, 3*(4), 334–337.

Limaye, V. S., Max, W., Constible, J., & Knowlton, K. (2020). Estimating the costs of inaction and the economic benefits of addressing the health harms of climate change: Commentary describes illuminates the costs of inaction on the climate crisis and the economic savings of addressing this problem. *Health Affairs, 39*(12), 2098–2104.

Lorenzoni, I., Nicholson-Cole, S., & Whitmarsh, L. (2007). Barriers perceived to engaging with climate change among the UK public and their policy implications. *Global Environmental Change, 17*, 445–459.

Lujala, P., Lein, H., & Rød, J. K. (2015). Climate change, natural hazards, and risk perception: The role of proximity and personal experience. *Local Environment, 20*, 489–509.

Macintosh, A. (2013). The Carbon Farming Initiative: removing the obstacles to its success. *Carbon Management, 4*(2), 185–202.

Mahbub, P., Goonetilleke, A., Ayoko, G. A., Egodawatta, P., & Yigitcanlar, T. (2011). Analysis of build-up of heavy metals and volatile organics on urban roads in Gold Coast, Australia. *Water Science and Technology, 63*(9), 2077–2085.

Malka, A., Krosnick, J. A., & Langer, G. (2009). The association of knowledge with concern about global warming: Trusted information sources shape public thinking. *Risk Analysis: An International Journal, 29*(5), 633–647.

Marquart-Pyatt, S. T., McCright, A. M., Dietz, T., & Dunlap, R. E. (2014). Politics eclipses climate extremes for climate change perceptions. *Global Environmental Change, 29*, 246–257.

McCright, A. M., & Dunlap, R. E. (2011). The politicization of climate change and polarization in the American public's views of global warming, 2001–2010. *The Sociological Quarterly, 52*(2), 155–194.

Melleuish, G. (2022). *Nationals vote holds steady. Will the Coalition become a party of the regions and outer suburbs?* Accessed on 25 July 2022 from https://theconversation.com/nationals-vote-holds-steady-will-the-coalition-become-a-party-of-the-regions-and-outer-suburbs-182856.

Menny, C., Osberghaus, D., Pohl, M., & Werner, U. (2011). General knowledge about climate change, factors influencing risk perception and willingness to insure. *ZEW-Centre for European Economic Research Discussion Paper, 11*, 60.

Metaxiotis, K., Carrillo, F., & Yigitcanlar, T. (2010). *Knowledge-based development for cities and societies: integrated multi-level approaches*. IGI Global.

Mortoja, M. G., & Yigitcanlar, T. (2020). Local drivers of anthropogenic climate change: quantifying the impact through a remote sensing approach in Brisbane. *Remote Sensing, 12*, 2270.

Mortoja, M. G., & Yigitcanlar, T. (2021a). Public perceptions of peri-urbanism triggered climate change: survey evidence from South East Queensland, Australia. *Sustainable Cities and Society, 75*, 103407.

Mortoja, M. G., & Yigitcanlar, T. (2021b). Why is determining peri-urban area boundaries critical for sustainable urban development? *Journal of Environmental Planning and Management*. https://doi.org/10.1080/09640568.2021.1978405

Mortoja, M. G., & Yigitcanlar, T., (2022a). Are climate change, urbanisation and political views correlated? Empirical evidence from South East Queensland. *Urban Climate, 41*(1), 101061.

Mortoja, M. G., & Yigitcanlar, T. (2022b). Factors influencing peri-urban growth: Empirical evidence from the Dhaka and Brisbane regions. *Remote Sensing Applications: Society and Environment, 26*, 100762.

Murtagh, F., & Legendre, P. (2014). Ward's hierarchical agglomerative clustering method: which algorithms implement Ward's criterion? *Journal of Classification, 31*, 274–295.

Myers, T. A., Maibach, E. W., Roser-Renouf, C., Akerlof, K., & Leiserowitz, A. A. (2013). The relationship between personal experience and belief in the reality of global warming. *Nature Climate Change, 3*(4), 343–347.

Nisbet, M. (2009). Communicating climate change: why frames matter for public engagement. *Environment: Science and Policy for Sustainable Development, 51*, 12–23.

Norton, A., & Leaman, J. (2004). *The day after tomorrow: public opinion on climate change*. MORI Social Research Institute.

O'Neill, S. J., & Boykoff, M. (2010). Climate denier, skeptic, or contrarian? *Proceedings of the National Academy of Sciences, 107*(39), E151.

Palutikof, J. P., Boulter, S. L., Stadler, F., & Vidaurre, A. C. (2019). Tracking the progress of climate change adaptation: An Australian case study. *Environmental Science & Policy, 101*, 126–135.

Perveen, S., Yigitcanlar, T., Kamruzzaman, M., & Agdas, D. (2020). How can transport impacts of urban growth be modelled? An approach to consider spatial and temporal scales. *Sustainable Cities and Society, 55*, 102031.

Poortinga, W., & Pidgeon, N. (2004). *Public perceptions of genetically modified food and crops, and the GM nation? Public debate on the commercialisation of agricultural biotechnology in the UK: main findings of a British survey*. Centre for Environmental Risk.

Queensland Parliament (2017). *Political representation, Fact Sheet 7.6*. https://www.parliament.qld.gov.au/documents/explore/education/factsheets/Factsheet_7.6_PoliticalRepresentation.pdf.

Shi, J., Visschers, V. H., & Siegrist, M. (2015). Public perception of climate change: The importance of knowledge and cultural worldviews. *Risk Analysis, 35*(12), 2183–2201.

Shi, J., Visschers, V. H., Siegrist, M., & Arvai, J. (2016). Knowledge as a driver of public perceptions about climate change reassessed. *Nature Climate Change, 6*(8), 759–762.

Skarbek, A., & Malos, A. (2022). *The teals and Greens will turn up the heat on Labor's climate policy. Here's what to expect.* Accessed on 25 July 2022 from https://theconversation.com/the-teals-and-greens-will-turn-up-the-heat-on-labors-climate-policy-heres-what-to-expect-183532.

Sotto, D., Philippi, A., Yigitcanlar, T., & Kamruzzaman, M. (2019). Aligning urban policy with climate action in the global south: Are Brazilian cities considering climate emergency in local planning practice? *Energies, 12*(18), 3418.

Stevenson, K. T., Peterson, M. N., Bondell, H. D., Moore, S. E., & Carrier, S. J. (2014). Overcoming skepticism with education: interacting influences of worldview and climate change knowledge on perceived climate change risk among adolescents. *Climatic Change, 126*(3), 293–304.

Sundblad, E. L., Biel, A., & Gärling, T. (2007). Cognitive and affective risk judgements related to climate change. *Journal of Environmental Psychology, 27*, 97–106.

Tangney, P. (2021). Are "climate deniers" rational actors? Applying weberian rationalities to advance climate policymaking. *Environmental Communication, 15*(8), 1077–1091.

Taylor, A., Dessai, S., & de Bruin, W. (2014). Public perception of climate risk and adaptation in the UK: A review of the literature. *Climate Risk Management, 4*, 1–16.

The Climate Council (2019). *Climate policies of majustralianian political parties.* https://www.climatecouncil.org.au/wp-content/uploads/2019/05/climate-policies-of-major-australian-political-parties-v2.pdf.

Tobler, C., Visschers, V. H., & Siegrist, M. (2012). Consumers' knowledge about climate change. *Climatic Change, 114*(2), 189–209.

Tranter, B. (2020). Does public knowledge of climate change really matter in Australia? *Environmental Communication, 14*(4), 537–554.

Tranter, B., & Lester, L. (2017). Climate patriots? Concern over climate change and other environmental issues in Australia. *Public Understanding of Science, 26*(6), 738–752.

Tyagi, A., Babcock, M., Carley, K. M., & Sicker, D. C. (2020). Polarizing tweets on climate change. In *International conference on social computing, behavioral-cultural modeling and prediction and behavior representation in modeling and simulation* (pp. 107–117). Springer.

Walsh, E. M., & Tsurusaki, B. K. (2014). Social controversy belongs in the climate science classroom. *Nature Climate Change, 4*(4), 259–263.

Weber, E. U. (2016). What shapes perceptions of climate change? New research since 2010. *Wiley Interdisciplinary Reviews: Climate Change, 7*(1), 125–134.

Whitmarsh, L. (2011). Scepticism and uncertainty about climate change: Dimensions, determinants and change over time. *Global Environmental Change, 21*(2), 690–700.

Wilby, R. L., Troni, J., Biot, Y., Tedd, L., Hewitson, B. C., Smith, D. M., & Sutton, R. T. (2009). A review of climate risk information for adaptation and development planning. *International Journal of Climatology, 29*(9), 1193–1215.

Williams, H. T., McMurray, J. R., Kurz, T., & Lambert, F. H. (2015). Network analysis reveals open forums and echo chambers in social media discussions of climate change. *Global Environmental Change, 32*, 126–138.

Wilson, K. M. (2013). Communicating climate change through the media: Predictions, politics and perceptions of risk. In *Environmental risks and the media* (pp. 217–233). Routledge.

Wong-Parodi, G., & Feygina, I. (2020). Understanding and countering the motivated roots of climate change denial. *Current Opinion in Environmental Sustainability, 42*, 60–64.

Yigitcanlar, T. (2010). *Sustainable urban and regional infrastructure development: Technologies, applications and management*. IGI Global.

Yigitcanlar, T., Fabian, L., & Coiacetto, E. (2008). Challenges to urban transport sustainability and smart transport in a tourist city: The Gold Coast, Australia. *Open Transportation Journal, 2*, 29–46.

Yigitcanlar, T., & Kamruzzaman, M. (2014). Investigating the interplay between transport, land use and the environment: A review of the literature. *International Journal of Environmental Science and Technology, 11*(8), 2121–2132.

Yin, R. K. (1992). The case study method as a tool for doing evaluation. *Current Sociology, 40*(1), 121–137.

Zihniah, N. (2021). The policy regime and shift of Emissions trading scheme by Australian Labour Party and Emissions Reduction Fund by the coalition in Australia year 2013–2018. *Earth and Environmental Science, 716*, 12002.

Appendix

APPENDIX 2.A

Views of the Respondents Regarding the Consequences of Climate Risk (n = 659)

Items	Level of Agreement	Cluster 1: Least Concerned/ Mostly Disagreed Group		Cluster 2: Highly concerned/mostly agreed group	
		Frequency	%	Frequency	%
Climate change is continuously degrading our valuable ecosystems and surrounding environment ($\chi^2 = 505.18$; df = 2; $p = 0.000$)	Disagree	133	56.8	4	0.9
	Neutral	83	35.5	17	4.0
	Agree	18	7.7	404	95.1
	Total	234	100.0	425	100.0
Extreme or unusual weather events—e.g., abrupt or extreme rainfall, drought, abrupt hot events in the summer, prolonged summer—are intensified by climate change impacts ($\chi^2 = 506.19$; df = 2; $p = 0.000$)	Disagree	143	61.1	3	0.7
	Neutral	74	31.6	20	4.7
	Agree	17	7.3	402	94.6
	Total	234	100.0	425	100.0
Sea-level-rise is intensified by climate change impacts ($\chi^2 = 460.31$; df = 2; $p = 0.000$)	Disagree	137	58.5	3	0.7
	Neutral	74	31.6	29	6.8
	Agree	23	9.8	393	92.5
	Total	234	100.0	425	100.0
The severity and frequency of natural disaster events—e.g., bushfires, floods—are intensified by climate change impacts ($\chi^2 = 501.20$; df = 2; $p = 0.000$)	Disagree	146	62.4	2	0.5
	Neutral	73	31.2	27	6.4
	Agree	15	6.4	396	93.2
	Total	234	100.0	425	100.0

APPENDIX 2.B

Views of the Respondents Regarding the Consequences of Climate Risk (n = 388, Excluding Politically Biased Group)

Items	Level of Agreement	Cluster 1: Least Concerned/Mostly Disagreed Group		Cluster 2: Highly Concerned/ Mostly Agreed Group	
		Frequency	%	Frequency	%
Climate change is continuously degrading our valuable ecosystems and surrounding environment ($\chi^2 = 289.74$; df = 2; $p = 0.000$)	Disagree	118	56.5	2	1.1
	Neutral	78	37.3	12	6.7
	Agree	13	6.2	165	92.2
	Total	209	100.0	179	100.0
Extreme or unusual weather events—e.g., abrupt or extreme rainfall, drought, abrupt hot events in the summer, prolonged summer—are intensified by climate change impacts ($\chi^2 = 277.61$; df = 2; $p = 0.000$)	Disagree	128	61.2	1	0.6
	Neutral	67	32.1	17	9.5
	Agree	14	6.7	161	89.9
	Total	209	100.0	179	100.0
Sea-level-rise is intensified by climate change impacts ($\chi^2 = 261.63$; df = 2; $p = 0.000$)	Disagree	123	58.9	1	0.6
	Neutral	68	32.5	18	10.1
	Agree	18	8.6	160	89.4
	Total	209	100.0	179	100.0
The severity and frequency of natural disaster events—e.g., bushfires, floods—are intensified by climate change impacts ($\chi^2 = 268$; df = 2; $p = 0.000$)	Disagree	128	61.2	2	1.1
	Neutral	66	31.6	18	10.1
	Agree	15	7.2	159	88.8
	Total	209	100.0	179	100.0

APPENDIX 2.C

Views of the Respondents Regarding the Consequences of Climate Risk (n = 271, Only Including Politically Biased Group)

Items	Level of Agreement	Cluster 1: Least Concerned/ Mostly Disagreed Group		Cluster 2: Highly Concerned/ Mostly Agreed Group	
		Frequency	%	Frequency	%
Climate change is continuously degrading our valuable ecosystems and surrounding environment ($\chi^2 = 161.59$; df = 2; $p = 0.000$)	Disagree	15	60.0	2	0.8
	Neutral	5	20.0	5	2.0
	Agree	5	20.0	239	97.2
	Total	25	100.0	246	100.0
Extreme or unusual weather events—e.g., abrupt or extreme rainfall, drought, abrupt hot events in the summer, prolonged summer—are intensified by climate change impacts ($\chi^2 = 189.46$; df = 2; $p = 0.000$)	Disagree	15	60.0	2	0.8
	Neutral	7	28.0	3	1.2
	Agree	3	12.0	241	98.0
	Total	25	100.0	246	100.0
Sea-level-rise is intensified by climate change impacts ($\chi^2 = 145.29$; df = 2; $p = 0.000$)	Disagree	14	56.0	2	0.8
	Neutral	6	24.0	11	4.5
	Agree	5	20.0	233	94.7
	Total	25	100.0	246	100.0
The severity and frequency of natural disaster events—e.g., bushfires, floods—are intensified by climate change impacts ($\chi^2 = 223.98$; df = 2; $p = 0.000$)	Disagree	18	72.0	0	0
	Neutral	7	28.0	9	3.7
	Agree	0	0	237	96.3
	Total	25	100.0	246	100.0

3

Public Perceptions of Peri-Urban Climate Risks

3.1 Introduction and Background

In accommodating growing urbanisation demands, especially in metropolitan areas, the gradual onslaughts of urban expansions to the most productive peri-urban greenfield sites have become a common strategy (Dizdaroglu & Yigitcanlar, 2014; Dur & Yigitcanlar, 2015; Kamruzzaman et al., 2018). While peri-urban areas are being perceived as important for agricultural production, scenic beauty, biodiversity conservation and carbon sequestration, they also serve as significant sources for ecosystem services (Sylla et al., 2020)—which are all critical for climate risk mitigation. Nevertheless, due to the global urbanisation drives, urban encroachment into peri-urban areas is perpetual (Liu et al., 2019). While all nations are experiencing climate change catastrophes, constant peri-urbanisation is a hindrance in mitigating climate change impacts and achieving long-term sustainability of cities—that house the majority of the world population (Aldieri et al., 2019, 2021; Dell'Anna, 2021; Sotto et al., 2019).

Nevertheless, the magnitude and dimension of peri-urbanisation are divergent across countries (Wandl & Magoni, 2016). Thus, their subsequent growth implications on climate risk mitigation vary across the globe, which in turn might generate new forms of climate risk (Kahn, 2009). For instance, following the 2011 Brisbane floods, insurance companies escalated their premium and became reluctant to insure properties and infrastructures that are located in peri-urban floodplains (Ma et al., 2012). Hence, dealing with constant peri-urbanisation and its subsequent growth implications has become a formidable urban planning challenge globally (Zhao, 2013), whereas global major cities are simultaneously becoming more exposed to climate change risk (Broto & Bulkeley, 2013; Llausas et al., 2016).

Few studies which investigated the implications of peri-urbanisation-driven climate risk observe that the implications of perpetual peri-urbanisation might pose severe impacts on climate risk mitigation. Such an observation is more critical to countries that are highly vulnerable to climate change impacts, in particular. For instance, Bangladesh is one of the most vulnerable countries to climate change risk (Ali, 1999; Biswas, 2013; Huq et al., 2019). In particular, at the metropolitan level, Dhaka's (the

DOI: 10.1201/9781003737049-4

capital city of Bangladesh) peri-urban growth areas are found to be one of the most vulnerable locations to climate change impacts (Mortoja & Yigitcanlar, 2020a).

Another major downfall of this rampant peri-urbanisation includes a decline in the valuable ecosystem and in the spaces of carbon sink, which in turn increases global warming potentials. For instance, due to indirect and direct anthropogenic climate change, Brisbane (Australia) and its surrounding areas experienced a net decline of nearly US$4.5 billion in ecosystem service value between 1989 and 2019 (Mortoja & Yigitcanlar, 2020b). According to Mortoja and Yigitcanlar (2020b), peri-urban areas are found with the highest losses in ecosystem service values and with the highest greenhouse gas (GHG) emission production levels when such figures are compared with the corresponding ecosystem service value losses and GHG in urban or rural areas. Long commuting from peri-urban areas is a significant contributor to both GHG and roadside pollution (Mahbub et al., 2011; Yigitcanlar & Kamruzzaman, 2019).

Consequently, such ongoing peri-urbanisation also endangers sustainable urban growth (McPhearson et al., 2015; Yigitcanlar, 2010). Henceforth, maintaining a balance between perpetual peri-urbanisation and preserving peri-urban areas' valuable climate risk mitigation elements appears to be a highly challenging task within the framework of sustainable development (Adedire, 2018; Shaw, 2005). Given these perceived challenges into consideration, the preparation of a set of policy documents in guiding and monitoring peri-urban areas' growth appears to be critical for achieving a region's long-term sustainability and mitigating climate change risk (Millward, 2002).

Nevertheless, peri-urban areas—which are neither purely urban nor rural—possess unique characteristics in terms of spatial settings, functionality, and governance (Rauws & de Roo, 2011). Thus, these peri urban areas are gaining increasing attention globally to be considered as an autonomous or independent jurisdictional space—which in turn deserves distinct policy guidelines of its own (Gallent & Shaw, 2007). In this way, with distinct policy guidelines, these peri-urban spaces individually exert specific implications in urban governance (Silva, 2019).

Besides, peri-urban areas, which hold the dichotomous characters of urban and rural land uses, always trigger controversies—e.g., land use conflicts, public discourse—with regard to further urban encroachment (Ruoso & Plant, 2018; Yigitcanlar & Kamruzzaman, 2014). Thus, the task of ensuring social justice becomes an important issue when a particular development activity takes place in peri-urban areas (Metaxiotis et al., 2010; Narain, 2014). However, social justice is often compromised by taking advantage of the weak opposition from the peri-urban dwellers (Zabik & Prytherch, 2013).

Such an occurrence usually happens when policymakers and planners are more convinced by the market-led speculative forces, and the long-term sustainability of peri-urban growth areas is undermined (Foelske et al., 2019; Friedberger, 2000; Merson et al., 2010). In this manner, despite being sufficiently aware of the long-term adverse consequences, policymakers and planners are more yielding to the market-led speculative forces underpinning peri-urban growth (Al-Mamun & Kim, 2020).

Thus, instead of making peri-urban dwellers sufficiently aware of the long-term adverse consequences of such an ongoing drive of peri-urbanisation, policymakers and planners generally provide indirect approval to such decisions on peri-urban growth areas (Yue et al., 2013). In this way, while simultaneously causing a decline in the reserve of peri-urban natural elements, the weak opposition of the peri-urban dwellers is capitalised, and thereby justice to these peri-urban dwellers is greatly compromised (Liu et al., 2019). Consequently, the concern of mitigating climate risk impacts and ensuring social justice with regard to peri-urban growth decisions appears to be highly critical.

Nonetheless, peri-urban dwellers themselves are also a part of the peri-urban growth elements. Thus, in order to ensure social justice, their level of awareness in foreseeing such peri-urbanisation-driven adverse consequences and commitment to preserve these peri-urban valuable resources appear to be important. Hence, a better understanding of the views of the general public on the interplay between urban growth and climate change, with particular interest to peri-urbanisation, is necessary.

Against this backdrop, the study reported in this chapter addresses the question of how the public perceives the interplay between peri-urban growth and climate change. The study selects the South East Queensland (SEQ) region of Australia as the case location, gathers data through an online survey, and analyses the collected data statistically.

3.2 Research Design

3.2.1 Case Study

In order to reveal public perceptions on the link between urban growth and climate change impacts, this chapter adopts a case study approach. For revealing public perceptions, the SEQ region of Australia is selected as a test-bed because of the following reasons: (a) SEQ is Australia's one of the most rapidly urbanised areas with continuing population increase trend (Perveen et al., 2017); (b) The peri-urban productive agrarian landscapes of SEQ—considered as significant for climate change mitigation circumstances—have been gradually encroached by urban sprawl (Guhathakurta & Stimson, 2007); (c) Urban expansions—taking place in SEQ—are predominantly occurring in the form of low-density dispersed peri-urban development (Bajracharya &

Hastings, 2018); (d) The major concentrations of these peri-urban expansions are directed towards the coastal areas of SEQ (Perveen et al., 2020); (e) SEQ's major climate change-induced incidents—e.g., bushfires, floods, drought, heat waves—are predicted to occur more frequently with an intensified degree (Kankanamge et al., 2020), and; (f) In terms of producing GHG—one of the prime catalysts of climate change—SEQ is responsible for a significant portion of the nation's GHG emissions derived from land use activities (Steffen & Dean, 2018).

3.2.2 Participants

By using the Key Survey Tool—an enterprise survey platform—an online questionnaire survey was carried out among the residents of urban, peri-urban, and rural dwellers of the SEQ region. This study adopted the simple random sample technique, and only people aged greater than or equal to 18 were selected as participants. The fuzzy-logic-driven peri-urban areas' mapping approach proposed by Mortoja and Yigitcanlar (2021) was used to distribute the online questionnaire survey forms among the respondents of urban areas, peri-urban areas and rural areas (Figure 3.1). The sample size is determined by using the sample size calculator of the Australian Bureau of Statistics (Figure 3.2)—available at https://www.abs.gov.au/websitedbs/D3310114.nsf/home/Sample+Size+Calculator. Based on the 2020 estimated resident population data of SEQ, the threshold sample size for this survey was calculated as 385 (with 95% confidence level).

In order to examine whether the selected sample is representative of the study area's population, the salient socioeconomic characteristics of the sample dataset were compared with the most recent census data of the Australian Bureau of Statistics (ABS)—i.e., 2016 census (Table 3.1). Such a comparison reveals that the selected sample does not demonstrate significant variation from the census data in terms of percentage distribution, and it comprises respondents with diversified strata of the public.

3.2.3 Data

After a thorough review of the literature on the public's knowledge and attitude on climate change impacts, a structured questionnaire was developed (Boykoff & Rajan, 2007; Hu et al., 2017; Kingwell, 2006; Leviston et al., 2011; Reser et al., 2012). This questionnaire comprised seven sections, concerning respondents': (a) Socioeconomic profile; (b) Attitudes towards sustainability practices; (c) Views on climate change; (d) Experience with climate change; (e) Views on urban growth and climate change; (f) Views on the federal, state and local government's strategies and actions in mitigating climate change impacts, and; (g) Views on the federal, state and local government's strategies and actions in directing urban growth. All the responses were closed-ended, including categorical options, and Likert-scale options were on a scale of

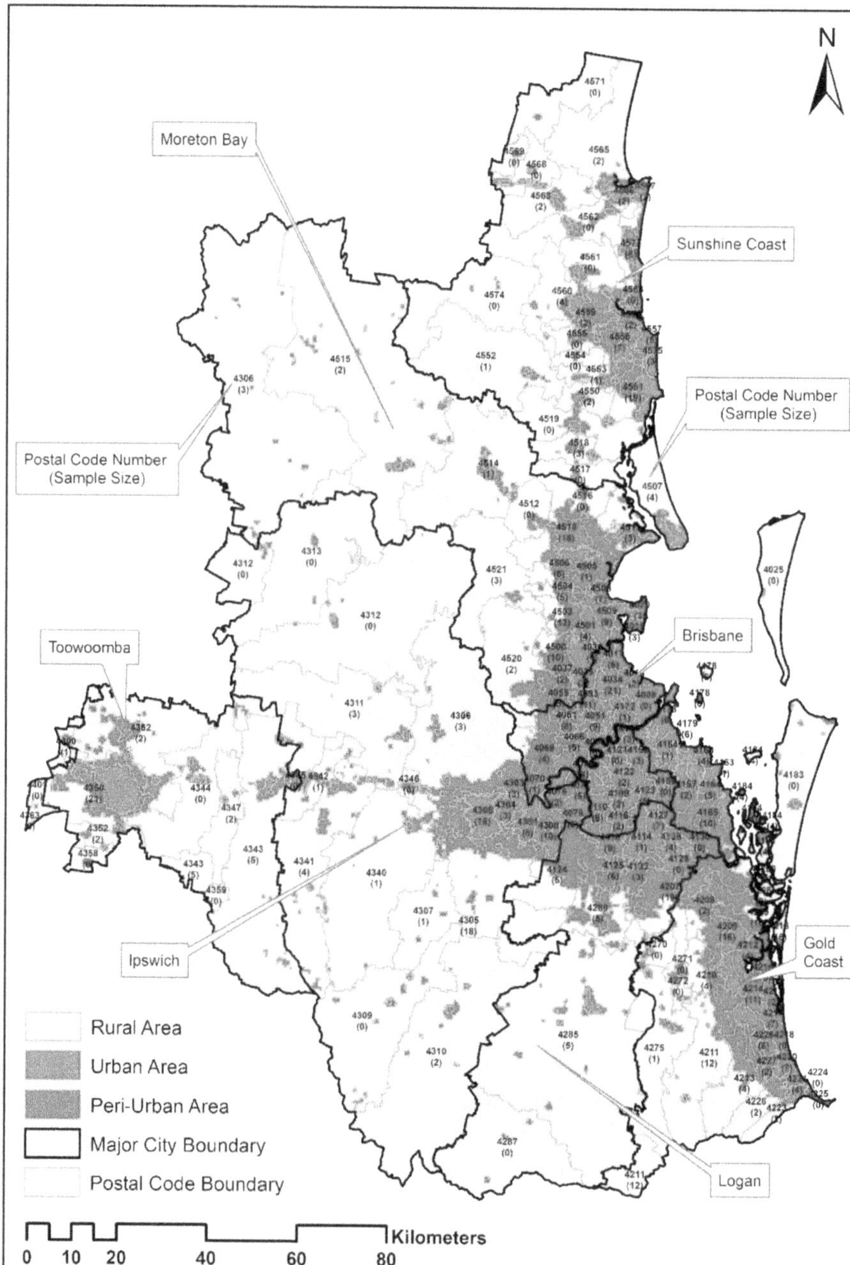

FIGURE 3.1
Urban, rural and peri-urban areas of SEQ.

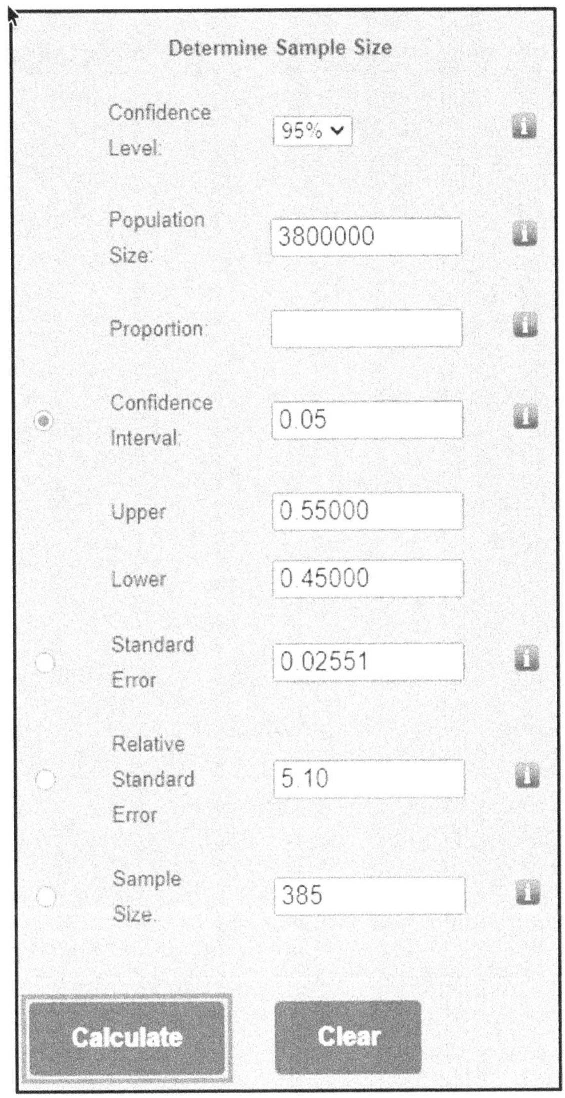

FIGURE 3.2
Sample size calculator. (From ABS, 2021.)

nine. To collect data from the targeted locations representing urban, peri-urban, and rural areas, the services of a reputed professional company were utilised. By sending the survey link to a small number (n = 20) of individuals, the survey work was first piloted. Later on, the questionnaire and corresponding responses were cross-checked, and further adjustments in the prescribed questionnaire were performed before the full-fledged launching of this survey activity in January 2021. With a response rate of 75.2%, a total

TABLE 3.1

Surveyed and Census Socioeconomic Profiles of the Investigated Peri-Urban Areas

Variables	Category	Frequency of Survey Data	Survey (%)	Census (%)
Gender	Female	320	48.6	50.60
	Male	335	50.8	49.40
Age	18–24	45	6.8	Not categorised according to census
	25–34	125	19	13.7
	35–44	119	18.1	13.4
	45–54	104	15.8	13.5
	55–64	97	14.7	11.7
	65–74	96	14.6	8.9
	75–84	65	9.9	4.5
	85 and over	8	1.2	1.8
Education	Year 10 or below	84	12.7	20.6
	Year 11	14	2.1	4.3
	Year 12	103	15.6	16.5
	Advanced diploma	97	14.7	8.7
	Bachelor	132	20	18.3
	Certificate	129	19.6	18.3
Profession	Professional	101	15.3	19.8
	Clerical or office worker	71	10.8	13.6
	Manager, executive or director	62	9.4	12.1
	Sales worker	35	5.3	9.7
	Skilled tradesperson	28	4.2	Not categorised according to census
	Unskilled or labourer	33	5.0	Not categorised according to census
	Consultant	10	1.5	Not categorised according to census
	Semi-skilled worker	19	2.9	Not categorised according to census
	Technology professional	18	2.7	Not categorised according to census
	Student	53	8.0	Not categorised according to census
	Manufacturer	11	1.7	Not categorised according to census
	Agriculture and fisheries employees	1	0.2	Not categorised according to census
	Business owner	17	2.6	Not categorised according to census
	Other	200	30.3	Not categorised according to census

of 659 responses were received. As the respondents were predominantly registered members of the survey company, the reported higher response rate as stated above appears to be reasonable. In order to ensure the consistency of the responses, all the responses were thoroughly checked and found to be reliable. Besides, as all the questions were closed-ended, there were no cases reported with missing data. Consequently, this study included all 659 responses for further analysis.

3.2.4 Statistical Analysis

Statistical analyses were carried out in the IBM SPSS Statistics platform. This study carried out two major types of statistical analyses: (a) Descriptive analyses; (b) Exploratory factor analysis (EFA). The descriptive analyses were carried out to reveal the nexus which lies between the respondents' socioeconomic characteristics and climate change perception. To test such a perceived nexus, the Chi-Square test was carried out. Whereas in order to carry out the EFA, by using a set of ordinal variables—which were all scaled from 1 to 9 (where 1 indicates strong disagreement, 5 indicates neutrality, and 9 indicates strong agreement), this study used the 'principal axis factoring' extraction method. This factor analysis was conducted to reveal the underlying factors that influence respondents' perception of the following sub-groups: (a) Climate change exposure factor; (b) Factor influencing urban growth and climate change; (c) Factor influencing climate change impacts driven by urban growth to peri-urban farmlands. In order to carry out the factor analysis, the methodological steps as proposed by Yong and Pearce (2013) were followed.

Under this factor analysis, first, this study derived the Cronbach's Alpha value in order to measure the internal consistency of ordinal datasets under each sub-group. For all these three sub-groups, the Cronbach's Alpha value was measured as 0.7 or greater than 0.7—meaning that all collected datasets taken for this factor extraction were reliable. Second, the presence of multicollinearity between the selected ordinal variables (also are called items) under each sub-group was examined by evaluating the 'determinant score'. A common rule of thumb—i.e., the determinant score of greater than 0.00001 implies the absence of multicollinearity (Yong & Pearce, 2013)—was employed. Through the use of the trial-and-error method, some variables were discarded to make the datasets free from the problem of multicollinearity.

Third, in order to investigate whether there is a patterned relationship within the selected variables, the Bartlett's Test of Sphericity was used. In all cases, the probability value under Bartlett's Test of Sphericity was derived as less than 0.001—meaning a patterned relationship between the selected variables has prevailed. Fourth, the Kaiser–Meyer–Olkin Measure of Sampling Adequacy (KMO MSA) was examined. For all three sub-groups, as stated earlier, the KMO MSA value was nearly 0.90—meaning that the factorability within the selected variables under each sub-group is achieved as 'marvellous' (Kaiser & Rice, 1974). Finally, this study used the oblique rotation technique (i.e., Promax with Kaiser Normalisation) in order to measure the correlation between the factors. Later on, the results of the factor analysis were interpreted based on the Scree plot test, rotated eigenvalues, and rotated factor loadings. Besides, coefficients smaller than 0.3 were suppressed from factor loadings. In this way, having derived the factors, reliability testing was later performed in order to validate the internal consistencies of the factorised variables under each extracted factor.

3.3 Analysis and Results

3.3.1 Descriptive Analysis

3.3.1.1 Socioeconomic Profiles

The participation of the two most common genders (i.e., female and male) was almost equal—i.e., male was 50.8%, and female was 48.6%. In terms of the age of the respondents, the respondents were chosen from assorted cohorts which predominantly comprise the age range of 25–74 by 82.1%. Whereas the least age cohort of participants was '85 and over' by 1.2%, and the age range of 18–24 by 6.8%. Besides, 85.1% of respondents have predominantly had an education level of Year 12 or equivalent to the postgraduate level. In terms of professional standing, there was no dominant group found—implying involvement of the varied professionals as survey participants. In terms of employment, 56.4% of respondents are involved with employment on a full-time or part-time basis. Besides, 61.2% of respondents dwell in their own houses.

This study observed a significant association in believing climate change impacts with age and education level. Respondents from relatively younger age cohort (i.e., 18–44) predominantly believe climate change has already taken place in the globe, whereas respondents who are agnostic about climate change, predominantly comprise the age range of 45 and above (Table 3.2). Besides, it is also revealed that respondents with higher educational attainment (i.e., bachelor's and postgraduate degrees) are more believers on climate change impacts than those of the moderate (i.e., certificate, advanced diploma/diploma) to lower (i.e., Year 10–Year 12) educational attainments (Table 3.2). Nevertheless, this study did not find any significant association in believing climate change impacts with gender, employment or income status (Table 3.3).

3.3.1.2 Sustainability Practices

The respondents' attitudes towards sustainability were tested in terms of practising a set of sustainability principles (Table 3.4)—that can be adopted

TABLE 3.2

Climate Change Views with Regard to Age Composition

	Do You Believe that the Climate Is Changing Now or Will Change Within Few Decades? (Pearson Chi-Square = 51.55; df = 21; p = 0.000)			
Age Group	**Yes, It Is Already Happening Now (%)**	**Not Now, but It Will Happen Within Few Decades (%)**	**Not Now, and It Will Not Happen Within Few Decades (%)**	**I Am Agnostic about Climate Change (%)**
18–24	77.8	11.1	4.4	6.7
25–34	82.4	8.8	4.0	4.8
35–44	77.3	14.3	5.0	3.4
45–54	72.1	2.9	7.7	17.3
55–64	71.1	3.1	9.3	16.5
65–74	69.8	6.3	8.3	15.6
75–84	60.0	9.2	4.6	26.2
85 and over	62.5	12.5	12.5	12.5
Total	73.6	7.9	6.4	12.1

TABLE 3.3

Climate Change Views with Regard to Educational Attainment

	Do You Believe that the Climate Is Changing Now or Will Change Within Few Decades? (Pearson Chi-Square = 36.38; df = 18; p = 0.006)			
Level of Education	**Yes, It Is Already Happening Now (%)**	**Not Now, but It Will Happen Within Few Decades (%)**	**Not now, and It Will Not Happen Within Few Decades (%)**	**I Am Agnostic about Climate Change (%)**
Year 10 or below	51.2	15.5	10.7	22.6
Year 11 or equivalent	78.6	0.0	7.1	14.3
Year 12 or equivalent	73.8	8.7	3.9	13.6
Certificate	74.4	4.7	6.2	14.7
Advanced diploma/ diploma	74.2	10.3	6.2	9.3
Bachelor's degree	79.5	6.1	5.3	9.1
Postgraduate degree	82.0	6.0	7.0	5.0
Total	73.6	7.9	6.4	12.1

TABLE 3.4

Sustainability Practices

Sustainability Practices Items	Number	%
Does your home have solar power?		
Yes	273	41.4
No	353	53.6
No, but considering to have in the next 5 years	33	5.0
Do you use any 'Energy Monitoring Device' in order to track electricity usage in your home?		
Yes	91	13.8
No	501	76.0
No, but considering to have in the next 5 years	67	10.2
Which type of car do you use at present?[a]		
Electric car	18	2.60
Hybrid car	18	2.60
LPG or CNG car	12	1.70
Petrol (Gasoline) car	511	72.50
Diesel car	95	13.50
Other	51	7.20
Which type of car do you prefer most to use in the future?		
Electric car	148	22.5
Hybrid car	133	20.2
LPG or CNG car	9	1.4
Petrol (Gasoline) car	264	40.1
Diesel car	74	11.2
Other	31	4.7
Do you have the provision of harvesting rainwater through the installation of rainwater tanks in your home?		
Yes	255	38.7
No	355	53.9
No, but considering to have in the next 5 years	49	7.4
Do you have the provision of 'composting' in your home?		
Yes	246	37.3
No	369	56.0
No, but considering to have in the next 5 years	44	6.7

[a] Total frequency exceeded the sample size, as multiple responses were acceptable.

at the household level and are linked to the respondents' cultural standings. Later on, these sustainability practices were linked to respondent's climate change perception.

Nevertheless, for this study area context, in particular, the usage of solar energy, composting, and rainwater harvesting are common at the household level. Other elements of sustainability practices that can be adopted at the household level—e.g., heating, cooling, cooking—were not considered under this study as they are more linked to technological innovation than to welcoming cultural changes. As Green and Vergragt (2002) rightfully stated that in order to cope with the upcoming sustainability challenges, technological innovations should be assisted by cultural changes too. Thus, the study reported in this chapter predominantly focused on examining respondents' attitudes towards commonly adopted sustainability practices that are endemic to the study area region. In this way, this study revealed how sustainability practices influence climate change perception.

Although 41.4% of respondents use solar energy, the analysis reveals that the trace of practising sustainability principles is not so significant at large. Besides, their intention to adopt sustainability practices soon—i.e., within the next five years—also appears to be insignificant. Nevertheless, an interesting contrast between their existing car usage type and preferred car choice types is observed. In terms of existing car usage type, petrol (gasoline) and diesel cars collectively accounted for 86% of the total usage type, whereas in terms of preferred car choice type, the preferences for petrol (gasoline) and diesel cars collectively accounted for 51.3%. Contrariwise, the preferences of electric car and hybrid car choices collectively reported as 42.7% against the existing usage type of electric car and hybrid car usage of 5.2%. In order to reveal whether such sustainability practices vary with the house ownership pattern, this study carried out a crosstab between these sustainability practices and the home ownership pattern (Table 3.5). Sustainability practices in connection to home ownership have been found highly statistically significant for all sustainability items except for the preferred car usage. In case of the preferred car usage, with respect to the home ownership pattern, the preferences of car choices were found statistically insignificant.

The analysis reveals that out of the respondents who already practice sustainability principles predominantly dwell in their own houses. Contrariwise, the respondents who replied 'No' to the sustainability practices are predominantly the rental dwellers (Table 3.5). Besides, out of the respondents who prefer to adopt sustainability practices within the next five years predominantly dwell in their own homes.

3.3.1.3 Climate Change Perceptions

Respondent climate change perceptions have been depicted in Table 3.6. About two-third (i.e., 73.6%) of the respondents believe that climate change is already happening, while only 6.4% of respondents opposed the occurrence

TABLE 3.5

Sustainability Practices with Relation to Home Ownership

	Sustainability Practices with Relation to the Home Ownership Pattern					
	Yes		No		No, but Considering to Have in the Next 5 Years	
Sustainability Practice Criteria	**Owned (%)**	**Rented (%)**	**Owned (%)**	**Rented (%)**	**Owned (%)**	**Rented (%)**
Does your home have solar power? (Pearson Chi-Square = 134.87; df = 2; $p = 0.000$)	86.10	21.90	40.80	59.20	72.70	27.30
Do you use any 'Energy Monitoring Device' in order to track electricity usage in your home? (Pearson Chi-Square = 14.41; df = 2; $p = 0.001$)	79.10	20.90	58.10	41.90	59.70	40.30
Do you have the provision of harvesting rainwater through the installation of rainwater tanks in your home? (Pearson Chi-Square = 58.41; df = 2; $p = 0.000$)	79.20	20.80	48.70	51.30	57.10	42.90
Do you have the provision of 'composting' in your home? (Pearson Chi-Square = 37.47; df = 2; $p = 0.000$)	76.00	24.00	53.10	46.90	45.50	54.50

TABLE 3.6

Climate Change Perceptions

Attributes	**Number**	**Percent**
From which source did you mainly learn what 'climate change' is?[a]		
Academic or professional journals/magazines	142	10.0
Colleagues/friends/family members	142	10.0
Movies	49	3.5
Newspapers	221	15.6
School	84	5.9
Social media/internet	236	16.6
TV/radio	394	27.8
University/courses	67	4.7
Workplace/training	40	2.8
I didn't know what 'Climate Change' was until this survey	11	0.8
Other	32	2.3

(Continued)

TABLE 3.6 *(Continued)*

Climate Change Perceptions

Attributes	Number	Percent
Do you believe that the climate is changing now or will change within few decades?		
Yes, it is already happening now	485	73.6
Not now, but it will happen within few decades	52	7.9
Not now, and it will not happen within few decades	42	6.4
I am agnostic about climate change	80	12.1
Please specify the particular climate change events that might have taken place in your local area.[a]		
Floods	292	16.5
Bushfires	284	16.0
Drought	290	16.4
Extreme heat events	355	20.0
Extreme rainfall events	240	13.5
More rainfall in the non-summer period	124	7.0
I am not sure whether the climate change has already taken place in my local area	72	4.1
No, the climate change has not yet taken place in my local area	116	6.5
What kind of effects climate change might generate?		
Nothing	55	8.3
Positive effects	21	3.2
Negative effects	221	33.5
Both of positive effects and negative effects	143	21.7
Negative effects will outweigh the positive effects	125	19.0
Positive effects will outweigh the negative effects	12	1.8
Not sure	82	12.4
Thinking about the causes of climate change, which of the following best describes your opinion?		
Climate change is entirely caused by natural processes	45	6.8
Climate change is mainly caused by natural processes	64	9.7
Climate change is partly caused by natural processes and partly by human activities	199	30.2
Climate change is mainly caused by human activities	212	32.2
Climate change is entirely caused by human activities	63	9.6
I think there is no such thing as climate change	41	6.2
Do not know	35	5.3
Thinking about climate change, which one best describes your views?		
The issue of climate change is generally overestimated	158	24.0
The issue of climate change is generally correct	216	32.8
The issue of climate change is generally underestimated	231	35.1
There is no climate change	54	8.2

[a] Total frequency exceeded the sample size, as multiple responses were acceptable.

of climate change at present or in the near future. Apropos of the kinds of effects that climate change might bring, one-third (i.e., 33.5%) of respondents believe that climate change will bring negative effects, while only 1.8% of respondents believe that climate change will bring positive effects. Besides, 89.4% of respondents believe that the incidents of climate change effects—e.g., floods, drought, bushfires, extreme rainfall events—have already taken place at the local level. With regard to the causes of climate change impacts, around one-third (i.e., 32.2%) of respondents believe that climate change is predominantly caused by human activities, while another one -third (i.e., 30.2%) of respondents believe that climate change is caused both by natural processes and anthropogenic disturbances. Surprisingly, more than one-third (i.e., 35.1%) of respondents believe that the issue of climate change is generally underestimated and another one-third (i.e., 32.8%) of respondents believe that the issue of climate change is generally correct.

Out of those (n = 212), who believe that climate change is either overestimated or non-existent, 59.4% of them are male, and the remaining fraction is female. However, there is no statistically significant correlation was found between such a particular belief with age, income, gender, educational standing or their willingness to participate in the climate change mitigation measures. Nonetheless, a strong correlation was found between such a belief with their views regarding what kind of effects climate change might generate (Pearson Chi-Square = 72.47; df = 6; p = 0.000). Only 5.7% of the respondents said that negative effects will outweigh the positive effects, while nearly one-third of them (24.5%) believe that it will create no effects indeed. Besides, 54.2% of the respondents disagreed with the statement that climate change is occurring because of human-activity-driven GHGs emissions, while 25% is neutral with this statement.

In addition, the analysis reveals that 60% of respondents came to know about climate change from media including newspapers, social media/internet, and TV/radio. Whereas only 0.8% respondents did not know what 'climate change' is until this survey.

3.3.1.4 Sustainability Practices and Climate Change Perceptions

A cross-tabulation linking respondents sustainability practices with climate change perception is depicted in Tables 3.7 and 3.8, where sustainability practices have been identified in terms of expressing preferred car choices. Out of the respondents who prefer to use electric car, 95.9% of them believe that climate change has already taken place or will take place soon (Table 3.7). Besides, respondents who use electric car predominantly (74.3%) perceive that climate change will bring negative effects or negative effects will outweigh the positive effects (Table 3.8). Whereas the respondents who use petrol or diesel cars predominantly believe that either climate change will bring negative effects or both positive and negative effects by 53.8 and 55.4%, respectively.

TABLE 3.7
Climate Change Views with Relation to the Preferred Car Choices

	Whether the Climate Is Changing Now or Will Change Within Few Decades? (Pearson Chi-Square = 59.33; df = 15; p = 0.000)			
Preferred Car Choices	**Yes, It Is Already Happening Now (%)**	**Not Now, but It Will Happen Within Few Decades (%)**	**Not Now, and It Will Not Happen Within Few Decades (%)**	**I Am Agnostic About Climate Change (%)**
Electric car	91.2	4.7	0.0	4.1
Hybrid car	82.0	7.5	4.5	6.0
LPG or CNG car	66.7	22.2	11.1	
Petrol (gasoline) car	62.1	9.8	9.8	18.2
Diesel car	64.9	8.1	9.5	17.6
Other	74.2	3.2	6.5	16.1

Nevertheless, this study did not find any statistically significant association between sustainability practices as populated in Table 3.5 with climate change belief except for the provision of having composting in their home. Thus, sustainability practices items—i.e., usage of solar energy and energy monitoring device or having the provision of rainwater harvesting—are not linked with believing the issue of climate change impacts. Nevertheless, out of the respondents who have the provision of composting in their home,

TABLE 3.8
Effects that Climate Change to Generate with Relation to the Preferred Car Choices

	Kind of Effects that Climate Change Might Generate (Pearson Chi-Square = 113.23; df = 30; p = 0.000)						
Preferred Car Choices	**Nothing (%)**	**Positive Effects (%)**	**Negative Effects (%)**	**Both of Positive Effects and Negative Effects (%)**	**Negative Effects Will Outweigh the Positive Effects (%)**	**Positive Effects Will Outweigh the Negative Effects (%)**	**Not Sure (%)**
Electric car	1.4	2.7	41.9	14.2	32.4	0.7	6.8
Hybrid car	5.3	7.5	42.1	15.8	21.1	2.3	6.0
LPG or CNG car	0.0	11.1	33.3	55.6	0.0	0.0	0.0
Petrol (Gasoline) car	13.3	1.1	26.1	27.7	10.6	2.3	18.9
Diesel car	10.8	4.1	28.4	27.0	17.6	2.7	9.5
Other	9.7	0.0	32.3	9.7	25.8	0.0	22.6

83.8% of them believe that climate change has already taken place or will take place soon (Pearson Chi-Square = 16.68; df = 6; p = 0.01). Thus, those who practice composting in their home are more believers on this climate change issue. Nevertheless, this study also did not find any statistically significant association between the home ownership pattern and climate change belief.

3.3.2 Factors Affecting Perceptions

3.3.2.1 Climate Change Exposure Factors

Climate change exposure factors are illustrated in Table 3.9. The Bartlett's Test of Sphericity (Chi-Square = 7329.416; df = 36; p = 0.000) indicates that there is a patterned relationship between the items which have been chosen under this sub-group for factor analysis. Besides, the KMO MSA was derived as 0.94—meaning that factorability between the selected items is marvellous. From the Scree plot test, this study derived three factors under this particular sub-group. Out of these three factors, two factors are observed with a minimum of at least three items, whereas one factor just comprised only one item. Besides, one cross-loading item was also observed. These three factors are classified as 'knowledge factor', 'moral factor', and 'livelihood factor'.

From the descriptive analysis, the highest disagreement is found in believing that climate change has had any income on their annual income or wellbeing. This item is labelled as a livelihood factor. This finding reveals that people do not believe that climate change is impacting their livelihoods. Besides, the item like 'respondents' willingness to pay more for the products that are produced with environmentally friendly materials and processes' is not supported and subsequently ranked as 2 in terms of disagreement. Contrariwise, 'the moral responsibility to use energy-efficient and clean technologies in reducing GHG' is found as the most agreed item. Besides, people's consideration to purchase environmentally friendly products is perceived with the liberal view—i.e., neither with higher agreement nor with higher disagreement. These all three items are labelled as 'moral factor'. The findings from this moral factor indicate that people are willing to adopt green technology in preventing climate change impacts, but their active participation in preventing climate change impacts appear to be limited. This inference is further supported by descriptive analysis as populated in Table 3.10. The findings show that only 7.7% of respondents ever participated in an event related to the mitigation of climate change impacts, and only 20.9% of respondents expressed their positive consent to participate in such events/activities/protests targeting the mitigation of climate change impacts in the future.

Another factor under this sub-group is the knowledge factor. The item which receives the highest agreement falls within this factor. Nevertheless, other items under this factor are perceived with liberal views—i.e., neither with higher agreement nor with higher disagreement. The highest agreed

TABLE 3.9

Climate Change Exposure Factors

Attributes	Descriptive Analysis					Factor Analysis (Pattern Matrix)		
	DA%	Neutral	A%	R-DA	R-A	Knowledge Factor	Moral Factor	Livelihood Factor
Climate change is occurring because of human-activity-driven GHG emissions	23.1	16.7	60.2	3	6	0.908		
Climate change is gradually deteriorating our environment and ecosystems	20.8	15.2	64	8	2	0.956		
The occurrences of extreme weather events are contributed by climate change impacts	22.2	14.3	63.5	5	3	0.968		
The occurrences of climate change are causing sea-level-rise and thereby submerge low-lying areas	21.2	15.6	63.2	7	4	0.85		
Climate change is causing an increase in the frequency and severity of natural disasters	22.5	15.2	62.3	4	5	0.875		
Climate change has had a negative impact on my annual income and/or wellbeing	50.2	26.3	23.5	1	9			0.489
It is our moral responsibility to use energy-efficient and clean technologies to reduce GHG emissions	15.8	16.1	68.1	9	1	0.714	0.341	
I do consider environmentally friendly products before any purchase is made	21.9	20.5	57.6	6	7		0.984	
I am willing to pay more for the products that are produced with environmentally friendly materials and processes	28.5	20.9	50.6	2	8		0.608	
Cronbach's Alpha = 0.96								
Determinant = 0.00001362								
Rotated sum of squared loadings						6.36	4.64	3.79
% of variance explained						73.067	6.35	1.801
Kaiser–Meyer–Olkin (KMO) = 0.94								
Extraction method = Principal Axis Factoring								
Rotation Method = Promax with Kaiser Normalisation								

TABLE 3.10

Willingness to Participate in Climate Change Mitigation Events

Attribute	Number	Percent
Did you ever participate in events/activities/protests related to the mitigation of climate change impacts?		
No	608	92.3
Yes	51	7.7
Are you willing to participate in such events/activities/protests targeting the mitigation of climate change impacts?		
Yes	138	20.9
No	340	51.6
Not sure	181	27.5

item, in fact, also was the only cross-loading item and falls within the moral factor too. Besides, this highest agreed item directly fits under the category of 'moral factor', whereas the remaining items predominantly describe the impacts and causes of climate change—indicating the respondents' knowledge about climate change. Thus, the highest agreed item was excluded from this factor and the remaining items are grouped as 'knowledge factor'. The items of knowledge factor include the causes of climate change—e.g., human-activity-driven GHG—and the consequences of climate change—e.g., climate change-induced natural disasters such as sea-level rise, floods, bushfires, droughts, extreme and abrupt rainfall, prolonged summer, more hot events in the summer, and having with deteriorated environment and ecosystems. Nevertheless, the findings from this particular factor analysis reveal that respondents provide qualified supports towards the views regarding the causes and consequences of climate change impacts.

3.3.2.2 ***Factors Influencing Urban Growth and Climate Change***

Factors influencing urban growth and climate change are illustrated in Table 3.11. By using the Bartlett's Test of Sphericity (Chi-Square = 6713.459; df = 55; p = 0.000), it has been assured that the items related to urban growth and corresponding consequences of climate change impacts exhibit a patterned relationship. The KMO MSA was derived as nearly 0.9 (i.e., 0.88 in specific)—again indicating that factorability within the selected items is marvellous. Under this particular factor analysis operation, the Scree plot diagram derives three factors: (a) Urban growth and policy considerations factor; (b) Urban growth controlling factor; (c) Urban growth action factor. Each factor comprises at least three items. Besides, one cross-loading item is also detected.

From the descriptive analysis, it is revealed that urban growth and policy considerations factor has been perceived with the highest agreement. The items

TABLE 3.11

Factors Influencing Urban Growth and Climate Change

	Descriptive Analysis					Factor Analysis (Pattern Matrix)		
Attributes	DA%	Neutral%	A%	R-DA	R-A	Urban Growth and Policy Considerations Factor	Urban Growth Controlling Factor	Direct Contribution Factor
Due to rapid urban growth, we are generating more GHG and exposed to climate change impacts	18.1	17.6	64.3	8	3	0.84		
Urban encroachment to the urban fringe limits ecosystem services and agricultural activities in these areas	15.3	20.3	64.4	11	2	1.04		
Urban encroachment to the urban fringe causes environmental degradation and biodiversity loss	16.1	18.5	65.4	10	1	1.02		
I am willing to pay more to rent/buy a home, if urbanisation predominantly is centred around the city in favour of preserving valuable ecosystems of the urban fringe	33.1	25	41.9	5	7			0.62
When developing policies and regulations for urban development, climate change should be an important consideration relative to other issues	17.3	21.2	61.5	9	4	0.73		
I am satisfied with the current urban development practices of my local government	36.9	31.9	31.2	4	9		0.93	
I am satisfied with the current urban development strategies of the state government	38.7	32	29.3	1	10		0.92	
I am satisfied with the current national urban policy of the federal government	38.4	32.8	28.8	3	11		0.9	
I am in favour of imposing a more stringent control on rapid urban growth	18.2	24.4	57.4	7	5	0.51		

(Continued)

TABLE 3.11 ***(Continued)***

Factors Influencing Urban Growth and Climate Change

	Descriptive Analysis					Factor Analysis (Pattern Matrix)		
Attributes	**DA%**	**Neutral%**	**A%**	**R-DA**	**R-A**	**Urban Growth and Policy Considerations Factor**	**Urban Growth Controlling Factor**	**Direct Contribution Factor**
I am willing to pay more tax to support a more climate-resilient urban growth practice	38.7	24.9	36.4	2	8			0.94
I have, or am keen to gain, knowledge on how to minimise my carbon footprint and contribute to sustainable development	19.6	26.6	53.8	6	6	0.52		0.4
Cronbach's Alpha = 0.90								
Determinant = 0.0000346								
Rotated sum of squared loadings						5.171	2.534	3.917
% of variance explained						49.189	23.108	5.253
Kaiser–Meyer–Olkin (KMO) = 0.88								
Extraction method = Principal Axis Factoring								
Rotation Method = Promax with Kaiser Normalisation								

of agreement—i.e., 'urban encroachment to the urban fringe causes environmental degradation and biodiversity losses', 'urban encroachment to the urban fringe limits ecosystem services and agricultural activities in these areas', and 'due to rapid urban growth, we are generating more GHG and exposed to climate change impacts'—are perceived with higher priorities and have been subsequently ranked as 1, 2, and 3 respectively. Besides, other two items—i.e., 'when developing policies and regulations for urban development, climate change should be an important consideration relative to other issues' and 'I am in favour of imposing a more stringent control on rapid urban growth'—which have been ranked as 4 and 5, respectively (in terms of the level of agreement) also are included within the urban growth factor.

The second factor under this particular operation is derived as the 'urban growth controlling factor'. This perceived urban growth controlling factor comprises three items—i.e., 'I am satisfied with the current urban development strategies of the state government', 'I am satisfied with the current national urban policy of the federal government', and 'I am satisfied with the current urban development practices of my local government'. These three items as mentioned above are subsequently ranked as 1, 3, and 4 in terms of the level of disagreement.

While respondents express their positive attitude towards mitigating the impacts of climate change through urban growth management, their direct contribution towards mitigating climate change impacts are observed with the higher disagreement. For instance, the items of disagreement—i.e., 'I am willing to pay more tax to support a more climate-resilient urban growth practice', 'I am willing to pay more to rent/buy a home, if urbanisation predominantly is centred around the city in favour of preserving valuable ecosystems of the urban fringe', and 'I have, or am keen to gain, knowledge on how to minimise my carbon footprint and contribute to sustainable development—are ranked as 2, 5, and 6 respectively'.

Besides, the analysis reveals that factors underpinning urban growth and preferred policies are negatively correlated with urban growth controlling factors (Table 3.12).

TABLE 3.12

Correlation between the Factors Influencing Urban Growth and Climate Change

Factors	Urban Growth Factor and Preferred Policy Considerations	Urban Growth Controlling Factor	Direct Contribution Factor
Urban growth factor and preferred policy considerations	1.000	−0.044	0.677
Urban growth controlling factor	−0.044	1.000	0.119
Direct contribution factor	0.677	0.119	1.000

3.3.2.3 Factors Influencing Climate Change Impacts Driven by Urban Growth to Farmlands

Factors influencing climate change impacts driven by urban growth to peri-urban farmlands are discussed in Table 3.13. The Bartlett's Test of Sphericity (Chi-Square = 3308.500; df = 45; p = 0.000) carried out for this particular section implies that a patterned relationship between the items influencing urban growth to the most productive peri-urban farmlands and consequent climate change impacts are evident. The KMO MSA was calculated as 0.84—indicating that factorability within the selected items is nearly marvellous. Through the Scree plot diagram, this particular section of factor analysis extracts three factors: (a) Evident climate change impact and mitigation factor; (b) Policy factor; and, (c) Adjustment factor. Each factor comprises at least three items. Nevertheless, one cross-loading item is also found.

The evident climate change impact factor comprises the items of impacts which are already evident due to destruction of peri-urban farmlands for urban growth accommodation—e.g., 'climate change is causing public health hazards', 'the seasons of vegetations and cropping already have been changed due to climate change', and 'the occurrences of climate change are generating a cause in the spread of infectious diseases'. In terms of the agreement, these particular three items as stated above are ranked as 4, 5, and 6. Nevertheless, another item underpinning the solution of climate change impact—i.e., 'sustainable urban development is critical for addressing and minimising climate challenge externalities'—is ranked with a higher order of agreement (i.e., ranked as 2).

The policy factor is perceived with a higher level of agreement. For instance, the items of policy factors—e.g., 'governments need to prevent the ongoing rapid urban encroachment to environmentally sensitive areas and agricultural farmlands' and 'governments should develop a policy for farmers not to convert their farms into large industrial or residential uses' are perceived with a higher agreement. In terms of the level of agreement, these two items as stated above get the rank of 1 and 3, respectively.

In order to investigate whether such perceptions vary with the framing activities, a cross-tabulation between the agricultural involvement of the respondents and the items as stated above was done (Table 3.14). The analysis revel that only the respondents who are involved with full-time framing activities predominantly (i.e., 78.30%) agreed to sub-dividing their farming lots to use for residential/commercial purposes in order to offset the losses from agricultural activities, whereas the remaining respondents predominantly opposed this idea. Nevertheless, the proposed policy for farmers not to convert their farms into large industrial or residential uses is predominantly agreed upon regardless of agricultural involvement.

TABLE 3.13

Factors Influencing Climate Change Impacts Driven by Urban Growth to the Farmland

	Descriptive Analysis					Factor Analysis (Pattern Matrix)		
Attributes	**DA%**	**Neutral %**	**A%**	**R-DA**	**R-A**	**Evident Climate Change Impact and Mitigation Factor**	**Policy Factor**	**Adjustment Factor**
Climate change is causing public health hazards	25.6	19.3	55.1	6	4	0.96		
The occurrences of climate change are generating a cause in the spread of infectious diseases	36.3	27.3	36.4	5	6	0.89		
I trust on the media coverage highlighting the issue of climate change	44.8	20.9	34.3	3	7	0.558		0.309
Climate change does not matter as I have a religious view on eternal afterlife	68.7	16.4	14.9	1	10			0.59
Sustainable urban development is critical for addressing and minimising climate challenge externalities	16.2	20.8	63	8	2	0.67		
Governments need to prevent the ongoing rapid urban encroachment to environmentally sensitive areas and agricultural farmlands	10.9	18.4	70.7	10	1		0.75	
The seasons of vegetations and cropping already have been changed due to climate change	18.1	28.8	53.1	7	5	0.609	0.318	
Farmers and residents are satisfied with the yearly return from agricultural yield and environmental services	37.2	38.5	24.3	4	8			0.68
Farmers should think of sub-dividing their farms to use for residential/commercial purposes in order to offset the losses from agricultural activities	48.1	29	22.9	2	9			0.62
Governments should develop policy for farmers not to convert their farms into large industrial or residential uses	14.7	26.1	59.2	9	3		0.72	

(Continued)

TABLE 3.13 ***(Continued)***

Factors Influencing Climate Change Impacts Driven by Urban Growth to the Farmland

	Descriptive Analysis					Factor Analysis (Pattern Matrix)		
Attributes	**DA%**	**Neutral %**	**A%**	**R-DA**	**R-A**	**Evident Climate Change Impact and Mitigation Factor**	**Policy Factor**	**Adjustment Factor**
Cronbach's Alpha = 0.84								
Determinant = 0.006								
Rotated sum of squared loadings						3.887	2.74	1.557
% of variance explained						40.987	13.209	5.401
Kaiser–Meyer–Olkin (KMO) = 0.84								
Extraction method = Principal Axis Factoring								
Rotation Method = Promax with Kaiser Normalisation								

TABLE 3.14

Cross-Tabulation of Farming Activities with the Policy Factor

		Farmers Should Think of Sub-Dividing Their Farms to Use for Residential/ Commercial Purposes In Order To Offset the Losses from Agricultural Activities (Pearson Chi-Square = 58.49; df = 6; p = 0.000)			Governments Should Develop Policy for Farmers Not to Convert their Farms into Large Industrial or Residential Uses (Pearson Chi-Square = 20.09; df = 6; p = 0.003)		
Attribute	**Involvement Type**	**Disagree (%)**	**Neutral (%)**	**Agree (%)**	**Disagree (%)**	**Neutral (%)**	**Agree (%)**
Do you engage with agricultural activities?	Yes, I work full-time in a farm	17.4	4.3	78.3	8.7	21.7	69.6
	Yes, I work part-time in a farm	43.8	18.8	37.5	31.3	25.0	43.8
	Yes, I grow my own (or some) fresh produce at my backyard or community garden	60.5	20.3	19.2	14.0	15.7	70.3
	No	45.1	33.9	21.0	14.7	30.4	54.9

The third factor under this particular section is the adjustment factor. This adjustment factor comprises four items and the items under this factor are ranked between 1 and 4 in terms of the level of disagreement.

While public perception of climate change impacts is highly evident through the study from different analyses, their trust in media is objected with a higher level of disagreement (ranked as 3)—implying that media is underestimating the climate change impacts with regard to peri-urbanisation. Such findings appear to be reasonable. While the climate change issue is gradually gaining increasing attention globally, anthropogenic climate change impacts due to the constant peri-urbanisation to the peri-urban productive farmlands yet did not get sufficient attention from the media. Nevertheless, in terms of disagreement, the highest-ranked item is a religious belief in eternal life—implying an altruistic and secular view of the respondents to promote sustainable urban development by obstructing urban encroachment to the productive peri-urban farmlands. Besides, the correlation matrix implies that policy factor and adjustment factor is negatively correlated (Table 3.15).

TABLE 3.15

Correlation between Climate Change Factors Driven by Urban Growth to Farmlands

	Evident Climate Change Impact and Mitigation Factor	Policy Factor	Adjustment Factor
Evident climate change impact and mitigation factor	1.000	0.620	0.243
Policy factor	0.620	1.000	−0.025
Adjustment factor	0.243	−0.025	1.000

3.4 Findings and Discussion

3.4.1 Descriptive Analysis

This study, to our best knowledge, is the first piece of academic literature that captures the views of general public on the interplay between urban growth and climate change with particular interest to peri-urbanisation. The study findings disclose that climate change impact, particularly the one which is driven by urban growth to the most productive peri-urban farmlands, garbs the attention from most of the public. This study also unveils the following findings.

First, media (e.g., TV/radio, social media/internet, and newspapers) is the dominant source of disseminating the information on climate change impacts—implying that respondents are exposed to the information on climate change during their everyday life activities. Other significant sources on the knowledge of climate change impacts include personal interactions such as interactions with colleagues, friends, and family members, and from scholarly resources and magazines. Thus, the sources of information on climate change impacts ranges from personal interactions to mass media. This outcome is similar to an earlier study finding carried out for the Australian context (Akompab et al., 2013). Such a finding has also confirmed the conjecture of Aitken et al. (2011) regarding possible knowledge sources on climate change for the New Zealand context. The positive role of media in generating increased climate risk perception is also globally recognised. For instance, Pasquaré and Oppizzi (2012) acknowledged the role of media in enhancing public perceptions on the link between climate change and geohazards for the Italian context. Whereas Boykoff and Rajan (2007) identified how media's portrayal of anthropogenic climate change is shaped within the framework of climate science and global climate politics for the contexts of UK and USA. Thus, it is revealed that with compared to other knowledge sources, the widespread media sources play more important role in disseminating the information on climate change. Hence, policymakers need to utilise the full potentials of these media sources in order to generate awareness creation and bring lifestyle and behavioural changes among the public in combating climate change impacts.

Second, age composition and academic standing play significant roles in believing on the issue of climate change impact. Other studies—e.g., Leviston et al.'s (2011) study carried out for the Australian context—also found significant association between academic standing and climate change belief. Nevertheless, while Leviston et al.'s (2011) study did not find any strong association between age and climate change belief, such an association reported in this chapter is observed as strong. In this study findings, it is disclosed that relatively younger respondents (aged less than 45) believe more on climate change impacts. This particular finding is similar to a recent study carried out for the Australian context (Chiw & Ling, 2019).

Nevertheless, while age and academic standing are perceived to be linked with climate risk perception for the Australian context, Hu et al.'s (2017) study carried out for the US context did not find any significant association between gender, age, educational standing, and climate risk perception. Some other studies carried out in Sweden, Germany, Norway, and US find that with regard to respondents' socioeconomic characteristics—e.g., age, gender, educational level, and income—gender is the only determinant that shapes climate risk perception, with female significantly worrying more than male concerning climate risk (Lujala et al., 2015; Marlon et al. 2019; Menny et al., 2011; Sundblad et al., 2007).

Leviston et al.'s (2011) study also observed that women are more prone to climate change belief. Nevertheless, such an observation was found statistically insignificant under the investigation of this particular study. This study observed that irrespective of gender issue, respondents predominantly believe on climate change impacts. Thus, this study findings are seen to be an improvement of climate change belief from the last decade of 2011. The higher level of educational attainments through either institutional education or normative development, widespread media campaign, and evident impacts of climate change incidents might have contributed to decline gender bias in believing climate change issue. Besides, like other Australian studies, this study also did not find any significant associations with income levels and climate change perception. These findings imply that issue of climate change is gaining momentum and become a common issue these days. Such an inference is similar to another study finding done for the Australian context (Reser et al., 2012).

Third, practising sustainability principles (e.g., the usages of solar energy and energy monitoring device, or having the provision of rainwater harvesting)—which are believed to have mitigating potentials in reducing climate change impacts—are not so linked with climate change perception, albeit sustainability practices are perceived to be linked predominantly with the home ownership pattern. The findings imply that the respondents who are not adapting sustainability principles do also believe on climate change impacts. Nevertheless, those who have the provision of composting in their home are observed to have higher level of climate impact perceptions. Nevertheless, these particular study findings as stated above had not been able to validate, because to our best knowledge, no studies till date investigated the nexus between sustainability practices and climate change perception. The overall findings of this particular section imply that the respondents' sense of belonging, and freedom of choice may be linked to sustainability practices with the home ownership pattern. Nevertheless, respondents' firm belief on climate change impacts is pervasive. The respondents' preferred car choices are found to be more linked with climate change perception, but their preference to have an electric car may be more linked to adapting the upcoming vogue or gaining higher social prestige than to climate change perception. Thus, these findings reveal that homeowners, particularly those who rent their home, are indifferent to provide with the provisions of sustainability practices to their rental dwellers. Therefore, federal and local governments should offer appropriate incentives to landlords in practising sustainability provisions in their home.

Fourth, respondents predominantly attributed human activities as one of the prime causes of climate change impacts. This particular finding is similar to another study conducted in Australia (Reser et al., 2012). However, neither their willingness to adapt behavioural changes or voluntary participation is so significant in mitigating the climate change impacts. Instead of offering voluntary participation from the respondents' end, they in fact suggested federal and local governments to innovate effective policy instruments in combating such perceived climate change impacts. These findings imply that federal and local governments should forge innovative ways to encourage more active and passive participations of the public on the issue of climate change impacts mitigation.

Nevertheless, Semenza et al.'s (2008) study carried out for the US context finds significant public motivations in offering voluntary participation for mitigating climate risk impacts. Notwithstanding, voluntary motivation towards mitigating climate risk largely depends on how the public perceives their level of vulnerabilities to be impacted by such climate risk events (Semenza et al., 2011). Thus, the finding implies that although respondents are predominantly aware of climate change impacts, they do not consider themselves as significantly vulnerable to these climate risks to date in bringing behavioural changes.

3.4.2 Factor Analysis

3.4.2.1 Climate Change Exposure Factors

The climate change exposure factors reported under this study include knowledge factor, moral factor, and livelihood factor. The livelihood factor comprises only one variable, whereas something to be regarded as a factor ideally should comprise minimum three variables. Nevertheless, this minimum threshold limit for factor consideration is not rigid and instead varies on the research design (Tabachnick & Fidell, 2007). This study only had one variable for identifying the impacts of climate change on income. Hence, this reported only one variable under the livelihood factor. Henceforth, such a finding is not shortcoming, instead is a well-fit in terms of explaining factorability among the selected variables.

The knowledge factor reveals that respondents provide qualified supports towards the views regarding the causes and consequences of climate change impacts. The moral factor indicates that people are willing to adopt green technology in preventing climate change impacts, but their active participation in preventing climate change impacts appear to be limited. This inference is further supported by descriptive analysis as stated in the earlier section.

Respondents' knowledge base and moral stance regarding the issue of climate change impacts suggest adopting innovative ways in ensuring passive participation of the public. The livelihood factor discloses that people are not aware of the long-term impacts of climate change on the overall livelihood pattern, or they did not realise the climate challenge externalities impacting on their livelihoods.

3.4.2.2 Factors Influencing Urban Growth and Climate Change

The climate change impacts related to urban growth consequences—which is factorised as urban growth and policy consideration factor—receive the highest level of agreement from the public. Whereas public perception demonstrates the highest level of disagreement towards the urban growth controlling factor—i.e., policies of urban growth management formulated by the Federal, State and Local government bodies of Australia. Consequently, the urban growth and policy consideration factor has been found negatively correlated with the controlling factor.

Nevertheless, the public expresses a bleak attitude towards playing active roles (i.e., the direct contribution factor) in mitigating the climate change impacts. The findings imply a clear dissatisfaction of the public towards government policies in addressing urbanisation-driven climate change impacts. This particular factor analysis section indicates that the public perceives the ongoing urban growth as ominous and are dissatisfied with the federal, state and local government policies in reducing the resultant climate change impacts. Thus, their perception of urbanisation-driven climate change impacts appears to be highly affirmative. Nevertheless, the public is also sceptical in assisting with their direct contributions. Thus, the Federal, State, and Local government bodies need to capitalise on these positive affirmations of the public and should formulate innovative policies which can involve passive participation of the public—a means of indirect involvement of the public in reducing urbanisation-driven climate change impacts.

3.4.2.3 Factors Influencing Climate Change Impacts Driven by Urban Growth to Farmlands

Under the evident climate change and mitigation factor, evident climate change is perceived with a moderate view—meaning neither with higher agreement nor with higher disagreement. Whereas the item illustrating the mitigation measures of climate change externalities are advocated with higher importance. The outcome from policy factor suggests that public perception is predominantly favourable to the peri-urban farmers—implying that the Government need to take proper actions in preventing the ongoing peri-urbanisation to the environmentally sensitive and most productive peri-urban farmlands.

The adjustment factor implies that farmers and residents are not satisfied with the yearly income from their agricultural yield and environmental services. However, the respondents are also strident opponents to sub-divide these peri-urban framing lots in order to offset the losses from agricultural activities. Besides, the negative correlation between policy factor and adjustment factor implies that the existing policies formulated by the Federal, State and Local government are not commensurate with the public's preferred coping mechanism.

3.5 Conclusion

While a highly marginal proportion of respondents were not familiar with the issue of climate change before the conducted survey, the remaining respondents being surveyed were aware of about climate change impacts beforehand. Besides, majority of the respondents are already exposed to the climate change induced incidents occurring in the SEQ region that include floods, drought, bushfire, and extreme heat and rainfall events. Thus, the findings reported in this chapter broadly represent the views of those who are already familiar with the impacts of climate change. Therefore, the study generates an evidence base for policymakers and planners and informs authorities in formulating climate change mitigation strategies and actions.

Nevertheless, in order to prepare appropriate climate risk mitigation and adaptation measures (where mitigation measures focus on neutralising or weakening the underlying causes of climate change risk, whereas adaptation measures focus on reducing vulnerability to the effects of such climate change risk), the extent of belief and knowledge about climate change appears to be critical. For instance, such measures might include changes in planning and development regulations as well as to bring changes in the management of natural resources. Thus, in order to ensure a successful implementation of climate change mitigation measures, along with the changes within intergovernmental organisations, bringing behavioural and motivational changes among the public appears to be equally important. As Bulkeley (2000) rightfully suggested that policy attention should be focused on encouraging more public involvement in combating climate change risks.

There are, nevertheless, only a few scholarly studies have been carried out so far for the Australian context in unveiling the public views about climate change and their extent of knowledge on climate change impacts. The climate change studies carried out for the Australian context either predominantly focused on the theoretical and practical perspectives of climate change impacts without much focus on the public views (Head et al., 2014; Hughes, 2011; Kingwell, 2006), or the limited studies concerned the public perceptions are now almost outdated (Henderson-Sellers, 1990). The same is correct for the international context as well, the public perception on climate change is an understudied area of research (Capstick et al., 2015; Sullivan & White, 2019; Yu et al., 2013).

A significant study that investigated public perception on climate change was carried out by Bulkeley (2000) for Newcastle (Australia). However, the time when this study was conducted, the issue of climate change was just burgeoning globally and therefore the lack of knowledge on climate change among the public was evident. Thus, following Bulkeley's (2000) study, investigating public perceptions on climate change could be an effective measure in determining the shifts in public views on climate change issues within the next two decades. Besides, few reports that are released so far from the different governmental organisations of Australia primarily capture the views of people on climate change impacts (Chiw & Ling, 2019; Gardner et al., 2010; Leviston et al., 2011).

Nevertheless, no studies to date focused on capturing the views of the general public with regard to peri-urbanisation led climate change impacts either in the Australian context or in the global context. Thus, the findings reported in this chapter possess both practical and theoretical implications. From the theoretical perspective, this chapter provides an important guideline in curbing down urbanisation to the cities' most productive peri-urban landscapes.

While peri-urbanisation always provokes land use conflicts, public perceptions connecting peri-urbanisation and climate change impacts provide an evidence base about how to manage these ongoing peri-urbanisation with the help of the public. With regard to practical implications, with a conviction that they have strong public support, the study findings enable policymakers in formulating effective strategies in order to preserve the most productive peri-urban farmland. These peri-urban productive farmlands shelter important ecosystems that are key to the mitigating circumstances of climate change impacts. In this manner, this research possesses significant implications to the countries that are biologically diverse and that shelter most of their ecosystem services in peri-urban areas.

In addition, the findings reported in this study unveil that the long-term consequences of climate change impacts on livelihoods are hardly perceived by the respondents. While the vast majority of peri-urban rural dwellers are heavily dependent upon agriculture, the livelihoods of peri-urban farmers are at an increasing threat to climate variability. Thus, the findings advocate developing appropriate climate change social communication policies in generating climate risk awareness to adapt their livelihoods. Such climate change social communication policies should highlight preserving the valuable regional ecosystems, developing a strong sense of communal belonging to this ecosystem services as well as recognising the simultaneous roles of these peri-urban spaces in securing their livelihoods and mitigating climate risk. In this way, any unwanted failure to climate risk mitigation can be contained and better adapted.

Nevertheless, the peri-urban growth pattern is geographically specific, as the determinants of peri-urban growth—e.g., societal characteristics, institutional arrangement, political system, economic drivers, and environmental settings—are geographically diverse (Mortoja et al., 2020). Thus, given the contextual nature of peri-urban growth, this research provokes carrying out similar studies with different geographical and continental settings. In this way, the variations in public perceptions on the link between climate risk and peri-urban growth with regard to a country's unique geographic, socioeconomic, environmental, and political settings can be better interpreted. Overall, the findings reported in this research highlight the distinct characteristics of peri-urban growth areas and recognises peri-urban areas as independent or autonomous spaces for climate risk governance in the city.

Acknowledgements

This chapter, with permission from the copyright holder, is a reproduced version of the following journal article: Mortoja, M., & Yigitcanlar, T., (2022). Public perceptions of peri-urbanism triggered climate change: survey evidence from South East Queensland, Australia. *Sustainable Cities and Society*, 75(1), 103407.

References

Adedire, F. M. (2018). Peri-urban expansion in Ikorodu, Lagos: Extent, causes, effects, and policy response. *Urban Forum, 29*, 259–275.

Aitken, C., Chapman, R., & McClure, J. (2011). Climate change, powerlessness and the commons dilemma: Assessing New Zealanders' preparedness to act. *Global Environmental Change, 21*, 752–760.

Akompab, D., Bi, P., Williams, S., Grant, J., Walker, I., & Augoustinos, M. (2013). Awareness of and attitudes towards heat waves within the context of climate change among a cohort of residents in Adelaide, Australia. *International Journal of Environmental Research and Public Health, 10*, 1–17.

Aldieri, L., Carlucci, F., Vinci, C. P., & Yigitcanlar, T. (2019). Environmental innovation, knowledge spillovers and policy implications. *Journal of Cleaner Production, 239*, 118051.

Aldieri, L., Grafstrom, J., & Vinci, C. P. (2021). The effect of Marshallian and Jacobian knowledge spillovers on jobs in the solar, wind and energy efficiency sector. *Energies, 14*(14), 4269.

Ali, A. (1999). Climate change impacts and adaptation assessment in Bangladesh. *Climate Research, 12*, 109–116.

Al-Mamun, M. M., & Kim, S. M. (2020). Stakeholder analysis matrix for buffer zone management in the peri-urban area of Chittagong, Bangladesh. *Environment, Development and Sustainability, 22*, 5503–5520.

Bajracharya, B., & Hastings, P. (2018). A regional, strategic growth-management approach to urban and peri-urban development in South East Queensland, Australia. *Journal of Regional and City Planning, 29*(3), 210–233.

Biswas, M. (2013). Climate change and its impacts on Bangladesh. Planned Decentralization: Aspired Development, 86–95.

Boykoff, M. T., & Rajan, S. R. (2007). Signals and noise: Mass-Media coverage of climate change in the USA and the UK. *EMBO Reports, 8*, 207–211.

Broto, V. C., & Bulkeley, H. (2013). A survey of urban climate change experiments in 100 cities. *Global Environmental Change, 23*, 92–102.

Bulkeley, H. (2000). Common knowledge? Public understanding of climate change in Newcastle, Australia. *Public Understanding of Science, 9*, 313–333.

Capstick, S., Whitmarsh, L., Poortinga, W., Pidgeon, N., & Upham, P. (2015). International trends in public perceptions of climate change over the past quarter century. *Wiley Interdisciplinary Reviews: Climate Change, 6*, 35–61.

Chiw, A., & Ling, H. (2019). *Young people of Australia and climate change: Perceptions and concerns*. Millennium Kids Inc.

Dell'Anna, F. (2021). Green jobs and energy efficiency as strategies for economic growth and the reduction of environmental impacts. *Energy Policy, 149*, 112031.

Dizdaroglu, D., & Yigitcanlar, T. (2014). A parcel-scale assessment tool to measure sustainability through urban ecosystem components: The MUSIX model. *Ecological Indicators, 41*, 115–130.

Dur, F., & Yigitcanlar, T. (2015). Assessing land-use and transport integration via a spatial composite indexing model. *International Journal of Environmental Science and Technology, 12*, 803–816.

Foelske, L., Van Riper, C. J., Stewart, W., Ando, A., Gobster, P., & Hunt, L. (2019). Assessing preferences for growth on the rural-urban fringe using a stated choice analysis. *Landscape and Urban Planning, 189*, 396–407.

Friedberger, M. (2000). The rural-urban fringe in the late twentieth century. *Agricultural History, 74*, 502–514.

Gallent, N., & Shaw, D. (2007). Spatial planning, area action plans and the rural-urban fringe. *Journal of Environmental Planning and Management, 50*, 617–638.

Gardner, J., Parsons, R., & Paxton, G. (2010). *Adaptation benchmarking survey: Initial report*. CSIRO.

Green, K., & Vergragt, P. (2002). Towards sustainable households: A methodology for developing sustainable technological and social innovations. *Futures, 34*, 381–400.

Guhathakurta, S., & Stimson, R. (2007). What is driving the growth of new "Sunbelt" metropolises? Quality of life and urban regimes in Greater Phoenix and Brisbane-South East Queensland region. *International Planning Studies, 12*, 129–152.

Head, L., Adams, M., McGregor, H., & Toole, S. (2014). Climate change and Australia. *Wiley Interdisciplinary Reviews: Climate Change, 5*, 175–197.

Henderson-Sellers, A. (1990). Australian public perception of the greenhouse issue. *Climatic Change, 17*, 69–96.

Hu, S., Jia, X., Zhang, X., Zheng, X., & Zhu, J. (2017). How political ideology affects climate perception: Moderation effects of time orientation and knowledge. *Resources, Conservation and Recycling, 127*, 124–131.

Hughes, L. (2011). Climate change and Australia: Key vulnerable regions. *Regional Environmental Change, 11*, 189–195.

Huq, S., Chow, J., Fenton, A., Stott, C., Taub, J., & Wright, H. (2019). *Confronting climate change in Bangladesh policy strategies for adaptation and resilience*. Springer.

Kahn, M. E. (2009). Urban growth and climate change. *Annual Review of Resource Economics, 1*, 333–350.

Kaiser, H., & Rice, J. (1974). Little jiffy, mark IV. *Educational and Psychological Measurement, 34*, 111–117.

Kamruzzaman, M., Deilami, K., & Yigitcanlar, T. (2018). Investigating the urban heat island effect of transit-oriented development in Brisbane. *Journal of Transport Geography, 66*, 116–124.

Kankanamge, N., Yigitcanlar, T., Goonetilleke, A., & Kamruzzaman, M. (2020). Determining disaster severity through social media analysis: Testing the methodology with South East Queensland Flood tweets. *International Journal of Disaster Risk Reduction, 42*, 101360.

Kingwell, R. (2006). Climate change in Australia: Agricultural impacts and adaptation. *Australasian Agribusiness Review, 14*, 1–29.

Leviston, Z., Leitch, A., Greenhill, M., Leonard, R., & Walker, I. (2011). *Australians' views of climate change*. CSIRO.

Liu, C., Valentine, G., Vanderbeck, R., Diprose, K., & McQuaid, K. (2019). Rural-urban inequality and the practice of promoting sustainability in contemporary China. *GeoJournal, 84*, 1187–1198.

Llausas, A., Buxton, M., & Beilin, R. (2016). Spatial planning and changing landscapes: A failure of policy in peri-urban Victoria, Australia. *Journal of Environmental Planning and Management, 59*, 1304–1322.

Lujala, P., Lein, H., & Rød, J. K. (2015). Climate change, natural hazards, and risk perception: The role of proximity and personal experience. *Local Environment, 20*, 489–509.

Ma, E., Guinery, M., McCarthy, P., & Shaw, R. (2012). Australian floods and their impact on insurance. In: The Geneva Reports: Extreme events and insurance, The Geneva Association (pp. 81–93).

Mahbub, P., Goonetilleke, A., Ayoko, G. A., Egodawatta, P., & Yigitcanlar, T. (2011). Analysis of build-up of heavy metals and volatile organics on urban roads in Gold Coast, Australia. *Water Science and Technology, 63*, 2077–2085.

Marlon, J. R., van der Linden, S., Howe, P. D., Leiserowitz, A., Woo, S. L., & Broad, K. (2019). Detecting local environmental change: The role of experience in shaping risk judgments about global warming. *Journal of Risk Research, 22*, 936–950.

McPhearson, T., Andersson, E., Elmqvist, T., & Frantzeskaki, N. (2015). Resilience of and through urban ecosystem services. *Ecosystem Services, 12*, 152–156.

Menny, C., Osberghaus, D., Pohl, M., & Werner, U. (2011). *General knowledge about climate change, factors influencing risk perception and willingness to insure.* ZEW-Centre for European Economic Research Discussion Paper, (11-060).

Merson, J., Attwater, R., Ampt, P., Wildman, H., & Chapple, R. (2010). The challenges to urban agriculture in the Sydney basin and lower Blue Mountains region of Australia. *International Journal of Agricultural Sustainability, 8*, 72–85.

Metaxiotis, K., Carrillo, F. J., & Yigitcanlar, T. (2010). *Knowledge-based development for cities and societies: Integrated multi-level approaches.* IGI Global.

Millward, H. (2002). Peri-urban residential development in the Halifax region 1960–2000: Magnets, constraints, and planning policies. *Canadian Geographer, 46*, 33–47.

Mortoja, M., & Yigitcanlar, T. (2020a). How does peri-urbanization trigger climate change vulnerabilities? An investigation of the Dhaka Megacity in Bangladesh. *Remote Sensing, 12*, 3938.

Mortoja, M., & Yigitcanlar, T. (2020b). Local drivers of anthropogenic climate change: Quantifying the impact through a remote sensing approach in Brisbane. *Remote Sensing, 12*, 2270.

Mortoja, M., & Yigitcanlar, T. (2021). Why is determining peri-urban area boundaries critical for sustainable urban development? Evidence from Brisbane, Australia. *Journal of Environmental Planning and Management.* https://doi.org/10.1080/09640568.2021.1978405.

Mortoja, M., Yigitcanlar, T., & Mayere, S. (2020). What is the most suitable methodological approach to demarcate peri-urban areas? A systematic review of the literature. *Land Use Policy, 95*, 104601.

Narain, V. (2014). Whose land? Whose water? Water rights, equity and justice in a peri-urban context. *Local Environment, 19*(9), 974–989.

Pasquaré, F. A., & Oppizzi, P. (2012). How do the media affect public perception of climate change and geohazards? An Italian case study. *Global and Planetary Change, 90*, 152–157.

Perveen, S., Kamruzzaman, M., & Yigitcanlar, T. (2017). Developing policy scenarios for sustainable urban growth management: A Delphi approach. *Sustainability, 9*, 1787.

Perveen, S., Yigitcanlar, T., Kamruzzaman, M., & Agdas, D. (2020). How can transport impacts of urban growth be modelled? An approach to consider spatial and temporal scales. *Sustainable Cities and Society, 55*, 102031.

Rauws, W. S., & de Roo, G. (2011). Exploring transitions in the peri-urban area. *Planning Theory & Practice, 12*, 269–284.

Reser, J., Bradley, G., Glendon, A., Ellul, M., & Callaghan, R. (2012). Public risk perceptions, understandings and responses to climate change in Australia and Great Britain. Griffith Climate Change Response Adaptation Facility, Gold Coast.

Ruoso, L. E., & Plant, R. (2018). A politics of place framework for unravelling peri-urban conflict: An example of peri-urban Sydney, Australia. *Journal of Urban Management, 7*, 57–69.

Semenza, J. C., Hall, D. E., Wilson, D. J., Bontempo, B. D., Sailor, D. J., & George, L. A. (2008). Public perception of climate change: Voluntary mitigation and barriers to behavior change. *American Journal of Preventive Medicine, 35*, 479–487.

Semenza, J. C., Ploubidis, G. B., & George, L. A. (2011). Climate change and climate variability: Personal motivation for adaptation and mitigation. *Environmental Health, 10*, 1–12.

Shaw, A. (2005). Peri-urban interface of Indian cities: Growth, governance and local initiatives. *Economic and Political Weekly, 40*, 129–136.

Silva, C. (2019). Auckland's urban sprawl, policy ambiguities and the peri-urbanisation to Pukekohe. *Urban Science, 3*, 1.

Sotto, D., Philippi, A., Yigitcanlar, T., & Kamruzzaman, M. (2019). Aligning urban policy with climate action in the global south: Are Brazilian cities considering climate emergency in local planning practice? *Energies, 12*, 3418.

Steffen, W., & Dean, A. (2018). *Land clearing and climate change: Risks and opportunities in the sunshine state*. The Climate Council of Australia.

Sullivan, A., & White, D. (2019). An assessment of public perceptions of climate change risk in three western US cities. *Weather, Climate and Society, 11*(2), 449–463.

Sundblad, E. L., Biel, A., & Gärling, T. (2007). Cognitive and affective risk judgements related to climate change. *Journal of Environmental Psychology, 27*, 97–106.

Sylla, M., Hagemann, N., & Szewrański, S. (2020). Mapping trade-offs and synergies among peri-urban ecosystem services to address spatial policy. *Environmental Science & Policy, 112*, 79–90.

Tabachnick, B., & Fidell, L. (2007). *Using multivariate statistics*. Allyn & Bacon.

Wandl, A., & Magoni, M. (2016). Sustainable planning of peri-urban areas: Introduction to the special issue. *Planning Practice & Research, 32*, 1–3.

Yigitcanlar, T. (2010). *Sustainable urban and regional infrastructure development: Technologies, applications and management*. IGI Global.

Yigitcanlar, T., & Kamruzzaman, M. (2014). Investigating the interplay between transport, land use and the environment: A review of the literature. *International Journal of Environmental Science and Technology, 11*, 2121–2132.

Yigitcanlar, T., & Kamruzzaman, M. (2019). Smart cities and mobility: Does the smartness of Australian cities lead to sustainable commuting patterns? *Journal of Urban Technology, 26*, 21–46.

Yong, A., & Pearce, S. (2013). A beginner's guide to factor analysis: Focusing on exploratory factor analysis. *Tutorials in Quantitative Methods for Psychology, 9*, 79–94.

Yu, H., Wang, B., Zhang, Y., Wang, S., & Wei, Y. (2013). Public perception of climate change in China: Results from the questionnaire survey. *Natural Hazards, 69*, 459–472.

Yue, W., Liu, Y., & Fan, P. (2013). Measuring urban sprawl and its drivers in large Chinese cities: The case of Hangzhou. *Land Use Policy, 31*, 358–370.

Zabik, M. J., & Prytherch, D. L. (2013). Challenges to planning for rural character: A case study from exurban Southern New England. *Cities, 31*, 186–196.

Zhao, P. (2013). Too complex to be managed? New trends in peri-urbanisation and its planning in Beijing. *Cities, 30*, 68–76.

Part II

Localised Climate Risk and Urban Exposure

This part of the book shifts attention from perception to exposure, examining how urban form, land use change, and human activity intensify localised climate risks. The chapters analyse peri-urbanisation as a key driver of vulnerability, where fragmented development patterns increase sensitivity to extreme heat and other climate hazards. They also investigate the role of anthropogenic climate change in reshaping urban landscapes, highlighting how built environments contribute to, and are affected by, rising temperatures. By focusing on place-based dynamics, this part demonstrates that climate risk is not uniform across cities but is produced through specific spatial, social, and developmental trajectories. It provides critical insights into how urban growth decisions shape present and future climate exposure.

DOI: 10.1201/9781003737049-5

4

Peri-Urbanisation and Climate Vulnerability Dynamics

4.1 Introduction and Background

Due to the global urbanisation drives, major metropolitan cities and regions across countries are gradually expanded by continuously encroaching their physical growth boundaries into adjoining peri-urban areas (Kamruzzaman et al., 2015, 2018; Yigitcanlar, 2006, 2010). Hence, urban growth at present predominantly is occurring in the form of peri-urbanisation globally (Mortoja & Yigitcanlar, 2020). Thus, rapid peri-urbanisation has already become an issue of increasing global concern (Yigitcanlar & Kamruzzaman, 2015). In addition, peri-urban areas are also vulnerable to climate change impacts (Carey & James, 2018; Nielsen et al., 2003). Consequently, these peri-urban growth pockets provoke considerable growth challenges to policymakers (James & O'Neill, 2016; Kennedy et al., 2016; Ruoso & Plant, 2018).

Although the rapidly occurring peri-urbanisation is commonly a global phenomenon, peri-urban growth factors vary across countries. Such variations in growth factors are deeply ingrained in a country's unique socio-economic settings, legal systems, institutional arrangements, and environmental conditions (Teriman et al., 2009; Woltjer, 2014). For instance, the in-migration—which is considered a major determinant of peri-urbanisation—varies across countries. Peri-urban in-migrants for the context of developed nations are predominantly amenity-led migrants (Argent et al., 2011; Zhao, 2012). Contrariwise, such in-migration in the developing countries context is largely attributed to forced factors such as poverty, lack of employment, and climate change impacts—e.g., sea level rise, river erosion, flooding, salinity intrusion, and drought (Datta, 2004; Pouliotte et al., 2009).

In addition, biased national policies—e.g., excessive capital city-oriented development tendencies—also accelerate this forced migration (Hossain, 2003). In this way, metropolitan cities in the developing countries' context are overwhelmed with unanticipated population growth. Hence, policymakers and planners are greatly hindered in maintaining a balance between economic development and social change (Todes, 2012), resulting in a gross failure in estimating the projected population and corresponding growth demands. While the cities of developing countries are frequently struggling to cope with the growing demand, the influx of additional migrants makes

DOI: 10.1201/9781003737049-6

the development control tasks extremely challenging (Mortoja et al., 2020). Subsequently, the unprecedented growth of informal economies occurs in their peripheries (ADB (Asian Development Bank), 2010).

While some cities are gradually expanding in response to this growing need, cities with geographical limitations are greatly obstructed from providing further avenues for expansion. For example, the expansion of Dhaka city—the capital of Bangladesh—is severely constrained by the lack of sufficient flood-free landscapes. The elevation of Dhaka city lies up to a maximum of 13 m above the mean sea level (Kabir & Parolin, 2012). Hence, Dhaka city is even more low-lying than its contemporaries—e.g., Mumbai (India), Kolkata (India), and Karachi (Pakistan).

Moreover, Dhaka's annual average rainfall is 2148 mm, out of which the monsoon rainfall that lasts from May to September accounts for nearly 78% (Weatherbase, 2020). Thus, Dhaka encounters frequent annual flooding in the monsoon due to heavy rainfall. While the portion of the city's annual flooding contributed by climate change impacts is as yet unknown, such frequently occurring flood events are widely claimed to be climate change-induced (Alam & Rabbani, 2007; Rabbani et al., 2011).

Furthermore, Bangladesh also tops the list of climate change impacts, and such impacts are already evident throughout the country. Due to climate change impacts—e.g., sea level rise, salinity intrusion, riverbank erosion, flooding, and drought—many people lose their livelihoods and properties, and eventually migrate to the major cities of Bangladesh (Siddiqui, 2010). Such internal migrants are globally defined as 'climate migrants' (Rigaud et al., 2018). So far, Bangladesh has two million climate migrants, and the capital city, Dhaka, individually hosts 68% of them (Ahmed, 2020). These climate migrants predominantly represent the poorest portion of the city. With an annual poverty growth rate of 4%, about 40% of Dhaka's urban population is poor (Nazrul & Salma, 2004), and are predominantly deprived of any sort of urban facilities (Baker, 2007). These poor people generally live in the peripheral areas of Dhaka city, which are highly flood-prone. A boom in the ready-made garments (RMG) industry has further made the population influx to these flood-prone areas unstoppable (DhakaTribune, 2020). Consequently, Dhaka's population is increasing at a rate of 0.4 million/year (WB (World Bank), 2007).

Thus, a more dense peri-urban growth occurs in their peri-peripheries, resulting in an intensified vulnerability to climate change's impact. These climate change impacts appear to be perpetual. As Bangladesh is one of the most vulnerable countries to climate change's impacts, incidents such as sea level rise, flooding, salinity intrusion, and their subsequent impacts on livelihoods and properties seem to persist. Thus, Dhaka will continue to absorb the direct and indirect impacts of climate change. While climate change has already become an issue of increasing concern for the government, policymakers are as yet unequipped to deal with the direct and indirect impacts of climate change at the metropolitan level.

Peri-urban areas—which are neither urban nor rural—are a distinct geographic space and have clear implications in urban governance (Silva, 2019, 2020). While the idea of independently operating peri-urban areas is globally gaining increasing concern (Gallent & Shaw, 2007; Rauws & de Roo, 2011), no study, to date, is reported in the literature investigating peri-urbanisation-driven climate change vulnerabilities based on an explicitly demarcated peri-urban growth boundary.

This study provides a remote sensing approach in demarcating peri-urban growth pockets, which are vulnerable to climate change impacts. In order to demarcate these peri-urban growth pockets, this chapter uses Mortoja et al.'s (2020) research findings apropos of 'what the most suitable methodological approach to demarcate peri-urban areas is'. The modus operandi of this investigation comprises carrying out change analyses with Landsat data and identifying peri-urban growth pockets with night-time light (NTL) data. By adapting the Greater Dhaka Area (Bangladesh)—or the Dhaka Megacity—as a testbed, the study undertakes the following four analyses. First, it maps out the changes in peri-urban growth boundaries that occurred in the peripheral areas of Dhaka over the last 24-year period (1992–2016). Second, it points out the consistencies of prevailing peri-urban growth with Dhaka's designated plan documents. Third, it spells out Dhaka's particular vulnerabilities to climate change impacts—i.e., flooding and groundwater table depletion. Lastly, it identifies the socioeconomic profile of the climate change victim groups from Dhaka. The insights generated from this study provide an evidence base in identifying the peri-urban growth pockets, and thereby enable policy makers to formulate area-specific growth policies and mitigate circumstances in dealing with increasing climate change vulnerabilities.

4.2 Research Design

4.2.1 Study Area

In order to reveal peri-urbanisation-related climate change vulnerabilities, this study chose one of the most vulnerable cities of global climate change impacts—i.e., Dhaka, the capital of Bangladesh. In terms of physical growth, Dhaka has become overwhelmingly saturated within its city corporation areas since the 2000s. Hence, further expansions of Dhaka city are protruding towards the adjoining peri-urban landscapes. The Capital Development Authority (a.k.a. Rajdhani Unnayan Kotripakhkha—RAJUK)—that is entitled as the prime organisation for guiding and monitoring Dhaka city's growth—projects the city's growth to nine adjoining sub-districts (a.k.a. upazilas), including Savar, Gazipur Sadar, Kaliganj, Rupganj, Sonargaon, Bandar, Narayanganj Sadar, and Keraniganj (Figure 4.1). These RAJUK-designated

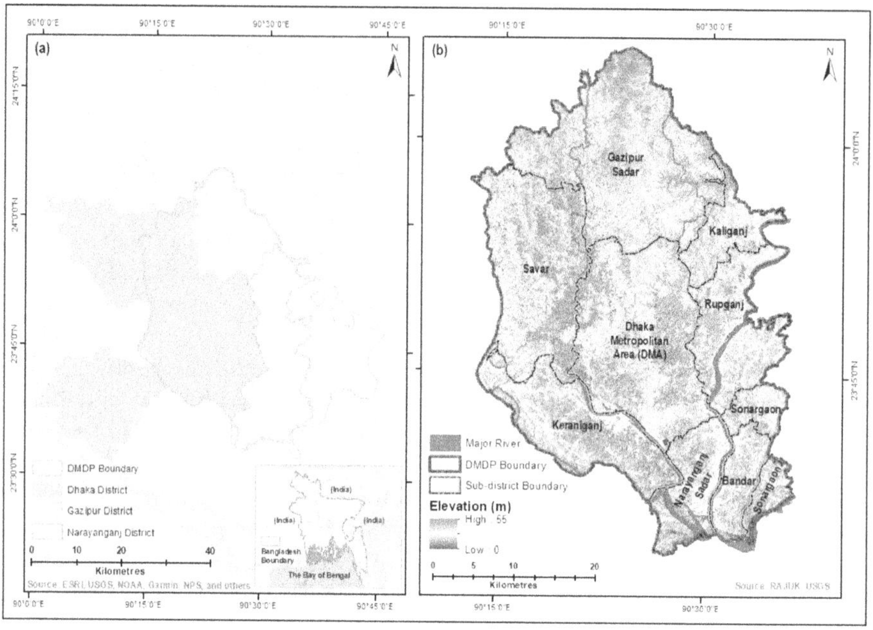

FIGURE 4.1
Location of the study area: (a) The Dhaka Metropolitan Development Plan (DMDP) area within the national context; (b) the Dhaka Metropolitan Development Plan (DMDP) area and its corresponding sub-districts' location.

sub-districts are altogether declared as the Dhaka Metropolitan Development Plan (DMDP) boundary. The study considered this DMDP boundary, comprising an aggregate area of 1530 km^2, as the testbed to demonstrate the peri-urbanisation-driven climate change vulnerabilities.

The compelling reasons for investigating the DMDP's peri-urban growth under this microscope entail: (a) The capital-city-centred biased national polices of agglomerating all major economic and infrastructural facilities around Dhaka city (Fan et al., 2020); (b) unprecedented flow of in-migrants to Dhaka city (Jahan, 2012); (c) rampant unsustainable development practices around Dhaka city (Hossain, 2008); (d) both in terms of the world least livable city index and global population density, Dhaka ranks third (BBS (Bangladesh Bureau of Statistics), 2011; DWUA (Demographia World Urban Areas), 2020; EIU (Economist Intelligence Unit), 2019a, b), and; (e) Dhaka is already facing climate change-induced frequent flooding (Rabbani et al., 2011).

4.2.2 Datasets

The datasets used for this study primarily comprise two types: (a) Level-one terrain-corrected (L1T) and cloud-free multispectral Landsat data for the years 1989, 1999, 2009, and 2019 collected from USGS (USGS (United States

TABLE 4.1

Characteristics of Landsat Data for the DMDP Area (USGS (United States Geological Survey), 2020)

Date	Path/Row	Sensor	Satellite	Resolution (m)
20 November 1989	137/44	TM	Landsat 5	30
9 March 1989	137/44	TM	Landsat 5	30
1 February 1999	137/44	TM	Landsat 5	30
1 February 1999	137/43	TM	Landsat 5	30
30 January 2010	137/44	TM	Landsat 5	30
28 February 2009	137/43	TM	Landsat 5	30
23 January 2019	137/44	OLI	Landsat 8	30
23 January 2019	137/43	OLI	Landsat 8	30

Geological Survey), 2020) (https://earthexplorer.usgs.gov) (Table 4.1), and; (b) NTL data for the years 1992 and 2016 collected from NCEI (NCEI (National Centers for Environmental Information), 2020) (http://ngdc.noaa.gov/eog/download.html).

The downloaded NTL Defense Meteorological Satellite Program's Operational Linescan System (DMSP-OLS) data of 1992 and NTL Visible Infrared Imaging Radiometer Suite (VIIRS) data of 2016 were 'F101992.v4b_web.stable_lights.avg_vis.tif' and 'SVDNB_npp_20160101-20161231_75N060E_v10_c201807311200', respectively.

While classified images of Landsat data over time were taken to derive the spatio-temporal dynamics of DMDP's growth, NTL data were used to detect the shifts in peri-urban growth boundaries corresponding to those spatial changes in DMDP's growth. As peri-urban areas take around 20–30 years to become completely urbanised, this study selected the NTL DMSP-OLS dataset of 1992 and the NTL VIIRS Day-Night Band (DNB) dataset of 2016 in order to investigate the changes in the peri-urbanisation pattern over a period of 24 years. This study selected the NTL dataset of 1992 and 2016, as the NTL data are available from 1992 onwards, and the latest year of NTL data which provides an annual composite is 2016.

In order to reveal the socioeconomic aspects of peri-urban growth, the UN-adjusted grid-level raster data on demography of the year 2001, 2016, and 2020, the average likelihood of poverty (ALP) (i.e., the average probability of living on less than $2.50/day), and standard deviation of the ALP of the year 2013 for the context of Bangladesh were collected from WorldPop and CIESIN (WorldPop, 2020) (https://www.worldpop.org/geodata/country?iso3=BGD). In addition, a continuous surface elevation map was generated for the DMDP area by using Shuttle Radar Topography Mission (SRTM) datasets of 2014 collected from USGS (USGS (United States Geological Survey), 2020).

Other relevant documents including administrative boundaries, data on flood plain areas (FPAs), proposed land use zone data, major river boundary, and reports of the City Region Development Plan (CRDP) 2015–2035 were collected from the RAJUK. The data on daily rainfall for the DMDP context from 2000 to 2018 were collected from the Bangladesh Water Development Board (BWDB).

4.2.3 Methods

The collected raster datasets and shapefiles were all projected to the universal coordinate system: 'WGS_1984_UTM_Zone_46N'. Figure 4.2 illustrates the overall methodology of this study, while the methodological steps are spelt out in the following sub-sections.

4.2.3.1 Preprocessing of Satellite Images

4.2.3.1.1 Preprocessing of Landsat Data

The collected Landsat images were carefully cross-checked to ensure that all collected images were geometrically corrected. Thus, no image-to-image spatial adjustments were performed thereafter.

Nevertheless, two accompanying Landsat scenes altogether cover the entire boundary of the DMDP area. However, due to the differences in sun angle as well as in the timing and date of image acquisition, the spectral signatures of each scene were notably different from its counterpart for each corresponding year. Thus, this study avoided image mosaicking before image classification. The radiometric calibration and atmospheric correction of these collected images were performed thereafter in the ENVI platform.

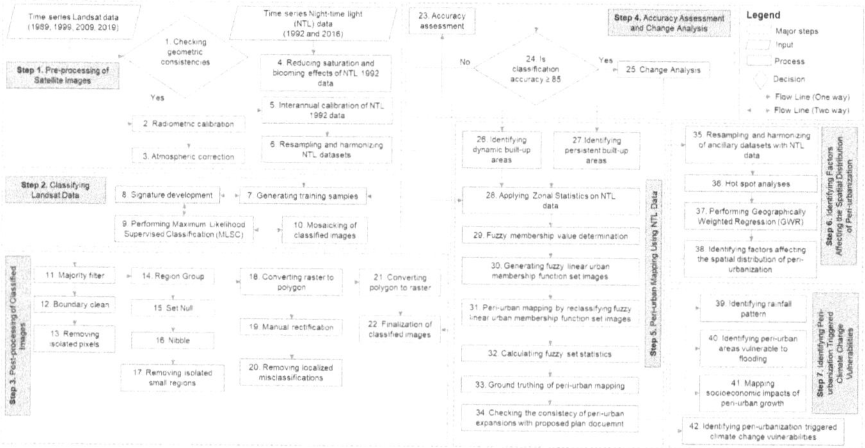

FIGURE 4.2
Methodological steps of this study.

4.2.3.1.2 Preprocessing of NTL Data

The NTL DMSP-OLS data inherently have two main problems (Bennett & Smith, 2017): (a) Saturation and blooming effects, and (b) inter-annual inconsistency, whereas such problems with the NTL VIIRS data are assumed to be minimal (Li & Zhou, 2017). Hence, in order to perform the spatio-temporal analysis using NTL data, producing temporally consistent DMSP-OLS data became necessary.

Reducing saturation and blooming effects of NTL DMSP-OLS 1992 data: In order to reduce the saturation and blooming effects of the NTL DMSP-OLS data of 1992, this study adapted Cao et al.'s (2019) self-adjusting model (SEAM).

The SEAM model was first applied in Beijing, China. Considering the compact land use development pattern and high-density sprawl development that are predominantly prevalent both in Dhaka and Beijing (Cox, 2020; Han et al., 2015), Dhaka's urban growth pattern and subsequent NTL illumination appear to be similar to the corresponding illuminations of Beijing. Thus, the chapter utilised this SEAM model for the DMDP area context by using the SEAM's script coded in MATLAB. In addition, this study chose the SEAM model because it can alleviate the saturation and blooming effects of the NTL DMSP-OLS images without the help of other auxiliary data.

According to the SEAM model, the saturation and blooming effects are estimated by pixel-based regression using pseudo light pixels (PLPs) and their neighbouring light sources. These PLPs are selected from the urban edges. The PLPs shown in Figure 4.3 represent weak brightness (i.e., Digital Number (DN) > 0), but one or more of their eight neighbours are dark (i.e., DN = 0). For each DMSP pixel with DN larger than 0, PLPs are selected within a radius of 150 km.

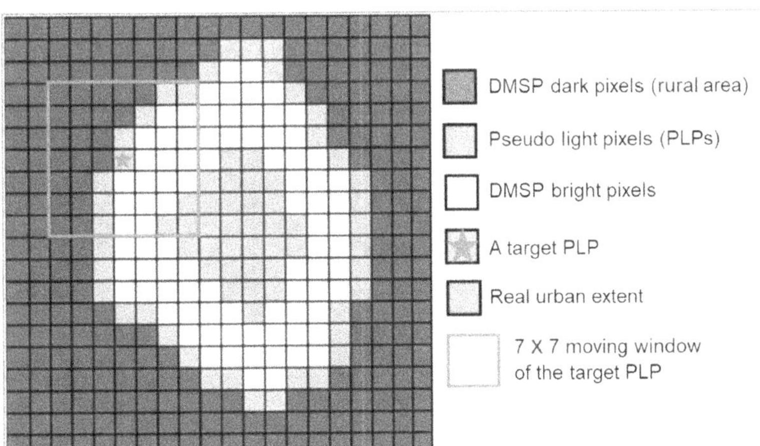

FIGURE 4.3
The selection of pseudo light pixels (PLPs) and their adjacent pixels within the frame of a 7 × 7 moving window (Cao et al., 2019) (p. 404).

Interannual Calibration of NTL 1992 data: The SEAM-corrected NTL DMSP-OLS data of 1992 (Figure 4.4b) were further intercalibrated by using Wu et al.'s (2013) study approach. The following equation was applied:

$$DN_c = a X \left(DN_m + 1\right)^b - 1 \tag{4.1}$$

where DN_c = intercalibrated NTL DMSP-OLS image of 1992; DN_m = SEAM-corrected NTL DMSP-OLS image of 1992; a = model coefficient = 0.8959; and b = model coefficient = 1.0310. The negative value of the NTL data of 1992 was later set to zero. This way, the NTL DMDP-OLS data of 1992 were made compatible with the NTL VIIRS data of 2016 to carry out spatio-temporal dynamics of peri-urban growth.

Resampling and harmonising NTL datasets: In the case of selecting the NTL dataset of 2016, this study chose the VIIRS annual average radiance composite vcm-orm-ntl data of 2016, which are cloud-free, outlier removed, stray-light corrected, ephemeral-lights eliminated, and geometrically corrected. Hence, no preprocessing on NTL VIIRS data was performed. These NTL data of 2016 were projected to the WGS_1984_UTM_Zone_46N reference system. The spatial resolution of the projected NTL VIIRS data was 451.52 m, and radiance values (a.k.a. DN values) within the selected DMDP area were between 0 and 69.86 nW/cm^2/sr. In order to bring harmony with the spatial resolution of the NTL DMSP-OLS data of 1992, the NTL data of 2016 were then resampled to 903.04 m spatial resolution by using the 'bilinear' resampling technique in the ArcGIS platform (Figure 4.4c).

4.2.3.2 Classifying Landsat Data

This study applied the most commonly used maximum likelihood supervised classification (MLSC) technique in the ENVI Platform to classify

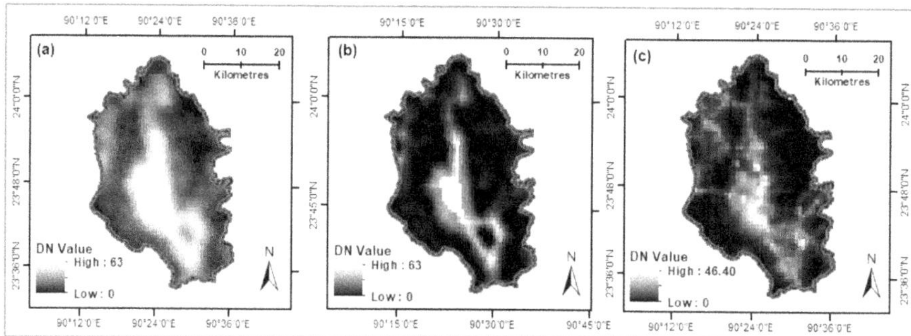

FIGURE 4.4
(a) The original DMSP-OLS image of 1992 for the DMDP area; (b) self-adjusting model (SEAM)-corrected image of 1992 for the DMDP area; (c) resampled night-time light (NTL) VIIRS image of 2016 for the DMDP area.

each selected Landsat image into five land cover categories: (a) Bare soil, (b) built-up, (c) vegetation, (d) water body, and (e) low-land. Other classification techniques—e.g., random forest algorithms, support vector machines, and decision tree algorithms—are also popular for classifying remotely sensed data. However, for classifying Landsat data, in particular, taking appropriate training samples produces more accurate classification outcomes than adapting any specific classification technique itself (Li et al., 2014; Phiri & Morgenroth, 2017). Given that the MLSC technique is robust and available with any remote sensing software package (Lu & Weng, 2007), this study selected this MLSC technique. While applying the MLSC technique, the training samples were repeatedly modified in order to ensure the selection of the most representative fraction for each land cover class, yielding the most accurate classification outcome. Later on, a post-classification mosaic for each selected year's image was performed in order to accommodate the entire boundary of the DMDP area within a single classified image frame.

4.2.3.3 Post-Processing of Classified Images

All classified images were found with 'salt-and-pepper' effects and some degree of localised misclassifications—e.g., 'bare soil' was misclassified as 'built-up areas', and 'built-up areas' was misclassified as 'bare soil'. In addition, in some places, low-density scattered settlements were misclassified as 'vegetation' due to the shade of vegetation coverage. Nonetheless, due to the lower resolution of Landsat images, such perceived misclassifications are common with Landsat image classification (Phiri & Morgenroth, 2017).

Hence, post-processing of classified images became necessary, which was subsequently carried out in the ArcGIS platform. First, in order to remove the 'salt-and-pepper' effects—i.e., removing the presence of isolated pixels from classified images, this study employed the 'Majority Filter' by using a three-by-three window. Second, the 'Boundary Clean' tool was used for smoothening the boundary of land cover classes. Third, for removing the small isolated regions, the study further generalised classified images by sequentially using the 'Region Group', 'Set Null', and 'Nibble' tools. Fourth, the classified raster images were further vectorised in order to manually rectify the localised misclassifications. After manual rectification of localised misclassification, these images were again rasterised to make them fit for accuracy assessment.

4.2.3.4 Accuracy Assessment of Classified Images and Change Analysis

The accuracy assessment of classified images was performed in the Google Earth platform. In the case of selecting the number of sampling points for accuracy assessment, if the study area's coverage is less than 1 million acres and the classified land cover categories are fewer than 12 land cover classes, a minimum of 50 sampling points for each land cover category is recommended (Congalton & Green, 2019). In this regard, the size of the study area was 377,842.53 acres, and the number of classified land cover categories was 5. Thus, the threshold sampling points for this accuracy assessment task were 250. However, the number of sampling points under the stratified random sampling technique is proportional to the area coverage of each land cover class. As some land cover classes (e.g., water bodies) were too rare, the formulation of a minimum of 50 stratified random sampling points for each land cover class was not possible. Finally, a total of 351 random sampling points was generated through a stratified random sampling technique in the ArcGIS spatial analyst platform. The image classification process and the corresponding post-processing tasks of classified images as mentioned above were repeatedly done until all classified images were found with a minimum classification accuracy of 85% (Rahman et al., 2017). Overall classification accuracies of the classified images for 1989, 1999, 2009, and 2019 were 90.00%, 84.90%, 84.90%, and 85.47% with Kappa coefficients of 0.849, 0.778, 0.775, and 0.769, respectively. Finally, post-classification change detection was carried out by using Land Change Modeller (LCM) in the TerrSet Geospatial Monitoring and Modelling platform.

4.2.3.5 Peri-Urban Mapping Using NTL Data

4.2.3.5.1 Recognising the Fuzzy Characters of Peri-Urban Areas

Peri-urban areas generally start from location proximity to an urban core and continue until a predominantly rural landscape is found. Thus, peri-urban areas, in general, comprise both the characteristics of urban and rural land uses, where some portions of peri-urban areas are more urban, and the remaining portions are more rural. Hence, peri-urban areas significantly possess the characteristics of the fuzzy-set theory, which limits peri-urban areas within a fuzzy membership value of 0 and 1, where '0 = predominantly rural', and '1 = predominantly urban'. Consequently, the values between 0 and 1 imply peri-urban areas, where higher membership values (e.g., 0.75) indicate more peri-urban—i.e., inclined to more urban—and lower values (e.g., 0.15) indicate less peri-urban—i.e., inclined to more rural. Henceforth, this study applied the fuzzy membership function to reveal these perceived fuzzy characteristics of peri-urban areas.

4.2.3.5.2 Identifying the Suitable Fuzzy Membership Function

This study primarily centred around selecting a suitable fuzzy membership function to reveal the fuzzy characteristics of peri-urban areas. Initially, fuzzy 'Gaussian' and 'Linear' membership functions on NTL data in the ArcGIS platform were run. By using multiple combinations as input values, it is observed that the fuzzy Linear membership function is more interpretable than the Gaussian one while using NTL data. Thus, this chapter chose the fuzzy Linear membership function to map the fuzzy characteristics of peri-urbanisation and named this the 'fuzzy linear urban membership function'.

4.2.3.5.3 *Selecting the Membership Value for the Fuzzy Linear Urban Membership Function*

In order to determine the minimum and maximum membership values of the fuzzy linear urban membership function, this study first extracted the persistent and dynamic built-up areas. The persistent built-up areas are those which remained unchanged between 1989 and 2019, while the dynamic built-up areas are the landscapes which were converted into built-up surfaces between 1989 and 2019. As peri-urban areas become urbanised over a 20- to 30-year period, the built-up areas showing persistence between 1989 and 2019 were hypothesised to be predominantly urban. Thus, this study assumes that the characteristics of peri-urban areas lie within the dynamic built-up surfaces. Henceforth, this study extracted the areas which were converted into built-up surfaces between 1989 and 2019.

Later on, this study carried out zonal statistics on the NTL datasets of 1992 and 2016 for the areas comprising persistent and dynamic built-up areas individually (Table 4.2). It is observed that the mean value between these two separate built-up zones is significantly different, while the differences in the maximum values between these two separate zones are not so conspicuous. Thus, the mean value of dynamic built-up areas appears to have significant potential to interpret the peri-urbanisation pattern of the DMDP area. Consequently, this study took the 'minimum value' and 'mean value' as the 'minimum value' and 'maximum value', respectively, to form the fuzzy linear urban membership function set images in the ArcGIS platform. Any values more than the mean estimate (which was subsequently considered the 'maximum value') indicated predominantly urban areas in the fuzzy linear urban membership function set images.

It is important to mention that this study only used the boundary of dynamic built-up areas in order to derive the membership values for the fuzzy linear urban membership function set images using NTL data. No pixel-to-pixel comparison was performed between 30 m resolution classified Landsat data and 903.04 m resolution NTL data. This way, the difficulty of handling the variations in spatial resolution of these two different image datasets was avoided.

TABLE 4.2

The Radiance Values of NTL Data Between Dynamic and Persistent Built-Up Areas

	Persistent Built-Up Areas			Dynamic Built-Up Areas		
Year	**Minimum**	**Maximum**	**Mean**	**Minimum**	**Maximum**	**Mean**
1992	0	64.22	29.22	0	64.22	9.0
2016	0	33.37	12.82	0	46.4	6.33

4.2.3.5.4 *Identification of Peri-Urban Areas Within the Fuzzy Linear Urban Membership Function Set Images*

The urban core areas were predominantly found within the values of 0.80–1.0, while the rural areas were predominantly found within the value range of 0–0.10 in the fuzzy linear urban membership function set images. Consequently, peri-urban areas are delimited within the value range of 0.11–0.79. Hence, by using these perceived value ranges, the fuzzy linear urban membership function set images of the years 1992 and 2016 are reclassified into three categories (Figure 4.5): (a) Predominantly urban (PURBAN), (b) predominantly rural (PRURAL), and (c) peri-urban (PU).

4.2.3.5.5 *Derivation of Fuzzy Set Statistics for the Study Area*

This study derived the overall level of urbanisation (LoU) and the degree of fuzziness (DoF) for the DMDP context. The LoU denotes the mean gross level of membership in the fuzzy linear urban membership function set. The following equation was used to calculate the LoU.

$$LoU_{year} = \frac{\sum_{i}^{n} F_{year}}{\sum n} \tag{4.2}$$

where LoU is the LoU of each corresponding year, and n is the total number of cells in the corresponding fuzzy linear urban membership function set image.

In order to determine the DoF, the following equation was used:

$$DoF_{year} = \frac{\left(\frac{F_{year} - F_{Predominantly\ Urban}}{F_{year}}\right)}{\sum n} \tag{4.3}$$

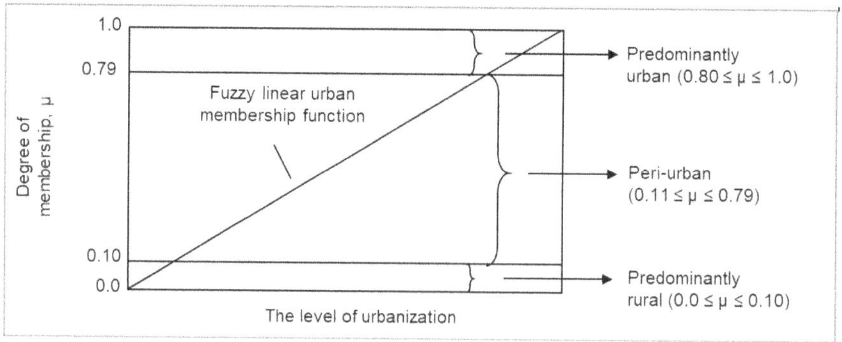

FIGURE 4.5
The level of urbanisation under the microscope of the fuzzy linear urban membership function set.

where DoF_{year} = the DoF for each corresponding year, and $F_{Predominantly\ Urban}$ = fuzzy linear membership function set for predominantly urban areas of each corresponding year, where the minimum and maximum values for the predominantly urban set image were assigned to 0.80 and 1.0, respectively.

4.2.3.6 Validation of Peri-Urban Mapping

4.2.3.6.1 Ground Truthing of Peri-Urban Mapping

In order to determine the accuracy of the peri-urban mapping approach, this study carried out ground-truthing by using 300 stratified random sampling points. Considering the availability of data for the purpose of ground-truthing using the Google Earth platform, the study chose NTL VIIRS data mapping of 2016 for this ground-truthing purpose. Nevertheless, as peri-urbanisation is local context-specific, a set of criteria was primarily formulated to distinguish the areas which are predominantly urban, peri-urban, or predominantly rural for the context of the DMDP area (Table 4.3). Later on, a ground-truthing exercise was performed on the Google Earth platform, and a confusion matrix was formed thereafter to derive the accuracy of the peri-urban mapping exercise of 2016.

TABLE 4.3

Criteria to Ground-Truth the NTL VIIRS Data Mapping of 2016

Category	Criteria
Predominantly urban	The ground-truthing point falls in places where the areas maintain a high-density (i.e., apparently, there is no distance between two settlements), continuous built-up development and surrounding areas are built-up.
	If the point does not fall within continuous built-up areas, the point falls in places where high-density continuous built-up areas are within 300 m.
Predominantly rural	The point falls in places where vegetation is more dominant, and a number of scattered rural homesteads are located within a 200 m radius
	The point falls within a 300 m radius of the nearby road
	Settlements are located in proximity to paddy lands, wetlands, or low-lying areas
Peri-urban	Located within 400 m of low-density continuous built-up development
	Located within 50 m of the nearby roads
	The point falls in places where vegetation is less dominant (i.e., built-up areas are not shaded by the vegetation) than in the presence of built-up areas
	Point falls in a location where settlements are dispersed but follows a linear development alongside the road

4.2.3.6.2 Checking the Consistency of Peri-Urban Expansions with the Proposed Plan Documents

In order to determine the changes in peri-urban growth corresponding to the proposed changes as stipulated in the DMDP's City Region Development Project (CRDP) 2015–2035, this study compared the NTL VIIRS data mapping of 2016 with the proposed land use zones of CRDP 2015–2035.

4.2.3.7 Identifying Factors Affecting the Spatial Distribution of Peri-Urbanisation

4.2.3.7.1 Preprocessing of Ancillary Datasets

The collected raster datasets on demography, the ALP and standard deviation of the ALP, and the continuous surface elevation dataset generated for the DMDP context were all resampled to 903.04 m spatial resolution (by using 'bilinear' resampling technique in the ArcGIS platform) in order to obtain consistency with the NTL data–driven peri-urban mapping for comparative analysis. This study considers the standard deviation of the ALP as a proxy variable of social stratification.

4.2.3.7.2 Identification of Statistically Significant Hot Spots for Interpreting Peri-Urbanisation

In order to reveal whether the spatial distribution of population growth, the standard deviation of the ALP, elevation, and corresponding changes in the spatial distribution of peri-urbanisation are clustered, random, or dispersed, the spatial autocorrelation was carried out to derive the Global Moran's I statistic. The positive Moran's Index implies a tendency towards spatial clustering, while the negative value of Moran's Index insinuates a tendency towards spatial dispersion. Based on the outcome of the spatial autocorrelation against each parameter as mentioned above, the null hypothesis 'spatial distribution of a given parameter is normally distributed' was tested.

Afterwards, hotspot analysis was carried out to compute the Getis-Ord Gi* in order to measure the intensity of such perceived clustering. While identifying the Getis-Ord Gi* statistic, a False Discovery Rate (FDR) correction was applied in order to resolve the issues of spatial dependency and multiple testing. Regardless of whether the z-score is positive or negative, a higher z-score value implies more spatial clustering. A positive z-score value indicates clustering of hot spots, while a z-score with a negative value indicates clustering of cold spots, and a z-score near zero implies clustering is 'not significant', meaning that no apparent spatial clustering is evident.

4.2.3.7.3 Performing Geographically Weighted Regression

The Geographically Weighted Regression (GWR) was carried out in order to reveal how peri-urbanisation is interpreted as a response variable to the spatial distribution of population growth, the standard deviation of the ALP, and the elevation pattern for the DMDP context.

4.2.3.8 Identifying Peri-Urbanisation Triggered Climate Change Vulnerabilities

4.2.3.8.1 Identifying Rainfall Pattern of the Study Area

The daily rainfall data from 2000 to 2018 (collected from the BWDB) were used to derive the statistics on total annual rainfall, annual monsoon rainfall, and annual non-monsoon rainfall. Each year's rainfall from May to September was considered the 'annual monsoon rainfall', and the rainfall of the remaining months for each year was considered the 'annual non-monsoon rainfall'.

4.2.3.8.2 Identification of Peri-Urban Growth Pockets Vulnerable to Flooding

The data on DMDP's FPA (collected from the RAJUK) were superimposed on NTL VIIRS data mapping of 2016 in order to find out the peri-urban growth pockets vulnerable to flooding.

4.2.3.8.3 Mapping the Socioeconomic Impacts of Peri-Urban Growth

The demographic data for the years 2001, 2016, and 2020, the ALP and the standard deviation of the ALP for the year 2013 for the DMDP context were all analysed and compared with the NTL VIIRS data mapping of 2016 in order to illustrate how the changes in the spatial distribution of demography and the standard deviation of ALP correspond to the changes in peri-urban growth.

4.3 Analysis and Results

4.3.1 Changes in Land Cover

The classified images are presented in Figure 4.6, while the aerial statistics of those classified images are presented in Table 4.4.

The results find that, except for built-up areas, the remaining land cover categories declined over time. While the highest decline of 22.18% is observed in the vegetation category, the 'bare soil' and 'low-land' categories also demonstrate a significant decline over this 30-year period. Thus, the losses in bare soil, vegetation, and lowlands contributed altogether to the development of built-up areas over time. Consequently, built-up areas quadrupled over time and constituted more than half (i.e., 52%) of the DMDP area in 2019, whereas bare soil and vegetation were more than halved within this 30-year period. Earlier studies also found a similar growth trend for bare soil, built-up, vegetation, and lowland areas (e.g., (Ahmed & Ahmed, 2012; Ahmed et al., 2013; Hassan & Southworth, 2018)).

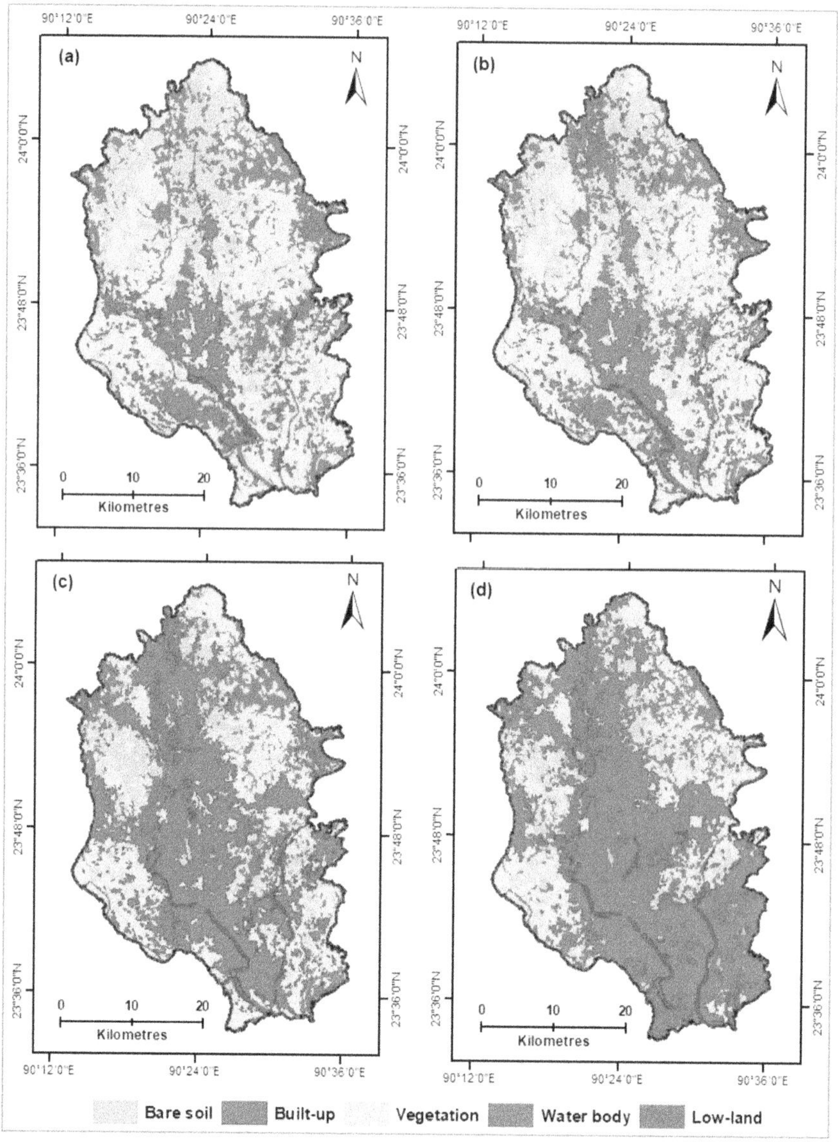

FIGURE 4.6
(a) Classified DMDP image from 1989; (b) classified DMDP image from 1999; (c) classified DMDP image from 2009; (d) classified DMDP image from 2019.

In addition, all land cover categories (except water bodies) demonstrate an exponential growth/decline trend with strong coefficient of determination values of R^2.

In the case of sub-district-wise net changes in bare soil, Gazipur Sadar encountered the highest decline (i.e., 26%) in bare soil, followed by the Dhaka

TABLE 4.4
Aerial Statistics of Classified Images for the DMDP Area

	1989		1999		2009		2019		
Land Cover/Year	**Area (km²)**	**%**	**Area (km²)**	**%**	**Area (km²)**	**%**	**Area (km²)**	**%**	**R^2 (Exponential)**
Bare soil	274	18	247	16	152	10	125	8	0.94
Built-up	199	13	329	22	632	41	790	52	0.97
Vegetation	659	43	636	42	435	28	320	21	0.92
Water body	38	3	40	3	47	3	42	3	0.00
Low land	359	23	277	18	262	17	252	16	0.83
Total	1529	100	1529	100	1529	100	1529	100	

Metropolitan Area (DMA), and Narayanganj Sadar by 13% and 8%, respectively, between 1989 and 2019 (Figure 4.7). While sub-districts predominantly lost bare soil over time, Kaliganj presented an increase of 7% for bare soil within this 30-year period. On the contrary, all sub-districts encountered significant losses in vegetation. As for the case of built-up areas, all sub-districts except for Kaliganj demonstrated significant increases over time, while the built-up areas of Kaliganj declined by 4% between 1989 and 2019.

4.3.2 Changes in Peri-Urban Boundary

The aerial statistics of peri-urban growth have been presented in Table 4.5, while the peri-urban mapping has been depicted in Figure 4.8. The analysis reveals that for the DMDP context, predominantly urban areas remained nearly unchanged over time, while the major changes occurred in transitioning predominantly rural areas into peri-urban areas (Table 5.5). Thus, major development within the DMDP context predominantly took place in the form of peri-urbanisation.

The LoU within the predominantly urban areas category was 0.94 and 0.96 in 1992 and 2016, respectively. Thus, predominantly urban areas had

TABLE 4.5
Aerial Statistics of Peri-Urban Growth Between 1992 and 2016

	Year 1992				Year 2016				% of Changes (1992–2016)	
Category	**Area (km²)**	**%**	**LoU**	**DoF**	**Area (km²)**	**%**	**LoU**	**DoF**	**LoU (%)**	**DoF (%)**
Predominantly rural	365	24	0.03	1	167	11	0.04	1	33.33	0.00
Peri-urban	715	47	0.3	0.98	874	58	0.4	0.94	33.33	(−)4.08
Predominantly urban	452	29	0.94	0.07	478	31	0.96	0.12	2.13	71.43

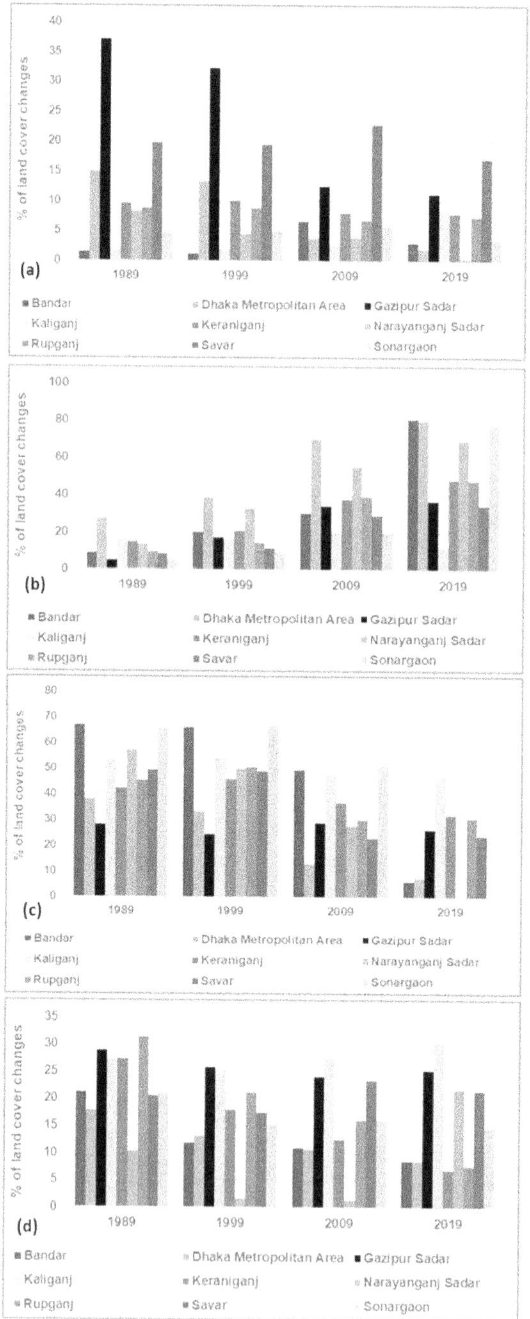

FIGURE 4.7
(a) % of net changes in bare soil; (b) % of net changes in built-up areas; (c) % of net changes in vegetation; (d) % of net changes in low-land areas.

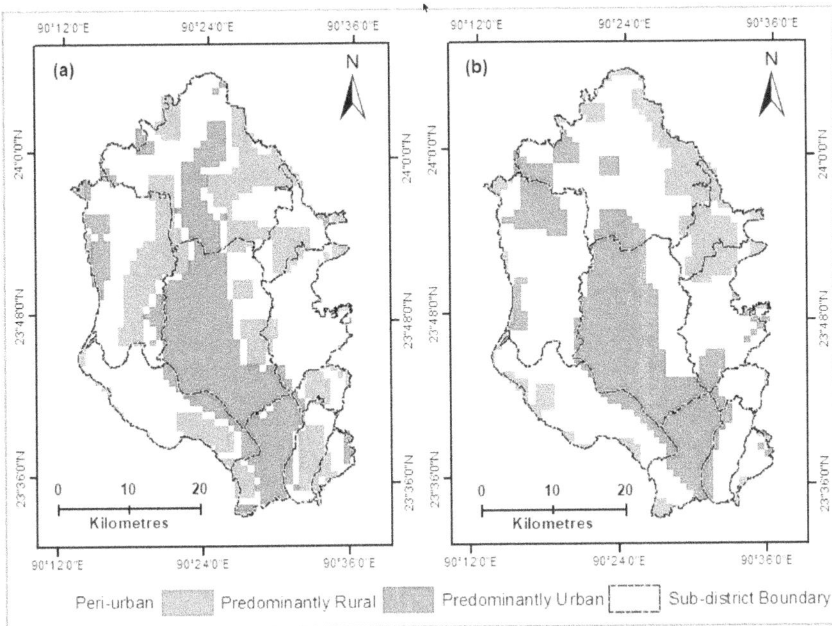

FIGURE 4.8
(a) Peri-urban mapping of 1992; (b) peri-urban mapping of 2016.

undergone the highest LoU within these 24 years (1992–2016), followed by peri-urban areas and predominantly rural areas. Nevertheless, in terms of the LoU, predominantly urban areas demonstrated an increase of 2.13% only, which resulted in an increase of 71.43% in the DoF in 2016 from the base year estimate of 1992. Such a finding reveals that, along with the pace of urbanisation, a less articulated pattern of urban land use practices becomes more evident.

In addition, the LoU within the predominantly rural areas category demonstrates the same thing, while the DoF within this zone is 1.00 in 1992 and 2016, meaning the land use practices embraced an unclear pattern in distinguishing whether the areas are predominantly rural or urban.

Furthermore, predominantly rural areas and peri-urban areas both demonstrated an increase of 33.33% in the LoU in 2016. However, while unquestionably rural areas demonstrated persistence in encountering the highest DoF in 1992 and 2016, peri-urban areas showed a slight decline of 4.08% in 2016, implying a nominal improvement in embracing articulated land use practices within peri-urban areas.

In the case of the sub-district-wise LoU and DoF, with the correlation coefficients of (–)0.97 and (–)0.95, respectively, for the years 1992 and 2016, the result reveals that the DoF is inversely correlated to the LoU, with Bandar being an exception (Figure 4.9). The Bandar sub-district encountered the

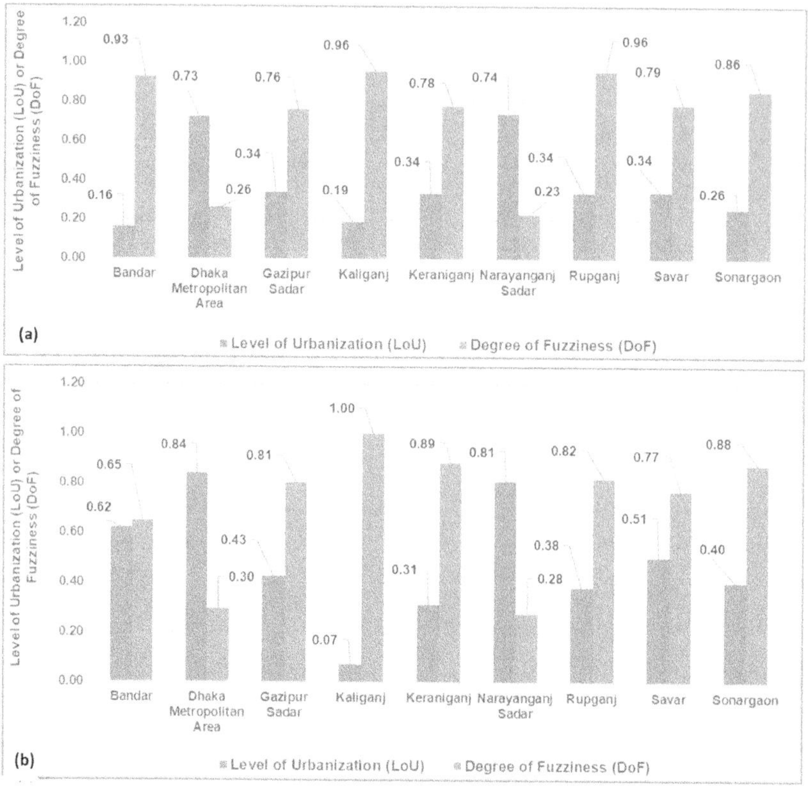

FIGURE 4.9
(a) Sub-district-wise level of urbanisation (LoU) and degree of fuzziness (DoF) statistics for 1992; (b) sub-district-wise level of urbanisation (LoU) and degree of fuzziness (DoF) statistics for 2016.

highest increase in the rate of the LoU (i.e., 277%) and experienced a reduction in the DoF, from 0.93 to 0.65 in 2016. Such a finding implies that scattered urban development in Bandar has been further consolidated through increased urbanisation, resulting in a decrease in the DoF thereafter. The DMA and Narayanganj Sadar continued the higher LoU both in 1992 and 2016, and hence scored the lowest in terms of the DoF.

The remaining sub-districts underwent less urbanisation and encountered more fuzziness, meaning that an intensified unarticulated peri-urbanisation predominantly took place within these (remaining) sub-districts.

4.3.3 Validation of Peri-Urban Mapping

4.3.3.1 Ground-Truthing of Peri-Urban Mapping

The spatial distribution of the ground-truthing points generated through the stratified random sampling technique is presented in Figure 4.10, while the

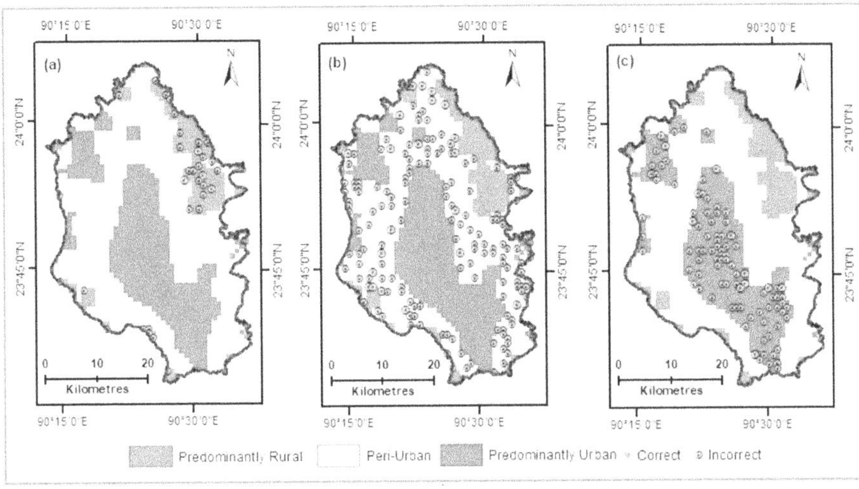

FIGURE 4.10
Spatial distribution of ground-truthing points on NTL VIIRS data mapping of 2016 for: (a) Predominantly rural areas; (b) peri-urban areas; (c) predominantly urban areas.

confusion matrix of this ground-truthing exercise is given in Table 4.6. With a Kappa coefficient of 0.75, the ground-truthing exercise yields an overall accuracy of 86%.

4.3.3.2 Consistency of Peri-Urban Expansions with the Proposed Plan Documents

The CRDP proposed land use zone of 2015–2035 was superimposed over the NTL VIIRS data mapping of 2016 in order to cross-check how the adapted

TABLE 4.6
Confusion Matrix of NTL VIIRS Data Mapping of 2016

Category	Predominantly Rural	Peri-Urban	Predominantly Urban	Total	Producer's Accuracy (%)	User's Accuracy (%)
Predominantly rural	30	4	0	34	55.56	88.24
Peri-Urban	24	147	1	172	89.09	85.47
Predominantly urban	0	14	80	94	98.77%	85.11
Total	54	165	81	300		
Overall Accuracy	86%					
Kappa Coefficient	0.75					

peri-urban mapping approach fits with the proposed land use zoning for the DMDP area.

The CRDP's 2015–2035 proposed land use zones comprise eight broad land use categories (Figure 4.11). The analysis reveals that the mixed-use zone (55%) and agricultural zone (29%) are the two most dominant proposed

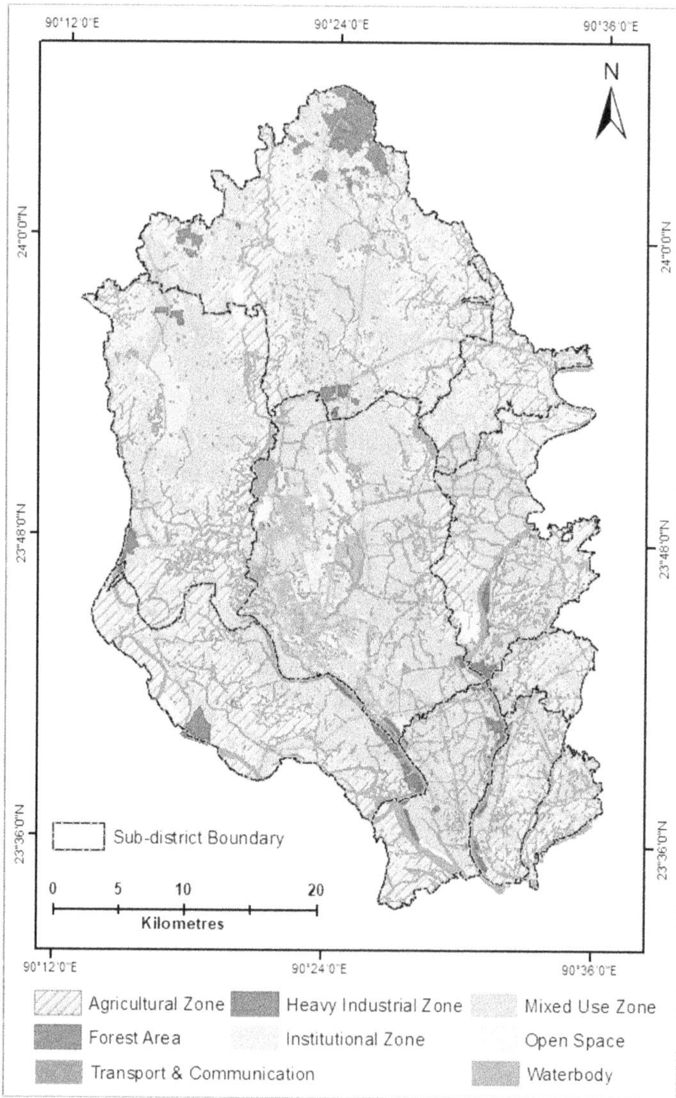

FIGURE 4.11
Spatial distribution of CRDP's 2015–2035 proposed land use zones (CRDP (City Regional Development Planning), 2014).

TABLE 4.7

Aerial Statistics of the Proposed Land Use Zone

	Predominantly Rural		Peri-Urban		Predominantly Urban			
Proposed Land Use Zoning	**Area (km²)**	**%**	**Area (km²)**	**%**	**Area (km²)**	**%**	**Total Area (km²)**	**Total %**
Agricultural Zone	86.36	5.81	325.23	21.87	19.46	1.31	431.05	29
Forest Area	4.71	0.32	14.42	0.97	0.08	0.01	19.21	1
Heavy Industrial Zone	0.21	0.01	6.60	0.44	23.11	1.55	29.93	2
Institutional Zone	0.41	0.03	19.81	1.33	38.88	2.61	59.10	4
Mixed-Use Zone	58.66	3.94	417.93	28.10	335.67	22.57	812.26	55
Open Space	1.46	0.10	4.72	0.32	7.00	0.47	13.18	1
Transport and Communication	0.94	0.06	7.47	0.50	14.77	0.99	23.18	2
Waterbody	7.40	0.50	56.17	3.78	35.60	2.39	99.16	7

Source: Derived from the CRDP 2015–2035 (CRDP (City Regional Development Planning), 2014).

land use zones in the DMDP area, which altogether account for 84% of the DMDP's area coverage (Table 4.7).

According to the NTL VIIRS data mapping of 2016, predominantly rural areas comprise 11% of the DMDP area, out of which the CRDP's 2015–2035 proposed agricultural zone and mixed-use (i.e., a blend of residential, commercial, and general industrial areas) zone altogether comprise around 10%. Similarly, in an aggregate, peri-urban areas comprise around 58% of the DMDP's area, whereas mixed-use zones (28.10%) and agricultural zones (21.87%) altogether comprise around 50% of the DMDP's peri-urban area. Thus, peri-urban areas are proposed to accommodate around three-fourths of the DMDP's proposed agricultural areas and more than half of the proposed mixed-use zone. Predominantly urban areas are proposed to be dominated by mixed-use zones, which comprise around two-fifths of the proposed mixed-use zone areas.

Among the other proposed land use zones, institutional zones comprise around 4% of the DMDP's area, and comprise 2.61% and 1.33% of the predominantly urban and peri-urban areas, respectively. In addition, more than three-fourths of the DMDP's proposed heavy industrial zone is located in predominantly urban areas. Meanwhile, among the total proposed forest areas, nearly three-fourths of the proposed forest zones are located in peri-urban areas, and the remaining portion was proposed to be accommodated by predominantly rural areas.

The analysis reveals that the proposed forest areas are observed as having the lowest values for LoU and the highest values for DoF (Figure 4.12). Contrariwise, the proposed heavy industrial zone appeared to be the

FIGURE 4.12
The level of urbanisation (LoU) and degree of fuzziness (DoF) under the CRDP's 2015–2035 proposed land use zones.

most urbanised in terms of LoU, and hence had the lowest values for DoF. Apparently, heavy industrial zones are proposed based on strict policy control, and hence, spontaneous development of these areas is not possible, resulting in an articulated land use pattern of these areas. Meanwhile, forested areas are spontaneous natural growth zones, and hence were observed as having the highest DoF.

In addition, in the case of each subdistrict, the higher the LoU, the lower the level of DoF in each proposed land use zone is (Figure 4.13). The Kaliganj sub-district, with the lowest LoU and an absolute DoF value of 1.0, provokes an immediate land use intervention by the city authorities (Figure 4.13d). Other sub-districts with higher DoF include Gazipur, Keraniganj, Rupganj, Sonargaon, and Savar. These areas are predominantly located in peri-urban areas, meaning that peri-urban growth appears to be dominated by unarticulated land use patterns.

Such general findings are also evident in the sub-district-wise LoU and DoF calculations that were derived from the NTL VIIRS data mapping of 2016 (Figure 4.9). Thus, it can be deduced that the proposed NTL data mapping exercise of 2016 yields substantive outcomes in distinguishing predominantly rural, peri-urban, and predominantly urban areas.

4.3.4 Factors Affecting the Spatial Distribution of Peri-Urbanisation

4.3.4.1 Identification of Statistically Significant Hot Spots for Interpreting Peri-Urbanisation

4.3.4.1.1 Hotspots of Population Growth

In the DMDP area, a highly statistically significant (Moran's Index: 0.69, p-value = 0, and hence the null hypothesis 'spatial distribution of the

FIGURE 4.13
The level of urbanisation (LoU) and degree of fuzziness (DoF) for CRDP's 2015–2035 proposed land use zones: (a) Bandar; (b) Dhaka Metropolitan Area (DMA); (c) Gazipur Sadar; (d) Kaliganj; (e) Keraniganj; (f) Narayanganj Sadar; (g) Rupganj; (h) Savar; and (i) Sonargaon.

population is normally distributed' was rejected) cluster pattern of population growth is evident. The hot spots of population increase are found predominantly in the Gazipur Sadar areas (Figure 4.14a), which bolsters the earlier findings on Gazipur Sadar's supremacy in attaining the highest rate of population growth. However, Savar areas are prevalent with both hot spots and cold spots. Indeed, the northern part of Savar in proximity to the Gazipur Sadar areas is observed as having hot spots, whereas the southern part of Savar in proximity to the DMA, is found to have cold spots for population growth. As Savar is found to have both cold and hot spots of population growth, the aggregate growth rate of population in Savar does not appear to be higher than its contemporaries, although the earlier findings claim that Savar is the most peri-urbanised sub-district for this study.

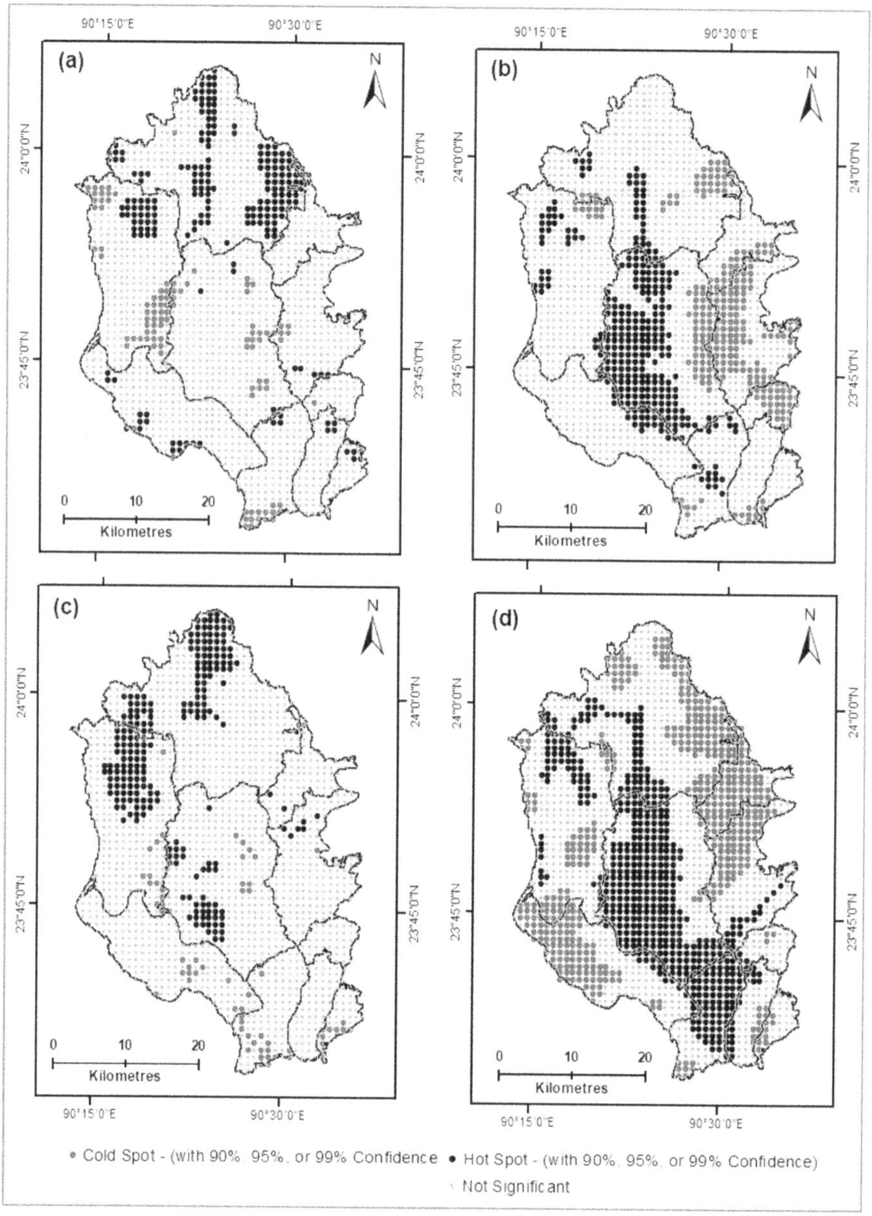

FIGURE 4.14
(a) Hot spot analysis of the aggregate population growth rate (APGR) mapping; (b) hot spot analysis of the standard deviation of the average likelihood of poverty (ALP); (c) hot spot analysis of elevation pattern; (d) hot spot analysis of the fuzzy linear urban membership function set image of 2016. ((c) Derived from the SRTM global DEM data (USGS (United States Geological Survey), 2020).)

4.3.4.1.2 *Hotspots Mapping of the Standard Deviation of the Average Likelihood of Poverty*

The hotspots mapping exercise on the standard deviation of the ALP demonstrates a highly statistically significant clustering pattern (Moran's Index = 0.78 and *p*-value = 0, and hence the null hypothesis 'spatial distribution of poverty deviation is normally distributed' was rejected). However, in this regard, hot spots are predominantly confined within the DMA, while the cold spots are predominantly observed near the eastern periphery of the DMA, which includes Rupganj, Sonargaon, and Kaliganj sub-districts (Figure 4.14b). Thus, the spatial distribution of the standard deviation of the ALP hot-spot mapping exercise does not appear to be commensurate with the hot-spot mapping of the population growth pattern.

4.3.4.1.3 *Hotspots of Elevation Pattern*

A statistically significant (Moran's Index: 0.56, *p*-value = 0, and hence the null hypothesis 'spatial distribution of elevation pattern is normally distributed' was rejected) clustering pattern is dominant in the spatial distribution of elevation for the DMDP area. The hot spots (i.e., areas with higher elevation) are predominantly concentrated in the Gazipur Sadar and Savar sub-districts—i.e., the areas which underwent the highest peri-urbanisation (Figure 4.14c).

4.3.4.1.4 *Hotspots of Peri-Urbanisation*

This study used the NTL VIIRS data mapping of 2016 in determining the spatial distribution of hotspots for peri-urbanisation (Figure 4.14d). The Moran's Index was computed as (+)0.90 with a z-score value of 54.51 and a *p*-value of 0. Thus, the null hypothesis 'spatial distribution of peri-urbanisation pattern is normally distributed' was rejected, implying that the spatial distribution of the fuzzy linear urban membership function set image of 2016 (i.e., NTL VIIRS data mapping of 2016) demonstrates a highly significant clustering pattern.

The analysis reveals that hot spots are predominantly urban, whereas the cold spots are predominantly rural (Figure 4.14d). On the contrary, the areas which are not statistically significant in the fuzzy linear urban membership function set image of 2016 are predominantly peri-urban. Indeed, peri-urban areas, which are neither urban nor rural, seem to possess more fuzzy characteristics, which in turn lead them to be statistically insignificant.

4.3.4.2 Carrying Out Geographically Weighted Regression for Interpreting Peri-Urbanisation

The findings from hot spot analysis reveal that the spatial distribution of peri-urban hot spots mapping is more linked to the corresponding distribution pattern of elevation and the growth of population. Thus, in order to illustrate how such perceived peri-urbanisation of the DMDP area varies across space corresponding to the changes in elevation and population growth pattern, GWR was carried out. The adjusted R^2 value of the GWR model was 0.87, indicating that peri-urbanisation is predominantly linked to higher elevation and higher population growth rate. Meanwhile, the R^2 value of the GWR model was 0.63 when peri-urbanisation was interpreted as a response variable of elevation, population growth rate, and the standard deviation of ALP. Thus, when the standard deviation of ALP is omitted, the GWR model interpreting DMDP's peri-urbanisation produces a better outcome, meaning that peri-urbanisation does not necessarily mean an over-concentration of poverty in the city's outskirts. Instead, such concentration is predominantly driven by the land scarcity in the city's core and an unrelenting influx of population to the capital city, Dhaka.

4.3.5 Identifying Peri-Urbanisation Triggered Climate Change Vulnerabilities

4.3.5.1 Identifying Rainfall Pattern of the Study Area

The annual rainfall pattern of the DMDP area is presented in Figure 4.15. The analysis reveals that from 2000 to 2018, total annual rainfall in the DMDP area ranged from 818.1 mm (in 2010) to 2873 mm (in 2017), with an annual average rainfall of 1824.91 mm within this 18-year period. In addition, it was calculated that the annual average rainfall in the monsoon—which lasts from September to May—comprised exactly 80% of the total annual rainfall within this 18-year period.

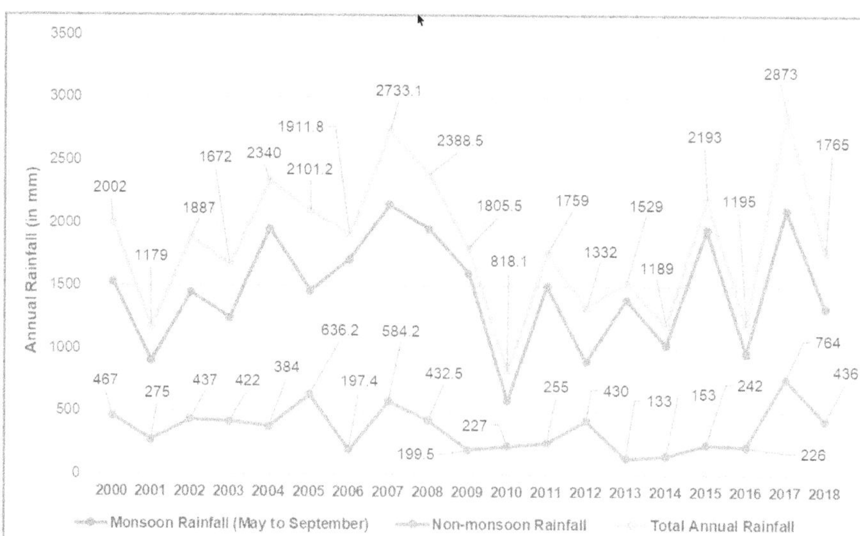

FIGURE 4.15
Rainfall pattern of the DMDP area.

Surprisingly, in 2012, the DMDP area had the lowest annual monsoon rainfall percentage of 67.72%, whereas in the following year (i.e., in the year 2013), the percentage of annual monsoon rainfall was the highest, at 91.30%. In addition, this rainfall pattern does not follow any significant growth trend over time. For example, although the highest annual rainfall occurred in the year 2017, annual monsoon rain within this year was 73.41%, which is below the annual average percentage of monsoon rainfall (i.e., 80%), whereas the highest annual rainfall in the non-monsoon period was observed in 2017.

However, this finding does not simplify with certainty that an occurrence of more rainfall events in the monsoon is linked with less rainfall in the non-monsoon period and vice versa, as the correlation coefficient of the rainfall pattern between the monsoon and non-monsoon period is 0.44 (which is not very strong), meaning that abrupt rainfall in the DMDP area has gradually become common. Consequently, events of frequent flooding due to heavy rainfall often occur in the DMDP area.

4.3.5.2 Identification of Peri-Urban Growth Pockets Vulnerable to Flooding

In general, DMDP is a flood-prone area. However, the northern part of the DMDP is at a higher elevation and is therefore comparatively less flood-prone (Figure 4.16). Thus, higher population concentration and subsequent peri-urbanisation are observed in the areas with higher elevation, which are predominantly concentrated in the Gazipur Sadar and Savar areas. The analysis reveals that around 44% of total peri-urban areas are located within the FPA, while such a percentage distribution for predominantly rural and urban areas is 35% and 33%, respectively, meaning that peri-urban areas are most vulnerable to frequent flooding. Such findings are relevant to Dewan et al.'s (2007) study, which claimed that the DMDP's peri-urban areas are moderately to highly vulnerable to annual flooding.

4.3.5.3 Mapping Socioeconomic Impacts of Peri-Urban Growth

4.3.5.3.1 *Changes in Demography*

The spatial distribution of population predominantly is concentrated within the DMA, Narayanganj Sadar, and along the Dhaka–Mymensingh and Tangail–Joydevpur Highways of the Gazipur Sadar area (Figure 4.17). Apparently, the spatial changes in demography between 2001 and 2016 are quite conspicuous, whereas such changes between 2016 and 2020 appear to be quite similar.

In addition, the percentage distribution of population between predominantly rural areas, peri-urban areas, and predominantly urban areas remained almost equal over time, whereas the density of population nearly doubled within this 19-year period (Table 4.8). Thus, this study finds a perpetual increase in the growth rate of the population over time for the DMDP

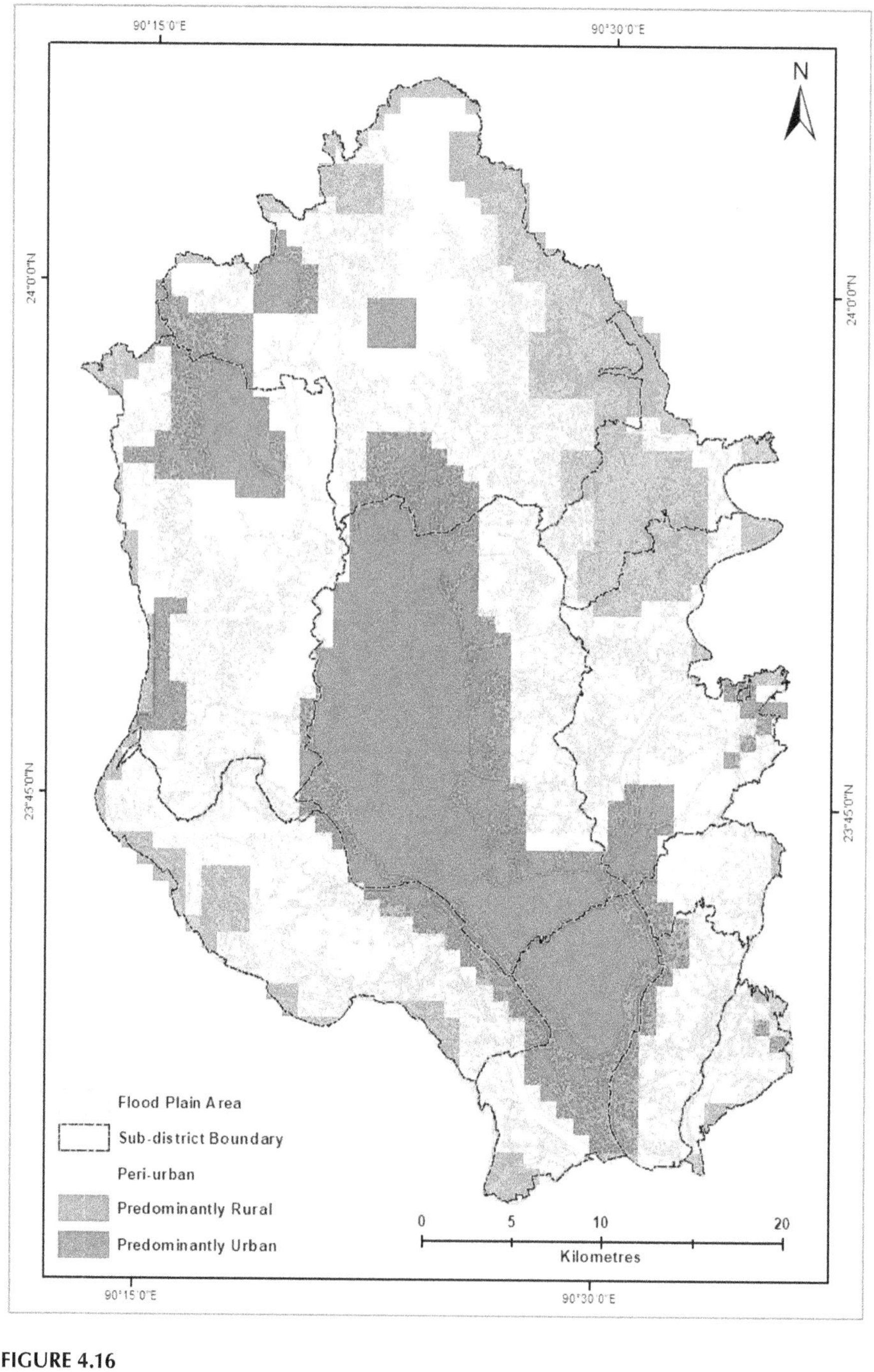

FIGURE 4.16
DMDP's flood plain area (FPA) superimposed on the VIIRTS NTL data mapping of 2016. (Derived from CRDP (CRDP (City Regional Development Planning), 2014).)

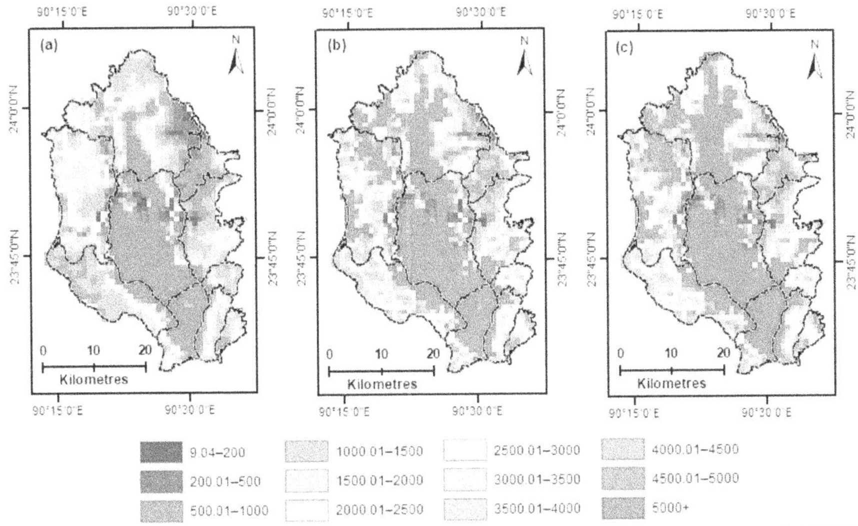

FIGURE 4.17
(a) Spatial distribution of 2001 population; (b) spatial distribution of 2016 population; (c) spatial distribution of 2020 population. (Derived from WorldPop and CIESIN (WorldPop, 2020).)

context. Although the density of population in DMDP's predominantly urban areas is unparalleled compared to the rest of the world, the population density of even DMDP's predominantly rural areas is much higher than that of some of the urban areas' population density in the developed world. For example, a country like Australia, which comprises more than 50 times the area coverage of Bangladesh, has an urban area population density of 903 persons/km^2 (DWUA (Demographia World Urban Areas), 2020).

Nevertheless, the population density of the Dhaka City Corporation (DCC) area is 52,000 persons/km^2 (BBS (Bangladesh Bureau of Statistics), 2011), whereas the estimated 2020 population density for predominantly urban areas is 31,666. Indeed, predominantly urban areas of 2016 cover an area of 478 km^2, whereas the area coverage of DCC is 134 km^2. Thus, NTL VIIRS 2016 data estimated that predominantly urban areas are more than 3.5 times larger than the DCC's area coverage. Hence, such perceived lower population density in predominantly urban areas is likely, as DMDP's predominantly urban areas extend far beyond the jurisdiction of the DCC boundary.

4.3.5.3.2 Population Densification and Peri-Urban Growth

In general, predominantly urban areas have a relatively lower growth of population, while predominantly rural areas and peri-urban areas have higher growth rates of population.

Figure 4.18a illustrates the spatial changes in peri-urban mapping between 1992 and 2016, while the aerial statistics of these changes are presented in

TABLE 4.8

Distribution of Population Between Predominantly Rural Areas, Peri-Urban Areas, and Predominantly Urban Areas Based on NTL VIIRS 2016 Data Mapping

Category	Area (km²)	Year 2001			Year 2016			Year 2020		
		Population	%	Density (km²)	Population	%	Density (km²)	Population	%	Density (km²)
Predominantly rural	167	148,583	1	890	246,185	1	1474	269,277	1	1612
Peri-urban	874	2,204,142	21	2522	3,512,173	20	4019	3,999,344	21	4576
Predominantly urban	478	8,073,684	77	16,891	13,436,218	78	28,109	15,136,359	78	31,666
Total	1519	10,426,409	100	6864	17,194,576	100	11,320	19,404,979	100	12,775

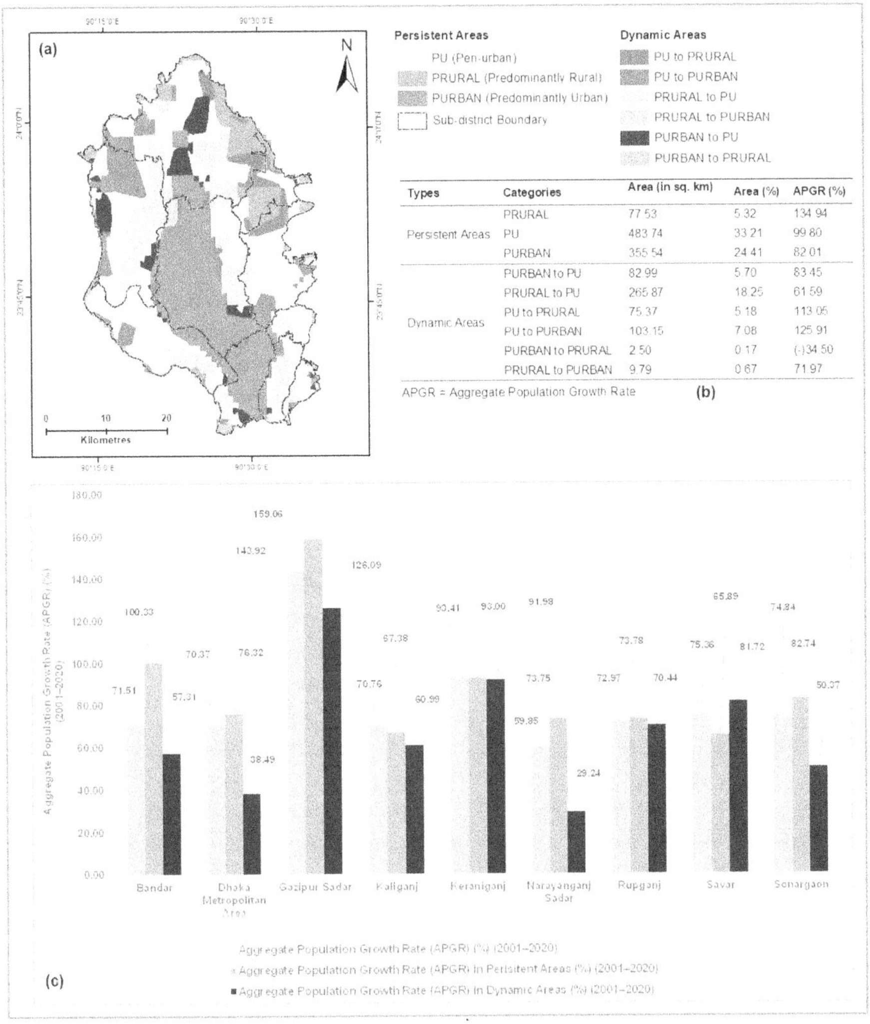

Types	Categories	Area (in sq. km)	Area (%)	APGR (%)
Persistent Areas	PRURAL	77.53	5.32	134.94
	PU	483.74	33.21	99.80
	PURBAN	355.54	24.41	82.01
Dynamic Areas	PURBAN to PU	82.99	5.70	83.45
	PRURAL to PU	265.87	18.25	61.59
	PU to PRURAL	75.37	5.18	113.05
	PU to PURBAN	103.15	7.08	125.91
	PURBAN to PRURAL	2.50	0.17	(-)34.50
	PRURAL to PURBAN	9.79	0.67	71.97

APGR = Aggregate Population Growth Rate

FIGURE 4.18
(a) Map showing the spatial distribution of predominantly rural (PRURAL), peri-urban (PU), and predominantly urban (PURBAN) areas; (b) table showing percentage distribution in the changes of PRURAL, PU, and PURBAN areas and corresponding changes in the aggregate population growth rate (APGR) between 2001 and 2020; (c) sub-district-wise aggregate population growth rate (APGR) between 2001 and 2020.

Figure 4.18b. The analysis reveals that the highest percentage distribution lies in the peri-urban areas category, which covers more than one-third of the DMDP's area. Surprisingly, in terms of spatial changes between the areas of predominantly rural, peri-urban, and predominantly urban, around two-thirds (i.e., 63%) of the DMDP area demonstrated persistence over time. However, the third and fourth highest categories are the transition of predominantly rural areas to peri-urban areas and the transition of peri-urban areas to predominantly urban areas, constituting 18.25% and 7.08% of the area, respectively.

In terms of APGR that occurred between 2001 and 2020, the analyses reveal that persistent predominantly rural areas encountered the highest pressure of population growth by around 135%—i.e., 7.10% per year. Among the areas which remained persistent over time, predominantly urban areas exhibited the lowest population growth rate. The second-highest growth rate of population was observed in the 'peri-urban to predominantly urban' areas category by 125.91%, implying a sequential transition of 'predominantly rural areas to peri-urban areas' and the subsequent transition of 'peri-urban areas to predominantly urban' areas.

Nevertheless, the areas demonstrating the transition of peri-urban to predominantly rural show the third highest concentration of population growth by 113.05%, which predominantly occurred near the persistent predominantly rural areas of Kaliganj sub-district. Such a finding, in fact, is not unlikely as these persistent predominantly rural areas conglomerate the highest percentage of population growth between 2001 and 2020.

Surprisingly, an unusual transition titled 'predominantly urban to predominantly rural' is observed in the Narayanganj Sadar area, which accounted for a negative growth rate of population. These areas of negative population growth are located in the highly flood-prone area of the Dhaleshwari River's catchment, and hence, frequent flooding displaces many people from this specific chunk of landscape.

In terms of the sub-district-wise population growth rate, the Gazipur Sadar area encountered the highest rate of population increase with a larger margin than its contemporaries (Figure 4.18c). The areas—i.e., Kaliganj, Keraniganj, Rupganj, which are predominantly peri-urban or predominantly rural in type—have nearly the same rate of population growth in dynamic and persistent areas. Meanwhile, the two most urbanised areas—i.e., the DMA and Narayanganj Sadar area—are observed as having the lowest rate of population increase. In addition, the population growth difference between persistent and dynamic areas is also found to be relatively higher in the Narayanganj Sadar and DMAs. However, the population growth rate in dynamic areas is relatively lower in each sub-district compared to the corresponding population growth rate of persistent areas, with Savar being an exception. The Savar sub-district possesses the largest chunk of landscapes which have been converted into peri-urban areas from predominantly rural, implying that Savar is the most peri-urbanised sub-district of this study. Such findings reveal that peri-urbanisation is more linked to population increase in dynamic areas. Thus, while the DMA and Narayanganj Sadar area have extremely limited scope for peri-urbanisation, the nearby sub-districts are potential areas for such peri-urban expansions.

4.3.5.3.3 Poverty and Peri-Urban Growth

The ALP is more dominant in the Kaliganj sub-district and trickles down to its nearby sub-districts of the Gazipur Sadar and Rupganj (Figure 4.19a), whereas

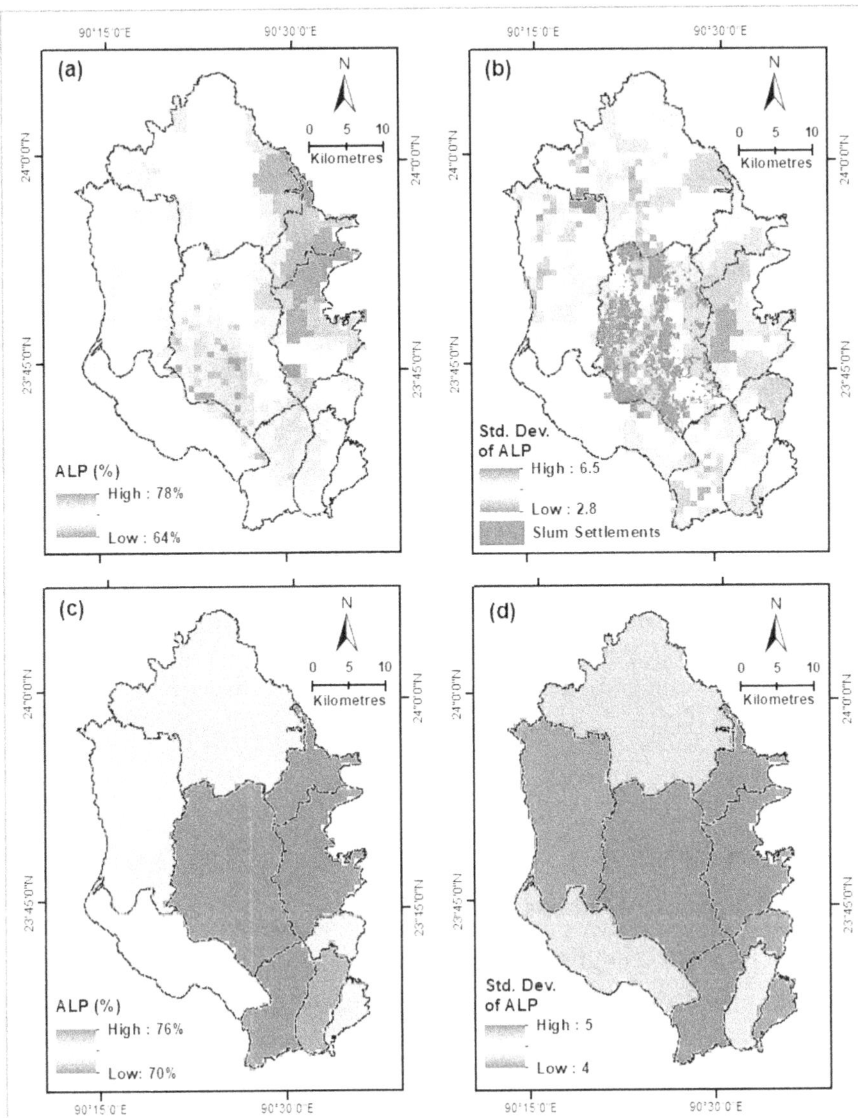

FIGURE 4.19
(a) Spatial distribution of the average likelihood of poverty (ALP) (i.e., the average probability of living on less than $2.50/day) of 2013 (Steele et al., 2017); (b) slum settlement areas superimposed on the standard deviation (Std. Dev.) of the average likelihood of poverty (ALP) mapping of 2013 (Steele et al., 2017); (c) sub-district-wise spatial distribution of the average likelihood of poverty (ALP); (d) sub-district-wise spatial distribution of the standard deviation (Std. Dev.) of the average likelihood of poverty (ALP). ((b) Derived from Gruebner et al. (2014)).)

such poverty concentration is relatively lower in the DMA, Narayanganj Sadar, and Bandar areas. The above-mentioned DMA, Narayanganj Sadar, and Bandar areas are also observed as having higher population growth difference between dynamic areas and persistent areas, meaning that the influx of poor migrants is predominantly confined to the dynamic land parcels of Kaliganj and its nearby areas.

However, the standard deviation mapping of the ALP depicts that the DMDP area is afflicted with higher variation in poverty concentration, which ranges from 2.8 to 6.5 (Figure 4.19b). Although a higher concentration of poverty is found in the Kaliganj, Rupganj and Gazipur Sadar areas, the DMA is observed with the highest deviation in the ALP, meaning that inequalities in social stratification within predominantly urban areas are comparatively higher.

Most parts of the DMA and Narayanganj Sadar area are already predominantly urbanised, while, through increasing peri-urbanisation, the Savar sub-district is in line to be predominantly urbanised shortly. Consequently, the DMA, Narayanganj Sadar area, and Savar are derived as the sub-districts of higher poverty deviation (Figure 4.19d). Thus, the deviation in poverty concentration appears to be correlated with urbanisation. For example, the DMA, as the most urbanised, ended up with the highest poverty deviation. Such a deviation in poverty intensification can further be illustrated by the abundance of slum settlements which are scattered throughout the DMDP area (Figure 4.19b).

4.4 Findings and Discussion

4.4.1 Factors Affecting Peri-Urbanisation

While the horizontal expansion of the DMDP area is remarkably constrained by the unavailability of flood-free landscapes, the northern part of the DMDP area encountered the highest growth rate of population and subsequent peri-urbanisation pressure. The DMDP's northern periphery, in general, lies at a higher elevation and is therefore naturally free from seasonal flooding. Thus, the findings reveal that peri-urbanisation for the DMDP context is interpreted as an interaction space of elevation and population growth rate. Consequently, in-migration of the population to the capital city Dhaka, plays a significant role in this perceived peri-urbanisation. Yet, such an in-migration pattern is profoundly interlinked to the unique socio-economic settings of this country.

For example, the Dhaka's share of the national gross domestic product (GDP) is USD 162 billion (i.e., 40%) (EIU (Economist Intelligence Unit), 2019), implying a heavily inclined concentration of economic activities towards the capital city Dhaka. Consequently, Dhaka is disproportionately equipped with nearly one third (i.e., 31.8%) of the country's total employment (ADB (Asian Development Bank), 2010). In general, Bangladesh's economic base is predominantly agricultural (i.e., 49%), and 45% of the country's labour force primarily relies on agriculture for employment (BBS (Bangladesh Bureau of Statistics), 2016). In addition, 84% of the rural population directly or indirectly depends on agriculture (Uddin, 2015). The disproportionate growth of agriculture-based industries coupled with the adaptation of more efficient farming technologies triggers an exodus of surplus agricultural labourers to the capital city Dhaka (CRDP (City Regional Development Planning), 2014). Consequently, 63% of Dhaka's population growth is due to the rural–urban migration (Razi & Dhrubo, 2020).

The majority of the migrants predominantly rely on the informal sector. Subsequently, the informal sector accounts for 84.30% of Dhaka's total employed population (ADB (Asian Development Bank), 2010). Thus, anticipating the social changes in response to economic growth becomes more difficult. Consequently, the task of addressing actual spatial growth patterns and accommodating informal economies becomes highly challenging.

In addition, Dhaka's per capita GDP is around three times higher than the national average. If Dhaka were an independent nation, its gross GDP would stand as the 50th largest economy in the world. Consequently, excessive agglomeration of the capital city-centred development paradigm makes the DMDP's peri-urban growth highly unpredictable and challenging. Thus, DMDP's physical growth predominantly implies a more dense peri-urban growth which agglomerates an overestimated population with inadequate infrastructural support. The scarcity of resources further hinders the timely upgrading of supporting infrastructures and provisions—e.g., land use–transport integration (Dur & Yigitcanlar, 2015; Yigitcanlar et al., 2007), utilities, and employment. This way, policymakers and planners are severely obstructed from enhancing planned development in their peripheries.

4.4.2 Implications for Growth Management and Natural Hazards

Bangladesh is one of the most vulnerable countries to the impact of climate change. Such impacts are already evident, as climate change-induced flooding—e.g., river flooding and stormwater flooding—often strikes in the DMDP area (Rabbani et al., 2011). Most parts of the predominantly urban areas (i.e., around 67%) are considered to be free from river flooding because of the proposed protective embankments, which surround them (CRDP (City Regional Development Planning), 2014). Nevertheless, the construction of the DMDP Urban Area Plan (1995–2005) proposed an eastern embankment, which as yet remains incomplete. In addition, the portion of predominantly urban areas—that is considered to be flood-free—severely lacks an efficient drainage system (Mowla & Islam, 2013). The analysis reveals that the DMDP's rainfall pattern from 2000 to 2018 is rather abrupt. Thus, heavy rainfall often happens, resulting in frequent stormwater flooding during each year, irrespective of the monsoon. Such a finding is also evident in other studies (Rabbani et al., 2011).

Nevertheless, due to the proximity to the nearby river, the remaining 23% of predominantly urban areas seem to be highly vulnerable to river flooding. The low-income people predominantly live in slums, and the spatial distribution of slums is scattered throughout predominantly urban areas. In addition, 80% of these slums are built on privately owned land and hence pose considerable institutional challenges in providing the basic urban facilities (Baker, 2007)—e.g., water supply, sanitation, sewerage, drainage, and electricity—in these slum-inflicted areas. On that very point, Baker (Baker, 2007) estimated that slum areas within a proximity of 50 m to nearby rivers accommodate around 76,000 households, which are at high risk of being frequently river flooded. Such river flooding usually occurs with a return period of every 10–40 years (Alam et al., 1996).

However, despite the geographical limitations to support further physical growth, Dhaka's unregulated population growth is not hindered by geographical limitations at all. For example, the analysis reveals that within these 19 years (2001–2020), the DMDP's population has increased at a rate of 0.5 million/year. Although the urban population as yet comprises less than 37.41% of the nation's population (WB (World Bank), 2020), such a portion reaches nearly 80% for the DMDP context. Nevertheless, the reported estimation of DMDP's annual population growth rate is 0.10 million/year higher than the earlier estimation by the World Bank (WB (World Bank), 2007). Nonetheless, from 2001 to 2020, the DMDP's population increased at an annual growth rate of 4.53%. Such a perceived growth rate in the population is much higher than the national average. For example, between 2000 and 2019, the national annual population growth rate was derived as 1.5% (UN (United Nations), 2020).

In terms of population size, Dhaka has secured the 19th position globally among the top 20 megacities, but stands first with a large margin in terms of the urban population density of 33,878 persons/km^2 (DWUA (Demographia World Urban Areas), 2020). Among the top 20 megacities, only six have constantly experienced a population growth rate of more than 3% over the last 20 years (UNFPA (United Nations Population Fund), 2007). Hence, the reported rate of DMDP's population growth is much higher than that of its contemporaries.

Consequently, due to an unabated growth in the DMDP's population, higher population concentration is also observed in the north-eastern periphery of the DMDP area (i.e., Rupganj and Kaliganj), which are predominantly flood-prone. In addition, these two areas are the hot spots of poverty with a lower standard deviation in the ALP index, meaning that these two areas predominantly accommodate the poorest portion of the DMDP's area. Thus, it can be inferred that poverty is forcing the poor migrants to live in these flood-prone areas. This way, the majority of the in-migrants, which also comprise climate migrants (Ahsan et al., 2011), are redundantly exposed to this climate change-induced frequent flooding. Such a finding poses a grave concern and brings forth the issue of climate justice for further consideration. On that very point, Ahsan (Ahsan, 2019) urged the adoption of area-specific growth policies in securing the basic rights for climate migrants of Bangladesh. Similarly, developing sound climate change mitigation policies is deemed critical to minimise climate migrant numbers across the globe (Sotto et al., 2019).

Furthermore, this unleashed population influx to the capital city Dhaka, worsens the prevailing water scarcity at large. For example, in 2005, the Dhaka Water Supply and Sanitation Authority (DWASA) was able to manage 1.6 mm^3/day groundwater extraction at maximum against the minimum demand of 2.1 mm^3/day (Akther et al., 2009). In addition, since 1986, the DMDP's groundwater table has been depleted at a rate of 2 m/year (Akther et al., 2009). While the DMDP area is suffering from evident water scarcity, the gradual groundwater depletion, coupled with the unprecedented population growth, further exacerbates this shortfall.

Nonetheless, due to the prevalence of wetlands, small water bodies, canals, and rivers, the groundwater level is relatively higher in the peripheral areas than in the central part of the DMDP area. Unfortunately, peri-urbanisation is predominantly taking place in the northern and north-eastern peripheries of the DMDP area, which are the hotspots of industrial clusters and agglomerate more than 2000 RMG industries (BWP (Bangladesh Water Partnership), 2019). This RMG sector has individually contributed to 11.17% of the total GDP and makes Bangladesh the second-largest global exporter in the RMG sector (DhakaTribune, 2020). The economic growth of this country is enormously linked to the development of the RMG sector (Islam, 2020). Hence, the country's RMG sector possesses great potential to grow further.

Nevertheless, this RMG sector heavily relies on groundwater extraction, because surface water is not available throughout the year. Thus, over-dependence on groundwater extraction coupled with the increased peri-urbanisation poses significant environmental threats to these peri-urban growth pockets, including land subsidence, increased earthquake vulnerability, and ecosystem degradation.

4.5 Conclusion

Although managing urban growth based on a demarcated peri-urban growth boundary remains rarely practised, explicitly defined peri-urban growth pockets appear to be a facilitating tool for promoting more rationalised land use practices in the periphery. Particularly in the context of developing countries, where rampant peri-urbanisation frequently occurs, quantifying the magnitude and DoF of peri-urban land parcels facilitates policymakers in identifying the areas with more urbanity and higher transition potentials. This way, policymakers and planners will be well equipped in dealing with the dual characteristics (i.e., urban or rural) of peri-urban areas and resultant growth directions, and thereby identify priority areas for immediate intervention.

Heikkila et al. (2003) first recognised the fuzzy characteristics of peri-urban areas, and statistically quantified the LoU and DoF in detecting the urban growth pattern of Ningbo, China by using Landsat data. However, they did not map the spatial distribution of peri-urban areas at all, whereas this chapter demarcated the spatial distribution of peri-urban growth pockets by unveiling the fuzzy characteristics of peri-urban areas. In doing so, this study developed fuzzy linear urban membership function set images using NTL data.

According to the findings of Mortoja et al.'s (2020) study, few studies used NTL data for peri-urban demarcation (e.g., Mustak et al., 2018; Peng et al., 2018; Sutton et al., 2006, 2010). While all the studies mentioned above used the crisp values of NTL data, none of the studies, to date, have investigated the potential of NTL data in unveiling the fuzzy characteristics of peri-urban areas. In addition, the conventional methodological approaches—which utilised the crisp values of NTL data for peri-urban demarcation—are claimed to be suitable to demarcate peri-urban areas at the national/global level only. Meanwhile, the ground-truthing outcome on NTL VIIRS data mapping of 2016 implies that the methodological approach reported in this study also appears to be a good fit at the regional level. Thus, quantifying the fuzzy characteristics of peri-urban areas and translating those findings into demarcating peri-urban growth pockets using NTL data provides a strong theoretical basis for this study.

Nevertheless, this study did not test other peri-urban demarcation approaches. Hence, how the methodological approach reported in this study prevails over others in terms of accuracy cannot be determined. However, the adapted methodological approach provides an easy way of demarcating peri-urban areas by using the readily available datasets at hand, which in turn facilitates prompt executions of peri-urban growth decisions.

In addition, this study explores how peri-urbanisation is interpreted as an interaction space of population growth, poverty, and increased vulnerability to natural hazards, particularly flooding. In this way, the research demonstrates the problem of rampant peri-urbanisation and subsequent fragilities—e.g., social and poverty problems, natural hazards, affecting large cities. Thus, the results are of interest to a wide audience and are not limited to the remote sensing community.

Acknowledgements

This chapter, with permission from the copyright holder, is a reproduction of the following journal article: Mortoja, M., & Yigitcanlar, T., (2020). How does peri-urbanisation trigger climate change vulnerabilities? An investigation of the Dhaka Megacity in Bangladesh. *Remote Sensing, 12*(23), 3938.

References

ADB (Asian Development Bank). (2010). *The informal sector and informal employment in Bangladesh, country report 2010.* Asian Development Bank.

Ahmed, B. (2020). Climate migrants in Bangladesh: A journey towards uncertainty! In: *Proceedings of the Seeking Refuge in the Climate Emergency, 2020 Global Shifts Colloquium*, University of Pennsylvania, Philadelphia, PA, USA, 14–16 September 2020; Perry World House: PA, USA, 2020.

Ahmed, B., & Ahmed, R. (2012). Modeling urban land cover growth dynamics using multi-temporal satellite images: A case study of Dhaka, Bangladesh. *ISPRS International Journal of Geo-Information, 1*, 3–31.

Ahmed, B., Kamruzzaman, M., Zhu, X., Rahman, M., & Choi, K. (2013). Simulating land cover changes and their impacts on land surface temperature in Dhaka, Bangladesh. *Remote Sensing, 5*, 5969–5998.

Ahsan, R. (2019). Climate-induced migration: Impacts on social structures and justice in Bangladesh. *South Asia Research, 39*, 184–201.

Ahsan, R., Karuppannan, S., & Kellett, J. (2011). Climate migration and urban planning system: A study of Bangladesh. *Environmental Justice, 4*, 163–170.

Akther, H., Ahmed, M. S., & Rasheed, K. B. S. (2009). Spatial and temporal analysis of groundwater level fluctuation in Dhaka city, Bangladesh. *Journal of Asian Earth Sciences, 2*, 49–57.

Alam, B. M., Damole, L. N., & Wickramanyake, E. Effects of flood mitigation measure: Lessons from Dhaka flood protection project, Bangladesh. *Asian Profile* 1996, *24*, 511–524.

Alam, M., & Rabbani, M. D. G. (2007). Vulnerabilities and responses to climate change for Dhaka. *Environment and Urbanization, 19*, 81–97.

Argent, N., Tonts, M., Jones, R., & Holmes, J. (2011). Amenity-led migration in rural Australia: A new driver of local demographic and environmental change? In G.W. Luck, D. Race, & R. Black (Eds.), *Demographic change in Australia's rural landscapes* (Vol. 12, pp. 23–44). Springer.

Baker, J. L. (2007). *Bangladesh-Dhaka: Improving living conditions for the urban poor* (Report No. 35824-BD). World Bank.

BBS (Bangladesh Bureau of Statistics). (2011). *Population and housing census 2011, community report: Dhaka.* Statistics and Informatics Division: Dhaka, Bangladesh. Available online: http://bbs.dhaka.gov.bd/sites/default/files/files/bbs.dhaka.

gov.bd/law_policy/6ed6b42c_2015_11e7_8f57_286ed488c766/Socio-Economic%20and%20demographic%20Report%202012.pdf (accessed on 15 November 2020).

BBS (Bangladesh Bureau of Statistics). (2016). *Yearbook of agricultural statistics of Bangladesh-2014 (26th series).* Bangladesh Bureau of Statistics: Dhaka, Bangladesh.

Bennett, M. M., & Smith, L. C. (2017). Advances in using multitemporal night-time lights satellite imagery to detect, estimate, and monitor socioeconomic dynamics. *Remote Sensing of Environment, 192,* 176–197.

BWP (Bangladesh Water Partnership) (2019). *Rapid assessment of greater Dhaka groundwater sustainability.* Global Water Partnership Organization (GWPO): Stockholm, Sweden.

Cao, X., Hu, Y., Zhu, X., Shi, F., Zhuo, L., & Chen, J. (2019). A simple self-adjusting model for correcting the blooming effects in DMSP-OLS nighttime light images. *Remote Sensing of Environment, 224,* 401–411.

Carey, R., & James, S. (2018). Peri-urban agriculture in Australia: Pressure on the urban fringe. In J. Zeunert & T. Waterman (Eds.), *Routledge handbook of landscape and food* (1st ed., pp. 213–227). Routledge.

Congalton, R. G., & Green, K. (2019). *Assessing the accuracy of remotely sensed data: Principles and practices* (3rd ed., pp. 1–346). CRC Press.

Cox, W. (2020). *The evolving urban form: Beijing.* Available online: http://www.newgeography.com/content/002406-the-evolving-urban-form-beijing (accessed on 17 October 2020).

CRDP (City Regional Development Planning). (2014). *Interim report (draft).* Rajdhani Unnayan Kartripakkha (RAJUK): Dhaka, Bangladesh,.

Datta, P. (2004). Push-pull factors of documented migration from Bangladesh to west bengal: A perception study. *Qualitative Report, 9,* 335–358.

Dewan, A. M., Humayun Kabir, M., Monirul Islam, M., Kumamoto, T., & Nishigaki, M. (2007). Delineating flood risk areas in greater Dhaka of Bangladesh using geoinformatics. *Georisk, 1,* 190–201.

DhakaTribune. (2020). Available online: https://www.dhakatribune.com/business/2019/04/26/apparel-sector-s-contribution-to-gdp-going-down-for-years (accessed on 20 August 2020).

Dur, F., & Yigitcanlar, T. (2015). Assessing land-use and transport integration via a spatial composite indexing model. *International Journal of Environmental Science and Technology, 12,* 803–816.

DWUA (Demographia World Urban Areas). (2020). *16th annual demography world urban areas.* Available online: http://www.demographia.com/db-worldua.pdf (accessed on 10 October 2020).

EIU (Economist Intelligence Unit) (2019a). Dhaka, Bangladesh GDP and income distribution. Economist Intelligence Unit: London, UK.

EIU (Economist Intelligence Unit). (2019b). *The Global Liveability Index 2019: A free overview;* Economist Intelligence Unit: London, UK.

Fan, Q., Rama, M., & Qian, J. (2020). *Seize the opportunity to make Dhaka a great, vibrant city.* End Poverty in South Asia, World Bank Blogs. Available online: https://blogs.worldbank.org/endpovertyinsouthasia/seize-opportunity-make-dhaka-great-vibrant-city# (accessed on 15 June 2020).

Gallent, N., & Shaw, D. (2007). Spatial planning, area action plans and the rural-urban fringe. *Journal of Environmental Planning and Management, 50,* 617–638.

Gruebner, O., Sachs, J., Nockert, A., Frings, M., Khan, M., Hossain, M., Lakes, T., & Hostert, P. (2014). Mapping the slums of Dhaka from 2006 to 2010. *Dataset Papers in Science, 2014,* 1–7.

Han, H., Yang, C., & Song, J. (2015). Scenario simulation and the prediction of land use and land cover change in Beijing, China. *Sustainability, 7,* 4260–4279.

Hassan, M. M., & Southworth, J. (2018). Analyzing land cover change and urban growth trajectories of the mega-urban region of Dhaka using remotely sensed data and an ensemble classifier. *Sustainability, 10,* 10.

Heikkila, E. J., Shen, T., & Yang, K. (2003). Fuzzy urban sets: Theory and application to desakota regions in China. *Environment and Planning B: Planning and Design, 30,* 239–254.

Hossain, M. (2003). Development through democratisation and decentralisation: The case of Bangladesh, South Asia. *Journal of South Asian Studies, 26,* 297–308.

Hossain, S. (2008). Rapid urban growth and poverty in Dhaka city. *Bangladesh e-Journal of Sociology, 5,* 1–24.

Islam, M. S. (2020). Ready-made garments exports earning and its contribution to economic growth in Bangladesh. *GeoJournal.* https://doi.org/10.1007/s10708-019-10131-0

Jahan, M. (2012). Impact of rural urban migration on physical and social environment: The case of Dhaka city. *International Journal of Sustainable Development, 1,* 186–194.

James, S. W., & O'Neill, P. M. (2016). Planning for peri-urban agriculture: A geographically-specific, evidence-based approach from Sydney. *Australian Geographic, 47,* 179–194.

Kabir, A., & Parolin, B. (2012). Planning and development of Dhaka—A story of 400 years. In *Proceedings of the Cities, Nations and Regions in Planning History, 15th International Planning History Society Conference,* São Paulo, Brazil, 15–18 July 2012; The International Planning History Society (IPHS): Delft, The Netherlands, 2012.

Kamruzzaman, M., Deilami, K., & Yigitcanlar, T. (2018). Investigating the urban heat island effect of transit-oriented development in Brisbane. *Journal of Transport Geography, 66,* 116–124.

Kamruzzaman, M., Hine, J., & Yigitcanlar, T. (2015). Investigating the link between carbon dioxide emissions and transport-related social exclusion in rural Northern Ireland. *International Journal of Environmental Science and Technology, 12,* 3463–3478.

Kennedy, M., Butt, A., & Amati, M. (2016). *Conflict and change in Australia's peri-urban landscapes* (1st ed., pp. 1–306). Routledge.

Li, C., Wang, J., Wang, L., Hu, L., & Gong, P. (2014). Comparison of classification algorithms and training sample sizes in urban land classification with Landsat thematic mapper imagery. *Remote Sensing, 6,* 964–983.

Li, X., & Zhou, Y. (2017). Urban mapping using DMSP/OLS stable night-time light: A review. *International Journal of Remote Sensing, 38,* 6030–6046.

Lu, D., & Weng, Q. (2007). A survey of image classification methods and techniques for improving classification performance. *International Journal of Remote Sensing, 28,* 823–870.

Mortoja, M., & Yigitcanlar, T. (2020). Local drivers of anthropogenic climate change: Quantifying the impact through a remote sensing approach in Brisbane. *Remote Sensing, 12,* 2270.

Mortoja, M. G., Yigitcanlar, T., & Mayere, S. (2020). What is the most suitable methodological approach to demarcate peri-urban areas? A systematic review of the literature. *Land Use Policy, 95,* 104601.

Mowla, Q. A., & Islam, M. S. (2013). Natural drainage system and water logging in Dhaka: Measures to address the problems. *Journal of Bangladesh Institute of Planners, 2075,* 9363.

Mustak, S., Baghmar, N. K., Srivastava, P. K., Singh, S. K., & Binolakar, R. (2018). Delineation and classification of rural–urban fringe using geospatial technique and onboard DMSP–Operational Linescan system. *Geocarto International, 33,* 375–396.

Nazrul, I., & Salma, A. S. (2004). Solid waste management and the urban poor in Dhaka. In *Proceedings of the Forum on Urban Infrastructure and Public Service Delivery for the Urban Poor Regional Focus,* New Delhi, India, 24–25 June 2004; Urban Sustainability Laboratory: New Delhi, India.

NCEI (National Centers for Environmental Information). *NOAA National Centers for Environmental Information.* Available online: https://ngdc.noaa.gov (accessed on 13 June 2020).

Nielsen, D. L., Brock, M. A., Rees, G. N., & Baldwin, D. S. (2003). Effects of increasing salinity on freshwater ecosystems in Australia. *Australian Journal of Botany, 51,* 655–665.

Peng, J., Liu, Y., Ma, J., & Zhao, S. (2018). A new approach for urban-rural fringe identification: Integrating impervious surface area and spatial continuous wavelet transform. *Landscape and Urban Planning, 175,* 72–79.

Phiri, D., & Morgenroth, J. (2017). Developments in Landsat land cover classification methods: A review. *Remote Sensing, 9,* 967.

Pouliotte, J., Smit, B., & Westerhoff, L. (2009). Adaptation and development: Livelihoods and climate change in Subarnabad, Bangladesh. *Climate and Development, 1,* 31–46.

Rabbani, G., Rahman, A. A., & Islam, N. (2011). Climate change implications for Dhaka City: A need for immediate measures to reduce vulnerability. In K. Otto-Zimmermann (Ed.), *Resilient cities* (Vol. 1, pp. 531–541). Springer.

Rahman, M. T., Aldosary, A. S., & Mortoja, M. G. (2017). Modeling future land cover changes and their effects on the land surface temperatures in the Saudi Arabian Eastern coastal City of Dammam. *Land, 6,* 36.

Rauws, W. S., & de Roo, G. (2011). Exploring transitions in the peri-urban area. *Planning Theory & Practice, 12,* 269–284.

Razi, A., & Dhrubo, A. (2020). *Rural-urban dichotomy: Dhaka.* Available online: http://www.futurarc.com/commentary/rural-urban-dichotomy-dhaka/ (accessed on 11 August 2020).

Rigaud, K. K., de Sherbinin, A., Jones, B., Bergmann, J., Clement, V., Ober, K., Schewe, J., Adamo, S., McCusker, B., & Heuser, S. (2018). *Groundswell: Preparing for internal climate migration.* World Bank.

Ruoso, L.-E., & Plant, R. (2018). A politics of place framework for unravelling peri-urban conflict: An example of peri-urban Sydney, Australia. *Journal of Urban Management, 7,* 57–69.

Siddiqui, T. (2010). *Impact of climate change: Migration as one of the adaptation strategies (Working Paper Series No. 18).* Refugee and Migratory Movements Research Unit (RMMRU): Dhaka, Bangladesh.

Silva, C. (2019). Auckland's urban sprawl, policy ambiguities and the peri-urbanization to Pukekohe. *Urban Science*, *3*, 1–20.

Silva, C. (2020). The rural lands of urban sprawl: Institutional changes and suburban rurality in Santiago de Chile. *Asian Geographer*, *37*, 117–144.

Sotto, D., Philippi, A., Yigitcanlar, T., & Kamruzzaman, M. (2019. Aligning urban policy with climate action in the global south: Are Brazilian cities considering climate emergency in local planning practice?. *Energies*, *12*, 3418.

Steele, J. E., Sundsøy, P. R., Pezzulo, C., Alegana, V. A., Bird, T. J., Blumenstock, J., Bjelland, J., Engø-Monsen, K., de Montjoye, Y.-A., & Iqbal, A. M. (2017). Mapping poverty using mobile phone and satellite data. *Journal of the Royal Society Interface*, *14*, 20160690.

Sutton, P. C., Cova, T. J., & Elvidge, C. D. (2006). Mapping "Exurbia" in the conterminous United States using nighttime satellite imagery. *Geocarto International*, *21*, 39–45.

Sutton, P. C., Goetz, A. R., Fildes, S., Forster, C., & Ghosh, T. (2010). Darkness on the edge of town: Mapping urban and peri-urban Australia using nighttime satellite imagery. *The Professional Geographer*, *62*, 119–133.

Teriman, S., Yigitcanlar, T., & Severine, M. (2009). Urban sustainability and growth management in south-east Asian city regions: The case of Kualalumpur and Hongkong. *Planning Malaysia*, *7*, 47–68.

Todes, A. (2012). New directions in spatial planning? Linking strategic spatial planning and infrastructure development. *Journal of Planning Education and Research*, *32*, 400–414.

Uddin, M. M. M. (2015). Causal relationship between agriculture, industry and services sector for GDP growth in Bangladesh: An econometric investigation. *Journal of Poverty, Investment and Development*, *8*, 124–129.

UN (United Nations). (2020). *World population prospects*. Department of Economic and Social Affairs Population Dynamics. Available online: https://population.un.org/wpp/ (accessed on 14 May 2020).

UNFPA (United Nations Population Fund. (2007). *State of world population 2007: Unleashing the potential of urban growth*. United Nations Population Fund: New York.

USGS (United States Geological Survey). (2020). *EarthExplorer*. Available online: https://earthexplorer.usgs.gov/(accessed on 11 June 2020).

WB (World Bank). (2007). *Improving living conditions for the urban poor (paper no. 17)*. World Bank: Washington DC.

WB (World Bank). (2020). *Bangladesh urban population 1960–2020*. Available online: https://www.macrotrends.net/countries/BGD/bangladesh/urban-population (accessed on 13 July 2020).

Weatherbase. (2020). Available online: http://www.weatherbase.com/weather/weather.php3?s=032914&refer=/ (accessed on 20 July 2020).

Woltjer, J. (2014). A global review on peri-urban development and planning. *Jurnal Perencanaan Wilayah dan Kota*, *25*, 1–16.

WorldPop (www.worldpop.org—School of Geography and Environmental Science, University of Southampton; Department of Geography and Geosciences, University of Louisville; Department de Geographie, Universite de Namur), CIESIN (Center for International Earth Science Information Network). (2020). *Global high-resolution population denominators project—Funded by the Bill and Melinda Gates Foundation (OPP1134076)*. Available online: https://www.worldpop.org/geodata/country?iso3=BGD (accessed on 11 June 2020).

Wu, J., He, S., Peng, J., Li, W., & Zhong, X. (2013). Intercalibration of DMSP-OLS night-time light data by the invariant region method. *International Journal of Remote Sensing, 34*, 7356–7368.

Yigitcanlar, T. (2006). Australian Local governments' practice and prospects with online planning. *URISA Journal, 18*, 7–17.

Yigitcanlar, T. (2010). *Rethinking sustainable development: Urban management, engineering, and design*. IGI Global.

Yigitcanlar, T., Dodson, J., Gleeson, B., & Sipe, N. (2007). Travel self-containment in master planned estates: Analysis of recent Australian trends. *Urban Policy and Research, 25*, 129–149.

Yigitcanlar, T., & Kamruzzaman, M. (2015). Planning, development and management of sustainable cities: A commentary from the guest editors. *Sustainability, 7*, 14677–14688.

Zhao, P. (2012). Urban-rural transition in China's metropolises: New trends in peri-urbanization in Beijing. *International Development Planning Review, 34*, 269–295.

5

Anthropogenic Climate Change in Urban Landscapes

5.1 Introduction and Background

Urban expansions to nearby peri-urban greenfield sites, particularly in metropolitan regions, have commonly become a global occurrence (Acheampong et al., 2016; Chen et al., 2014; Yigitcanlar et al., 2015; Zhou et al., 2019). These greenfield sites primarily host numerous ecosystem services—e.g., places for scenic beauty, biodiversity conservation, sources of natural resources, crop production, and recreational values (Buxton & Low-Choy, 2007). Nevertheless, urban growth stimuli—e.g., economic growth, increasing population influx, market-led speculative forces, consumer preferences, lack of alternative urban growth corridors, and lack of policy instruments (Friedberger, 2000; Kijas, 2002; Merson et al., 2010; Yue et al., 2013)—make urban encroachments to these greenfield sites almost inevitable. Consequently, sooner or later, a significant portion of these greenfield sites usually become urbanised (Dur & Yigitcanlar, 2015; Sutton et al., 2006), and ecosystem services inherent with these spaces are lost permanently (Sinclair et al., 2003).

Besides, peri-urban greenfield sites also serve as a carbon reservoir. Hence, the conversion of greenfield sites to built-up areas contributes to the release of substantial greenhouse gas (GHG) emissions, implying the changing role of peri-urban greenfield sites from a carbon reservoir to a carbon source. Such increased GHG emissions intensify global warming potential (GWP) (Yousefi et al., 2014). Thus, urban encroachment on nearby peri-urban greenfield sites poses two major environmental challenges, namely, (a) ecosystem service losses and (b) GWP intensification.

The losses in ecosystem services are quantified in terms of the losses in ecosystem service values (ESV) (Costanza et al., 1997; Martínez et al., 2009). These ESV losses are the total sum attributable to the losses in both economic (e.g., agricultural production) and non-economic values (e.g., aesthetic, recreational, cultural) (Martínez et al., 2009; Sutton & Costanza, 2002). The dynamics of land use and land cover (LULC) are strongly linked to ESV (Kindu et al., 2016; Polasky et al., 2011; Styers et al., 2010). On that point, Costanza et al. (2014) demonstrate that globally LULC changes from 1997 to 2011 resulted in a decline in ESV of between USD 4.3 trillion/annum and USD 20.2 trillion/annum. Such losses already accounted for a total of 30% decline

DOI: 10.1201/9781003737049-7

in global ESV (Sutton et al., 2016). The losses in ESV appear to be perpetual and are predicted to encounter further decrease by USD 51 trillion/annum (Kubiszewski et al., 2017).

While allowing urban expansions to nearby peri-urban greenfield sites is a commonly adopted global strategy in accommodating increasing urbanisation needs (Mortoja et al., 2020), such peri-urbanisation appears to have dire consequences for countries with unique biodiversity habitats. For example, Australia has inherited one of the most unique biodiversity habitats over the centuries, where per 1 km^2 woodland clearing results in the destruction of 45,000 trees and death casualties of approximately 20,000 reptiles and 3000 birds (Cogger, 2003). Due to constant peri-urbanisation, over the last 200 years, this country has encountered the worst decline in biodiversity habitats than the rest of the continents in the globe (ABS (Australian Bureau of Statistics), 2010). Such reported decline results in a higher rate of ESV losses (i.e., USD 6.8 trillion/annum) in Australia than the global average loss in ESVs of USD 6.3 trillion/annum (Sutton et al., 2016).

Consequently, urban encroachment on nearby greenfield sites in Australia is viewed as a significant threat to farmland ecosystem services (Australian State of the Environment, 2011; Harman & Choy, 2011). Nonetheless, while performing the conversion of Australia's greenfield sites to other land use activities—e.g., residential, industrial, commercial, and hobby farming—ecosystem services remain less prioritised (Rothwell et al., 2015). Several factors—e.g., high land values, land tenures, and the prevalence of diversified stakeholders—hinder implementing the effective mitigation measures of urban encroachment to ecosystem services in the Australian context (Bekessy & Gordon, 2007). The lack of research related to the impacts of urban encroachment on natural ecosystem services makes the mitigation practices more complicated (Miller & Hobbs, 2002). By using peri-urban Sydney as a testbed, Merson et al. (2010) evaluate the role of the country's greenfield sites in preserving ecosystem services. They observe that a broad range of intangible values of ecosystem services—that cannot be converted into conventional market prices—are not adequately addressed in the policy documents.

Besides, land clearing policy exerts significant climate change implications (Steffen & Dean, 2018; Zhao, 2011). Along with the significant losses in ESV, about 10% of Australia's net GHG emissions stem from LULC-related activities (Hatfield-Dodds et al., 2015), and the country is currently vulnerable to climate change impacts, with this vulnerability becoming intolerable by the year 2050 (IPCC (Intergovernmental Panel on Climate Change), 2007). Despite this, the impacts of climate change—recently the term 'climate emergency' is being used instead—on Australia's ecosystem services are still unidentified (Alamgir et al., 2014; Goonetilleke et al., 2014).

To this end, by adopting a case study approach, this chapter provides evidence on the anthropogenic climate change impacts stemming from the direct and indirect implications of LULC-driven ESV losses and GHG emissions. In doing so, the study brings forth the case of Brisbane and adjoining local government areas (LGAs) from Australia under the microscope. A change analysis of over a 30-year period (1989–2019) is carried out to examine how these anthropogenic activities intensify climate change impacts over time corresponding to the changes in greenfield sites in the case study area. The study also evaluates the implications of government policies—i.e., urban growth policies and land-clearing policies—that regulate those changes.

5.2 Research Design

5.2.1 Case Study

The LGA of Brisbane and its surrounding LGAs covering an area of about 8556 km^2 are chosen as the case study area in order to determine the impacts of anthropogenic activities on climate change. The study area comprises LGAs of Brisbane, Gold Coast, Moreton Bay, Redland, Logan, and Ipswich (Figure 5.1). The reasons that motivate us to select this region as the testbed are as follows: (a) The selected LGAs are located in the state of Queensland, where the state was subjected to nearly 60% of the nation's land clearing over

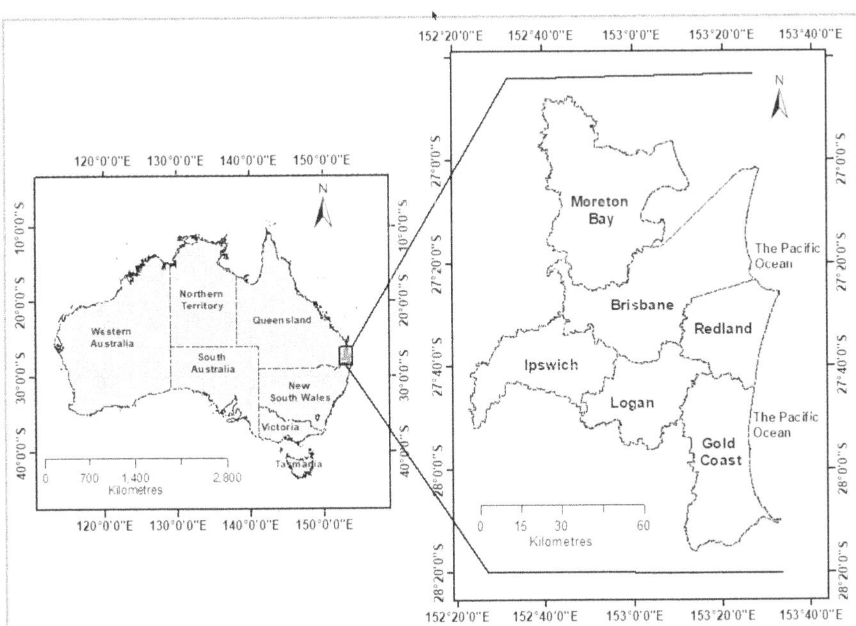

FIGURE 5.1
Location of the study area.

the last four decades (Steffen & Dean, 2018); (b) Queensland has been generating about 80% of the nation's land use emissions (Steffen & Dean, 2018), which has substantial impacts on local climate change; (c) the selected LGAs are the most urbanised part of the state (Yigitcanlar, 2006); (d) Brisbane ranks as the most sprawling city in Australia (Sutton et al., 2010; Yigitcanlar et al., 2007); (e) while Brisbane's urban areas only accommodate 6% of the national population (Spencer et al., 2015), its low-density urban expansion to nearby peri-urban areas places a great risk on the ecosystem services (Van Delden, 2017); and (f) Brisbane is claimed to be one of the most sustainable and smart cities in Australia (Yigitcanlar, 2010a, 2010b; Yigitcanlar et al., 2019).

5.2.2 Data

Four Landsat images corrected at the level-one terrain (L1T) level were used to detect the climate change impacts triggered by anthropogenic activities for a period of 30 years with 10-year intervals (USGS, 2020) (Table 5.1).

The estimated resident population (ERP) data of the Australian Bureau of Statistics (ABS) were used to investigate the socioeconomic aspects of anthropogenic activities. In the case of calculating GHG emissions due to anthropogenic activities, this study used the national GHG inventory database of the Australian government—AGEIS (AGEIS (Australian Greenhouse Emissions Information System), 2020). In order to reveal the changes in precipitation and temperature over time, data were taken from the Australian Bureau of Meteorology (BoM).

5.2.3 Methods

5.2.3.1 Landsat Image Classification and Performing Accuracy Assessment

Multispectral Landsat images processed at L1T level were classified into four LULC categories, namely, bare soil (i.e., exposed soil, vacant lands, sand), built-up (i.e., transportation networks, industrial, residential, institutional settlements, commercial), vegetation (i.e., playground, parks, greeneries, trees), and water body (i.e., lakes, reservoirs, retention ponds, coastal water, rivers) (Anderson et al., 1976). In order to derive the abovementioned LULC

TABLE 5.1

Characteristics of Landsat Data

Date	Path/Row	Sensor	Satellite	Resolution (m)
16 September 1989	89/79	TM	Landsat 5	30
19 August 1999	89/79	ETM+	Landsat 7	30
23 September 2009	89/79	TM	Landsat 5	30
3 September 2019	89/79	OLI	Landsat 8	30

classes, the maximum likelihood supervised classification (MLSC) technique was applied. Nearmap data with 5.8–7.5 cm spatial resolution, Google Earth imageries, and field survey data were utilised collectively to develop training samples required to perform the MLSC technique in the ENVI image analysis software.

For carrying out accuracy assessments of the classified images, 350 stratified random sampling points were generated in the ArcGIS software. Afterwards, ground-truthing of the classified images was performed using field surveys, Nearmap data, and Google Earth images. Then, error matrices were produced to check the accuracy of each classified image. The image classification process was repeated by using different training samples until the threshold accuracy of 85% for each classified image was achieved (Jiang & Eastman, 2000). The error matrices of each classified image are presented in Tables 5.2–5.5.

TABLE 5.2

The Error Matrix for the Classified Image of 1989

LULC Category	Ground Truth Points	LULC Data Points	Number Correct	Producer's Accuracy (%)	User's Accuracy (%)
Bare soil (total count)	77	72	65	90.28	84.42
Built-up (total count)	50	51	45	88.24	90.00
Vegetation (total count)	168	174	160	91.95	95.24
Water body (total count)	55	53	49	92.45	89.09
Grand total points count	350	350	319		
Overall accuracy	91.14%				
Kappa coefficient	0.87				

TABLE 5.3

The Error Matrix for the Classified Image of 1999

LULC Category	Ground Truth Points	LULC Data Points	Number Correct	Producer's Accuracy (%)	User's Accuracy (%)
Bare soil (total count)	54	53	47	88.68	87.04
Built-up (total count)	69	69	61	88.41	88.41
Vegetation (total count)	170	173	158	91.33	92.94
Water body (total count)	57	55	49	89.09	85.96
Grand total points count	350	350	315		
Overall accuracy	90%				
Kappa coefficient	0.85				

TABLE 5.4
The Error Matrix for the Classified Image of 2009

LULC Category	Ground Truth Points	LULC Data Points	Number Correct	Producer's Accuracy (%)	User's Accuracy (%)
Bare soil (total count)	61	58	52	89.66	85.25
Built-up (total count)	84	89	78	87.64	92.86
Vegetation (total count)	148	148	139	93.92	93.92
Water body (total count)	57	55	52	94.55	91.23
Grand total points count	350	350	321		
Overall accuracy	91.71%				
Kappa coefficient	0.88				

TABLE 5.5
The Error Matrix for the Classified Image of 2019

LULC Category	Ground Truth Points	LULC Data Points	Number Correct	Producer's Accuracy (%)	User's Accuracy (%)
Bare soil (total count)	75	72	64	88.89	85.33
Built-up (total count)	79	81	71	87.65	89.87
Vegetation (total count)	126	129	118	91.47	93.65
Water body (total count)	70	68	63	92.65	90.00
Grand total points count	350	350	316		
Overall accuracy	90.29%				
Kappa coefficient	0.87				

5.2.3.2 Estimating Changes due to Anthropogenic Activities

5.2.3.2.1 Identifying Major LULC Changes

This study investigated urbanisation-led major LULC changes which exert a significant influence on climate change. In the case of determining the major LULC changes, the study considered the LULC transitions as major, which were more than 50 km^2 between 1989 and 2019. Such changes comprise four major LULC transitions in particular—i.e., vegetation to bare soil (VEG_BS), vegetation to built-up (VEG_BU), bare soil to built-up (BS_BU), and bare soil to vegetation (BS_VEG). By using TerrSet, this study quantifies these four major transitions.

5.2.3.2.2 Calculating Ecosystem Service Value Losses

In order to derive the impact of urban expansions on ecosystem services, the study quantified the losses in ESV over time. In this connection, this study used Sandhu et al.'s (2013) ESV estimate of the New Zealand context in order to calculate the losses in ESV for the Australian context.

Besides, this study used the consumer price index (CPI) value to adjust Sandhu et al.'s (2013) estimation of ESV. The following equation was used for this adjustment:

$$ESV_{2019} = ESV_{2012} \times \left(CPI_{2019} / CPI_{2012}\right) \tag{5.1}$$

where ESV_{2019} = adjusted ESV of the year 2019; ESV_{2012} = Sandhu et al.'s (2013) estimation of ESV; CPI_{2019} = CPI value of the year 2019; CPI_{2012} = CPI value of the year 2012.

Thus, $ESV_{2019} = 3224 \times (258.68/229.59)$ = USD 3632.40 (CPIIC (CPI Inflation Calculator), 2020). This adjusted ESV_{2019} was used to calculate the net losses in ESV that occurred in the decades of 1989–1999, 1999–2009, and 2009–2019.

The net deforestation that took place for each ten-year interval—i.e., 1989–1999, 1999–2009, 2009–2019—was calculated by using the following equation:

$$\begin{aligned} \textit{Net deforestation} &= \left(\textit{Vegetation to bare soil} + \textit{Vegetation to built} - \textit{up}\right) \\ &\quad - \textit{Bare soil to vegetation} \end{aligned} \tag{5.2}$$

5.2.3.2.3 Calculating Greenhouse Gas Emissions

To quantify urbanisation-led GHG emissions, this study utilised 'Australian GHG information system' (AGEIS) data to derive GHG emissions per capita, which was later used to derive GHG emissions/ha due to urbanisation-led major LULC changes.

While calculating GHG emissions, the study hypothesised the following:

- The conversion of vegetation to built-up areas undergoes two consecutive steps: First, the conversion of vegetation to bare soil (i.e., deforestation); second, the conversion of bare soil to built-up areas.
- Total GHG emissions from each major LULC conversion comprise two parts: (a) Conversion-related emissions; (b) GHG emissions from the end LULC category. For example, in the case of the conversion from vegetation to built-up areas, GHG is emitted from two sources: (i) The emitted GHG due to the conversion of vegetation to built-up areas, and (ii) the emitted GHG from the end LULC category—i.e., existing built-up areas.
- The specific time frame within which an LULC transition occurred and the time frame since subsequent GHG is emitted from the end LULC category are unknown. Hence, in estimating the rate of GHG emissions/ha, conversion-related GHG emissions and subsequent GHG emissions from the end LULC category were calculated altogether.

- The GHG emissions related to the conversion of bare soil to vegetation (that can be termed as 'revegetation') were ignored, as such a transition-related emission was insignificant. For example, in 2016, Queensland's revegetation-related GHG emissions were only 0.009 tonnes/capita (AGEIS (Australian Greenhouse Emissions Information System), 2020). Hence, only the end LULC GHG sink from the conversion of bare soil to vegetation—i.e., 'space of carbon sink'—is considered, which is 5 tonnes/hectare/annum (ACS (Australia's Chief Scientist), 2009).
- Deforestation (that is, due to the conversion of vegetation to bare soil (VEG_BS)) and urbanisation-led conversions—i.e., bare soil to built-up areas (BS_BU) and vegetation to built-up areas (VEG_BU)—were considered as sources of GHG emissions. On the other hand, the conversion of bare soil to vegetation was considered a source of carbon sink.
- Bare soil was considered as no land use activity, and hence, no emission from bare soil was calculated.

The GHG emissions data were gathered from AGEIS (AGEIS (Australian Greenhouse Emissions Information System), 2020). Queensland's ERP of each corresponding year was used to derive per capita GHG emissions for each selected LULC transition. The average of per capita GHG emissions of the Queensland state between each ten-year period (e.g., between 1999 and 2009) was later used to derive GHG emissions originating from the urbanisation-led LULC changes in the context of the study area.

In order to calculate GHG emissions stemming from the conversion of vegetation to built-up areas, emissions from deforestation, construction emissions, and household emissions (such as from energy usage and solid waste disposal), direct non-agricultural (NA) residential emissions, and indirect emissions (i.e., emissions from purchased electricity) were considered. While calculating GHG emissions due to the conversion of bare soil to built-up areas, all the above-mentioned sources of GHG emissions except for deforestation were used. In order to calculate GHG emissions due to the conversion of vegetation to bare soil, only the emissions attributable to deforestation were considered.

Finally, the aggregate GHG emissions figure was calculated by estimating the emissions released from the LULC, which did not change over time (i.e., 'built-up areas' and 'vegetation' showing persistence) and the emissions from major LULC changes over the investigated 30-year period.

5.2.3.2.4 *Checking the Reliability of GHG Estimation Approaches*

In order to cross-check the reliability of our GHG estimation approaches, we specifically calculated GHG emissions for the LGA of Brisbane. The Brisbane LGA was selected for this examination because such relevant data for comparison were available for this LGA from secondary resources.

5.2.3.2.5 Calculating the Variations in GHG Emissions Between Unquestionably Urban, Peri-Urban, and Rural Areas

In order to determine the spatial distribution of GHG emissions within different urban growth land use areas, this chapter adopts Mortoja et al.'s (2020) study approach to demarcate unquestionably urban, peri-urban, and rural areas. They applied fuzzy logic on night-time light (NTL) data to demarcate these urban growth areas. This study applied its methodological approach to demarcate different urban growth areas for the year 2019 by using NTL data of 2019 (Figure 5.2). Later, this study estimated GHG emissions within the demarcated urban growth areas between 1989 and 2019. While estimating per capita GHG emissions for each LULC transition between 1989 and 2019, the study took the average value of per capita GHG emissions of the three decades.

5.2.3.3 Changes in Atmospheric Temperature and Precipitation

The study at hand did not directly measure the fraction of climate change attributable to major LULC changes. Nevertheless, we looked at the general trend of temperature and precipitation pattern between 2000 and 2020, which appears to help in anticipating the anthropogenic climate change impacts indirectly.

5.2.3.4 Shifts in State Polices and Corresponding Deforestation

In order to illustrate the policy impacts influencing deforestation, this study looked at how the state policies governed deforestation, and how these incurring deforestations resulted in increased GHG emissions over time.

5.3 Analysis and Results

5.3.1 Changes in Land Use and Land Cover

The classified LULC images are presented in Figure 5.3, while the percentage of LGA share of total LULC changes for 1989, 1999, 2009, and 2019 is depicted in Figure 5.4. Between 1989 and 2019, built-up areas increased by nearly 1711 km^2 (20% of the study area), bare soil declined by approximately 513 km^2 (6% of the area), and vegetation decreased by about 1027 km^2 (12%) (Figure 5.3e). In 1999, built-up areas experienced the highest growth of 9% from the base year estimate of 1989, whereas from 1999 to 2019, built-up areas collectively gained a growth of 11%. Interestingly, between 1989 and 1999, bare soil declined by 9%, and such a decline in vegetation within this time frame was only 2%, indicating the period was undergoing brownfield development.

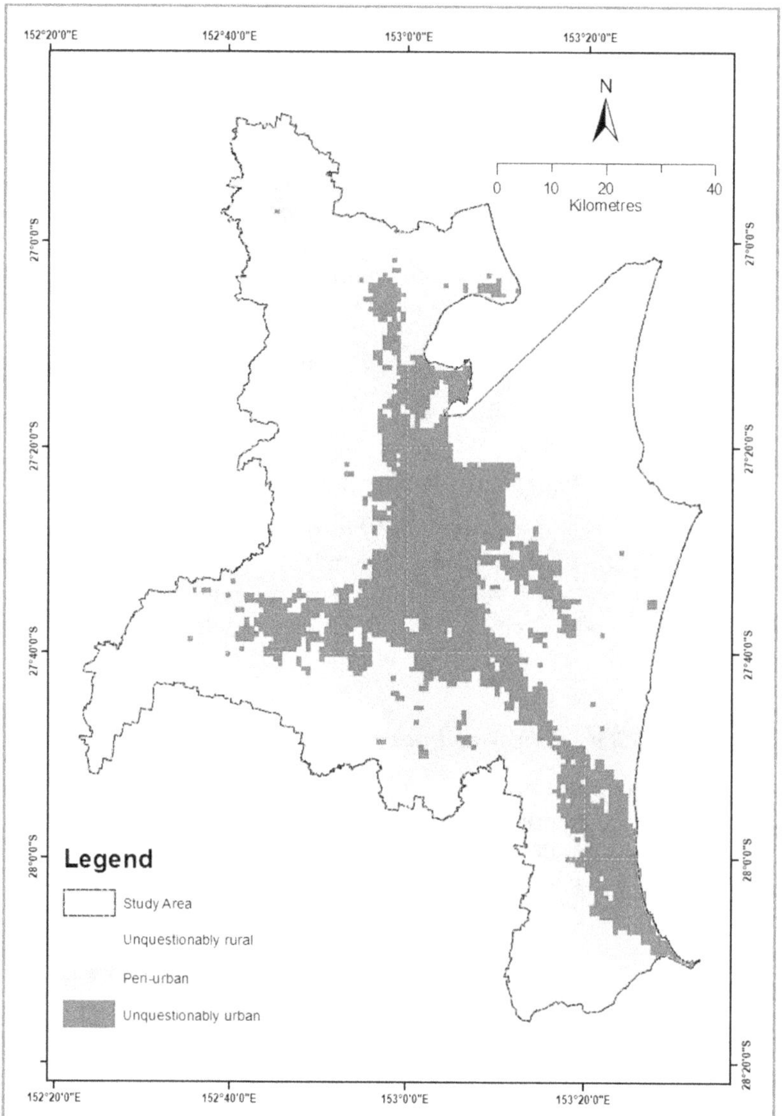

FIGURE 5.2
Mapping peri-urban areas of 2019 by fuzzy logic on night-time light (NTL) data.

Between 1999 and 2009, built-up areas increased by 6%, and vegetation declined by 7% with little change in bare soil (i.e., only 2%), implying that this decade had undergone greenfield development. Between 2009 and 2019, vegetation and bare soil declined by 4% and 2%, respectively, while built-up areas increased by 5%, indicating that this period had undergone both brownfield and greenfield development.

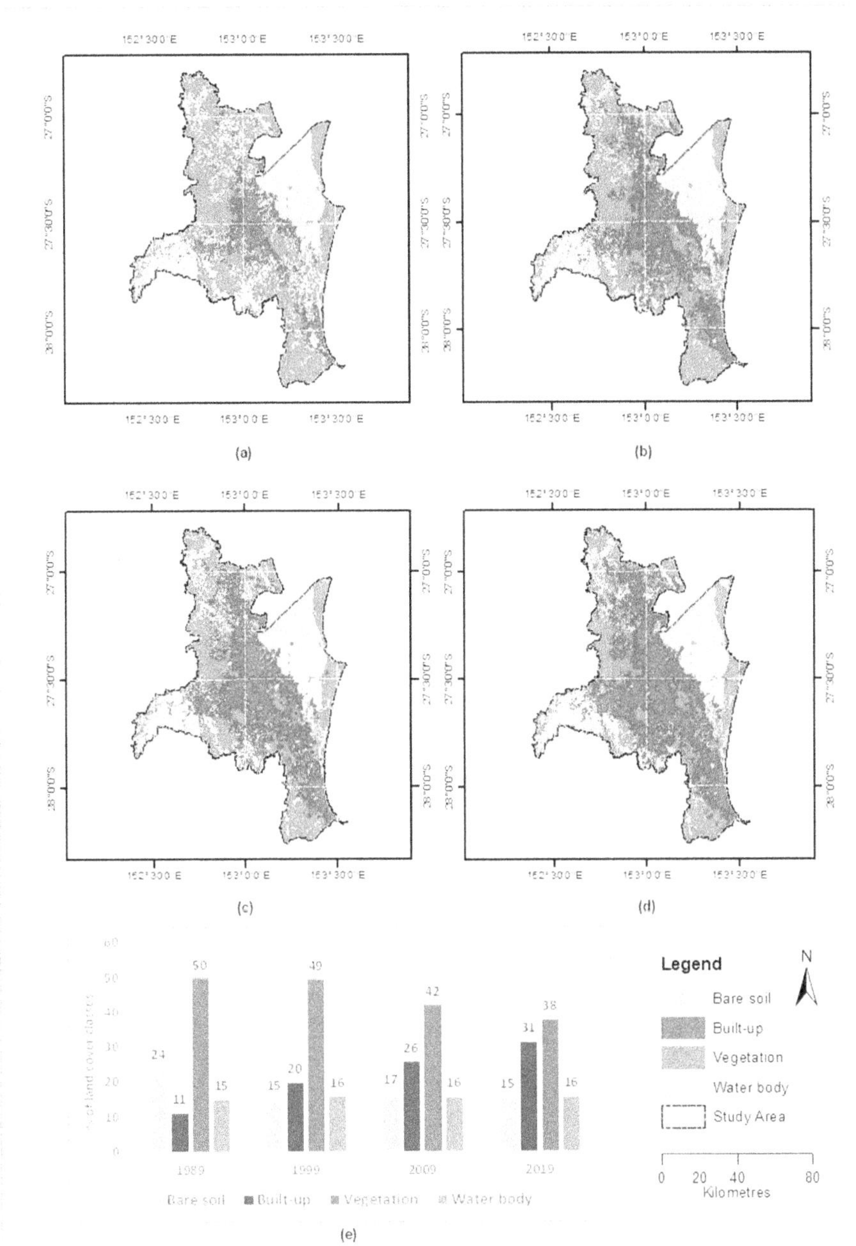

FIGURE 5.3
(a) Classified image of 1989; (b) Classified image of 1999; (c) Classified image of 2009; (d) Classified image of 2019; (e) % of changes in LULC between 1989, 1999, 2009, and 2019.

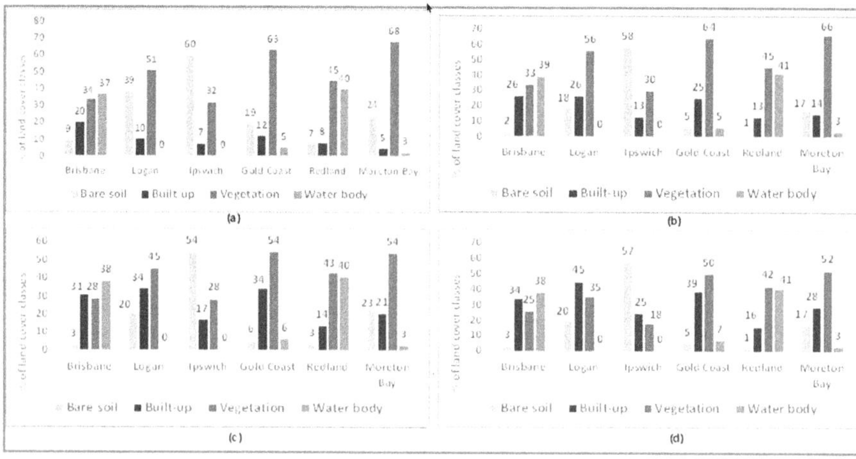

FIGURE 5.4
(a) % of local government area (LGA) share of total LULC changes for 1989; (b) % of LGA share of total LULC changes for 1999; (c) % of LGA share of total LULC changes for 2009; (d) % of LGA share of total LULC changes for 2019.

In terms of built-up areas growth within LGAs, Logan experienced the highest growth of 35%, followed by Gold Coast (27%), and Moreton Bay (23%) between 1989 and 2019 (Figure 5.4). Within this 30-year period, Logan and Moreton Bay both also encountered the highest loss in vegetation (16%), followed by Ipswich (14%) and Gold Coast (13%). Again, in terms of bare soil, Logan experienced the most by 19%. Subsequently, throughout the 30-year period, Logan, Gold Coast, and Moreton Bay experienced higher growth than Brisbane.

Brisbane comprises the highest amount of water body areas, and its growth to the East is constrained by the Pacific Ocean. Consequently, Brisbane's growth trickles down towards Moreton Bay in the North, Logan in the South, and Ipswich in the West. Nevertheless, 1% net increase in Brisbane's water body between 1989 and 2019 does not imply sea-level rise. Instead, such an increase might be attributed to the underlying error in the image classification process.

Besides, Ipswich covers the highest amount of bare soil, which only declined by 3% over time. In contrast, vegetation in Ipswich declined by 14%, meaning bare soil in Ipswich is not suitable for development due to the presence of some natural constraints. In reality, a large chunk of bare soil in Ipswich remains undeveloped, as these are declared as grazing areas, and some portions are conserved as ecologically sensitive areas. Thus, overgrazing might lead to more bare soil formation in Ipswich, which warrants further research.

5.3.2 Changes in Land Use and Land Cover and Ecosystem Service Values

The 1999–2009 period demonstrated the highest deforestation (i.e., 71,543 ha) and ESV losses (i.e., 2.6 billion)—which is around 7.5 times higher than that of the 1989–1999 period (Table 5.6). More than 57% of ESV were lost in the 1999–2009 decade. From 1999 to 2009, the transition from vegetation to built-up area was 54,283 ha, which is around 1.8 times higher than that which occurred in the 1989–1999 and 2009–2019 periods. On the contrary, the conversion from bare soil to vegetation in the 1989–1999 decade was 44,877 ha, which is around three to four times higher than that of the 1999–2009 and 2009–2019 decades, respectively. The 1989–1999 period is also the lowest in terms of converting vegetation to bare soil, which led to the smallest value in deforestation (i.e., 9541 ha) and subsequent losses in ESV (0.35 billion) in the 1989–1999 period.

5.3.3 Changes in Land Use and Land Cover and Greenhouse Gas Emissions

Queensland's per capita GHG emissions for each selected LULC transition over time are presented in Table 5.7, while the estimated GHG emissions/ha/annum for these selected LULC transitions are highlighted in Table 5.8. Our analysis reveals that GHG emissions/ha/annum related to the conversion of built-up areas from bare soil and vegetation increased over time, with a more than two-fold increase in the 1999–2009 decade than those of the 1989–1999 decade. Nevertheless, deforestation-related GHG emissions/ha/annum were highest (i.e., 4.63 tonnes/ha/annum) in the 1999–2009 decade, which is around 1.7 times higher than in its former and later decades.

In terms of per-hectare aggregate GHG emissions, the analysis shows that such an emission was lowest (i.e., 2.34 tonnes/ha/annum) in the 1989–1999

TABLE 5.6

LULC Transitions and Ecosystem Service Value Changes Over the 30-Year Period

Major Land Cover Changes	1989–1999	1999–2009	2009–2019
Bare soil to built-up (BS_BU) (ha)	53,681	16,697	32,481
Bare soil to vegetation (BS_VEG) (ha)	44,877	14,787	11,242
Vegetation to bare soil (VEG_BS) (ha)	24,184	32,047	24,318
Vegetation to built-up (VEG_BU) (ha)	30,234	54,283	30,776
Net deforestation (ha)	9541	71,543	43,852
Net ESV losses (USD billion)	0.35	2.60	1.59
Percentage of ESV losses	7.7	57.27	35.02

TABLE 5.7

Estimating urbanisation-led built-up greenhouse gas (GHG) emissions of Queensland. (Derived from AGEIS (Australian Greenhouse Emissions Information System) (2020) and Queensland Government (2020).)

	Direct GHG Emissions (1000 tonnes)										
Year	Energy	Solid Waste Disposal	Deforestation	Construction	Non-Agricultural Residential	Indirect GHG Emission (1000 tonnes)	Total GHG Emission (1000 tonnes)	Estimated Resident Population (ERP)	GHG (Per Capita) for 'Bare Soil to Built-Up'	GHG (Per Capita) for 'Vegetation to Built-Up'	GHG (Per Capita) for 'Vegetation to Bare Soil'
1990	48,542	3324	31,123	1865	7449	33.38	97,770	2,928,713	22.76	33.38	10.63
1999	71,218	3010	51,016	1814	8560	40.98	143,312	3,497,147	26.39	40.98	14.59
2009	98,636	3165	29,740	1716	11,563	34.81	154,431	4,436,882	28.10	34.81	6.70
2016	112,689	2214	16,224	2165	13,502	31.90	155,795	4,883,821	28.58	31.90	3.32

TABLE 5.8

Estimating GHG Emissions for the Changes of 'Bare Soil to Built-Up (BS_BU)', 'Vegetation to Built-Up (VEG_BU)', and 'Vegetation to Bare Soil (VEG_BS)' for the Study Area

Indicator	1989–1999	1999–2009	2009–2019
Average GHG/per capita (tonnes/per capita) (for BS_BU)	24.57	27.25	28.34
Average GHG/per capita (tonnes/per capita) (for VEG_BU)	37.18	37.9	33.36
Average GHG/per capita (tonnes/per capita) (for VEG_BS)	12.61	10.65	5.01
Changes in population	324,100	512,054	541,354
Bare soil to built-up (BS_BU) (ha)	53,681	16,697	32,481
Vegetation to built-up (VEG_BU) (ha)	30,234	54,283	30,776
Vegetation to bare soil (VEG_BS) (ha)	24,184	32,047	24,318
Bare soil to vegetation (BS_VEG) (ha)	44,877	14,787	11,242
Sum of major LU transitions	152,976	117,814	98,817
Ratio of BS_BU = BS_BU/Sum of major LULC transitions	0.35	0.14	0.33
Ratio of VEG_BU = VEG_BU/Sum of major LULC transitions	0.20	0.46	0.31
Ratio of VEG_BS = VEG_BS/Sum of major LULC transitions	0.16	0.27	0.25
Increased GHG for BS_BU= (changes in population × average GHG/capita for BS_BU X ratio of BS_BU)	2,794,805	1,977,337	5,043,018
Increased GHG for VEG_BU = (changes in population × average GHG/capita for VEG_BU X ratio of VEG_BU)	2,381,556	8,941,737	5,624,552
Increased GHG for VEG_BS = (changes in population × average GHG/capita for VEG_BS X ratio of VEG_BS)	645,962	1,482,759	667,777
GHG emission for BS_BU/ha/annum	5.21	11.84	15.53
GHG emission for VEG_BU/ha/annum	7.88	16.47	18.28
GHG emission for VEG_BS/ha/annum	2.67	4.63	2.75

decade but was quadrupled in the 1999–2009 decade (Table 5.9). Nonetheless, although per-hectare aggregate GHG emissions increased over time, such an increase did not demonstrate significant variation between the periods of 1999–2009 and 2009–2019.

As for the case of aggregate GHG emissions/ha/annum, due to the prevalence of vegetation coverage, aggregate GHG emissions/ha/annum were found to be negative (i.e., –1.72 tonnes/ha/annum) in the 1989–1999 decade, indicating that aggregate GHG emissions were offset by the amount of carbon sink. Nonetheless, aggregate GHG emissions/ha/annum were highest (i.e., 4.49 tonnes/ha/annum) in the period of 2009–2019, which is more than two times higher than the former ten-year period of 1999–2009.

TABLE 5.9

Calculating Aggregate GHG Emissions in the Study Area

	LULC Category/ Transition	Area (ha) 1989–1999	Area (ha) 1999–2009	Area (ha) 2009–2019	GHG Emissions (tonnes) 1989–1999	GHG Emissions (tonnes) 1999–2009	GHG Emissions (tonnes) 2009–2019
Persistent	Built-up (≈ BS_BU) (+)	84,135	146,933	201,475	4,383,429	17,396,838	31,289,063
	Vegetation (carbon sink) (−)	367,552	333,902	300,892	18,377,616	16,695,117	15,044,608
	Sub-total (persistent)	451,687	480,835	502,367	(−)13,994,187	701,721	16,244,455
Major transitions	Bare soil to built-up (ha) (+)	53,681	16,697	32,481	2,796,785	1,976,874	5,044,278
	Bare soil to vegetation (ha) (i.e., carbon sink) (−)	44,877	14,787	11,242	2,243,849	739,331	562,099
	Vegetation to bare soil (ha) (+)	24,184	32,047	24,318	645,720	1,483,770	668,741
	Vegetation to built-up (ha) (+)	30,233	54,283	30,776	2,382,398	8,940,384	5,625,844
	Sub-total (major LULC transitions)	152,976	117,813	98,817	3,581,054	11,661,697	10,776,763
Grand total		604,663	598,648	601,184	(−)10,413,132	12,363,417	27,021,218
Per unit GHG calculation	GHG/ha/annum (for major LULC transitions)				2.34	9.90	10.91
	Aggregate GHG/ha/annum				(−)1.72	2.07	4.49
	Total carbon sink	412,429	348,689	312,134	20,621,465	17,434,448	15,606,707
	Carbon source to sink ratio (in terms of area)	0.47	0.72	0.93			
	Carbon source to sink ratio (in terms of aggregate GHG emissions)				0.50	1.71	2.73

The carbon source to sink ratio (in terms of area coverage) increased over time, but always remained below 1.0, indicating that the coverage of vegetation was always higher than that of carbon sources throughout the decades. Contrastingly, the carbon source to sink ratio, in terms of aggregate GHG emissions, was below 1.0 in the 1989–1999 decade, but increased to 1.71 and 2.73 by the periods of 1999–2009 and 2009–2019, respectively. Therefore, the period of 1989–1999 released the least aggregate GHG emissions/ha/annum.

5.3.4 Checking the Reliability of GHG Estimation Approaches

Our estimated GHG emissions figures, which originated from the persistent built-up area category and major LULC transitions for the LGA of Brisbane

TABLE 5.10

Calculating Brisbane's GHG Emissions That Originated from the Persistent Built-Up Area Category and Major LULC Transitions

LULC Category/Transition	Area (in ha) 2009–2019	GHG Emissions (tonnes) 2009–2019
Built-up (persistent ≈ BS_BU)	63,235	9,820,398
Bare soil to built-up	2571	399,298
Vegetation to bare soil	2440	67,105
Vegetation to built-up	5655	1,033,747
Total GHG emissions		11,320,549
Population increases from 2009 to 2019 (UN, United Nations, 2020)		410,000
GHG emissions/capita/annum		2.8

between 2009 and 2019, are calculated as 2.8 tonnes/annum (Table 5.10). On the other hand, Brisbane's overall per capita GHG emissions are indicated in the literature as 14 tonnes/annum (Chen et al., 2016). As household emissions for the Australian context contribute to 12% of total emissions (DoISER, 2020), Brisbane's household emissions can be derived as (14) × (0.12) = 1.68 tonnes/capita/annum. Thus, our estimated GHG figures appear to be a bit higher than household emissions. Such an outcome is reasonable as we considered the construction- and deforestation-related emissions along with calculating household emissions.

5.3.5 Variations in GHG Emissions Between Unquestionably Urban, Peri-Urban, and Rural Areas

In terms of urban growth land use classification, which sources GHG emissions, peri-urban areas comprise the biggest chunk (i.e., 36.17%), followed by unquestionably rural areas, and unquestionably urban areas account for 34.67% and 29.16%, respectively (Table 5.11). Conversely, the persistence in built-up areas is mostly confined within unquestionably urban areas and peri-urban areas, accounting for 89.57% and 9.4%, respectively.

As for the case of major LULC transitions, the highest amount of transitions occurred in the category of peri-urban areas (i.e., 44.98%), followed by unquestionably urban areas (33.75%) and unquestionably rural areas (21.28%). Additionally, transitions to built-up areas—i.e., 'bare soil to built-up' and 'vegetation to built-up'—are found to be the highest in peri-urban areas.

In terms of total GHG emissions between different urban growth areas, in unquestionably rural areas' net GHG emissions are negative (i.e., –32.31%), indicating that the presence of vegetation coverage largely offsets the

TABLE 5.11

Calculating GHG Emissions Within Different Urban Growth Land Use Areas Between 1989 and 2019

		Unquestionably Urban			Peri-Urban			Unquestionably Rural		
	LULC Category/Transition	**Area (ha)**	**%**	**GHG (tonnes)**	**Area (ha)**	**%**	**GHG (tonnes)**	**Area (ha)**	**%**	**GHG (in tonnes)**
Persistent	Built-up	79,861	89.57	26,018,830	8385	9.40	2,731,773	911	1.02	296,957
	Vegetation	20,166	6.93	3,024,974	108,365	37.25	16,254,815	162,368	55.82	24,355,240
	Sub-total (persistence)	100,028	26.32	22,993,856	116,750	30.72	(−)13,523,042	163,280	42.96	(−)24,058,284
Major transitions	Bare soil to Built-up	39,289	43.74	12,800,292	42,988	47.86	14,005,532	7536	8.39	2,455,337
	Bare soil to vegetation	2760	13.07	413,963	9432	44.67	1,414,814	8924	42.26	1,338,610
	Vegetation to bare soil	1597	4.18	160,498	11,497	30.11	1,155,469	25,085	65.70	2,521,031
	Vegetation to built-up	35,707	41.51	15,221,899	41,840	48.63	17,836,510	8482	9.86	3,615,920
	Sub-total (major transitions)	79,353	33.75	27,768,725	105,758	44.98	31,582,697	50,027	21.28	7,253,678
Grand total (GHG release—carbon sink)		179,380		50,762,581	222,508		18,059,655	213,307		(−)16,804,606
% of grand total		29.16		97.59	36.17		34.72	34.67		(−)32.31
Rate (tonnes/ha/annum)		9.43			2.71			(−)2.63		

LULC-related GHG emissions. Besides, the results show that unquestionably urban areas emitted the highest emissions, followed by peri-urban areas, and per-hectare GHG emissions in unquestionably urban areas are about 3.5 times higher than those of peri-urban areas.

5.3.6 Changes in Atmospheric Temperature and Precipitation

The increase in global temperature between the years 1900 and 2000 was approximately 0.7°C, whereas the mean temperature increase value in Australia between the years 1910 and 2009 was more than 1°C (Braganza & Church, 2011). Australia is expected to experience 1°C and 1.8°C increases in atmospheric temperature by the year 2030 and 2050, respectively, from the base year estimate of 1990 (IPCC, (Intergovernmental Panel on Climate Change), 2007). Such an increasing trend of temperature has also been depicted from our study area, i.e., Brisbane and adjoining LGAs. While the study area experienced a slight increase in annual mean maximum temperature (MMT) between 2000 and 2020 (Figure 5.5), such temperature steadily increased during the last ten-year period (Figure 5.6). Although the trend line of annual MMT shows a weak predictive level of acceptance ($R^2 = 0.27$), the substantial acceptance level of MMT for the non-summer season was found ($R^2 = 0.79$). Furthermore, the moderate predictive level of acceptance of MMT ($R^2 = 0.57$) was observed for the summer season.

Moreover, Australia is also predicted to experience a decline in annual precipitation, and the rate of decrease in annual precipitation from the year 1970 to 2011 was 5 mm/annum (Alamgir et al., 2014). In the context of our study area, this precipitation pattern showed a declining trend from 2010 to 2020 (Figure 5.7). While the trend of decline in average precipitation is weak

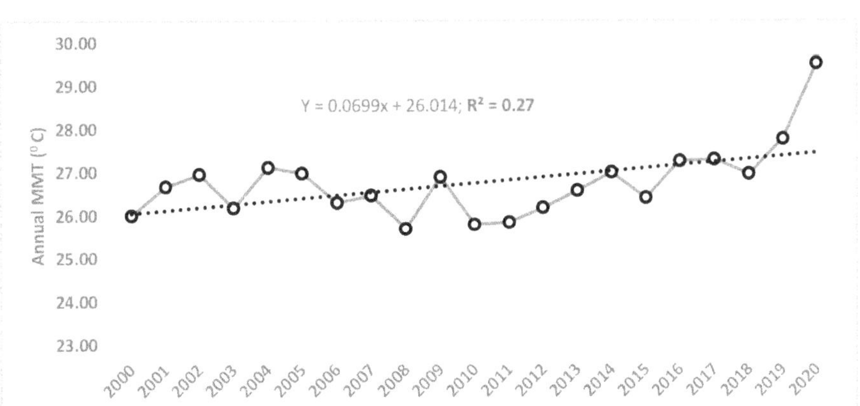

FIGURE 5.5
Trend of annual mean maximum temperature from 2000 to 2020.

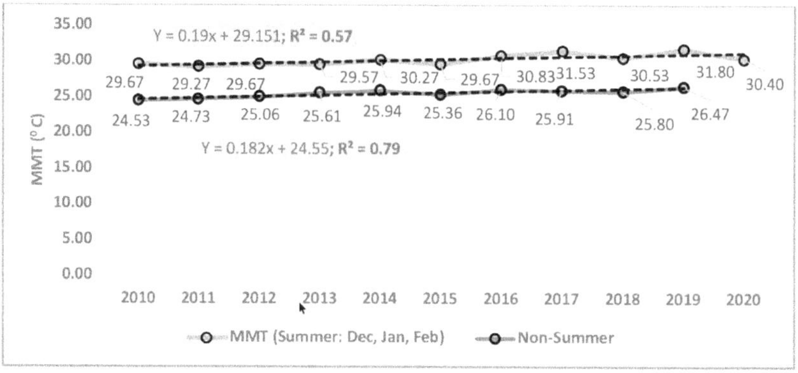

FIGURE 5.6
Variations in the trend of mean maximum temperature between summer and non-summer seasons from 2010 to 2020.

(R^2 = 0.12) during the monsoon period, such a declining trend during the non-monsoon period is moderately significant (R^2 = 0.41).

Interestingly, in the Australian context, the monsoon period comes in the summer, which includes the months of December, January, and February due to the Southern Hemisphere positioning. Hence, these three months simultaneously represent both the summer and rainy seasons. Nonetheless, the increasing trend of MMT coupled with the decreasing trend of average precipitation appears to pose a detrimental effect on the growth of summer and non-summer agricultural products, e.g., fruits and vegetables.

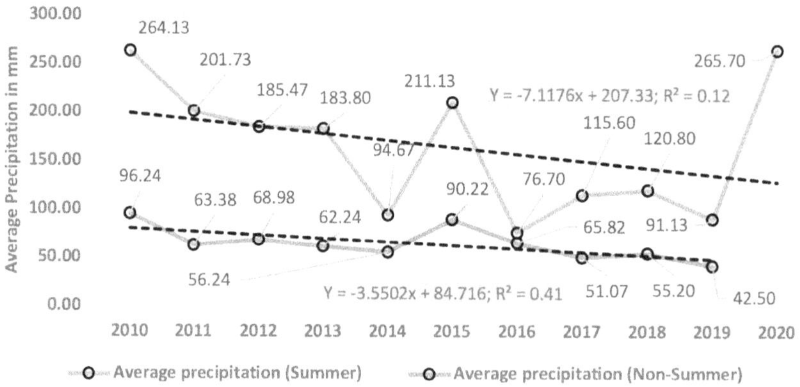

FIGURE 5.7
Variations in the trend of average precipitation between summer and non-summer seasons from 2010 to 2020.

5.3.7 Deforestation Scenarios Correspond to the Shifts in State Policies

The state polices of Queensland concerning deforestation play a significant role in land clearing. (Figure 5.8). The amount of deforestation declined over time when stringent policy control on land clearing was imposed, and later gained a sudden momentum when such strictly defined regulations were slackened.

For example, from 1990 to 1995, deforestation declined to 331,000 ha from 637,000 ha. Then, again in 2000, deforestation reached its peak of 758,000 ha. The Vegetation Management Act (VMA) 1999 curbed this amount to 375,000 ha in 2006. Later in 2006, a ban on broad-scale vegetation clearing was imposed, which resulted in a further decline to 100,000 ha in 2009. In 2009, the VMA 1999 was further reinforced by giving additional emphasis to the protection of high-value regrowth vegetation. Such a reinforcement brought down the amount of vegetation clearing to a minimum of 79,000 ha in 2010. Nonetheless, the Vegetation Management Amendment Act 2013 weakened the 2006 and 2009 regulations, leading to increased deforestation to 396,000 ha in 2016.

In terms of GHG emissions due to deforestation, Queensland shares the highest portion within Australia. For example, from 1990 to 2016, the percentage of Queensland's deforestation-related share of national GHG emissions was always more than 50%, while such a share accounted for a minimum of 50.12% in 2011 (Figure 5.9). Yet, from 1990 to 1999, Queensland's share increased to 72.04% from the base year estimate of 50.35% in 1990. In the periods of 1999–2009 and 2009–2016, such shares declined to 56.78% and 55.73%,

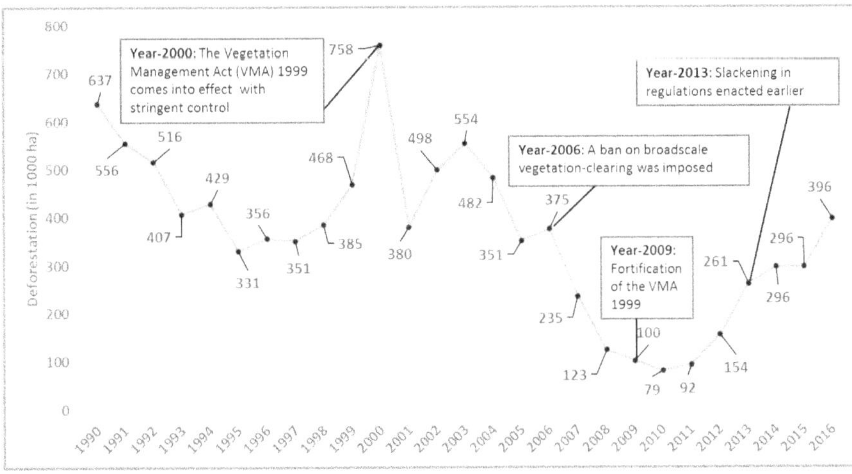

FIGURE 5.8
Queensland's deforestation between 1990 and 2016. (Derived from The Wilderness Society (2020) and Steffen and Dean (2018).)

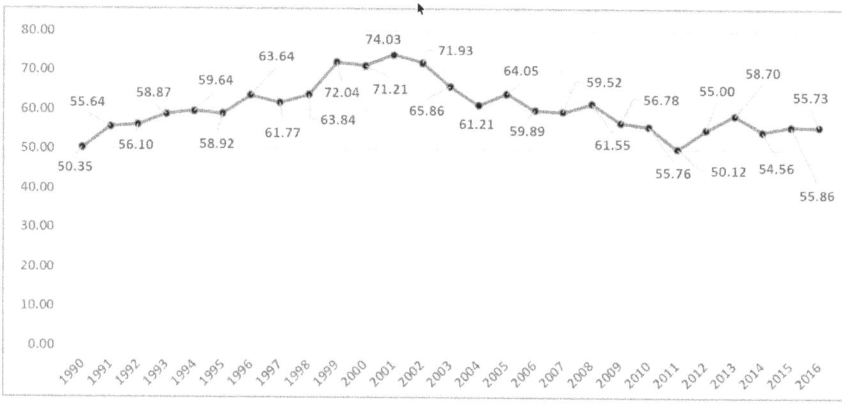

FIGURE 5.9
Percentage of Queensland's share of national GHG emissions between 1990 and 2016 due to deforestation, derived from AGEIS (AGEIS (Australian Greenhouse Emissions Information System), 2020).

respectively. In sum, from 1990 to 2001, such shares steadily increased and later started to decrease from 2002 to 2016.

5.4 Findings and Discussion

The study reported in this chapter mainly focuses on peri-urbanisation-related LULC changes and their impacts on ESV losses and GHG emissions. Nonetheless, ecosystem services, which lie within peri-urban farmlands, are the agroecosystem services. These agroecosystem services are undermined in the conventional ESV estimation process. The conventional ESV estimation approaches assign more weight to wetlands and assign nominal dollar values (i.e., USD 92/ha/annum) to farmland ecosystem services inherent with peri-urban greenfield sites (Costanza et al., 1997; van der Ploeg et al., 2010).

In general, more than 80% of the ecosystem service benefits cannot be monetised (De Groot et al., 2002; Dizdaroglu et al., 2012). Thus, farmland ecosystem services are rich in value, but are highly discredited due to the inability to appropriately monetise these intangible service benefits. Furthermore, the global estimation approaches of ESV are backed up by secondary data, which appears to be inappropriate in evaluating greenfield sites' farmland ecosystem services at the regional/local context. Henceforth, Wu et al. (2013) rightfully contend that peri-urbanisation (which triggers the conversion of greenfield sites to bare soil and built-up areas) results in a significant decline in regional ESV. Thus, while conventional studies evaluate the loss of ESV due to the conversion of forested lands and wetlands to agricultural uses, our study analyses the loss of ESV due to the conversion of peri-urban farmlands to urban land uses.

The study at hand adopted Sandhu et al.'s (2013) ESV estimation on agroecosystem services. They carried out a regional estimation of ESV based on primary data and calculated ESV to farmland as USD 3224/ha/annum—more than 35 times higher than that of Costanza et al.'s, (1997) global ESV estimation, which is commonly adopted in many studies (De Groot et al., 2002).

Our ESV estimation approach relied on (Sandhu et al., 2013) ESV of the Canterbury Plains of New Zealand. Countries like New Zealand and Australia are both native vegetation dominated, and a large proportion of this vegetation is nonexistent elsewhere on the planet (EIANZ (The Environment Institute of Australia and New Zealand), 2009). Besides, these two countries were separated from the rest of the globe for millions of years (ABS (Australian Bureau of Statistics), 2010). Thus, because of their evolutionary and geographical detachments, both New Zealand and Australia have become biologically diverse from the rest of the globe (Williams & West, 2000). Hence, ESV to farmlands for the contexts of New Zealand and Australia cannot be generalised with the global farmland estimation of ESV. Furthermore, New Zealand and Australia also share similar socioeconomic and cultural settings. Henceforth, ESV to farmlands for the Australian context appears to be equivalent to Sandhu et al.'s (2013) estimation of farmlands' ESV.

Our study area has undergone a significant brownfield development between 1989 and 1999, and then encountered a considerable greenfield development between 1999 and 2019. Such development trends reflected their corresponding consequences in ESV losses and GHG emissions, which collectively might induce the impact of anthropogenic climate change.

Nonetheless, the losses in ESV were conspicuous in the period of 1999–2009, and were lower during the 1989–1999 period. This is due to the extensive brownfield developments taking place between 1989 and 1999. Such a development pattern indicates a significant amount of vegetation coverage. In contrast, in the 1999–2009 period, the study area underwent greenfield development, which was more than three times higher than the brownfield development that occurred between 1989 and 1999.

Such a development trend was also intensified by the mining boom in Australia in general. Between 2002 and 2012, Australia's mining exports had tripled, while spending contributed by the mining sector increased from 2% of gross domestic product (GDP) to 8% between this ten-year period (Downes et al., 2014).

Deforestation, which is a major catalyst of declining ESV, seems to be impacted by the Millennium Drought (2001–2009) event in Australia. However, anthropogenic influences on the drought event are not yet possible to determine, meaning that if there was an impact, the proportion of anthropogenic influences on the drought event remains immeasurable. Consequently, the proportion of anthropogenic ESV losses attributable to the drought event also cannot be measured.

Nevertheless, ESV is determined based on four ecosystem services, namely, regulating services (e.g., climate regulation such as regulating the emissions of GHG), provisioning services (e.g., food production), supporting services (e.g., biodiversity habitat), and cultural services (e.g., amenity landscapes) (Costanza et al., 1997). Additionally, climate change-induced extreme heat events are strongly linked to human-led increased GHG emissions. The changes in LULC and ESV losses result in increased GHG emissions in the atmosphere.

Several studies quantified the impacts of human-led increased GHG emissions on extreme heat events. For example, researchers claimed that, along with several past records of extreme heat events, the record-breaking heat events in Australia that occurred in the summer of 1997–1998 and 2012–2013 were largely attributed to anthropogenic influences (King et al., 2016). On that very point, the record-breaking Australian hot summers of 1997–1998 and 2012–2013 are estimated to have a minimum of seven-fold and five-fold increase, respectively, caused by anthropogenic climate change (Lewis & Karoly, 2013).

Furthermore, our analysis reveals that during the last decade (2010–2020), along with the declining precipitation trend, extreme heat events have also intensified and surpassed the previous record of extreme hot events. An increase in atmospheric temperature coupled with a decrease in annual precipitation leads to a drought event, which eventually impacts ecosystem services in the long run (Australia State of the Environment, 2011a, b). Besides, the climate impacts on ecosystem services are not uniform throughout Australia (Alamgir et al., 2014)—they vary across the regions both temporarily and spatially (Ding & Nunes, 2014).

In the context of our study area, between 1989 and 1999, the gross emissions of GHG were negative, but 79,64,483 tonnes of GHG were collectively released from persistent built-up areas, and due to the emissions from major LULC changes. In unquestionably urban areas, the percentage of vegetation coverage is insignificant, implying that extreme heat events are likely to be perceived more in unquestionably urban areas. On the other hand, the presence of a large stock of vegetation within unquestionably rural areas appears to lessen human-induced climate change impact within this zone, as the presence of vegetation is strongly linked to lowering the heat island effect (Ahmed et al., 2013; Deilami & Kamruzzaman, 2017; Deilami et al., 2018; Kamruzzaman et al., 2018). However, the presence of vegetation can only lower the heat island effect by reducing the land surface temperature; that is, it cannot discernibly lessen the atmospheric temperature, which is substantially influenced by the impact of GHG emission-led anthropogenic climate change. Furthermore, the highest transition in LULC and corresponding GHG emissions occurred in peri-urban areas, meaning that peri-urbanisation-related GHG emissions play a major role in overall GHG emissions.

Queensland's deforestation scenarios and corresponding GHG emissions over time have been linked to policy control. Still, our study area experienced the highest ESV losses and GHG emissions between 1999 and 2009. Such losses in ESV and corresponding GHG emissions were apparently due to the mining boom that lasted from 2002 to 2012 and resulted in a massive urbanisation drive. Hence, though Queensland's land clearing regulations were tightened from 2001, investment spending via the mining boom outweighed land-clearing regulations in the context of our study area. However, this mining-led urbanisation drive was later subdued due to the aftermath of the global financial crisis of 2007–2008. Consequently, we reported fewer LULC changes in the following period of 2009–2019 than those of the preceding ten-year period of 1999–2009.

In summary, the study discloses the following main findings: (a) The utilised remote sensing method is a useful technique in quantifying the impacts of climate change; (b) over the last 30-year period, Brisbane and its adjoining areas encountered a total loss of about USD 4.5 billion in ecosystem services due to direct and indirect anthropogenic climate change; (c) peri-urban areas encountered the biggest losses in ESV; (d) peri-urban areas experienced the highest GHG emission production levels; and (e) ecosystem services should be backed up by robust urban management policies—this is critical for mitigating climate change.

5.5 Conclusion

As growth in urban areas in most of the metropolitan cities and regions around the globe is already saturated, peri-urban areas are now the potential pockets in accommodating further growth (Metaxiotis et al., 2010; Yigitcanlar & Kamruzzaman, 2015). For that reason, peri-urbanisation-driven ESV losses and GHG emissions should be taken into consideration.

The mix of urban and rural land uses, accompanied by ambiguous land use patterns in peri-urban areas, appears to quantify GHG emissions, and their corresponding impacts on GWP are challenging. Hence, quantifying peri-urbanisation-driven GHG emissions remains unincorporated in the Intergovernmental Panel on Climate Change (IPCC)'s estimated climate change scenarios (Sotto et al., 2019; Stocker et al., 2013). Still, globally occurring constant peri-urbanisation to nearby greenfield sites and the corresponding changes in ESV and GHG emissions most likely will have an increasing climate change impact (Van Delden et al., 2016). These peri-urban areas simultaneously serve both as a source of carbon sink and a carbon source. At inception, the vegetation dominating peri-urban areas serves as an ample source of ecosystem services, including the space of a carbon sink. As urbanisation progresses, carbon sink areas become minor over time, and looming anthropogenic climate impacts become more visible.

Accordingly, identifying the locations experiencing constant losses in ESV and the locations shifting their role from a carbon reservoir to a carbon source within a peri-urban context is necessary to quantify the subsequent impacts of anthropogenic activities on climate change. Such impact identification will provide an evidence base to policymakers in formulating the most appropriate climate change mitigation strategies within peri-urban settings.

This study quantifies the spaces of carbon sources and carbon sinks over time. Simultaneously, converting peri-urban areas hinders policymakers in identifying the GHG emissions emanating from anthropogenic sources, as differentiating carbon source areas is always difficult due to the dynamic nature of peri-urban areas (Dodman, 2009; Satterthwaite, 2008). Our study approach overcomes this limitation and helps estimate anthropogenic GHG emissions with increased precision. This way, our findings enable policymakers to decide the level of vegetation coverage to be preserved in order to offset anthropogenic climate change impact due to increased GHG emissions.

The IPCC estimates mainly focus on the climate change scenarios, while peri-urbanisation-related GHG emission figures still remain uncovered. The unique contribution of this study is that it calculates peri-urbanisation-related GHG emissions and discloses that they are very high to the extent that they endanger the sustainability of our cities and regions (Arbolino et al., 2018; Van Vuuren et al., 2017; Yigitcanlar & Kamruzzaman, 2015). This finding also suggests that peri-urbanisation-related GHG emissions are critical for building reliable climate change scenarios.

Nevertheless, it is important to note that the study reported in this chapter mainly focused on how peri-urbanisation triggers anthropogenic climate change in terms of ESV losses and GHG emissions. In the context of the study area, i.e., Brisbane (Australia), such peri-urbanisation was mostly the conversions of bare soil and vegetation to built-up areas. Other sectors of emissions—e.g., agriculture, forestry and fishing, mining, manufacturing, transport, postal and warehousing, and commercial services—were not included in our GHG calculation. Thus, our GHG estimation outcome only provides a fraction of the GHG that originated from the peri-urbanisation-led major LULC transitions. This study neither carried out any GHG modelling nor simulation. Instead, we used GHG data from the available and credible sources—i.e., AGEIS—in order to derive per-hectare GHG emissions due to the peri-urbanisation-led major LULC transitions.

Besides, the study at hand did not calculate GHG emissions from the persistent bare soil category—i.e., the bare soil, which did not change within a decade—as we considered bare soil as 'no land use activity'. Nevertheless, bare soil, indeed, emits a substantial amount of GHG, which ranges from 0.57 to 0.86 μmol $CO_2/(m^2\ s)$ (Lai et al., 2012). Consequently, bare soil contributes to 10% of the global CO_2 emissions (Oertel et al., 2016). Nonetheless, such emissions are related to natural atmospheric GHG emissions, whereas the focus of our study was to estimate LULC-driven GHG emissions. Hence, we omitted the calculation of GHG emissions emanating from the persistent bare soil category.

Another scope of concern in our GHG estimation approach lies in the inability to know the exact time frame when an LULC transition occurred. Such an inability compels us to generalise the GHG estimation. Thus, we calculated LULC-related major GHG emissions and subsequent GHG emissions from the end LULC category altogether.

Nevertheless, the estimated GHG outcome does not represent the gross scenarios of GHG emissions—instead, such an outcome appears to fit well with regard to peri-urbanisation-driven real-estate emissions. Thus, with regard to permitting housing projects (Ingrao et al., 2018; Lee et al., 2008), this research provides an evidence base for policymakers to construct more realistic GHG estimation scenarios in line with a nation's targeted GHG emissions under the international agreements, e.g., the Kyoto Protocol and Paris Agreement.

Acknowledgements

This chapter, with permission from the copyright holder, is a reproduced version of the following journal article: Mortoja, M., & Yigitcanlar, T. (2020). Local drivers of anthropogenic climate change: Quantifying the impact through a remote sensing approach in Brisbane. *Remote Sensing, 12*(14), 2270.

References

ABS (Australian Bureau of Statistics). (2010). *Year book Australia*. Australian Bureau of Statistics: Canberra, ACT, Australia. Available online: https://www.abs.gov.au/ausstats/abs@.nsf/Previousproducts/1301.0Feature%20Article12009%E2%80%9310?opendocument&tabn (accessed on 15 March 2020).

Acheampong, R. A., Agyemang, F., & Abdul-Fatawu, M. (2016). Quantifying the spatio-temporal patterns of settlement growth in a metropolitan region of Ghana. *GeoJournal, 82*, 823–840. http://dx.doi.org/10.1007/s10708-016-9719-x

ACS (Australia's Chief Scientist). (2009). *Which plants store more carbon in Australia: Forests or grasses?* Australian Government: Canberra, ACT, Australia.

AGEIS (Australian Greenhouse Emissions Information System). (2020). *National greenhouse gas inventory – Kyoto Protocol classifications*. Department of the Environment and Energy: Australia.

Ahmed, B., Kamruzzaman, M., Zhu, X., Rahman, S., & Choi, K. (2013). Simulating land cover changes and their impacts on land surface temperature in Dhaka, Bangladesh. *Remote Sensing, 5*, 5969–5998. http://dx.doi.org/10.3390/rs5115969

Alamgir, M., Pert, P. L., & Turton, S. M. (2014). A review of ecosystem services research in Australia reveals a gap in integrating climate change and impacts on ecosystem services. *International Journal of Biodiversity Science, Ecosystem Services & Management, 10*, 112–127. http://dx.doi.org/10.1080/21513732.2014.919961

Anderson, J. R., Hardy, E. E., Roach, J. T., & Witmer, R. E. (1976). *A land use and land cover classification system for use with remote sensor data*. US Government Printing Office: Washington DC.

Arbolino, R., De Simone, L., Carlucci, F., Yigitcanlar, T., & Ioppolo, G. (2018). Towards a sustainable industrial ecology: Implementation of a novel approach in the performance evaluation of Italian regions. *Journal of Cleaner Production, 178*, 220–236. http://dx.doi.org/10.1016/j.jclepro.2017.12.183

Australia State of the Environment. (2011). *Drivers of Australia's environment*. Observatory Hill: Sydney, Australia.

Australian State of the Environment. (2011). *Committee*. Canberra, Australia: Australia State of the Environment. Available online: http://www.environment.gov.au/soe/2011/report/built-environment/index.html (accessed on 15 March 2020).

Bekessy, S., & Gordon, A. (2007). Nurturing nature in the city. In *Steering sustainability in an urbanising world: Policy, practice and performance* (pp. 227–238). Routledge.

Braganza, K., & Church, J. A. (2011). *Observations of global and Australian climate* (pp. 1–14). CSIRO Publishing.

Buxton, M., & Low-Choy, D. (2007). Change in peri-urban Australia: Implications for land use policies. In *Proceedings of the State of Australian Cities Conference*; Adelaide, Australia, 28–30 November 2007; pp. 291–302.

Chen, G., Wiedmann, T., Hadjikakou, M., & Rowley, H. (2016). City carbon footprint networks. *Energies, 9*, 602. http://dx.doi.org/10.3390/en9080602

Chen, M., Zhang, H., Liu, W., & Zhang, W. (2014). The global pattern of urbanization and economic growth: Evidence from the last three decades. *PLoS ONE, 9*, e103799. http://dx.doi.org/10.1371/journal.pone.0103799

Cogger, H. G. (2003). *Impacts of land clearing on Australian wildlife in Queensland*. World-Wide Fund for Nature Australia: Sydney, Australia.

Costanza, R., D'Arge, R., De Groot, R., Farber, S., Grasso, M., Hannon, B., Limburg, K., Naeem, S., O'Neill, R. V., Paruelo, J., & Raskin, R. G. (1997). The value of the world's ecosystem services and natural capital. *Nature, 387*, 253–260. http://dx.doi.org/10.1038/387253a0

Costanza, R., De Groot, R., Sutton, P., Van Der Ploeg, S., Anderson, S., Kubiszewski, I., Farber, S., & Turner, R. K. (2014). Changes in the global value of ecosystem services. *Global Environmental Change, 26*, 152–158. http://dx.doi.org/10.1016/j.gloenvcha.2014.04.002

CPIIC (CPI Inflation Calculator). (2020). Available online: https://www.in2013dollars.com/us/inflation/2012?amount=1 (accessed on 27 March 2020).

De Groot, R. S., A Wilson, M., & Boumans, R. M. (2002). A typology for the classification, description and valuation of ecosystem functions, goods and services. *Ecological Economics, 41*, 393–408. http://dx.doi.org/10.1016/s0921-8009(02)00089-7

Deilami, K., & Kamruzzaman, M. (2017). Modelling the urban heat island effect of smart growth policy scenarios in Brisbane. *Land Use Policy, 64*, 38–55. http://dx.doi.org/10.1016/j.landusepol.2017.02.027

Deilami, K., Kamruzzaman, M., & Liu, Y. (2018). Urban heat island effect: A systematic review of spatio-temporal factors, data, methods, and mitigation measures. *International Journal of Applied Earth Observation and Geoinformation, 67*, 30–42. http://dx.doi.org/10.1016/j.jag.2017.12.009

Ding, H., & Nunes, P. (2014). Modeling the links between biodiversity, ecosystem services and human wellbeing in the context of climate change: Results from an econometric analysis of the European forest ecosystems. *Ecological Economics, 97*, 60–73. http://dx.doi.org/10.1016/j.ecolecon.2013.11.004

Dizdaroglu, D., Yigitcanlar, T., & Dawes, L. (2012). A micro-level indexing model for assessing urban ecosystem sustainability. *Smart and Sustainable Built Environment, 1*, 291–315. http://dx.doi.org/10.1108/20466091211287155

Dodman, D. (2009). Blaming cities for climate change? An analysis of urban greenhouse gas emissions inventories. *Environment and Urbanization, 21*, 185–201. http://dx.doi.org/10.1177/0956247809103016

DoISER (Department of Industry, Science, Energy and Resources). (2020). Australian Government. Available online: https://publications.industry.gov.au/publications/climate-change/climate-change/individuals-andhouseholds.html#:~:text=Australia's%20households%20are%20responsible%20for,to%20emissions%20through%20electricity%20consumption (accessed on 1 July 2020).

Downes, P. M., Hanslow, K., & Tulip, P. (2014). *The effect of the mining boom on the Australian economy*. Reserve Bank of Australia Research Discussion Paper No. 2014-08. Brisbane, QLD, Australia, 22 August 2014. https://ssrn.com/abstract=2701080 or http://dx.doi.org/10.2139/ssrn.2701080

Dur, F., & Yigitcanlar, T. (2015). Assessing land-use and transport integration via a spatial composite indexing model. *International Journal of Environmental Science and Technology, 12*, 803–816. http://dx.doi.org/10.1007/s13762-013-0476-9

EIANZ (The Environment Institute of Australia and New Zealand). (2009). *Position statement on biodiversity*. Available online: https://www.eianz.org/resources/position-statements (accessed on 18 June 2020).

Friedberger, M. (2000). The rural-urban fringe in the late twentieth century. *Agricultural History, 74*, 502–514.

Goonetilleke, A., Egodawatta, P., Yigitcanlar, T., & Ayoko, G. (2014). *Sustainable urban water environment*. Edward Elgar Publishing.

Harman, B. P., & Choy, D. L. (2011). Perspectives on tradable development rights for ecosystem service protection: Lessons from an Australian peri-urban region. *Journal of Environmental Planning and Management, 54*, 617–635. http://dx.doi.org/10.1080/09640568.2010.526405

Hatfield-Dodds, S., Schandl, H., Adams, P.D., Baynes, T.M., Brinsmead, T.S., Bryan, B.A., Chiew, F.H., Graham, P.W., Grundy, M., Harwood, T. & McCallum, R. (2015). Australia is 'free to choose' economic growth and falling environmental pressures. *Nature, 527*, 49–53. http://dx.doi.org/10.1038/nature16065

Ingrao, C., Messineo, A., Beltramo, R., Yigitcanlar, T., & Ioppolo, G. (2018). How can life cycle thinking support sustainability of buildings? Investigating life cycle assessment applications for energy efficiency and environmental performance. *Journal of Cleaner Production, 201*, 556–569. http://dx.doi.org/10.1016/j.jclepro.2018.08.080

IPCC (Intergovernmental Panel on Climate Change). (2007). *Climate change 2007: Synthesis report*. Intergovernmental Panel on Climate Change: Geneva, Switzerland.

Jiang, H., & Eastman, J. R. (2000). Application of fuzzy measures in multi-criteria evaluation in GIS. *International Journal of Geographical Information Science, 14*, 173–184. http://dx.doi.org/10.1080/136588100240903

Kamruzzaman, M., Deilami, K., & Yigitcanlar, T. (2018). Investigating the urban heat island effect of transit oriented development in Brisbane. *Journal of Transport Geography, 66*, 116–124. http://dx.doi.org/10.1016/j.jtrangeo.2017.11.016

Kijas, J. (2002). A place at the coast: Internal migration and the shift to the coastal countryside. *Transformations, 2*, 1–12.

Kindu, M., Schneider, T., Teketay, D., & Knoke, T. (2016). Changes of ecosystem service values in response to land use/land cover dynamics in Munessa–Shashemene landscape of the Ethiopian highlands. *Science of the Total Environment, 547*, 137–147. http://dx.doi.org/10.1016/j.scitotenv.2015.12.127

King, A., Black, M., Min, S., Fischer, E., Mitchell, D., Harrington, L., & Perkins-Kirkpatrick, S. (2016). Emergence of heat extremes attributable to anthropogenic influences. *Geophysical Research Letters, 43*, 3438–3443. http://dx.doi.org/10.1002/2015gl067448

Kubiszewski, I., Costanza, R., Anderson, S., & Sutton, P. (2017). The future value of ecosystem services: Global scenarios and national implications. *Ecosystem Services, 26*, 289–301. http://dx.doi.org/10.1016/j.ecoser.2017.05.004

Lai, L., Zhao, X., Jiang, L., Wang, Y., Luo, L., Zheng, Y., Chen, X., & Rimmington, G. M. (2012). Soil respiration in different agricultural and natural ecosystems in an arid region. *PLoS ONE, 7*, e48011. http://dx.doi.org/10.1371/journal.pone.0048011

Lee, S., Yigitcanlar, T., Han, J., & Leem, Y. (2008). Ubiquitous urban infrastructure: Infrastructure planning and development in Korea. *Innovation, 10*, 282–292. http://dx.doi.org/10.5172/impp.453.10.2-3.282

Lewis, S. C., & Karoly, D. J. (2013). Anthropogenic contributions to Australia's record summer temperatures of 2013. *Geophysical Research Letters, 40*, 3705–3709.

Martínez, M. L., Pérez-Maqueo, O., Vazquez, G., Castillo-Campos, G., García-Franco, J. G., Mehltreter, K., Equihua, M., & Landgrave, R. (2009). Effects of land use change on biodiversity and ecosystem services in tropical montane cloud forests of Mexico. *Forest Ecology and Management, 258*, 1856–1863. http://dx.doi.org/10.1016/j.foreco.2009.02.023

Merson, J., Attwater, R., Ampt, P., Wildman, H., & Chapple, R. (2010). The challenges to urban agriculture in the Sydney basin and lower Blue Mountains region of Australia. *International Journal of Agricultural Sustainability, 8*, 72–85. http://dx.doi.org/10.3763/ijas.2009.0464

Metaxiotis, K., Carrillo, J., & Yigitcanlar, T. (2010). *Knowledge-based development for cities and societies: Integrated multi-level approaches*. IGI Global.

Miller, J. R., & Hobbs, R. J. (2002). Conservation where people live and work. *Conservation Biology, 16*, 330–337. http://dx.doi.org/10.1046/j.1523-1739.2002.00420.x

Mortoja, G., Yigitcanlar, T., & Mayere, S. (2020). What is the most suitable methodological approach to demarcate peri-urban areas? A systematic review of the literature. *Land Use Policy, 95*, 104601. http://dx.doi.org/10.1016/j.landusepol.2020.104601

Oertel, C., Matschullat, J., Zurba, K., Zimmermann, F., & Erasmi, S. (2016). Greenhouse gas emissions from soils: A review. *Geochemistry, 76*, 327–352. http://dx.doi.org/10.1016/j.chemer.2016.04.002

Polasky, S., Nelson, E., Pennington, D., & Johnson, K. A. (2011). The impact of land-use change on ecosystem services, biodiversity and returns to landowners: A case study in the state of Minnesota. *Environmental and Resource Economics, 48,* 219–242. http://dx.doi.org/10.1007/s10640-010-9407-0

Queensland Government. (2020). *Queensland Government Statistician's Office.* Queensland Treasury: Australia.

Rothwell, A., Ridoutt, B., Page, G., Bellotti, W., & Ridoutt, B. (2015). Feeding and housing the urban population: Environmental impacts at the peri-urban interface under different land-use scenarios. *Land Use Policy, 48,* 377–388. http://dx.doi.org/10.1016/j.landusepol.2015.06.017

Sandhu, H., Porter, J., & Wratten, S. D. (2013). Experimental assessment of ecosystem services in agriculture. In *Ecosystem services in agricultural and urban landscapes* (pp. 122–135). John Wiley & Sons.

Satterthwaite, D. (2008). Cities' contribution to global warming: Notes on the allocation of greenhouse gas emissions. *Environment and Urbanization, 20,* 539–549. http://dx.doi.org/10.1177/0956247808096127

Sinclair, I., Bunker, R., & Holloway, D. (2003). From the outside looking in – Planning and land management in Sydney's fringe. In *Proceedings of the State of Australian Cities National Conference*; Sydney, Australia, 3–5 December 2003.

Sotto, D., Philippi, J. A., Yigitcanlar, T., & Kamruzzaman, M. (2019). Aligning urban policy with climate action in the global South: Are Brazilian cities considering climate emergency in local planning practice? *Energies, 12,* 3418. http://dx.doi.org/10.3390/en12183418

Spencer, A., Gill, J., & Schmahmann, L. (2015). Urban or suburban? Examining the density of Australian cities in a global context. In *The Proceedings of the State of Australian Cities Conference*; 9–11 December 2015, Gold Coast, Australia; pp. 9–11.

Steffen, W., & Dean, A. (2018). *Land clearing and climate change: Risks and opportunities in the sunshine state.* The Climate Council of Australia: NSW, Australia.

Stocker, T. F., Qin, D., Plattner, G. K., Tignor, M., Allen, S. K., Boschung, J., & Midgley, P. M. (2013). Climate change 2013: The physical science basis. In *Contribution of Working Group I to the Fifth Assessment Report of the Intergovernmental Panel on Climate Change (IPCC)*; Geneva, Switzerland.

Styers, D. M., Chappelka, A. H., Marzen, L. J., & Somers, G. L. (2010). Developing a land-cover classification to select indicators of forest ecosystem health in a rapidly urbanizing landscape. *Landscape and Urban Planning, 94,* 158–165. http://dx.doi.org/10.1016/j.landurbplan.2009.09.006

Sutton, P., Anderson, S., Costanza, R., & Kubiszewski, I. (2016). The ecological economics of land degradation: Impacts on ecosystem service values. *Ecological Economics, 129,* 182–192. http://dx.doi.org/10.1016/j.ecolecon.2016.06.016

Sutton, P., & Costanza, R. (2002). Global estimates of market and non-market values derived from nighttime satellite imagery, land cover, and ecosystem service valuation. *Ecological Economics, 41,* 509–527. http://dx.doi.org/10.1016/s0921-8009(02)00097-6

Sutton, P., Goetz, A. R., Fildes, S., Forster, C., & Ghosh, T. (2010). Darkness on the edge of town: Mapping urban and peri-urban Australia using nighttime satellite imagery. *The Professional Geographer, 62,* 119–133. http://dx.doi.org/10.1080/00330120903405006

Sutton, P. C., Cova, T. J., & Elvidge, C. D. (2006). Mapping "exurbia" in the conterminous United States using night-time satellite imagery. *Geocarto International, 21,* 39–45.

The Wilderness Society. (2020.). *Towards zero deforestation.* Australia.

UN (United Nations). (2020). *Brisbane, Australia metro area population*. Available online: https://www.macrotrends.net/cities/206170/brisbane/population (accessed on 1 July 2020).

USGS. (2020). *EarthExplorer*. Available online: https://earthexplorer.usgs.gov/ (accessed on 26 February 2020).

Van Delden, L. (2017). *Implications of urbanization related land use change on the carbon and nitrogen cycle from subtropical soils* [Ph.D. thesis, Queensland University of Technology, Brisbane, QLD, Australia].

Van Delden, L., Larsen, E., Rowlings, D., Scheer, C., & Grace, P. (2016). Establishing turf grass increases soil greenhouse gas emissions in peri-urban environments. *Urban Ecosystems, 19*, 749–762. http://dx.doi.org/10.1007/s11252-016-0529-1

van der Ploeg, S., De Groot, R. S., & Wang, Y. (2010). *The TEEB valuation database: Overview of structure, data and results*. Foundation for Sustainable Development.

Van Vuuren, D., Stehfest, E., Gernaat, D. E., Doelman, J. C., Berg, M. V. D., Harmsen, M., De Boer, H. S., Bouwman, A. F., Daioglou, V., Edelenbosch, O. Y., & Girod, B. (2017). Energy, land-use and greenhouse gas emissions trajectories under a green growth paradigm. *Global Environmental Change, 42*, 237–250. http://dx.doi.org/10.1016/j.gloenvcha.2016.05.008

Williams, J. A., & West, C. J. (2000). Environmental weeds in Australia and New Zealand: Issues and approaches to management. *Austral Ecology, 25*, 425–444.

Wu, K. Y., Ye, X. Y., Qi, Z. F., & Zhang, H. (2013). Impacts of land use/land cover change and socioeconomic development on regional ecosystem services: The case of fast-growing Hangzhou metropolitan area, China. *Cities, 31*, 276–284. http://dx.doi.org/10.1016/j.cities.2012.08.003

Yigitcanlar, T. (2006). Australian Local governments' practice and prospects with online planning. *URISA Journal, 18*, 7–17.

Yigitcanlar, T. (2010a). *Rethinking sustainable development: Urban management, engineering, and design*. IGI Global.

Yigitcanlar, T. (2010b). *Sustainable urban and regional infrastructure development: Technologies, applications and management*. IGI Global.

Yigitcanlar, T., Dodson, J., Gleeson, B., & Sipe, N. (2007). Travel self-containment in master planned estates: Analysis of recent Australian trends. *Urban Policy and Research, 25*, 129–149. http://dx.doi.org/10.1080/08111140701255823

Yigitcanlar, T., Han, J. H., Kamruzzaman, M., Ioppolo, G., & Sabatini-Marques, J. (2019). The making of smart cities: Are Songdo, Masdar, Amsterdam, San Francisco and Brisbane the best we could build? *Land Use Policy, 88*, 104187. http://dx.doi.org/10.1016/j.landusepol.2019.104187

Yigitcanlar, T., & Kamruzzaman, M. (2015). Planning, development and management of sustainable cities: A commentary from the guest editors. *Sustainability, 7*, 14677–14688. http://dx.doi.org/10.3390/su71114677

Yigitcanlar, T., Kamruzzaman, M., & Teriman, S. (2015). Neighborhood sustainability assessment: Evaluating residential development sustainability in a developing country context. *Sustainability, 7*, 2570–2602. http://dx.doi.org/10.3390/su7032570

Yousefi, M., Khoramivafa, M., & Mondani, F. (2014). Integrated evaluation of energy use, greenhouse gas emissions and global warming potential for sugar beet (*Beta vulgaris*) agroecosystems in Iran. *Atmospheric Environment, 92*, 501–505. http://dx.doi.org/10.1016/j.atmosenv.2014.04.050

Yue, W., Liu, Y., & Fan, P. (2013). Measuring urban sprawl and its drivers in large Chinese cities: The case of Hangzhou. *Land Use Policy*, *31*, 358–370. http://dx.doi.org/10.1016/j.landusepol.2012.07.018

Zhao, X. (2011). Is global warming mainly due to anthropogenic greenhouse gas emissions? *Energy Sources Part A: Recovery, Utilization, and Environmental Effects*, *33*, 1985–1992. http://dx.doi.org/10.1080/15567030903515013

Zhou, D., Xiao, J., Bonafoni, S., Berger, C., Deilami, K., Zhou, Y., Frolking, S., Yao, R., Qiao, Z., & Sobrino, J. A. (2019).0 Satellite remote sensing of surface urban heat islands: Progress, challenges, and perspectives. *Remote Sensing*, *11*, 48. http://dx.doi.org/10.3390/rs11010048

Part III

Methods and Mapping Urban Heat Vulnerability

This final part of the book focuses on the tools and methods used to assess, analyse, and visualise urban heat vulnerability. It reviews established and emerging assessment approaches, traces key trends in heat vulnerability research, and introduces advanced techniques using machine learning, remote sensing, and artificial intelligence. The chapters demonstrate how these methods enable finer-grained, data-driven insights into spatial inequities, revealing who is most exposed and why. By integrating technological innovation with equity-focused analysis, this part shows how mapping can move beyond description to inform more targeted, inclusive, and actionable climate adaptation strategies. This part positions AI-enabled urban analytics as a critical asset for evidence-based and just climate decision-making.

DOI: 10.1201/9781003737049-8

6

Assessment Methods for Urban Heat Vulnerability

6.1 Introduction and Background

The United Nations' (UN) projections suggest the world population will reach about 9.7 billion inhabitants by the mid-century, where this figure is to exceed 11 billion by the end of the century (As Nações Unidas No Brasil). Along with this rapid population growth, urbanisation levels across the globe are also on the rise due to accelerated migration from rural to urban areas. In 1960, the world population residing in urban centres was only 33.6%; in 2018, this number increased to 55.7% (Dong et al., 2022). The UN estimates that by 2050, 60% of the world's population will reside in urban areas (ONU News). In many developing and developed regions of the world, such as Latin America and Europe, this number has already exceeded 75% (Albino et al., 2015; População Rural e Urbana|Educa|Jovens-IBGE).

Along with population growth and urbanisation, cities face problems such as sprawling urban expansion, inequitable distribution of infrastructure and amenities, environmental degradation and pollution, urban immobility, increased demand for energy, uneven distribution of water and food, and lack of basic sanitation, among others (Hussain et al., 2021; Mahbub et al., 2011; Yigitcanlar et al., 2007). Given the current situation and future projections, the need for cities to seek smart and sustainable solutions to deal with these challenges is imminent, especially in the era that gave rise to smart cities (Albino et al., 2015), where these offer an urban development blueprint in line with sustainable and knowledge-based urban development principles (Ballas, 2013; Yigitcanlar & Dur, 2013; Yigitcanlar et al., 2021).

Energy, sustainability, and smart cities are associated concepts that require an integrated approach in promoting the quality of life and sustainability for the world's population that today lives concentratedly in cities (Dur & Yigitcanlar, 2015; Kim et al., 2021; Mattoni et al., 2015). For humanity to continue to develop and prosper into the next generations, sustainable urban development with increased energy efficiency is necessary (Rosales Carreón & Worrell, 2018). In this context, energy efficiency can be understood as the rational and efficient use of energy in all stages of the process, from primary form to final consumption (Lovins, 2004). It is an essential tool with a relevant impact to reduce energy costs, but it depends on implementing public

DOI: 10.1201/9781003737049-9

policies with fiscal and financial incentives for its application in local management (Costa et al., 2021).

In the world energy perspective, energy consumption is an essential indicator of development for countries. However, a significant constraint on the economic development of smart cities is the cost of electricity (Gates, 2021), where the demand for energy increases as the population and its living conditions increase. The global energy consumption trend, between 1990 and 2017, is illustrated with data from the International Energy Agency (IEA) (Data & Statistics-IEA) in Figure 6.1. The financial crisis caused by the COVID-19 pandemic showed us how financial factors influence access to energy (Energy Agency). The IEA (Data & Statistics-IEA) estimated that, in 2020, 592 million people across the world did not have access to electricity. In 2019 (just before the beginning of the pandemic), this figure was 572 million, which shows a significant increase in inequality, going against the UN's Sustainable Development Goals (SDGs) (Energy Agency; Organização das Nações Unidas). This situation shows that the supply of affordable electricity is a requirement for cities' socioeconomic development (Gates, 2021; United Nations).

The solution to this imminent increase in the demand for electric energy cannot be given only by expanding energy production from the traditional fossil fuel resources. UN's SDG 7 deals with access to different energy sources, especially renewable, efficient, and non-polluting ones. SDG 7 highlights that development requires electricity to be clean and affordable (Organização das Nações Unidas). In addition to being affordable, electricity needs to be sustainable (Yigitcanlar, 2022; Yuksel & Kaygusuz, 2011). The COP 26 (2021 United Nations Conference on Climate Change) proposed reducing 50% of greenhouse gas emissions by 2030 (United Nations-Climate Change COP26

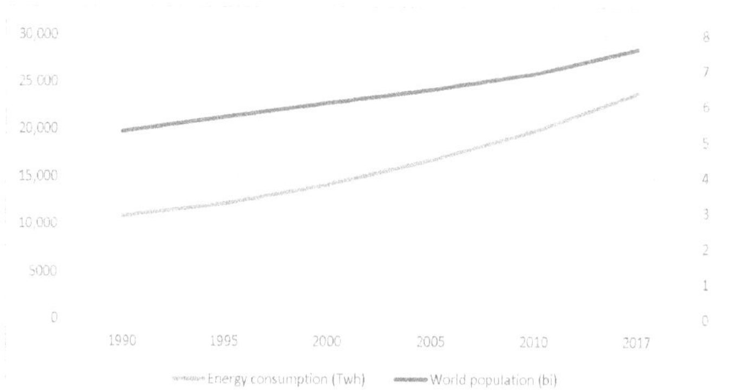

FIGURE 6.1
Global population (billion) and energy consumption (TWh) by year (1990–2017). (Derived from Data & Statistics-IEA.)

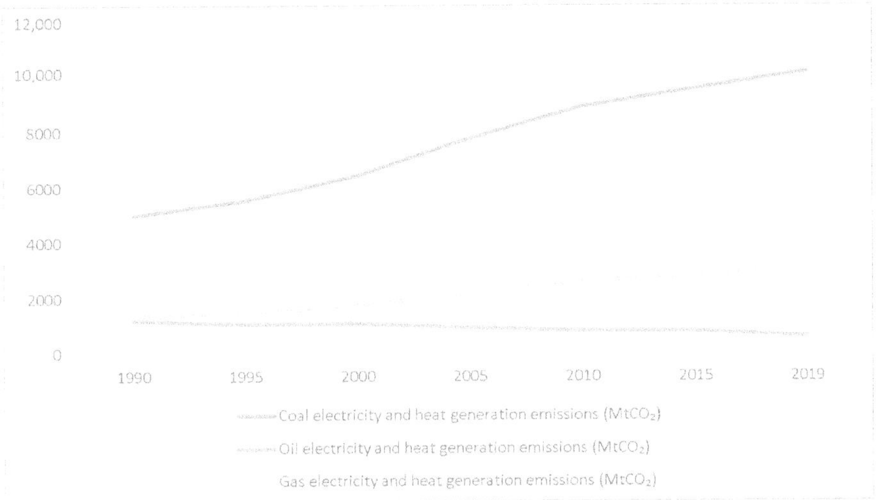

FIGURE 6.2
Global CO_2 emissions ($MTCO_2$) from electricity and heat by energy sources and year (1990–2019). (Derived from Data & Statistics-IEA.)

EXPLAINED). One of the industrial sectors that requires attention in this emission reduction plan is the energy production sector, which currently accounts for more than a quarter (27%) of total carbon equivalent emissions (Gates, 2021; United Nations-Climate Change COP26 EXPLAINED), illustrated in Figure 6.2.

6.1.1 Theoretical Background

The sixth report of the Intergovernmental Panel on Climate Change (IPCC) disclosed the increase in the planet's temperature caused by the emission of greenhouse gases, also highlighting the relationship of this warming with extreme climatic events (IPCC AR6). Scientific evidence alerts the public to the need to seek sustainable and less polluting solutions. One of the actions to reduce emissions in the energy production sector is to accelerate the replacement of coal, which represents 36% of the world's energy matrix, with clean energy, such as solar, wind, and tidal (United Nations-Climate Change COP26 EXPLAINED).

Replacing traditional fossil fuel with sustainable resources is not the only proposal to reduce the emission of greenhouse gases (Policy Solutions-Electricity; United Nations-Climate Change COP26 EXPLAINED). Another way of promoting energy sustainability is by applying technological resources to obtain energy efficiency (EP100|Climate Group the World's Energy-Smart Companies Committed to Doing More with Less Energy), a technique that does not change production but improves how energy is handled and utilised. These techniques also reduce losses in the distribution of electric energy, smart grids, the Internet of Things, and energy reduction for heating and cooling of buildings (EP100|Climate Group the World's Energy-Smart Companies Committed to Doing More with Less Energy; Tostar et al.).

The relationship between smart cities and sustainability has been examined recently. One of the leading studies for the topic answered the question 'can cities become smart without being sustainable?' and revealed that cities cannot be truly smart without being sustainable in the first place (Yigitcanlar et al., 2019). On a higher level, sustainability or its varied forms can become key indicators for measuring cities' smartness, such as conservation of natural resources (Jurevičienė & Biekšaitė, 2020) and environmental impact from energy use (Skvarciany et al., 2021). Smart cities are designed to promote sustainability, ensuring access to essential human development and well-being (Yigitcanlar et al., 2019). Smart cities and sustainability are intertwining concepts that can be infused in many aspects of urban living, such as in urban mobility (Yigitcanlar, 2022), urban planning (Pezzutto et al., 2017), and energy management (Sayah et al., 2020).

6.1.2 Aim and Contributions

Considering the growing demand for electric energy accompanied by the tremendous environmental impacts caused by its production, this study aims to seek the ways smart cities promote sustainability in the electricity sector. For this purpose, the study employed the systematic literature review method to map the scientific knowledge produced so far on the topic (Tranfield et al., 2003).

The main contributions of this study include the following:

- Being a pioneering systematic review that focuses on energy sustainability in smart cities;
- Offering the depiction of the academic publication landscape for smart city and energy sustainability research (Section 3.1);
- Providing a critical evaluation of geographical continents' energy use intensity vs. smart cities' energy sustainability research outcomes (Section 3.2);
- Identifying major research gaps in energy and urban planning areas (Section 3.2.3).

Following this introduction, Section 6.2 details the methodological approach of the chapter. Next, Section 6.3 reveals the results and offers a discussion on the results. Lastly, Section 6.4 presents the key findings and contributions of the study.

6.2 Research Design

In this study, a systematic literature review method is used to tackle the question of 'How do smart cities promote sustainability in the electricity sector?' (Tranfield et al., 2003). The PRISMA (Preferred Reporting Items for Systematic Reviews and Meta-Analyses) protocol is adopted to offer the replicability of the study (Page et al., 2021). The review is structured in three stages according to the methodology suggested by Yigitcanlar et al. (2019) and Regona et al. (2022). The three stages are: (a) Planning for developing the research aim, question, and search criteria; (b) conducting the review; and (c) reporting and dissemination stage.

The review study used the Web of Science and Scopus databases to collect documents, as they are the two databases with the most significant number of indexed journals on the subject studied (Martín-Martín et al., 2018). In the first stage, the study made a review planning by creating the inclusion and exclusion criteria. The keywords 'smart cities' and 'energy' were defined (considering all the terms with and without plural inflection) to construct the search expression to collect the materials. In this step, the inclusion criteria were defined: Only complete articles, available online, written in English, and relevant to the study. Finally, the exclusion criteria were defined: As described in Table 6.1, books, conference reports, editorials, articles in languages other than English, and grey literature (government documents, industrial reports, non-academic documents, or any other material that did not pass the inclusion criteria).

In the second stage, the study conducted a systematic literature review by following the PRISMA protocol. In this step, the materials were collected from the selected search bases, for which the keywords gave rise to the search expression (Title or Publication Title contains the terms ('smart city' OR 'smart cities') AND ('energy" OR "energies')). For the collection of

TABLE 6.1

Inclusion and Exclusion Criteria

Primary Data		Secondary Data	
Inclusionary	**Exclusionary**	**Inclusionary**	**Exclusionary**
Journal articles	Duplicate records	Energy in smart cities	Not related to energy sustainability
Peer-reviewed	Books and chapters	Energy sustainability	Not related to cities
Full-text available online	Industry reports	Relevant to the research objective	Irrelevant research objectives
Published in English	Government reports		
	Conferences		

TABLE 6.2
Selection Criteria

Selection Criteria
• Identify which continents promote energy sustainability.
• Determine how energy sustainability is promoted.
• Quantify the methodological approach of the analysed works.
• Analyse whether the theme is promising through a temporal analysis.
• Relevant categories are distributed and selected under the most pertinent categories.

materials, the authors did not establish an initial date, so that all materials published until 31 December 2021 were collected. The results of this search were tabulated; this initial sample contained 928 documents. After removing documents that were duplicated, in languages other than English, and non-academic documents, a sample of 226 articles was identified. The abstracts of the articles in this sample were read to separate only the articles that had elements capable of contributing to the answer to the guiding question according to the criteria in Table 6.2.

After the above step, 154 articles were selected (listed in Table 6.A1) to compose the final sample, which were read in full and had their information tabulated, and the articles were grouped according to their objective/central theme. The process for literature screening and selection is illustrated in Figure 6.3 following the PRISMA model (Page et al., 2021).

The results were presented in a literature review format in the third stage. In addition to the articles that made up the sample used in this work, complementary materials were selected for the composition, introduction, contextualisation, discussion, and conclusion. The results of this review are presented and discussed in Section 6.3.

6.3 Analysis and Results

6.3.1 General Observations

The research showed that despite being a relatively new topic, it continues to proliferate in academia, showing a scientific drive to seek solutions and innovation in the energy sustainability sector in cities. The initially selected literature pieces contained 928 documents. After applying the inclusion and exclusion criteria, a final list was created containing 154 articles in English published in scientific journals for analysis. The final list showed that 81% of the studies on energy sustainability in smart cities were published in the last four years (13% in 2018, 16% in 2019, 20% in 2020, and 32% in 2021). It was also

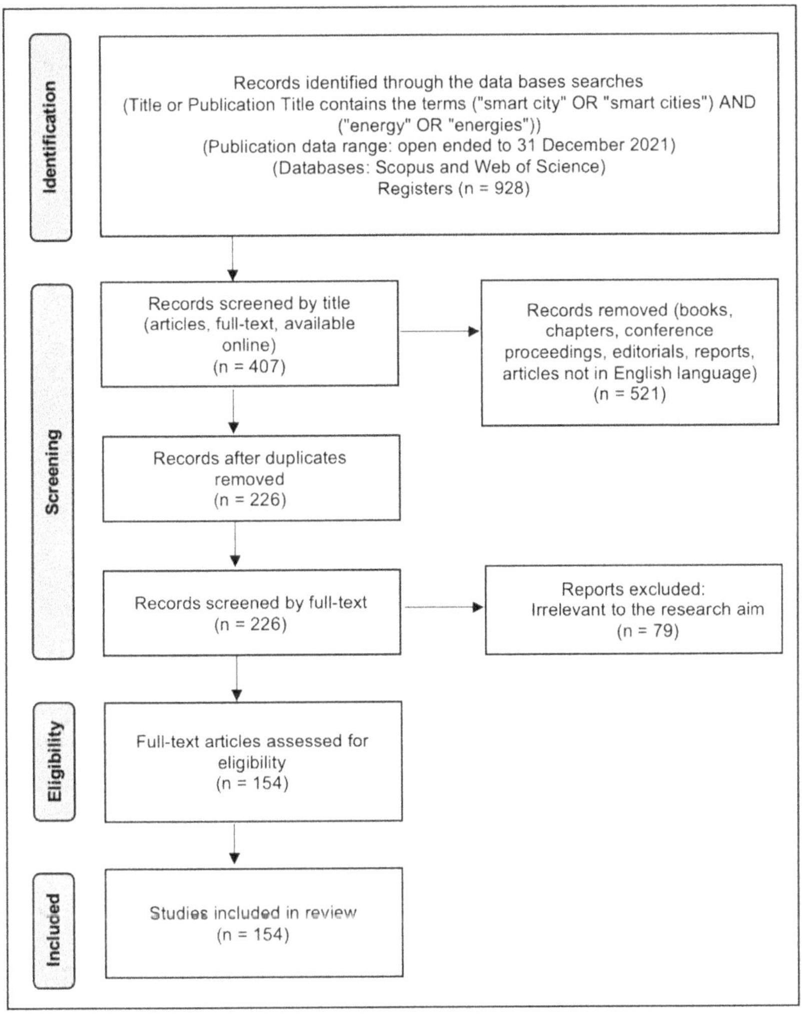

FIGURE 6.3
The PRISMA selection process of relevant literature.

notable that in 2012, there were two publications on the topic, and that the cumulative distribution of these publications until the end of 2016 was only 11%, as shown in Figure 6.4.

The analysis disclosed that the 154 selected articles are distributed among 97 journals, of which only 20 have two or more publications on the topic, grouping 51% of these articles, as shown in Table 6.1. Of these works, 13.6% were published in the journals 'Sustainable Energy Technologies and Assessments' and 'Energies', followed by the journals 'Sustainability'

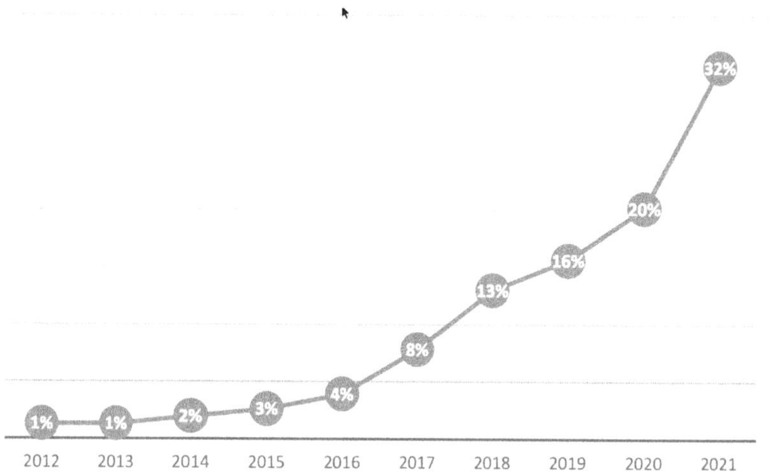

FIGURE 6.4
Distribution of publications by year (2012–2021).

and 'Sustainable Cities and Society' with 5 and 4%, respectively. The journals 'Smart Cities', 'IEEE Access', 'Sensors', 'IEEE Transactions on Industrial Informatics', 'Applied Energy', 'Journal of Cleaner Production', and 'Green Energy and Technology' published together 16% of the articles found on the theme. Despite having a distribution of materials in 97 journals, the research showed that the topic is still segmented with greater concentration in only 13% of the sample journals (12 journals with three or more articles about the subject since 2012) as shown in Table 6.3.

To understand whether the amount of academic scientific production is associated with the energy demand of a specific region, the number of publications per region was determined. This indicator was related to each continent's amount of electricity consumed (Data & Statistics-IEA). This analysis showed that Asia, the continent with the highest energy consumption, is the second continent that produces the most scientific research on the subject, behind Europe in terms of scientific production, representing 44 and 45% of the total sample, respectively. Next in energy consumption comes North America. However, the number of academic publications on the continent on the subject represents only 5% of the sample, and only a little more than half of these studies are of practical content. Europe is the continent that ranks third in terms of energy consumption. Next comes Oceania, Latin and Central America, and Africa, in that order. The energy consumption of these continents together represents less than the total consumption of North America, and the sum of the academic publications of these continents on the subject represents only 5% of the sample, as shown in Table 6.4. Electricity consumption figures are gathered from IEA statistics (Data & Statistics-IEA).

TABLE 6.3
Distribution of Publications by Journals and Years

Journal	2012	2013	2014	2015	2016	2017	2018	2019	2020	2021	Total
Energies	1	0	0	0	0	0	1	1	3	5	11
Sustainable Energy Technologies and Assessments	0	0	0	0	0	0	0	0	0	10	10
Sustainability	0	0	1	0	0	0	0	0	3	3	7
Sustainable Cities and Society	0	0	0	0	0	0	0	0	1	4	5
Smart Cities	0	0	0	0	0	0	0	0	2	2	4
IEEE Access	0	0	0	0	0	0	1	1	2	0	4
Sensors	0	0	0	0	2	1	0	0	0	0	3
IEEE Transactions on Industrial Informatics	0	0	0	0	0	1	3	3	0	0	7
Applied Sciences	0	0	0	0	0	0	0	0	1	1	2
Journal of Cleaner Production	0	0	0	0	0	0	1	1	0	1	3
Green Energy and Technology	0	0	1	0	1	1	0	0	0	0	3
Techne-Journal of Technology for Architecture and Environment	0	0	0	0	0	0	3	3	0	0	6
Energy and Buildings	0	0	0	0	0	0	0	0	1	1	2
Techne	0	0	0	0	0	0	1	1	0	1	3
Future Generation Computer Systems	0	0	0	0	0	0	0	0	1	0	1
Energy Policy	0	0	0	0	0	1	0	0	0	0	1
IEEE Communications Magazine	0	0	0	0	0	1	0	0	0	0	1
Energy, Sustainability and Society	0	0	0	0	1	1	0	0	0	0	2
Applied Energy	0	1	0	0	0	0	0	0	0	0	1
Expert Systems	0	0	0	0	0	0	0	0	1	1	2
Other	1	1	1	4	2	6	10	14	16	21	76
Total	2	2	3	4	6	12	20	24	31	50	154

6.3.2 Classification of Categories

With the increase in urbanisation and population, the energy demand also grows. Estimates from the UN predict that by 2050, the world population will reach 9.7 billion (Organização das Nações Unidas). This challenge demands the implementation of urban planning considering three categories

TABLE 6.4

Number of Publications by Continent, Methodology, and Energy Consumption

Continent	Number of Studies	Empirical	Theoretical	Electricity Consumption (TWh)
Asia	69	49	20	11,985.5
Europe	70	38	32	3837.9
North America	7	4	3	5056.2
Africa	3	2	1	732.4
Latin and Central America	3	3	0	1109.5
Oceania	2	1	1	1912.7

built from this systematic literature review: The energy issue (Haarstad & Wathne, 2019; Maltese et al., 2016; Pilipczuk, 2020), adoption of renewable energy sources (Abdullah et al., 2018Hoang & Nguyen, 2021), and actions on energy efficiency (Chui et al., 2018; Maltese et al., 2016; Yu & Zhang, 2019) for better handling and utilisation of available energy.

During the systematic literature review, the articles are grouped and analysed according to their central theme and research question, resulting in the construction of three categories. Together, these three categories are an integral part of energy and urban planning. Figure 6.5 illustrates the union of these categories, where energy sustainability is the central theme and underlines the main theme of this chapter, i.e., urban energy sustainability. This

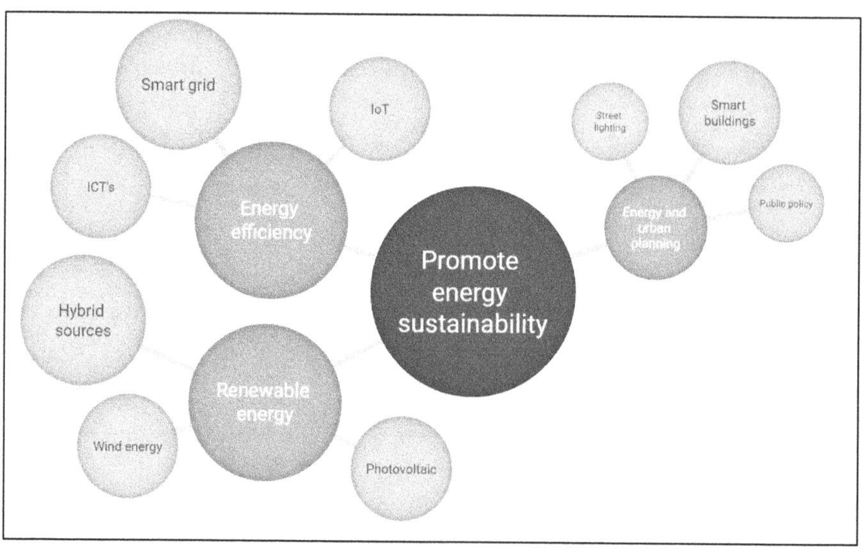

FIGURE 6.5
Main categories concerning sustainable energy in the context of smart cities.

theme is connected to the three observed categories: (a) Energy efficiency; (b) renewable energy; and (c) energy and urban planning. Figure 6.5 represents the connection between the central theme, energy sustainability, followed by the three themes that appeared the most in the articles, which, for categorisation, were treated as categories, together with their subcategories converging towards the established objective.

Figure 6.5 represents the proposition of UN, which declared the decade from 2014 to 2023 as the decade of 'Sustainable energy for all", contemplating universal access to energy, renewable energy, and energy efficiency as primary objectives (Sustainable Energy for All is Funded by Sustainable Energy for All). The stimulus is intense to expand the supply of renewable energy in the world energy matrix and ensure that energy promotes economic development in all countries. In this sense, sustainability and energy are directly related topics, and, as most of the world's population lives in cities, this research agenda integrates with the challenges within the scope of smart and sustainable cities (Philippi Cortese et al., 2017).

6.3.2.1 Energy Efficiency

From reading the articles, the review sought to understand the focus that each one had. The first category identified was 'Energy efficiency' due to the number of articles published (86 documents), equivalent to 55.8% of the selected literature pieces. Energy efficiency is characterised by actions and projects that use technology to reduce energy consumption or make this consumption process more efficient (Carrera et al., 2021; Shen et al., 2022).

Technology stands out as one of smart cities' leading and central dimensions (Caragliu et al., 2011; Batista & Storopoli). This dimension is also a protagonist in the energy efficiency category. Some of the main topics identified in these studies are smart grids, the Internet of Things (IoT), and information and communication technologies (ICTs).

Smart grids use technological resources for the efficient production and distribution of electricity, promoting energy security and diversifying production sources, enabling the use of more than one source of energy production in the same distribution network (Manshahia, 2018; Sato et al., 2019). In addition to promoting energy efficiency, smart grids have emerged as elements for creating resilient cities, as they ensure diversity in energy production sources and diversification of distribution routes (Konstantinou, 2022).

Due to the wide range of power transmission and distribution lines, IoT resources emerge as vital allies in promoting energy efficiency, enabling the monitoring of quality of service, and assisting in decision-making (Zhang & Yue, 2021; Zhang et al., 2021). Associated with the use of IoT, the use of ICTs also appears as a necessary resource for the implementation of IoT (Duan et al., 2021; Kumar Garg et al., 2020). The emergence of 5G technology favours the use of IoT in the energy efficiency sector, enabling the connection of the

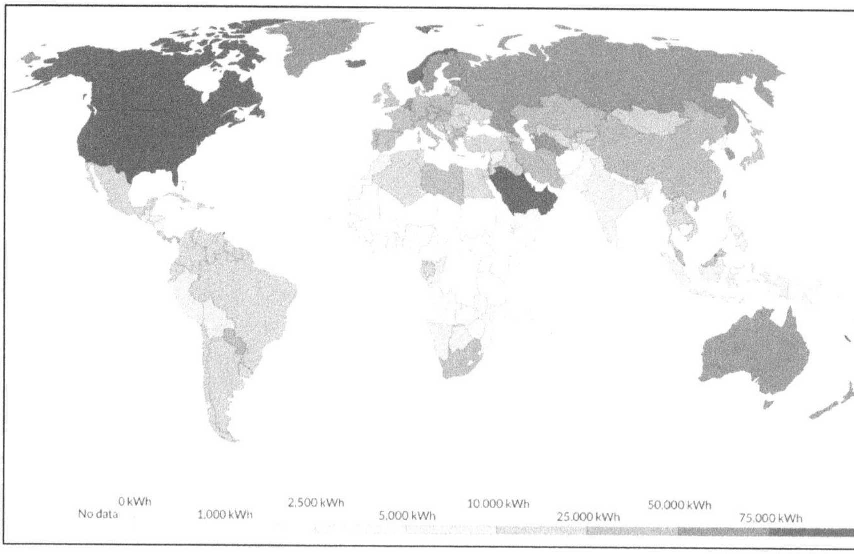

FIGURE 6.6
Per capita global energy consumption (kWh) in 2019. (Derived from Data & Statistics-IEA.)

resource in more remote areas and with better connection quality and speed (Sánchez-Cano et al., 2021; Yigitcanlar et al., 2020).

An IEA (Data & Statistics-IEA) survey pointed out a jump in electricity consumption in the world, which went from 2.1 MWh/per capita in 1990 to 3.3 MWh/per capita in 2018, an increase of 57% in just 28 years. With the population increase projected for the next decades (United Nations), there is an imminent need to produce more electricity to meet this demand—energy consumption figures are illustrated in Figure 6.6.

This research unveiled that the continent with the highest per capita consumption is not necessarily the most advanced academic research on the subject. North America, a continent with an average consumption of 10.3 MWh/capita, published only seven articles in this category, representing 5% of the total sample. On the other hand, Asia and Europe, which occupy the second and third positions of the continents with the highest consumption of electric energy per capita, are the ones that produced the most research on the subject. Asia published 70 articles, representing 45% of the total sample. Europe appears to be tied with Asia in the number of publications. Most of the works developed by Asian and European authors are of an empirical, methodological nature, representing 71 and 54%, respectively, whereas 57% North American publications are of an empirical nature.

On the energy efficiency issue, on the one hand, energy efficiency has contributed to the promotion of sustainability in smart cities through its innovations and the use of technology; on the other hand, its specific actions and independent projects alone are not sufficient to guarantee significant changes in the consumption patterns or behaviour of the entire society. Another critical point of this category is the concentration of studies on the technology dimension, which demonstrates that smart cities remain 'stuck' in their stereotype without considering socioeconomic issues that directly affect the life and behaviour of the population.

In sum, promoting energy sustainability is possible by applying appropriate technological solutions to improve energy efficiency, but this should be implemented along with other measures, such as investing in renewable energy resources.

6.3.2.2 Renewable Energy

The second category is 'Renewable energy', which is present in 36 documents, equivalent to 23.3% of the selected literature. Renewable energy corresponds to natural energy sources; they regenerate naturally in each cycle (Deakin & Reid, 2018; Govindarajan & Ganesh, 2021). Despite being a global commitment established in international agreements (Kylili & Fokaides, 2015; United Nations-Climate Change COP26 EXPLAINED; UNFCCC Paris Agreement Text English), the production of scientific research on renewable energy in smart cities is still limited. Asia is the continent that produced the most research on the topic, with 23 articles in the sample, followed by Europe with 15 and North America and Oceania, with only one article each.

Sustainability is one of the structuring axes of research on smart cities (Alawadhi et al., 2012; Batista & Storopoli; Caragliu et al., 2011). In this context, environmental sustainability has been the focus of discussions and commitments by several countries due to the urgency of measures to restore ecosystems and mitigate climate change (Organização das Nações Unidas). Renewable energy contributes significantly to environmental sustainability. Among the main topics identified in the sample, studies are mainly in the areas of solar energy, wind energy, tidal energy, and biomass energy.

The world energy matrix is composed of different resources to produce electric energy, using the most viable resource for each region and the combination of these resources (Gates, 2021; Kuznetsov et al., 2021; Mahmood et al., 2021), as illustrated in Figure 6.7. Despite a growth in the use of renewable resources such as photovoltaic (PV) and wind energy, the electricity generated from fossil fuels is still the majority in the world. They are also the ones that release the most greenhouse gases into the atmosphere, as illustrated in Figure 6.8 (Data & Statistics-IEA).

In this category, most studies refer to hybrid systems, which use two or more sources for energy production. This predominance occurs due to the natural variation of renewable resources, such as fluctuations in solar irradiance and wind speeds (Gates, 2021; Kuznetsov et al., 2021; Mahmood et al., 2021). With the increased installation of PV panels at homes and buildings, locally generated renewable electricity has increased for local use, making

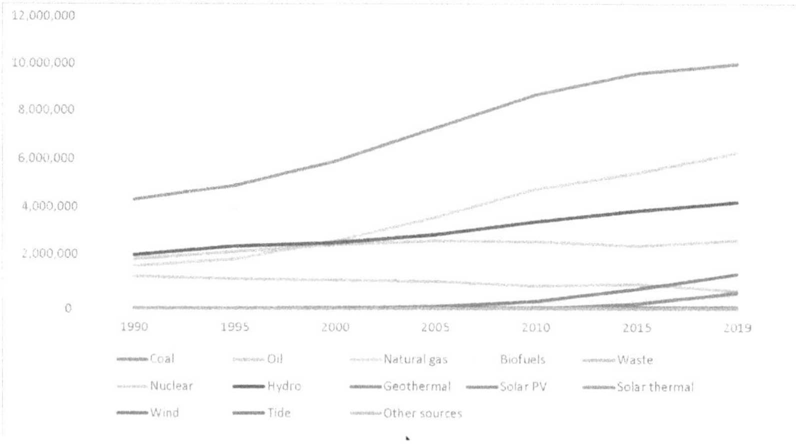

FIGURE 6.7
Global electricity generation by source (GWh) and year (1990–2019). (Derived from Data & Statistics-IEA.)

it possible to adopt a renewable source of energy without depending solely on utility and network companies responsible for the supply of electricity (Meenakshi et al., 2021). With the popularisation of electric vehicles, the concern of some cities also arises in providing clean energy for recharging these vehicles (Lotfi et al., 2022).

On the renewable energy issue, we consider this category the most urgent and promising of all. Nonetheless, renewable energies still lack investments for their development and expansion. Likewise, the results show that research on this topic is still incipient. Even in smart cities, the lack of investments, incentives, and public policies makes switching to renewable energies very

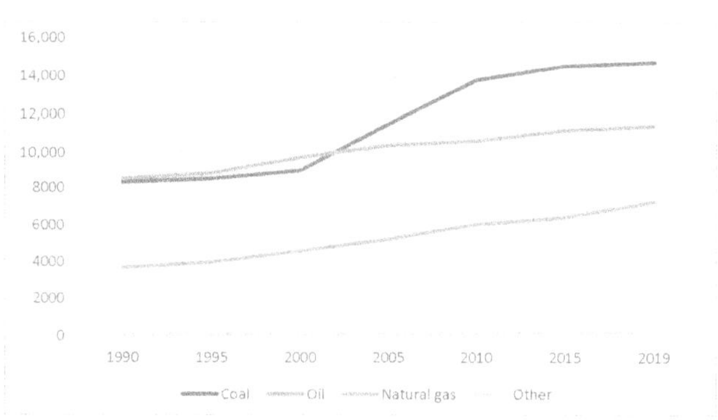

FIGURE 6.8
Global CO_2 emissions in electric energy production by source and year (1990–2019). (Derived from Data & Statistics-IEA.)

slow. Another critical point related to this category is the economic aspect, traditional companies in the energy sector that use polluting sources represent true economic powers and profit a lot from inefficient and destructive energy generation.

In sum, the generation of clean energy from renewable energy resources has been gaining popularity in recent years. Renewable energy is a critical and integral part of the smart and sustainable cities agenda. Nonetheless, renewable energy with respect to urban planning is still an area for further growth for smart cities to deliver their desired sustainable energy goals.

6.3.2.3 ***Energy and Urban Planning***

The 'Energy and urban planning' category was the third identified and consisted of planning, management, creation, and development of public policies, actions, and strategies focused on the development of urban energy sustainability (Haarstad & Wathne, 2019; Krutwig et al., 2018; Maltese et al., 2016).

The issue of energy and urban planning was identified as a significant gap in this study, present in 32 of the studies analysed, representing 20.7% of the sample. With only a few studies as a reference, this category should be the basis or the starting point for energy sustainability in smart cities. Without adequate planning, SDGs and the agreements signed at the COP26 in 2021 on the energy matrix would be lost in goals and promises that will not be achievable for several countries (Organização das Nações Unidas).

Promoting energy sustainability also comes from creating public policies for urban planning, sustainable urban development, and management instruments. These policies should focus on access to electricity and energy security, ensuring the availability and reliability of the resource, and sustainability in production (Biancardi et al., 2021; Martín et al., 2021). For this, urban planning must consider elements such as efficient and sustainable public lighting (Chen et al., 2022) and the construction of smart homes, buildings, and neighbourhoods capable of promoting energy efficiency and renewable energy sources.

One of the essential strategies to reduce energy consumption and increase efficiency is to implement technological innovation, such as automation of lighting systems using presence sensors and timers, materials with thermal comfort consideration, and automation of temperature control systems for air conditioners. In new constructions, this discussion needs to become mandatory in the design phase. Furthermore, retrofitting is an effective and economical option to combine innovation and sustainability for existing and old buildings (Philippi Cortese et al., 2017).

On the energy and urban planning aspect, we consider that smart cities demand the improvement of the energy systems, regarding production, distribution, and consumption in public urban planning policies to achieve significant changes and proceed with the behavioural transformation of the whole society to a sustainable model (Lewandowska et al., 2020). The great challenge is how to maintain energy supply and promote energy sustainability in these cities to allow continued economic development, ensuring an ecologically balanced environment (Hoang & Nguyen, 2021). Another critical point for this change is the integration and commitment of different actors in the society, such as governments, public institutions, non-governmental institutions, private companies, and civil societies (Nasr et al., 2020).

In sum, energy, sustainability, and smart cities are associated concepts and require an integrated approach in promoting urban sustainability and quality of life of the world's population that today live concentratedly in cities. Urban planning could provide such integration. Nevertheless, there is a high concentration on the technology dimension of energy sustainability and efficiency, and renewables in the reviewed literature, but much less attention is paid to the most needed symbiosis between energy and urban planning (Lenhart et al., 2015).

6.3.3 A Conceptual Model and Further Discussion

Based on the systematic literature review discoveries, a conceptual model for energy sustainability in smart cities was developed, and the model is presented in Figure 6.9. Furthermore, this section extends the discussion and presents potential implications for researching energy sustainability in smart cities.

As shown in Figure 6.9, energy efficiency, renewable energy, and energy and urban planning are pillars of energy sustainability in smart cities. They are integral components and together support the sustainable development of smart cities. Energy efficiency runs through the whole energy cycle, from generation to distribution to utilisation at the users' end. Renewable

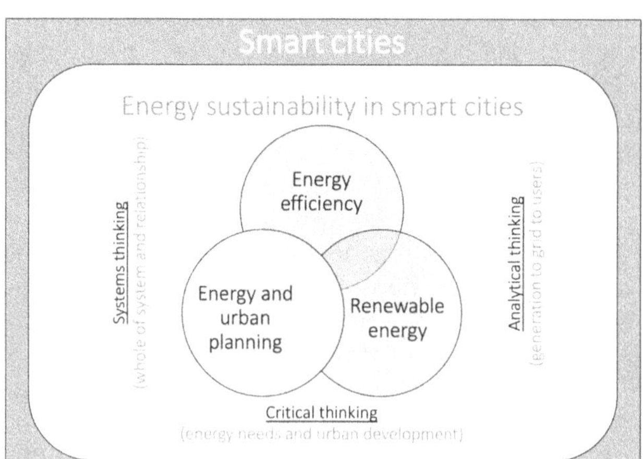

FIGURE 6.9
Conceptual model for energy sustainability in smart cities.

energy supports sustainability by generating clean and sustainable energy for smart cities. In the context of smart cities, energy and urban planning can be enabling and overarching factors to ensure energy sustainability. Implications and examples of the three components are discussed in the following paragraphs.

Implications for energy efficiency in smart cities: Energy efficiency is often of the highest priority in terms of resource conservation because it often has a permanent impact and cuts down the need to use energy, hence reducing the need to consume resources (Miller et al., 2012; Palensky & Dietrich, 2011). When sustainability becomes an inherent part of smart cities, urban energy efficiency would also become a key enabler and indicator for smart cities, for example, passive design for buildings (Miller et al., 2012) and innovative technology implementation for the urban environment (Seifhashemi et al., 2018). However, as Section 3.2.1 revealed, energy efficiency may not be the silver bullet for all energy sustainability needs. There is a need to enable more renewable energy in smart cities. This is critical as the forecasts are for the future of urban mobility being electric, hence there is investment needed for clean energy–fed electromobility (Vilathgamuwa et al., 2022).

Implications for renewable energy in smart cities: There are various renewable energy forms that can be utilised in the urban environment. One of the most common renewable energy technologies is a solar PV system. In smart and sustainable cities, PV can provide renewable energy in a distributed manner, such as on rooftops, carparks, and building façades (Liu et al., 2020; Miller et al., 2018). Another common form of renewable energy is wind, which needs more focused study on how wind resources can be integrated in smart cities for energy generation and cooling the built environment (He et al., 2020; Shahizare et al., 2016).

Implication for energy and urban planning for smart cities: In smart cities, urban planning may lay the foundation to enable energy sustainability without consuming extra resources and avoid pitfalls that may happen if planning has not first been confirmed. A real-world example can be given as follows: No dark roof is allowed in the western region of Sydney (Australia Net Zero), which is a low to no-risk option to mitigate the impact of urban heat islands and reduce electric energy use by natural ways of lowering solar heat gain for buildings.

A wide area study of 34 cities in Poland confirmed that sustainable development is at the core of urban management (Hajduk, 2016). This study did not focus on the energy or electricity sector; however, it highlighted that urban planning is a crucial factor to enable sustainable development.

Urban planning can offer more opportunities in enabling energy sustainability in cities, such as population density planning for net zero goals, building height limits for utilising solar energy, and urban forms for wind utilisation (Perera et al., 2019; Wang et al., 2022). On the other hand, energy sustainability supports sustainable urban development in the context of smart cities.

6.4 Conclusion

The trend of rapidly increasing human population creates a high demand for energy, particularly in large metropolitan cities, which poses a major threat to our future energy security (Lantz et al., 2021; Ingrao et al., 2018; Li et al., 2021; Madlener & Sunak, 2011). Given that this trend is already coupled with anthropogenic climate change externalities, the magnitude of this problem is colossal, which makes it a serious challenge for the authorities to tackle (Mortoja & Yigitcanlar, 2020). This sustainable energy challenge is of the utmost importance in meeting basic human needs, improving the quality of life, and avoiding contributing to further climate catastrophes (Gunningham, 2019).

Smart cities, during the last decade and a half, have offered some invaluable directions (at least at the theoretical level) for addressing this problem (Orecchini et al., 2019). The smart city energy solutions focus on establishing: (a) Energy efficiency; (b) energy sustainability, including utilisation of renewable energy resources; and (c) energy and urban planning, including the utilisation of smart grids, smart homes/buildings, smart urban and energy planning, electric vehicles, and so on (Liu et al., 2021; Lund et al., 2017; Maier et al., 2019; Yigitcanlar & Cugurullo, 2020; Yigitcanlar & Kamruzzaman, 2019).

While there is substantial literature with solutions in each of these areas, there are only limited comprehensively conducted review studies that shed light on the big picture for understanding sustainable energy in the context of smart cities (Kamyab et al., 2020; Lewandowska et al., 2020; Mosannenzadeh et al., 2017). To address this gap in the literature, this chapter contributed to mapping out the academic studies that investigate the ways smart cities promote sustainability in the electricity sector. For this purpose, the study adopted a PRISMA protocol for the systematic literature review for the transparency and replicability of the review.

Energy, sustainability, and smart cities are associated concepts and require an integrated approach in promoting urban sustainability and quality of life of the world's population that today lives concentratedly in cities. The sustainable energy issue is a critical and integral part of the smart and sustainable cities agenda. Promoting energy sustainability is possible by applying appropriate technological solutions to obtain energy efficiency, but this should be implemented along with investing in renewable energy resources.

The major findings of this review include: (a) The academic publication landscape for smart city and energy sustainability research; (b) unbalanced publications when critically evaluating geographical continents' energy use intensity vs. smart cities' energy sustainability research outcomes; and (c) heavy concentration on the technology dimension of energy sustainability and efficiency, and renewables topics in the literature, but much less attention is paid to the energy and urban planning issues.

These insights inform urban and energy authorities in understanding sustainable energy in the context of smart cities and provide urban and energy scholars with directions for prospective research to tackle the wicked sustainable energy problems of our cities and societies. The focus of this chapter is limited to smart cities and energy research. Potential future research may include energy in practical smart city cases, smart communities, smart homes, or specific areas, such as connectivity and digital features.

Acknowledgements

This chapter, with permission from the copyright holder, is a reproduced version of the following journal article: Li, F., Yigitcanlar, T., Nepal, M., Thanh, K., & Dur, F. (2022). Understanding urban heat vulnerability assessment methods: a PRISMA review. *Energies*, 15(19), 6998.

References

Abbas, S., Khan, M. A., Eduardo Falcon-Morales, L., Rehman, A., Saeed, Y., Zareei, M., Zeb, A., & Mohamed, E. (2020). Modeling, simulation and optimization of power plant energy for IoT enabled smart cities empowered with deep extreme machine. *IEEE Access, 8*, 39982–39997. https://doi.org/10.1109/ACCESS.2020.2976452.

Abdel-Basset, M., Hawash, H., Chakrabortty, R. K., & Ryan, M. (2021). Energy-net: a deep learning approach for smart energy management in IoT-based smart cities. *IEEE Internet of Things Journal, 8*, 12422–12435. https://doi.org/10.1109/JIOT.2021.3063677.

Abdullah, A., Yusoff, S. H., Zaini, S. A., Midi, N. S., & Mohamad, S. Y. (2019). Energy efficient smart street light for smart city using sensors and controller. *Bulletin of Electrical Engineering and Informatics, 8*, 558–568. https://doi.org/10.11591/eei.v8i2.1527.

Abdullah, M. A., Al-Hadhrami, T., Tan, C. W., & Yatim, A. H. (2018). Towards green energy for smart cities: particle swarm optimization based MPPT approach. *IEEE Access, 6*, 58427–58438. https://doi.org/10.1109/ACCESS.2018.2874525.

Abu-Rayash, A., & Dincer, I. (2021). Development and analysis of an integrated solar energy system for smart cities. *Sustainable Energy Technologies and Assessments, 46*, 101281. https://doi.org/10.1016/j.seta.2021.101170.

Ahuja, K., & Khosla, A. (2019). Network selection criterion for ubiquitous communication provisioning in smart cities for smart energy system. *Journal of Network and Computer Applications, 127*, 82–91. https://doi.org/10.1016/j.jnca.2018.11.011.

Alawadhi, S., Aldama-Nalda, A., Chourabi, H., Gil-Garcia, J. R., Leung, S., Mellouli, S., Nam, T., Pardo, T. A., Scholl, H. J., & Walker, S. *Building understanding of smart City initiatives*; Springer: Berlin/Heidelberg, Germany, 2012; pp. 40–53.

Albino, V., Berardi, U., & Dangelico, R. M. (2015). Smart cities: Definitions, dimensions, performance, and initiatives. *Journal of Urban Technology, 22*, 3–21. https://doi.org/10.1080/10630732.2014.942092.

Al-Nory, M. T. (2019). Optimal decision guidance for the electricity supply chain integration with renewable energy: aligning smart cities research with sustainable development goals. *IEEE Access, 7*, 74996–75006. https://doi.org/10.1109/ACCESS.2019.2919408.

Andreucci, M. B. (2018). Linking future energy systems with heritage requalification in smart cities. On-going research and experimentation in the City of Trento (IT). *Techne, 1*, 87–91. https://doi.org/10.13128/Techne-22824.

Angelakoglou, K., Kourtzanidis, K., Giourka, P., Apostolopoulos, V., Nikolopoulos, N., & Kantorovitch, J. (2020). From a comprehensive pool to a project-specific list of key performance indicators for monitoring the positive energy transition of smart cities—An experience-based approach. *Smart Cities, 3*, 705–735. https://doi.org/10.3390/smartcities3030036.

Anthony Jnr, B. (2020). Smart city data architecture for energy prosumption in municipalities: Concepts, requirements, and future directions. *International Journal of Green Energy, 17*, 827–845. https://doi.org/10.1080/15435075.2020.1791878.

Anthony Jnr, B., Abbas Petersen, S., Ahlers, D., & Krogstie, J. (2020). API deployment for big data management towards sustainable energy prosumption in smart cities—A layered architecture perspective. *International Journal of Sustainable Energy, 39*, 263–289. https://doi.org/10.1080/14786451.2019.1684287.

Arif, A., Barrigon, F. A., Gregoretti, F., Iqbal, J., Lavagno, L., Lazarescu, M. T., Ma, L., Palomino, M., & Segura, J. L. L. (2020). Performance and energy-efficient implementation of a smart city application on FPGAs. *Real-Time Image Processing, 17*, 729–743.

As Nações Unidas No Brasil. (2020). População Mundial Deve Chegar a 9,7 Bilhões de Pessoas Em 2050, Diz Relatório Da ONU. As Nações Unidas No Brasil. Available online: https://brasil.un.org/pt-br/83427-populacao-mundial-deve-chegar-97-bilhoes-de-pessoas-em-2050-diz-relatorio-da-onu (accessed on 12 February 2022).

Ashwin, M., Alqahtani, A. S., & Mubarakali, A. (2021). IoT based intelligent route selection of wastage segregation for smart cities using solar energy. *Sustainable Energy Technologies and Assessments, 46*. https://doi.org/10.1016/j.seta.2021.101281.

Aslam, S., Hasan, N. U., Jang, J. W., & Lee, K. G. (2016). Optimized energy harvesting, cluster-head selection and channel allocation for IoTs in smart cities. *Sensors, 16*. https://doi.org/10.3390/s16122046.

Aujla, G. S., Kumar, N., Singh, M., & Zomaya, A. Y. (2019). Energy trading with dynamic pricing for electric vehicles in a smart city environment. *Journal of Parallel and Distributed Computing, 127*, 169–183. https://doi.org/10.1016/j.jpdc.2018.06.010.

Australia Net Zero. (2021). NSW to Ditch Dark-Coloured Roofs to Prevent Heat Islands. Available online: https://www.afr.com/property/commercial/nsw-to-ditch-dark-coloured-roofs-to-prevent-heat-islands-20211117-p599ok (accessed on 1 March 2022).

Avotins, A., Adrian, L. R., Porins, R., Apse-Apsitis, P., & Ribickis, L. (2021). Smart City Street lighting system quality and control issues to increase energy efficiency and safety. *The Baltic Journal of Road and Bridge Engineering, 16*, 28–57. https://doi.org/10.7250/bjrbe.2021-16.538.

Aymen, F., & Mahmoudi, C. (2019). A novel energy optimization approach for electrical vehicles in a smart City. *Energies, 12*, 929. https://doi.org/10.3390/en12050929.

Babar, M., Khattak, A. S., Jan, M. A., & Tariq, M. U. (2021). Energy aware smart city management system using data analytics and internet of things. *Sustainable Energy Technologies and Assessments, 44,* 100992. https://doi.org/10.1016/j.seta.2021.100992.

Bachanek, K. H., Tundys, B., Wiśniewski, T., Puzio, E., & Maroušková, A. (2021). Intelligent street lighting in a smart city concepts—A direction to energy saving in cities: an overview and case study. *Energies, 14,* 3018. https://doi.org/10.3390/en14113018.

Ballas, D. (2013). What makes a 'Happy City'? *Cities, 32,* S39–S50. https://doi.org/10.1016/J.CITIES.2013.04.009.

Barresi, A. (2018a). Urban densification and energy efficiency in smart cities: The verge project (Switzerland). *Advances in Modelling and Analysis A, 55,* 173–176.

Barresi, A. (2018b). Urban densification and energy efficiency in smart cities—The VerGe (Switzerland). *Journal of Technology for Architecture and Environment,* 28–32. https://doi.org/10.13128/Techne-22713.

Batista, G., & Storopoli, E. (2021). Planejamento de Cidades Inteligentes: Uma Análise Dos Planos Diretores Dos Municípios Brasileiros. Available online: http://bibliotecatede.uninove.br/handle/tede/2713 (accessed on 14 March 2022).

Battista, G., Evangelisti, L., Guattari, C., Basilicata, C., & de Lieto Vollaro, R. (2014). Buildings energy efficiency: Interventions analysis under a smart cities approach. *Sustainability, 6,* 4694–4705. https://doi.org/10.3390/su6084694.

Bhati, A., Hansen, M., & Chan, C. M. (2017). Energy conservation through smart homes in a smart city: A lesson for Singapore households. *Energy Policy, 104,* 230–239. https://doi.org/10.1016/j.enpol.2017.01.032.

Biancardi, M., di Bari, A., & Villani, G. (2021). R&D investment decision on smart cities: Energy sustainability and opportunity. *Chaos Solitons Fractals, 153,* 111554. https://doi.org/10.1016/j.chaos.2021.111554.

Boeri, A., Boulanger, S. O. M., Turci, G., & Pagliula, S. (2021). Enabling strategies for mixed-used PEDs: Energy efficiency between smart cities and Industry 4.0. *Techne, 22,* 170–180. https://doi.org/10.36253/techne-10594.

Brundu, F. G., Patti, E., Osello, A., del Giudice, M., Rapetti, N., Krylovskiy, A., Jahn, M., Verda, V., Guelpa, E., & Rietto, L. et al (2017). IoT software infrastructure for energy management and simulation in smart cities. *IEEE Transactions on Industrial Informatics, 13,* 832–840. https://doi.org/10.1109/TII.2016.2627479.

Calvillo, C. F., Sanchez-Miralles, A., & Villar, J. (2018). Synergies of electric urban transport systems and distributed energy resources in smart cities. *IEEE Transactions on Intelligent Transportation Systems, 19,* 2445–2453. https://doi.org/10.1109/TITS.2017.2750401.

Caponio, G., Massaro, V., Mossa, G., & Mummolo, G. (2015). Strategic energy planning of residential buildings in a smart City: a system dynamics approach. *International Journal of Engineering Business Management, 7,* 61768. https://doi.org/10.5772/61768.

Caragliu, A., del Bo, C., & Nijkamp, P. (2011). Smart cities in Europe. *Journal of Urban Technology, 18,* 65–82. https://doi.org/10.1080/10630732.2011.601117.

Carli, R., Dotoli, M., & Pellegrino, R. (2017). A hierarchical decision-making strategy for the energy management of smart cities. *IEEE Transactions on Automation Science and Engineering, 14,* 505–523. https://doi.org/10.1109/TASE.2016.2593101.

Carrera, B., Peyrard, S., & Kim, K. (2021). Meta-regression framework for energy consumption prediction in a smart: a case study of songdo in South Korea. *Sustainable Cities and Society, 72,* 103025. https://doi.org/10.1016/j.scs.2021.103025.

Causone, F., Sangalli, A., Pagliano, L., & Carlucci, S. (2018). Assessing energy performance of smart cities. *Building Services Engineering Research & Technology, 39,* 99–116. https://doi.org/10.1177/0143624417725220.

Chen, L., & Han, P. (2021). The construction of a smart city energy efficiency management system oriented to the mobile data aggregation of the internet of things. *Complexity, 2021,* 9988282. https://doi.org/10.1155/2021/9988282.

Chen, Y., Ardila-Gomez, A., & Frame, G. (2017). Achieving energy savings by intelligent transportation systems investments in the context of smart cities. *Transportation Research Part D: Transport and Environment, 54,* 381–396. https://doi.org/10.1016/j.trd.2017.06.008.

Chen, Z., Sivaparthipan, C. B., & Muthu, B. A. (2022). IoT based smart and intelligent smart city energy optimization. *Sustainable Energy Technologies and Assessments, 49,* 101724. https://doi.org/10.1016/j.seta.2021.101724.

Chithaluru, P., Al-Turjman, F., Kumar, M., & Stephan, T. (2020). I-AREOR: An energy-balanced clustering protocol for implementing green IoT in smart cities. *Sustainable Cities and Society, 61,* 102254. https://doi.org/10.1016/j.scs.2020.102254.

Cho, K., Yang, J., Kim, T., & Jang, W. (2021). Influence of building characteristics and renovation techniques on the energy-saving performances of EU smart city projects. *Energy Build, 252,* 111477. https://doi.org/10.1016/j.enbuild.2021.111477.

Chui, K. T., Lytras, M. D., & Visvizi, A. (2018). Energy sustainability in smart cities: artificial intelligence, smart, and optimization of energy consumption. *Energies, 11,* 2869. https://doi.org/10.3390/en11112869.

Cioara, T., Anghel, I., Salomie, I., Antal, M., Pop, C., Bertoncini, M., Arnone, D., & Pop, F. (2019). Exploiting data centres energy flexibility in smart cities: Business scenarios. *Information Sciences, 476,* 392–412. https://doi.org/10.1016/j.ins.2018.07.010.

Colmenar-Santos, A., Molina-Ibáñez, E. L., Rosales-Asensio, E., & López-Rey, Á (2018). Technical approach for the inclusion of superconducting magnetic energy storage in a smart City. *Energy, 158,* 1080–1091. https://doi.org/10.1016/j.energy.2018.06.109.

Condotta, M., & Borga, G. (2018). Urban energy performance monitoring for smart city decision support. *TECHNE-Journal of Technology for Architecture and Environment,* 73–80. https://doi.org/10.13128/Techne-22688.

Corsini, F., Certomà, C., Dyer, M., & Frey, M. (2019). Participatory energy: Research, imaginaries and practices on people' contribute to energy systems in the smart city. *Technological Forecasting and Social Change, 142,* 322–332. https://doi.org/10.1016/j.techfore.2018.07.028.

Costa, J., dos, S., & Andrade Junior, L. M. L. (2021). de Eficiência Energética Aplicada Ao Consumo de Eletricidade: Um Estudo de Revisão Bibliográfica. *Research, Society and Development, 10,* e26210414085. https://doi.org/10.33448/rsd-v10i4.14085.

Cui, J., Yoon, H., & Youn, B. D. (2018). An omnidirectional biomechanical energy harvesting (OBEH) sidewalk block for a self-generative power grid in a smart city. *International Journal of Precision Engineering and Manufacturing, 5,* 507–517. https://doi.org/10.1007/s40684-018-0054-1.

Data & Statistics-IEA. (2021). Available online: https://www.iea.org/data-and-statistics/data-browser?country=WORLD&fuel=Energy%20consumption&indicator=TotElecCons (accessed on 12 February 2022).

Deakin, M., & Reid, A. (2018). Smart cities: Under-gridding the sustainability of city-districts as energy efficient-low carbon zones. *Journal of Cleaner Production, 173*, 39–48. https://doi.org/10.1016/j.jclepro.2016.12.054.

Deakin, M., Reid, A., & Mora, L. (2020). Smart cities: The metrics of future internet-based developments and renewable energies of urban and regional innovation. *Journal of Urban Technology, 27*, 59–78. https://doi.org/10.1080/10630732.2020.1868738.

Dong, F., Li, Y., Li, K., Zhu, J., & Zheng, L. (2022). Can smart city construction improve urban ecological total factor energy efficiency in China? Fresh evidence from generalized synthetic control method. *Energy, 241*, 119976. https://doi.org/10.1016/J.ENERGY.2021.122909.

Duan, P., Askari, M., Hemat, K., & Ali, Z. M. (2021). Optimal operation and simultaneous analysis of the electric transport systems and distributed energy resources in the smart City. *Sustainable Cities and Society, 75*, 103306. https://doi.org/10.1016/j.scs.2021.103306.

Dur, F., & Yigitcanlar, T. (2015). Assessing land-use and transport integration via a spatial composite indexing model. *International Journal of Environmental Science and Technology, 12*, 803–816. https://doi.org/10.1007/s13762-013-0476-9.

Ejaz, W., Naeem, M., Shahid, A., Anpalagan, A., & Jo, M. (2017). Efficient energy management for the internet of things in smart cities. *IEEE Communications Magazine, 55*, 84–91. https://doi.org/10.1109/MCOM.2017.1600218CM.

Elkamel, M., Ahmadian, A., Diabat, A., & Zheng, Q. P. (2021). Stochastic optimization for price-based unit commitment in renewable energy-based personal rapid transit systems in sustainable smart cities. *Sustainable Cities and Society, 65*, 346–354. https://doi.org/10.1016/j.scs.2020.102618.

Energy Agency, I. (2020). World Energy Outlook 2020. Available online: www.iea.org/weo (accessed on 14 March 2022).

EP100|Climate Group the World's Energy-Smart Companies Committed to Doing More with Less Energy. (2020). Available online: https://www.theclimategroup.org/ep100 (accessed on 10 December 2021).

Fadeyi, O., Krejcar, O., Maresova, P., Kuca, K., Brida, P., & Selamat, A. (2020). Opinions on sustainability of smart cities in the context of energy posed by cryptocurrency mining. *Sustainability, 12*, 169. https://doi.org/10.3390/su12010169.

Farid, A. M., Alshareef, M., Badhesha, P. S., Boccaletti, C., Cacho, N. A. A., Carlier, C. I., Corriveau, A., Khayal, I., Liner, B., & Martins, J. S. B. et al (2021). Smart city drivers and challenges in energy and water systems. *IEEE Potentials, 40*, 6–10. https://doi.org/10.1109/MPOT.2020.3011378.

Gaska, K., & Generowicz, A. (2020). SMART computational solutions for the optimization of selected technology processes as an innovation and progress in improving energy efficiency of smart cities—A case study. *Energies, 13*, 3338. https://doi.org/10.3390/en13133338.

Gates, B. *How to avoid a climate disaster: The solutions we have and the breakthroughs we need*; Companhia das Letras: São Paulo, 2021; Volume 1; ISBN 978-85-359-3427-4.

Ghadami, N., Gheibi, M., Kian, Z., Faramarz, M. G., Naghedi, R., Eftekhari, M., Fathollahi-Fard, A. M., Dulebenets, M. A., & Tian, G. (2021). Implementation of solar energy in smart cities using an integration of artificial neural network, photovoltaic system and classical delphi methods. *Sustainable Cities and Society, 74*, 103149. https://doi.org/10.1016/j.scs.2021.103149.

Giourka, P., Apostolopoulos, V., Angelakoglou, K., Kourtzanidis, K., Nikolopoulos, N., Sougkakis, V., Fuligni, F., Barberis, S., Verbeek, K., Costa, J. M. et al (2020). The nexus between market needs and value attributes of smart city solutions towards energy transition. An empirical evidence of two European Union (EU) smart cities, Evora and Alkmaar. *Smart Cities, 3,* 604–641. https://doi.org/10.3390/smartcities3030032.

Gitelman, L. D., Kozhevnikov, M. v., Starikov, E. M., & Gamburg, A. v. (2020). Technology entrepreneurship as a factor of sustainable energy in smart cities. *WIT Transactions on Ecology and the Environment.* 246, 101–112. https://doi.org/10.2495/EPM200101.

Gitelman, L. D., Kozhevnikov, M. V., & Adam, L. A. (2019). Sustainable energy for smart City. *International Journal of Energy Production and Management, 4,* 273–286. https://doi.org/10.2495/EQ-V4-N4-343-353.

Gonçalves, D., Sheikhnejad, Y., Oliveira, M., & Martins, N. (2020). One step forward toward smart City utopia: Smart building energy management based on adaptive surrogate modelling. *Energy Build, 223,* 110146. https://doi.org/10.1016/j.enbuild.2020.110146.

Gorla, P., & Chamola, V. (2021). Battery lifetime estimation for energy efficient telecommunication networks in smart cities. *Sustainable Energy Technologies and Assessments, 46,* 101205. https://doi.org/10.1016/j.seta.2021.101205.

Govindarajan, H. K., & Ganesh, L. S. (2021). Renewable energy for electricity use in India: Evidence from India's smart cities mission. *Renewable Energy Focus, 38,* 36–43. https://doi.org/10.1016/j.ref.2021.05.005.

Gunningham, N. (2019). Averting climate catastrophe: environmental activism, extinction rebellion and coalitions of influence. *King's Law Journal, 30,* 194–202. https://doi.org/10.1080/09615768.2019.1645424.

Haarstad, H., & Wathne, M. W. (2019. Are smart city projects catalyzing urban energy sustainability?. *Energy Policy, 129,* 918–925. https://doi.org/10.1016/j.enpol.2019.03.001.

Hadidi, L. A., Rahman, S. M., & Maghrabi, A. T. (2021). Smart city—A sustainable solution for enhancing energy efficiency and climate change mitigation in Saudi Arabia. *International Journal of Global Warming, 24,* 91–107.

Hajduk, S. (2016). The concept of a smart city in urban management. *Business, Management and Economics Engineering, 14,* 34–49. https://doi.org/10.3846/bme.2016.319.

Hajduk, S., & Jelonek, D. (2021). A decision-making approach based on Topsis method for ranking smart cities in the context of urban energy. *Energies, 14,* 2691. https://doi.org/10.3390/en14092691.

Hati, S., Dey, P., & De, D. (2019). WLAN based energy efficient smart City design. *Microsystem Technologies, 25,* 1599–1612. https://doi.org/10.1007/s00542-017-3530-6.

Hayashi, Y., Fujimoto, Y., Ishii, H., Takenobu, Y., Kikusato, H., Yoshizawa, S., Amano, Y., Tanabe, S. I., Yamaguchi, Y., & Shimoda, Y. et al (2018). Versatile modeling platform for cooperative energy management systems in smart cities. *Proceedings of the IEEE, 106,* 594–612. https://doi.org/10.1109/JPROC.2018.2809468.

He, B.-J., Ding, L., & Prasad, D. (2020). Wind-sensitive urban planning and design: precinct ventilation performance and its potential for local warming mitigation in an open midrise gridiron precinct. *Journal of Building Engineering, 29,* 101145. https://doi.org/10.1016/j.jobe.2019.101145.

Hoang, A., & Nguyen, X. (2021). Integrating renewable sources into energy system for smart city as a sagacious strategy towards clean and sustainable process. *Journal of Cleaner Production, 305*, 127–161.

Huang, X., Huang, P., & Huang, T. (2020). Multi-objective optimization of digital management for renewable energies in smart cities. *Journal Europeen des Systemes Automatises, 53*, 893–902. https://doi.org/10.18280/jesa.530615.

Hung, P., & Peng, K. (2017). Green-energy, water-autonomous greenhouse system: An alternative-technology approach towards sustainable smart-green vertical greening in smart cities. *International Review for Spatial Planning and Sustainable Development, 5*, 55–70. https://doi.org/10.14246/irspsd.5.1_55.

Hussain, M. M., Akram, R., Memon, Z. A., Nazir, M. H., Javed, W., & Siddique, M. (2021). Demand side management techniques for home energy management systems for smart cities. *Sustainability, 13*, 11740. https://doi.org/10.3390/su132111740.

Ingrao, C., Messineo, A., Beltramo, R., Yigitcanlar, T., & Ioppolo, G. (2018). How can life cycle thinking support sustainability of buildings? Investigating life cycle assessment applications for energy efficiency and environmental performance. *Journal of Cleaner Production, 201*, 556–569. https://doi.org/10.1016/j.jclepro.2018.08.080.

IPCC AR6. (2021). Climate Change 2021: The Physical Science Basis. Available online: https://www.ipcc.ch/assessment-report/ar6/ (accessed on 20 December 2021).

Jablonski, I. (2015). Smart transducer interface-from networked on-site optimization of energy in research-demonstrative office building to smart city. *IEEE Sensors Journal, 15*, 2339135. https://doi.org/10.1109/JSEN.2014.2339135.

Jettanasen, C., Songsukthawan, P., & Ngaopitakkul, A. (2020). Development of micromobility based on piezoelectric energy harvesting for smart city applications. *Sustainability, 12*, 2933. https://doi.org/10.3390/su12072933.

Jurevičienė, D., & Biekšaitė, A. (2020). Valuation of Lithuanian cities' smartness. *Economics and Culture, 17*, 104–115. https://doi.org/10.2478/jec-2020-0010.

Kai, C., Li, H., Xu, L., Li, Y., & Jiang, T. (2018). Energy-efficient device-to-device communications for green smart cities. *IEEE Transactions on Industrial Informatics, 14*, 1542–1551. https://doi.org/10.1109/TII.2017.2789304.

Kalli, M. (2017). Energy solutions for future smart cities. *Euroheat and Power (English Edition), 14*, 24–26.

Kamyab, H., Klemeš, J. J., Fan, Y., & van; Lee, C. T. (2020). Transition to sustainable energy system for smart cities and industries. *Energy, 207*, 118104. https://doi.org/10.1016/j.energy.2020.118104.

Khalil, M. I., Jhanjhi, N. Z., Humayun, M., Sivanesan, S. K., Masud, M., & Hossain, M. S. (2021). Hybrid smart grid with sustainable energy efficient resources for smart cities. *Sustainable Energy Technologies and Assessments, 46*, 101211. https://doi.org/10.1016/j.seta.2021.101211.

Khattak, H. A., Tehreem, K., Almogren, A., Ameer, Z., Din, I. U., & Adnan, M. (2020). Dynamic pricing in industrial internet of things: Blockchain application for energy management in smart cities. *Journal of Information Security and Applications, 55*, 102615. https://doi.org/10.1016/j.jisa.2020.102615.

Kim, D., Kwon, D., Park, L., Kim, J., & Cho, S. (2021). Multiscale LSTM-based deep learning for very-short-term photovoltaic power generation forecasting in smart city energy management. *IEEE Systems Journal, 15*, 346–354. https://doi.org/10.1109/JSYST.2020.3007184.

Kim, H., Choi, H., Kang, H., An, J., Yeom, S., & Hong, T. (2021). A systematic review of the smart energy conservation system: From smart homes to sustainable smart cities. *Renewable and Sustainable Energy Reviews, 140,* 110755. https://doi.org/10.1016/j.rser.2021.110755.

Konstantinou, C. (2022). Toward a secure and resilient all-renewable energy grid for smart cities. *IEEE Consumer Electronics Magazine, 11,* 33–41. https://doi.org/10.1109/MCE.2021.3055492.

Krozer, Y. (2017). Innovative offices for smarter cities, including energy use and energy-related carbon dioxide emissions. *Energy, Sustainability and Society, 7,* s13705–s13717.

Krutwig, M., Starosta, K., & Tantau, A. Comparative study on quality assurance for mandatory energy audits in Romania and Germany. In Proceedings of the International Conference on Business Excellence, Bucharest, Romania, 22–23 March 2018.

Kumar, D. (2020). Urban energy system management for enhanced energy potential for upcoming smart cities. *Energy Exploration & Exploitation, 38,* 1968–1982. https://doi.org/10.1177/0144598720937529.

Kumar Garg, A., Janyani, V., & Batagelj, B. (2020). Ring based latency-aware and energy-efficient hybrid WDM TDM-PON with ODN interconnection capability for smart cities. *Optical Fiber Technology, 58,* 102242. https://doi.org/10.1016/j.yofte.2020.102242.

Kuznetsov, P., Rimar, M., Yakimovich, B., Kulikova, O., Lopusniak, M., Voronin, D., & Evstigneev, V. (2021). Parametric optimization of combined wind-solar energy power plants for sustainable smart city development. *Applied Sciences, 11,* 10351. https://doi.org/10.3390/app112110351.

Kylili, A., & Fokaides, P. A. (2015). European smart cities: The role of zero energy buildings. *Sustainable Cities and Society, 15,* 86–95. https://doi.org/10.1016/j.scs.2014.12.003.

Lanini, L., & Barsanti, E. A. (2018). Hybrid building as social and energy hub for smart cities: Unite 2.0, A. *TECHNE-Journal of Technology for Architecture and Environment,* 49–55. https://doi.org/10.13128/Techne-22706.

Lantz, T. L., Ioppolo, G., Yigitcanlar, T., & Arbolino, R. (2021). Understanding the correlation between energy transition and urbanization. *Environmental Innovation and Societal Transitions, 40,* 73–86. https://doi.org/10.1016/j.eist.2021.06.002.

Le, L. T., Nguyen, H., Dou, J., & Zhou, J. (2019). A comparative study of PSO-ANN, GA-ANN, ICA-ANN, and ABC-ANN in estimating the heating load of buildings' energy efficiency for smart city planning. *Applied Sciences, 9,* 2630. https://doi.org/10.3390/app9132630.

Lenhart, J., van Vliet, B., & Mol, A. P. J. (2015). New roles for local authorities in a time of climate change: The Rotterdam energy approach and planning as a case of urban symbiosis. *Journal of Cleaner Production, 107,* 593–601. https://doi.org/10.1016/j.jclepro.2015.05.026.

Lewandowska, A., Chodkowska-Miszczuk, J., Rogatka, K., & Starczewski, T. (2020). Smart energy in a smart city: Utopia or reality? evidence from Poland. *Energies, 13,* 5795.

Li, L., Zheng, Y., Zheng, S., & Ke, H. (2020). The new smart city programme: evaluating the effect of the internet of energy on air quality in China. *Science of the Total Environment, 714,* 136380. https://doi.org/10.1016/j.scitotenv.2019.136380.

Li, W., Yigitcanlar, T., Erol, I., & Liu, A. (2021). Motivations, barriers and risks of smart home adoption: From systematic literature review to conceptual framework. *Energy Research & Social Science, 80,* 102211. https://doi.org/10.1016/j.erss.2021.102211.

Liu, A., Miller, W., Chiou, J., Zedan, S., Yigitcanlar, T., & Ding, Y. (2021). Aged care energy use and peak demand change in the COVID-19 year: Empirical evidence from Australia. *Buildings, 11,* 570. https://doi.org/10.3390/buildings11120570.

Liu, J., Wang, M., Peng, J., Chen, X., Cao, S., & Yang, H. (2020). Techno-economic design optimization of hybrid renewable energy applications for high-rise residential buildings. *Energy Conversion and Management, 213,* 112868. https://doi.org/10.1016/j.enconman.2020.112868.

Liu, Y., Yang, C., Jiang, L., Xie, S., & Zhang, Y. (2019). Intelligent edge computing for IoT-based energy management in smart cities. *IEEE Network, 33,* 111–117. https://doi.org/10.1109/MNET.2019.1800254.

Lotfi, M., Almeida, T., Javadi, M. S., Osório, G. J., Monteiro, C., & Catalão, J. P. S. (2022). Coordinating energy management systems in smart cities with electric vehicles. *Applied Energy, 307,* 118241. https://doi.org/10.1016/j.apenergy.2021.118241.

Lovins, A. B. (2004). Energy efficiency, taxonomic overview. In: *Encyclopedia of energy* (pp. 383–401). Elsevier.

Lu, C. W., Huang, J. C., Chen, C., Shu, M. H., Hsu, C. W., & Tapas Bapu, B. R. (2021). An energy-efficient smart City for sustainable green tourism industry. *Sustainable Energy Technologies and Assessments, 47,* 101494. https://doi.org/10.1016/j.seta.2021.101494.

Lu, W., Gong, Y., Liu, X., Wu, J., & Peng, H. (2018). Collaborative energy and information transfer in green wireless sensor networks for smart cities. *IEEE Transactions on Industrial Informatics, 14,* 1585–1593. https://doi.org/10.1109/TII.2017.2777846.

Lund, H., Østergaard, P. A., Connolly, D., & Mathiesen, B. V. (2017). smart energy and smart energy systems. *Energy, 137,* 556–565. https://doi.org/10.1016/j.energy.2017.05.123.

Luo, H., Cai, H., Yu, H., Sun, Y., Bi, Z., & Jiang, L. (2019). A short-term energy prediction system based on edge computing for smart. *Future Generation Computer Systems: The International Journal of eScience, 101,* 444–457. https://doi.org/10.1016/j.future.2019.06.030.

Lützenberger, M., Masuch, N., Küster, T., Freund, D., Voß, M., Hrabia, C. E., Pozo, D., Fähndrich, J., Trollmann, F., & Keiser, J. et al (2015). A common approach to intelligent energy and mobility services in a smart city environment. *Journal of Ambient Intelligence and Humanized Computing, 6,* 337–350. https://doi.org/10.1007/s12652-015-0263-1.

Madlener, R., & Sunak, Y. (2011. Impacts of urbanization on urban structures and energy demand: what can we learn for urban energy planning and urbanization management?. *Sustainable Cities and Society, 1,* 45–53. https://doi.org/10.1016/j.scs.2010.08.006.

Mahapatra, C., Moharana, A. K., & Leung, V. C. M. (2017). energy management in smart cities based on internet of things: peak demand reduction and energy savings. *Sensors, 17,* 2812. https://doi.org/10.3390/s17122812.

Mahbub, P., Goonetilleke, A., Ayoko, G., Egodawatta, P., & Yigitcanlar, T. (2011). Analysis of build-up of heavy metals and volatile organics on urban roads in gold coast, Australia. *Water Science and Technology, 63,* 2077–2085.

Mahmood, D., Javaid, N., Ahmed, G., Khan, S., & Monteiro, V. (2021). A review on optimization strategies integrating renewable energy sources focusing uncertainty factor—Paving path to eco-Friendly smart cities. *Sustainable Computing: Informatics and Systems, 30,* 100559. https://doi.org/10.1016/j.suscom.2021.100559.

Maier, S. (2016). Smart energy systems for smart city districts: Case study Reininghaus district. *Energy, Sustainability and Society, 6,* s13705–s016.

Maier, S., Narodoslawsky, M., Borell-Damián, L. et al (2019). Theory and practice of European co-operative education and training for the support of energy transition. *Energy, Sustainability and Society, 9,* 29. https://doi.org/10.1186/s13705-019-0213-4.

Maltese, I., Mariotti, I., & Boscacci, F. (2016). Smart city, urban performance and energy. In R. Papa & R. Fistola (Eds.), *Smart energy in the smart City* (pp. 25–42). Springer. https://doi.org/10.1007/978-3-319-31157-9_2.

Manshahia, M. S. (2018). Swarm intelligence-based energy-efficient data delivery in WSAN to virtualise IoT in smart cities. *IET Wireless Sensor Systems, 8,* 256–259. https://doi.org/10.1049/iet-wss.2018.5143.

Martín, C., Castillo-Calzadilla, T., Zabala, K., Arrizabalaga, E., Hernández, P., Mabe, L., López, J. R., Casado, J. M., Santos, M. N., Guardo, J., et al. (2021). The opportunity for smart city projects at municipal scale: Implementing a positive energy district in Zorrozaurre. Ekonomiaz, *99,* 119–149.

Martín-Martín, A., Orduna-Malea, E., Thelwall, M., & Delgado López-Cózar, E. (2018). Google scholar, web of science, and scopus: A systematic comparison of citations in 252 subject categories. *Journal of Informetrics, 12,* 1160–1177. https://doi.org/10.1016/j.joi.2018.09.002.

Martins, F., Patrão, C., Moura, P., & de Almeida, A. T. (2021). A review of energy modeling tools for energy efficiency in smart cities. *Smart Cities, 4,* 1420–1436.

Mattoni, B., Gugliermetti, F., & Bisegna, F. (2015). A multilevel method to assess and design the renovation and integration of smart cities. *Sustainable Cities and Society, 15,* 105–119. https://doi.org/10.1016/j.scs.2014.12.002.

Meenakshi, Mothi Kumar, K. E., & Kumari, N. (2021). A geo-spatial approach for quantifying rooftop photovoltaic energy potential in Karnal smart City, Haryana—A case study. *Journal of Applied and Natural Science, 13,* 512–519. https://doi.org/10.31018/jans.v13i2.2665.

Mekhum, W. (2020). Smart cities: impact of renewable energy consumption, information and communication technologies and e-governance on CO2 emission. *Journal of Security and Sustainability Issues, 9,* 785–795.

Miller, W., Buys, L., & Bell, J. (2012). Performance evaluation of eight contemporary passive solar homes in subtropical Australia. *Building and Environment, 56,* 57–68. https://doi.org/10.1016/j.buildenv.2012.02.023.

Miller, W., Liu, L. A., Amin, Z., & Gray, M. (2018). Involving occupants in net-zero-energy solar housing retrofits: An Australian sub-tropical case study. *Solar Energy, 159,* 390–404. https://doi.org/10.1016/j.solener.2017.10.008.

Miyasawa, A., Akira, S., Fujimoto, Y., & Hayashi, Y. (2021). Spatial demand forecasting based on smart meter data for improving local energy self-sufficiency in smart cities. *IET Smart Cities, 3,* 107–120. https://doi.org/10.1049/smc2.12011.

Moreno, M., Zamora, M. A., & Skarmeta, A. F. (2014). User-centric smart buildings for energy sustainable smart cities. *Transactions on Emerging Telecommunications Technologies, 25,* 41–55. https://doi.org/10.1002/ett.2771.

Mortoja, M. G., & Yigitcanlar, T. (2020). Local drivers of anthropogenic climate change: Quantifying the impact through a remote sensing approach in Brisbane. *Remote Sensing, 12,* 2270. https://doi.org/10.3390/rs12142270.

Mosannenzadeh, F., Bisello, A., Vaccaro, R., D'alonzo, V., Hunter, G. W., & Vettorato, D. (2017). Smart energy city development: A story told by urban planners. *Cities, 64,* 54–65. https://doi.org/10.1016/j.cities.2017.02.001.

Muhammad, K., Lloret, J., & Baik, S. W. (2019). Intelligent and energy-efficient data prioritization in green smart cities: Current challenges and future directions. *IEEE Communications Magazine, 57,* 60–65. https://doi.org/10.1109/MCOM.2018.1800371.

Mutule, A., Domingues, M., Ulloa-Vásquez, F., Carrizo, D., García-Santander, L., Dumitrescu, A. M., Issicaba, D., & Melo, L. (2021). Implementing smart city technologies to inspire change in consumer energy behaviour. *Energies, 14,* 4310. https://doi.org/10.3390/en14144310.

Mutule, A., Teremranova, J., & Antoskovs, N. (2018). Smart city through a flexible approach to smart energy. *Latvian Journal of Physics and Technical Sciences, 55,* 3–14.

Nagy, Z., Sebestyén Szép, T., & Szendi, D. (2019). Regional disparities in Hungarian urban energy consumption-a link between smart cities and successful cities. *Geographia Technica, 14,* 102.

Nanni, S., Benetti, E., & Mazzini, G. (2017). Indoor monitoring in public buildings: workplace wellbeing and energy consumptions. An example of IoT for smart cities application. *Advances in Science, Technology and Engineering Systems, 2,* 884–890. https://doi.org/10.25046/aj0203110.

Naseer, S., Liu, W., & Sarkar, N. I. (2019). Energy-efficient massive data dissemination through vehicle mobility in smart cities. *Sensors, 19,* 4735. https://doi.org/10.3390/s19214735.

Nasr, A., Kashan, M., Maleki, A., Jafari, N., & Hashemi, H. (2020). Assessment of barriers to renewable energy development using stakeholders approach. *Entrepreneurship and Sustainability Issues, 7,* 2526.

Nesmachnow, S., Colacurcio, G., Rossit, D. G., Toutouh, J., & Luna, F. (2020). Optimizing household energy planning in smart cities: A multiobjective approach. *Revista Facultad de Ingeniería Universidad de Antioquia, 101,* 8–19. https://doi.org/10.17533/udea.redin.20200587.

Nuvvula, R., Devaraj, E., & Srinivasa, K. T. (2021). A comprehensive assessment of large-scale battery integrated hybrid renewable energy system to improve sustainability of a smart city. *Energy Sources Part A: Recovery, Utilization, and Environmental Effects.* https://doi.org/10.1080/15567036.2021.1905109.

O'Dwyer, E., Pan, I., Acha, S., & Shah, N. (2019). Smart energy systems for sustainable smart cities: Current developments, trends and future directions. *Applied Energy, 237,* 581–597.

Oldenbroek, V., Smink, G., Salet, T., & van Wijk, A. J. M. (2020). Fuel cell electric vehicle as a power plant: Techno-economic scenario analysis of a renewable integrated transportation and energy system for smart cities in two climates. *Applied Sciences, 10,* 143. https://doi.org/10.3390/app10010143.

Oldenbroek, V., Verhoef, L. A., & van Wijk, A. J. M. (2017). Fuel cell electric vehicle as a power plant: fully renewable integrated transport and energy system design and analysis for smart city areas. *International Journal of Hydrogen Energy, 42,* 8166–8196. https://doi.org/10.1016/j.ijhydene.2017.01.155.

ONU News. (2021). ONU Prevê Que Cidades Abriguem 70% Da População Mundial Até 2050. Available online: https://news.un.org/pt/story/2019/02/1660701 (accessed on 12 February 2022).

Orecchini, F., Santiangeli, A., Zuccari, F., Pieroni, A., & Suppa, T. (2019). Blockchain technology in smart city: A new opportunity for smart environment and smart mobility. In P. Vasant, I. Zelinka, & G.W. Weber (Eds.), *Intelligent computing & optimization* (pp. 346–354). Springer Volume 1. https://doi.org/10.1007/978-3-030-00979-3_36.

Organização das Nações Unidas. (2020). Glasgow Concentra Esperança Sobre Temas Centrais da Ação Climática BR. Available online: https://news.un.org/pt/story/2021/10/1768762 (accessed on *10* December 2021).

Organização das Nações Unidas. (2021). Objetivos de Desenvolvimento Sustentável. Available online: https://brasil.un.org/pt-br/sdgs/7 (accessed on 14 March 2022).

Page, M. J., McKenzie, J. E., Bossuyt, P. M., Boutron, I., Hoffmann, T. C., Mulrow, C. D., Shamseer, L., Tetzlaff, J. M., Akl, E. A., & Brennan, S. E. et al (2021). The PRISMA 2020 statement: An updated guideline for reporting systematic reviews. *BMJ, 88*, 105906. https://doi.org/10.1136/bmj.n71.

Palanca, J., Jordán, J., Bajo, J., & Botti, V. (2020). An energy-aware algorithm for electric vehicle infrastructures in smart cities. *Future Generation Computer Systems, 108*, 454–466. https://doi.org/10.1016/j.future.2020.03.001.

Palensky, P., & Dietrich, D. (2011). Demand side management: Demand response, intelligent energy systems, and smart loads. *IEEE Transactions on Industrial Informatics, 7*, 381–388. https://doi.org/10.1109/TII.2011.2158841.

Palomar, E., Chen, X., Liu, Z., Maharjan, S., & Bowen, J. (2016). Component-based modelling for scalable smart city systems interoperability: A case study on integrating energy demand response systems. *Sensors, 16*, 1810. https://doi.org/10.3390/s16111810.

Parks, D. (2019). Energy efficiency left behind? Policy assemblages in Sweden's most climate-smart city. *European Planning Studies, 27*, 318–335. https://doi.org/10.1080/09654313.2018.1455807.

Parks, D. (2020). Promises and techno-politics: Renewable energy and Malmö's vision of a climate-smart city. *Scientific Culture, 29*, 388–409. https://doi.org/10.1080/09505431.2019.1705274.

Pei, P., Huo, Z., Martínez, O. S., & Crespo, R. G. (2020). Minimal green energy consumption and workload management for data centers on smart city platforms. *Sustainability, 12*, 3140. https://doi.org/10.3390/SU12083140.

Perera, A. T. D., Coccolo, S., & Scartezzini, J.-L. (2019). The influence of urban form on the grid integration of renewable energy technologies and distributed energy systems. *Scientific Reports, 9*, 17756. https://doi.org/10.1038/s41598-019-53653-w.

Petritoli, E., Leccese, F., Pizzuti, S., & Pieroni, F. (2019). Smart lighting as basic building block of smart city: An energy performance comparative case study. *Measurement: Journal of the International Measurement Confederation, 136*, 466–477. https://doi.org/10.1016/j.measurement.2018.12.095.

Petrović, N., & Kocić, D. (2020). Data-driven framework for energy-efficient smart cities. *Serbian Journal of Electrical Engineering, 17*, 41–63. https://doi.org/10.2298/SJEE2001041P.

Pezzutto, S., Mosannenzadeh, F., Grilli, G., & Sparber, W. (2017). European Union research and development funding on smart cities and their importance on climate and energy goals. In: *Smart and sustainable planning for cities and regions* (pp. 421–435). Springer International Publishing. ISBN 9783319448992.

Philippi Cortese, T. T., Kniess, C. T., & Maccari, E. A. (2017). *Cidades inteligentes e sustentáveis*; Editora Manole. Volume 1. ISBN 9788520451403.

Pieroni, A., Scarpato, N., di Nunzio, L., Fallucchi, F., & Raso, M. (2018). Smarter city: Smart energy grid based on blockchain technology. *International Journal on Advanced Science, Engineering and Information Technology, 8*, 298–306. https://doi.org/10.18517/ijaseit.8.1.4954.

Pilipczuk, O. (2020). Sustainable smart cities and energy management: The labor market. *Energies, 13*, 6084. https://doi.org/10.3390/en13226084.

Pirisi, A., Grimaccia, F., Mussetta, M., & Zich, R. E. (2012). Novel speed bumps design and optimization for vehicles' energy recovery in smart cities. *Energies, 5*, 4624–4642. https://doi.org/10.3390/en5114624.

Pol, O., Palensky, P., Kuh, C., Leutgöb, K., Page, J., & Zucker, G. (2012). Integration of centralized energy monitoring specifications into the planning process of a new urban development area: A step towards smart cities. *Elektrotechnik und Informationstechnik 129*, 258–264. https://doi.org/10.1007/s00502-012-0010-7.

Policy Solutions-Electricity. (2020). Available online: https://www.breakthroughenergy.org/us-policy-overview/electricity (accessed on 10 December 2021).

População Rural e Urbana|Educa|Jovens-IBGE. (2021). Available online: https://educa.ibge.gov.br/jovens/conheca-o-brasil/populacao/18313-populacao-rural-e-urbana.html (accessed on 12 February 2022).

Regona, M., Yigitcanlar, T., Xia, B., & Li, R. Y. M. (2022). Opportunities and adoption challenges of AI in the construction industry: A PRISMA review. *Journal of Open Innovation: Technology, Market, and Complexity, 8*, 45. https://doi.org/10.3390/joitmc8010045.

Rodríguez-Molina, J., Martínez, J. F., Castillejo, P., & de Diego, R. (2013). SMArc: A proposal for a smart, semantic middleware architecture focused on smart city energy management. *International Journal of Distributed Sensor Networks, 13*, 560418. https://doi.org/10.1155/2013/560418.

Rosales Carreón, J., & Worrell, E. (2018). Urban energy systems within the transition to sustainable development. A research agenda for urban metabolism. *Resources, Conservation and Recycling, 132*, 258–266. https://doi.org/10.1016/j.resconrec.2017.08.004.

Rostirolla, G., Righi, R. D. R., Barbosa, J. L. V., & da Costa, C. (2018). ElCity: An elastic multilevel energy saving model for smart cities. *IEEE Transactions on Sustainable Computing, 3*, 30–43. https://doi.org/10.1109/TSUSC.2017.2749880.

Sabory, N. R., Senjyu, T., Danish, M. S. S., Hosham, A., Noorzada, A., Amiri, A. S., & Muhammdi, Z. (2021). Applicable smart City strategies to ensure energy efficiency and renewable energy integration in poor cities: Kabul case study. *Sustainability, 13*, 11984. https://doi.org/10.3390/su132111984.

Sami, M. S., Abrar, M., Akram, R., Hussain, M. M., Nazir, M. H., Khan, M. S., & Raza, S. (2021). Energy management of microgrids for smart cities: A review. *Energies, 14*, 5976. https://doi.org/10.3390/en14185976.

Sánchez-Cano, J. E., García-Quilachamin, W. X., Pérez-Véliz, J., Herrera-Tapia, J., & Fuentes, K. A. (2021). Review of methods to reduce energy consumption in a smart city based on IoT and 5G technology. *International Journal of Online and Biomedical Engineering, 17*, 4–21. https://doi.org/10.3991/ijoe.v17i08.23671.

Sanchez-Miralles, A., Calvillo, C., Martín, F., & Villar, J. (2014). Use of renewable energy systems in smart cities. *Green Energy Technology*, 2014, 341–370. https://doi.org/10.1007/978-3-319-03224-5_10.

Sanseverino, E. R., Scaccianoce, G., Vaccaro, V., Carta, M., & Sanseverino, R. R. (2015). Smart cities and municipal building regulation for energy effciency. *International Journal of Agricultural and Environmental Information Systems*, *6*, 56–82. https://doi.org/10.4018/IJAEIS.2015100104.

Sato, M., Fukuyama, Y., El-Abd, M., Iizaka, T., & Matsui, T. (2019a). Total optimization of energy networks in smart city by cooperative coevolution using global-best brain storm optimization. 2019 IEEE Congress on Evolutionary Computation (CEC), pp. 681–688, doi: 10.1109/CEC.2019.8790288.

Sato, M., Fukuyama, Y., Iizaka, T., & Matsui, T. (2019b). Total optimization of energy networks in a smart city by multi-swarm differential evolutionary particle swarm optimization. *IEEE Transactions on Sustainable Energy*, *10*, 2186–2200. https://doi.org/10.1109/TSTE.2018.2882203.

Sato, M., Fukuyama, Y., Iizaka, T., & Matsui, T. (2019c). Total optimization of energy networks in a smart city by multi-population global-best modified brain storm optimization with migration. *Algorithms*, *12*, 15. https://doi.org/10.3390/a12010015.

Sayah, Z., Kazar, O., Lejdel, B., Laouid, A., & Ghenabzia, A. (2020). An intelligent system for energy management in smart cities based on big data and ontology. *Smart and Sustainable Built Environment*, *10*, 169–192. https://doi.org/10.1108/SASBE-07-2019-0087.

Seifhashemi, M., Capra, B. R., Milller, W., & Bell, J. (2018). The potential for cool roofs to improve the energy efficiency of single storey warehouse-type retail buildings in Australia: A simulation case study. *Energy and Buildings*, *158*, 1393–1403. https://doi.org/10.1016/j.enbuild.2017.11.034.

Serban, A. C., & Lytras, M. D. (2020). Artificial intelligence for smart renewable energy sector in europe-smart energy infrastructures for next generation smart cities. *IEEE Access*, *8*, 77364–77377. https://doi.org/10.1109/ACCESS.2020.2990123.

Shahizare, B., Nik-Ghazali, N., Chong, W. T., Tabatabaeikia, S., Izadyar, N., & Esmaeilzadeh, A. (2016). Novel investigation of the different omni-direction-guide-vane angles effects on the urban vertical axis wind turbine output power via three-dimensional numerical simulation. *Energy Conversion and Management*, *117*, 206–217. https://doi.org/10.1016/j.enconman.2016.03.034.

Shen, X., Lu, Y., & Zhang, Y. et al (2022). An innovative data integrity verification scheme in the internet of things assisted information exchange in transportation systems. *Cluster Computing*. https://doi.org/10.1007/s10586-021-03471-5.

Shen, X., Yu, H., Liu, X., Bin, Q., Luhach, A. K., & Saravanan, V. (2021). The optimized energy-efficient sensible edge processing model for the internet of vehicles in smart cities. *Sustainable Energy Technologies and Assessments*, *47*, 101477. https://doi.org/10.1016/j.seta.2021.101477.

Shin, C. E. S., & Cho, Y. (2021). Efficient energy consumption prediction model for a data analytic-enabled industry building in a smart city. *Building Research and Information*, *49*, 127–143. https://doi.org/10.1080/09613218.2020.1809983.

Shu, M., Wu, S., Wu, T., Qiao, Z., Wang, N., Xu, F., Shanthini, A., & Muthu, B. A. (2020). Efficient energy consumption system using heuristic renewable demand energy optimization in smart city. *Computational Intelligence*. https://doi.org/10.1111/coin.12412.

Skvarciany, V., Jurevičienė, D., Žitkienė, R., Lapinskaitė, I., & Dudė, U. (2021). A different approach to the evaluation of smart cities' indicators. *TalTech Journal of European Studies, 11*, 130–147. https://doi.org/10.2478/bjes-2021-0018.

Strielkowski, W., Veinbender, T., Tvaronaviciene, M., & Lace, N. (2020). Economic efficiency and energy security of smart cities. *Economic Research-Ekonomska Istraživanja, 33*, 788–803. https://doi.org/10.1080/1331677X.2020.1734854.

Sustainable Energy for All is Funded by Sustainable Energy for All. Available online: Sustainable Energy for All is funded by (accessed on 17 February 2022).

Swain, A., Salkuti, S. R., & Swain, K. (2021). An optimized and decentralized energy provision system for smart cities. *Energies, 14*. https://doi.org/10.3390/en14051451.

Tantau, A., & Şanta, A. M. I. (2021). New energy policy directions in the European Union developing the concept of smart cities. *Smart Cities, 4*, 241–252. https://doi.org/10.3390/smartcities4010015.

Tanwar, S., Popat, A., Bhattacharya, P., Gupta, R., & Kumar, N. (2021). A taxonomy of energy optimization techniques for smart cities: Architecture and future directions. *Expert Systems*. https://doi.org/10.1111/exsy.12703.

Tostar, A., Ferrand, L., Lee, T., & Roaf, L. (2021). Energy Efficiency: The Unsung Hero of Net Zero Carbon|Climate Group. Available online: https://theclimategroup.prod.acquia-sites.com/our-work/news/energy-efficiency-unsung-hero-net-zero-carbon (accessed on 10 December 2021).

Tranfield, D., Denyer, D., & Smart, P. (2003). Towards a methodology for developing evidence-informed management knowledge by means of systematic review. *British Journal of Management, 14*, 207–222. https://doi.org/10.1111/1467-8551.00375.

UNFCCC Paris Agreement Text English. (2021). Available online: https://unfccc.int/sites/default/files/english_paris_agreement.pdf (accessed on 14 March 2022).

United Nations. World Population Projected to Reach 9.8 Billion in 2050, and 11.2 Billion in 2100. United Nations. Available online: https://www.un.org/en/desa/world-population-projected-reach-98-billion-2050-and-112-billion-2100 (accessed on 10 December 2021).

United Nations-Climate Change COP26 EXPLAINED. (2021). Available online: https://ukcop26.org/wp-content/uploads/2021/07/COP26-Explained.pdf (accessed on 14 March 2022).

Uspenskaia, D., Specht, K., Kondziella, H., & Bruckner, T. (2021). Challenges and barriers for net-zero/positive energy buildings and districts—Empirical evidence from the smart city project sparcs. *Buildings, 11*, 1–25. https://doi.org/10.3390/buildings11020078.

Vázquez-Canteli, J. R., Ulyanin, S., Kämpf, J., & Nagy, Z. (2019). Fusing TensorFlow with building energy simulation for intelligent energy management in smart cities. *Sustainable Cities and Society, 45*, 243–257. https://doi.org/10.1016/j.scs.2018.11.021.

Verma, S. (2021). Energy-efficient routing paradigm for resource-constrained internet of things-based cognitive smart city. *Expert Systems*, e12905. https://doi.org/10.1111/exsy.12905.

Vilathgamuwa, M., Mishra, Y., Yigitcanlar, T., Bhaskar, A., & Wilson, C. (2022). Mobile-Energy-as-a-service (MEaaS): Sustainable electromobility via integrated energy–transport–urban infrastructure. *Sustainability, 14*, 2796. https://doi.org/10.3390/su14052796.

Vrabie, C. (2021). Converting municipal waste to energy through the biomass chain, a key technology for environmental issues in (smart) cities. *Sustainability, 13,* 4633. https://doi.org/10.3390/su13094633.

Vukovic, N., Koriugina, U., Illarionova, D., Pankratova, D., Kiseleva, P., & Gontareva, A. (2021). Towards smart green cities: Analysis of integrated renewable energy use in smart cities. *Strategic Planning for Energy and the Environment, 40,* 75–94. https://doi.org/10.13052/spee1048-5236.4015.

Wang, B., Geoffroy, S., & Bonhomme, M. (2022). Urban form study for wind potential development. *Environment and Planning B: Urban Analytics and City Science, 49,* 76–91. https://doi.org/10.1177/2399808321994449.

Wang, C., Gu, J., Sanjuán Martínez, O., & González Crespo, R. (2021a). Economic and environmental impacts of energy efficiency over smart cities and regulatory measures using a smart technological solution. *Sustainable Energy Technologies and Assessments, 47,* 101422. https://doi.org/10.1016/j.seta.2021.101422.

Wang, S., Wang, X., Wang, Z. L., & Yang, Y. (2016). Efficient scavenging of solar and wind energies in a smart city. *ACS Nano, 10,* 5696–5700. https://doi.org/10.1021/acsnano.6b02575.

Wang, X., Chen, Q., & Wang, J. (2021b). Fuzzy rough set based sustainable methods for energy efficient smart city development. *Journal of Intelligent & Fuzzy Systems, 40,* 8173–8183. https://doi.org/10.3233/JIFS-189640.

Wei, L., & Hu, Y. (2019). Research on new and old kinetic energy transformation supported by smart city construction in big data era. *Journal of Advanced Computational Intelligence and Intelligent Informatics, 23,* 102–106. https://doi.org/10.20965/jaciii.2019.p0102.

Xia, X., Wu, X., BalaMurugan, S., & Marimuthu, K. (2021). Effect of environmental and social responsibility in energy-efficient management models for smart cities infrastructure. *Sustainable Energy Technologies and Assessments, 47,* 101525. https://doi.org/10.1016/j.seta.2021.101525.

Xiaoyi, Z., Dongling, W., Yuming, Z., Manokaran, K. B., & Benny Antony, A. (2021). IoT driven framework based efficient green energy management in smart cities using multi-objective distributed dispatching algorithm. *Environmental Impact Assessment Review, 88,* 106567. https://doi.org/10.1016/j.eiar.2021.106567.

Yamagata, Y., & Seya, H. (2013). Simulating a future smart city: An integrated land use-energy model. *Applied Energy, 112,* 1466–1474. https://doi.org/10.1016/j.apenergy.2013.01.061.

Ye, H., Li, F. M., Liu, Z. X., & Deng, X. D. (2020). A green energy consumption policy of bluetooth mobile devices for smart cities. *Computing, 102,* 1077–1091. https://doi.org/10.1007/s00607-019-00765-w.

Yigitcanlar, T. (2022). Towards smart and sustainable urban electromobility: An editorial commentary. *Sustainability, 14,* 2264.

Yigitcanlar, T., & Cugurullo, F. (2020). The sustainability of artificial intelligence: an urbanistic viewpoint from the lens of smart and sustainable cities. *Sustainability, 12,* 8548. https://doi.org/10.3390/su12208548.

Yigitcanlar, T., Dodson, J., Gleeson, B., & Sipe, N. (2007). Travel self-containment in master planned estates: Analysis of recent Australian trends. *Urban Policy and Research, 25,* 129–149.

Yigitcanlar, T., & Dur, F. (2013). Making space and place for knowledge communities: Lessons for Australian practice. *Australasian Journal of Regional Studies, 19,* 36–63.

Yigitcanlar, T., Kamruzzaman, M., Foth, M., Sabatini-Marques, J., da Costa, E., & Ioppolo, G. (2019). Can cities become smart without being sustainable? A systematic review of the literature. *Sustainable Cities and Society, 45*, 348–365.

Yigitcanlar, T., & Kamruzzaman, M. (2019). Smart cities and mobility: does the smartness of Australian cities lead to sustainable commuting patterns?. *Journal of Urban Technology, 26*, 21–46. https://doi.org/10.1080/10630732.2018.1476794.

Yigitcanlar, T., Kankanamge, N., Regona, M., Ruiz Maldonado, A., Rowan, B., Ryu, A., Desouza, K., Corchado, J., Mehmood, R., & Li, R. (2020). Artificial intelligence technologies and related urban planning and development concepts: how are they perceived and utilized in Australia?. *Journal of Open Innovation: Technology, Market, and Complexity, 6*, 168.

Yigitcanlar, T., Kankanamge, N., & Vella, K. (2021). How are smart city concepts and technologies perceived and utilized? A systematic geo-twitter analysis of smart cities in Australia. *Journal of Urban Technology, 28*, 135–154.

Yu, Y., & Zhang, N. (2019). Does smart city policy improve energy efficiency? Evidence from a quasi-natural experiment in China. *Journal of Cleaner Production, 229*, 501–512. https://doi.org/10.1016/j.jclepro.2019.04.316.

Yuksel, I., & Kaygusuz, K. (2011). Renewable energy sources for clean and sustainable energy policies in Turkey. *Renewable and Sustainable Energy Reviews, 15*, 4132–4144.

Zambon, I., Monarca, D., Cecchini, M., Bedini, R., Longo, L., Romagnoli, M., & Marucci, A. (2016). Alternative energy and the development of local rural contexts: An approach to improve the degree of smart cities in the Central-Southern Italy. *Contemporary Engineering Sciences, 9*, 1371–1386. https://doi.org/10.12988/ces.2016.68143.

Zekić-Sušac, M., Mitrović, S., & Has, A. (2021). Machine learning based system for managing energy efficiency of public sector as an approach towards smart cities. *International Journal of Information Management, 58*, 391–402. https://doi.org/10.1016/j.ijinfomgt.2020.102074.

Zhang, W., & Yue, M. (2021). The application of building energy management system based on IoT technology in smart city. *International Journal of System Assurance Engineering and Management, 12*, s13198–s021.

Zhang, X., Manogaran, G., & Muthu, B. A. (2021). IoT enabled integrated system for green energy into smart cities. *Sustainable Energy Technologies and Assessments, 46*, 101208. https://doi.org/10.1016/j.seta.2021.101208.

Zhou, L., Wu, D., Chen, J., & Dong, Z. (2018). Greening the smart cities: Energy-efficient massive content delivery via D2D communications. *IEEE Transactions on Industrial Informatics, 14*, 1626–1634. https://doi.org/10.1109/TII.2017.2784100.

Appendix

TABLE 6.A1

Publications Selected for Systematic Literature Review

Ref.	Authors	Title	Year	Categories	Aim	Approach	Continent
Pol et al. (2012)	Pol, O., Palensky, P., Kuh, C., Leutgöb, K., Page, J., Zucker, G.	Integration of centralized energy monitoring specifications into the planning process of a new urban development area: a step towards smart cities	2012	Energy and urban planning	Other	Theoretical	Oceania
Pirisi et al. (2012)	Pirisi, A., Grimaccia, F., Mussetta, M., Zich, R.E.	Novel speed bumps design and optimization for vehicles' energy recovery in smart cities	2012	Renewable energy	Energy harvesting devices	Empirical	Europe
Rodríguez-Molina et al. (2013)	Rodríguez-Molina, J., Martínez, J.-F., Castillejo, P., De Diego, R.	Smarc: a proposal for a smart, semantic middleware architecture focused on smart city energy management	2013	Energy efficiency	Smart grid	Empirical	Europe
Yamagata and Seya (**2013**)	Yamagata, Y., Seya, H.	Simulating a future smart city: an integrated land use-energy model	2013	Renewable energy	Hybrid energy system	Theoretical	Asia
Sanchez-Miralles et al. (2014)	Sanchez-Miralles, A., Calvillo, C., Martín, F., Villar, J.	Use of renewable energy systems in smart cities	2014	Renewable energy	Hybrid energy system	Theoretical	Europe
Battista et al. (2014)	Battista, G., Evangelisti, L., Guattari, C., Basilicata, C., de Lieto, Vollaro, R.	Building's energy efficiency: interventions analysis under a smart cities approach	2014	Energy efficiency	Smart buildings	Empirical	Europe
Moreno et al. (2014)	Moreno, M.V., Zamora, M.A., Skarmeta, A.F.	User-centric smart buildings for energy sustainable smart cities	2014	Energy efficiency	Smart buildings	Empirical	Europe

(Continued)

TABLE 6.A1 ***(Continued)***

Publications Selected for Systematic Literature Review

Ref.	Authors	Title	Year	Categories	Aim	Approach	Continent
Caponio et al. (2015)	Caponio, G., Massaro, V., Mossa, G., Mummolo, G.	Strategic energy planning of residential buildings in a smart city: a system dynamics approach	2015	Energy efficiency	Energy saving	Empirical	Europe
Sanseverino et al. (2015)	Sanseverino, E.R., Scaccianoce, G., Vaccaro, V., Carta, M., Sanseverino, R.R.	Smart cities and municipal building regulation for energy efficiency	2015	Energy efficiency	Smart buildings	Theoretical	Europe
Lützenberger et al. (2015)	Lützenberger, M., Masuch, N., Küster, T., Freund, D., Voß, M., Hrabia, C.-E., Pozo, D., Fähndrich, J., Trollmann, F., Keiser, J., Albayrak, S.	A common approach to intelligent energy and mobility services in a smart city environment	2015	Energy efficiency	Electric vehicle	Empirical	Europe
Jablonski (2015)	Jablonski, I.	Smart transducer interface-from networked on-site optimization of energy balance in research-demonstrative office building to smart city conception	2015	Energy efficiency	Smart buildings	Empirical	Europe
Aslam et al. (2016)	Aslam, S., Hasan, N.U., Jang, J.W., Lee, K.-G.	Optimized energy harvesting, cluster-head selection and channel allocation for IoTs in smart cities	2016	Energy efficiency	Smart grid	Empirical	Asia
Maier(2016)	Maier, S.	Smart energy systems for smart city districts: case study Reininghaus district	2016	Renewable energy	Hybrid energy system	Empirical	Oceania

Palomar et al. (2016)	Palomar, E., Chen, X., Liu, Z., Maharjan, S., Bowen, J.	Component-based modelling for scalable smart city systems interoperability: a case study on integrating energy demand response systems	2016	Energy efficiency	Smart buildings	Empirical	Europe
Wang et al. (2016)	Wang, S., Wang, X., Wang, Z.L., Yang, Y.	Efficient scavenging of solar and wind energies in a smart city	2016	Renewable energy	Hybrid energy system	Empirical	Asia
Maltese et al. (2016)	Maltese, I., Mariotti, I., Boscacci, F.	Smart city, urban performance and energy	2016	Energy and urban planning	Other	Theoretical	Europe
Zambon et al. (2016)	Zambon, I., Monarca, D., Cecchini, M., Bedini, R., Longo, L., Romagnoli, M., Marucci, A.	Alternative energy and the development of local rural contexts: an approach to improve the degree of smart cities in the central-southern Italy	2016	Renewable energy	Hybrid energy system	Empirical	Europe
Mahapatra et al. (2017)	Mahapatra, C., Moharana, A.K., Leung, V.C.M.	Energy management in smart cities based on internet of things: peak demand reduction and energy savings	2017	Energy efficiency	Smart grid	Empirical	North America
Krozer(2017)	Krozer, Y.	Innovative offices for smarter cities, including energy use and energy-related carbon dioxide emissions	2017	Energy efficiency	Smart buildings	Theoretical	Europe
Chen et al. (2017)	Chen, Y., Ardila-Gomez, A., Frame, G.	Achieving energy savings by intelligent transportation systems investments in the context of smart cities	2017	Energy efficiency	Smart mobility	Theoretical	North America
Carli et al. (2017)	Carli, R., Dotoli, M., Pellegrino, R.	A hierarchical decision-making strategy for the energy management of smart cities	2017	Energy and urban planning	Other	Theoretical	Europe

(Continued)

TABLE 6.A1 *(Continued)*

Publications Selected for Systematic Literature Review

Ref.	Authors	Title	Year	Categories	Aim	Approach	Continent
Brundu et al. (2017)	Brundu, F.G., Patti, E., Osello, A., Giudice, M.D., Rapetti, N., Krylovskiy, A., Jahn, M., Verda, V., Guelpa, E., Rietto, L., Acquaviva, A.	IoT software infrastructure for energy management and simulation in smart cities	2017	Energy efficiency	Smart grid	Empirical	Europe
Oldenbroek et al. (2017)	Oldenbroek, V., Verhoef, L.A., van Wijk, A.J.M.	Fuel cell electric vehicle as a power plant: fully renewable integrated transport and energy system design and analysis for smart city areas	2017	Renewable energy	Hybrid energy system	Empirical	Europe
Nanni et al. (2017)	Nanni, S., Benetti, E., Mazzini, G.	Indoor monitoring in public buildings: workplace wellbeing and energy consumptions. An example of IoT for smart cities application	2017	Energy efficiency	Smart grid	Empirical	Europe
Kalli(2017)	Kalli, M.	Energy solutions for future smart cities	2017	Energy and urban planning	Other	Theoretical	North America
Hung and Peng (**2017**)	Hung, P., Peng, K.	Green-energy, water-autonomous greenhouse system: an alternative-technology approach towards sustainable smart-green vertical greening in smart cities	2017	Energy and urban planning	Sustainable environment/ vertical greening	Empirical	Asia
Bhati et al. (2017)	Bhati, A., Hansen, M., Chan, C.M.	Energy conservation through smart homes in a smart city: a lesson for Singapore households	2017	Energy efficiency	Energy saving	Empirical	Asia

Ejaz et al. (2017)	Ejaz, W., Naeem, M., Shahid, A., Anpalagan, A., Jo, M.	Efficient energy management for the internet of things in smart cities	2017	Energy efficiency	Smart grid	Empirical	Asia
Pezzutto et al. (2017)	Pezzutto, S., Mosannenzadeh, F., Grilli, G., Sparber, W.	European union research and development funding on smart cities and their importance on climate and energy goals	2017	Energy and urban planning	Other	Theoretical	Europe
Manshahia(2018)	Manshahia, M.S.	Swarm intelligence-based energy-efficient data delivery in WSAN to virtualise IoT in smart cities	2018	Energy efficiency	Smart grid	Empirical	Asia
Chui et al. (2018)	Chui, K.T., Lytras, M.D., Visvizi, A.	Energy sustainability in smart cities: artificial intelligence, smart monitoring, and optimization of energy consumption	2018	Energy efficiency	Smart grid	Empirical	Asia
Barresi (2018a)	Barresi, A.	Urban densification and energy efficiency in smart cities: the verge project (Switzerland)	2018	Renewable energy	Solar energy	Empirical	Europe
Colmenar-Santos et al. (2018)	Colmenar-Santos, A., Molina-Ibáñez, E.-L., Rosales-Asensio, E., López-Rey, Á.	Technical approach for the inclusion of superconducting magnetic energy storage in a smart city	2018	Energy efficiency	Smart grid	Empirical	Europe
Cui et al. (2018)	Cui, J., Yoon, H., Youn, B.D.	An omnidirectional biomechanical energy harvesting (OBEH) sidewalk block for a self-generative power grid in a smart city	2018	Renewable energy	Omnidirectional biomechanical energy harvesting	Empirical	Asia

(Continued)

TABLE 6.A1 ***(Continued)***

Publications Selected for Systematic Literature Review

Ref.	Authors	Title	Year	Categories	Aim	Approach	Continent
Calvillo et al. (2018)	Calvillo, C.F., Sanchez-Miralles, A., Villar, J.	Synergies of electric urban transport systems and distributed energy resources in smart cities	2018	Energy efficiency	Smart grid	Empirical	Europe
Hayashi et al. (2018)	Hayashi, Y., Fujimoto, Y., Ishii, H., Takenobu, Y., Kikusato, H., Yoshizawa, S., Amano, Y., Tanabe, S.-I., Yamaguchi, Y., Shimoda, Y., Yoshinaga, J., Watanabe, M., Sasaki, S., Koike T., Jacobsen H.-A., Tomsovic K.	Versatile modeling platform for cooperative energy management systems in smart cities	2018	Energy efficiency	Smart grid	Empirical	Asia
Kai et al. (2018)	Kai, C., Li, H., Xu, L., Li, Y., Jiang, T.	Energy-efficient device-to-device communications for green smart cities	2018	Energy efficiency	Smart grid	Empirical	Asia
Zhou et al. (2018)	Zhou, L., Wu, D., Chen, J., Dong, Z.	Greening the smart cities: energy-efficient massive content delivery via D2D communications	2018	Energy efficiency	Smart grid	Empirical	Asia
Lu et al. (2018)	Lu, W., Gong, Y., Liu, X., Wu, J., Peng, H.	Collaborative energy and information transfer in green wireless sensor networks for smart cities	2018	Energy efficiency	Smart grid	Empirical	Asia

Deakin and Reid (**2018**)	Deakin, M., Reid, A.	Smart cities: under-gridding the sustainability of city-districts as energy efficient-low carbon zones	2018	Energy and urban planning	Other	Theoretical	Europe
Rostirolla et al. (2018)	Rostirolla, G., Righi, R.D.R., Barbosa, J.L.V., Da Costa, C.A.	Elcity: an elastic multilevel energy saving model for smart cities	2018	Energy efficiency	Smart grid	Empirical	Latin America
Andreucci (2018)	Andreucci, M.B.	Linking future energy systems with heritage requalification in smart cities. On-going research and experimentation in the city of Trento (IT)	2018	Energy efficiency	Smart grid	Empirical	Europe
Abdullah et al. (2018)	Abdullah, M.A., Al-Hadhrami, T., Tan, C.W., Yatim, A.H.	Towards green energy for smart cities: particle swarm optimization based MPPT approach	2018	Renewable energy	Wind energy	Empirical	Asia
Pieroni et al. (2018)	Pieroni, A., Scarpato, N., Di Nunzio, L., Fallucchi, F., Raso, M.	Smarter city: smart energy grid based on blockchain technology	2018	Energy efficiency	Smart grid	Theoretical	Europe
Causone et al. (2018)	Causone, F., Sangalli, A., Pagliano, L., Carlucci, S.	Assessing energy performance of smart cities	2018	Energy and urban planning	Other	Theoretical	Europe
Mutule et al. (2018)	Mutule, A.; Teremranova, J.; Antoskovs, N.	Smart city through a flexible approach to smart energy	2018	Energy and urban planning	Other	Theoretical	Europe
Condotta and Borga (**2018**)	Condotta, M.; Borga, G.	Urban energy performance monitoring for smart city decision support environments	2018	Energy efficiency	Smart grid	Empirical	Europe

(Continued)

TABLE 6.A1 ***(Continued)***

Publications Selected for Systematic Literature Review

Ref.	Authors	Title	Year	Categories	Aim	Approach	Continent
Barresi (2018)	Barresi, A.	Urban densification and energy efficiency in smart cities—the verge project (Switzerland)	2018	Renewable energy	Solar energy	Empirical	Europe
Lanini and Barsanti (**2018**)	Lanini, L.; Barsanti, E.A.	Hybrid building as social and energy hub for smart cities: Unite 2.0, a prototype	2018	Energy efficiency	Smart buildings	Empirical	Europe
Luo et al. (2019)	Luo, H., Cai, H., Yu, H., Sun, Y., Bi, Z., Jiang, L.	A short-term energy prediction system based on edge computing for smart city	2019	Energy efficiency	Monitoring and energy efficiency	Empirical	Asia
Naseer et al. (2019)	Naseer, S., Liu, W., Sarkar, N.I.	Energy-efficient massive data dissemination through vehicle mobility in smart cities	2019	Energy efficiency	Network transmission	Empirical	Europe
Sato et al. (2019b)	Sato, M., Fukuyama, Y., Iizaka, T., Matsui, T.	Total optimization of energy networks in a smart city by multi-swarm differential evolutionary particle swarm optimization	2019	Energy efficiency	Smart grid	Empirical	Asia
Yu and Zhang (**2019**)	Yu, Y., Zhang, N.	Does smart city policy improve energy efficiency? Evidence from a quasi-natural experiment in China	2019	Energy efficiency	Conceptualization of energy efficiency	Empirical	Asia
Le et al. (2019)	Le, L.T., Nguyen, H., Dou, J., Zhou, J.	A comparative study of PSO-ANN, GA-ANN, ICA-ANN, and ABC-ANN in estimating the heating load of buildings' energy efficiency for smart city planning	2019	Energy efficiency	Energy planning	Empirical	Asia

Abdullah et al. (2019)	Abdullah, A., Yusoff, S.H., Zaini, S.A., Midi, N.S., Mohamad, S.Y.	Energy efficient smart street light for smart city using sensors and controller	2019	Energy efficiency	LED lighting	Empirical	Asia
Haarstad and Wathne (**2019**)	Haarstad, H., Wathne, M.W.	Are smart city projects catalysing urban energy sustainability?	2019	Renewable energy	Hybrid energy system	Theoretical	Europe
Corsini et al. (2019)	Corsini, F., Certomà, C., Dyer, M., Frey, M.	Participatory energy: research, imaginaries and practices on people' contribute to energy systems in the smart city	2019	Energy and urban planning	Network transmission	Theoretical	Europe
Aujla et al. (2019)	Aujla, G.S., Kumar, N., Singh, M., Zomaya, A.Y.	Energy trading with dynamic pricing for electric vehicles in a smart city environment	2019	Energy efficiency	Electric vehicles	Empirical	Asia
Hati et al. (2019)	Hati, S., Dey, P., De, D.	WLAN based energy efficient smart city design	2019	Energy efficiency	Energy saving	Empirical	Asia
Liu et al. (2019)	Liu, Y., Yang, C., Jiang, L., Xie, S., Zhang, Y.	Intelligent edge computing for IoT-based energy management in smart cities	2019	Energy efficiency	Smart grid	Empirical	Asia
Petritoli et al. (2019)	Petritoli, E., Leccese, F., Pizzuti, S., Pieroni, F.	Smart lighting as basic building block of smart city: an energy performance comparative case study	2019	Energy efficiency	LED lighting	Empirical	Europe
Muhammad et al. (2019)	Muhammad, K., Lloret, J., Baik, S.W.	Intelligent and energy-efficient data prioritization in green smart cities: current challenges and future directions	2019	Energy efficiency	Smart grid	Theoretical	Asia
Ahuja and Khosla (**2019**)	Ahuja, K., Khosla, A.	Network selection criterion for ubiquitous communication provisioning in smart cities for smart energy system	2019	Energy efficiency	Smart grid	Empirical	Asia

(Continued)

TABLE 6.A1 ***(Continued)***

Publications Selected for Systematic Literature Review

Ref.	Authors	Title	Year	Categories	Aim	Approach	Continent
Vázquez-Canteli et al. (2019)	Vázquez-Canteli, J.R., Ulyanin, S., Kämpf, J., Nagy, Z.	Fusing TensorFlow with building energy simulation for intelligent energy management in smart cities	2019	Energy efficiency	Smart grid	Empirical	North America
Cioara et al. (2019)	Cioara, T., Anghel, I., Salomie, I., Antal, M., Pop, C., Bertoncini, M., Arnone, D., Pop, F.	Exploiting data centres energy flexibility in smart cities: business scenarios	2019	Energy efficiency	Smart grid	Empirical	Europe
Parks(2019)	Parks, D.	Energy efficiency left behind? Policy assemblages in Sweden's most climate-smart city	2019	Energy efficiency	Smart grid	Theoretical	Europe
Gitelman et al. (2019)	Gitelman, L.D., Kozhevnikov, M.V., Adam, L.A.	Sustainable energy for smart city	2019	Energy and urban planning	Other	Theoretical	Europe
Al-Nory (2019)	Al-Nory, M.T.	Optimal decision guidance for the electricity supply chain integration with renewable energy: aligning smart cities research with sustainable development goals	2019	Renewable energy	Electricity supply chain	Theoretical	Asia
Nagy et al. (2019)	Nagy, Z., Sebestyén Szép, T., Szendi, D.	Regional disparities in Hungarian urban energy consumption—a link between smart cities and successful cities	2019	Energy and urban planning	Other	Theoretical	Europe

Aymen and Mahmoudi (**2019**)	Aymen, F., Mahmoudi, C.	A novel energy optimization approach for electrical vehicles in a smart city	2019	Energy efficiency	Smart grid	Empirical	Africa
Wei and Hu (**2019**)	Wei, L., Hu, Y.	Research on new and old kinetic energy transformation supported by smart city construction in big data era	2019	Energy and urban planning	Urban experimentation	Empirical	Asia
Sato et al. (2019)	Sato, M., Fukuyama, Y., Iizaka, T., Matsui, T.	Total optimization of energy networks in a smart city by multi-population global-best modified brain storm optimization with migration	2019	Energy efficiency	Smart grid	Empirical	Asia
O'Dwyer et al. (2019)	O'Dwyer, E.; Pan, I.; Acha, S.; Shah, N.	Smart energy systems for sustainable smart cities: current developments, trends and future directions	2019	Energy and urban planning	Other	Theoretical	Europe
Huang et al. (2020)	Huang, X., Huang, P., Huang, T.	Multi-objective optimization of digital management for renewable energies in smart cities	2020	Renewable energy	Hybrid energy system	Empirical	Asia
Khattak et al. (2020)	Khattak, H.A., Tehreem, K., Almogren, A., Ameer, Z., Din, I.U., Adnan, M.	Dynamic pricing in industrial internet of things: blockchain application for energy management in smart cities	2020	Energy efficiency	Energy saving	Theoretical	Asia
Pilipczuk (2020)	Pilipczuk, O.	Sustainable smart cities and energy management: the labor market perspective	2020	Energy and urban planning	Energy manager profession	Theoretical	Europe

(Continued)

TABLE 6.A1 ***(Continued)***

Publications Selected for Systematic Literature Review

Ref.	Authors	Title	Year	Categories	Aim	Approach	Continent
Lewandowska et al. (2020)	Lewandowska, A., Chodkowska-Miszczuk, J., Rogatka, K., Starczewski, T.	Smart energy in a smart city: utopia or reality? Evidence from Poland	2020	Renewable energy	Hybrid energy system	Theoretical	Europe
Anthony Jnr (2020)	Anthony Jnr, B.	Smart city data architecture for energy presumption in municipalities: concepts, requirements, and future directions	2020	Energy efficiency	Energy distribution	Theoretical	Asia
Chithaluru et al. (2020)	Chithaluru, P., Al-Turjman, F., Kumar, M., Stephan, T.	I-AREOR: an energy-balanced clustering protocol for implementing green IoT in smart cities	2020	Energy efficiency	Energy planning	Empirical	Asia
Gonçalves et al. (2020)	Gonçalves, D., Sheikhnejad, Y., Oliveira, M., Martins, N.	One step forward toward smart city utopia: smart building energy management based on adaptive surrogate modelling	2020	Energy efficiency	Smart building	Empirical	Europe
Giourka et al. (2020)	Giourka, P., Apostolopoulos, V., Angelakoglou, K., Kourtzanidis, K., Nikolopoulos, N., Sougkakis, V., Fuligni, F., Barberis, S., Verbeek, K., Costa, J.M., Formiga, J.	The nexus between market needs and value attributes of smart city solutions towards energy transition. An empirical evidence of two European Union (EU) smart cities, Evora and Alkmaar	2020	Energy and urban planning	Urban energy planning	Empirical	Europe

Angelakoglou et al. (2020)	Angelakoglou, K., Kourtzanidis, K., Giourka, P., Apostolopoulos, V., Nikolopoulos, N., Kantorovitch, J.	From a comprehensive pool to a project-specific list of key performance indicators for monitoring the positive energy transition of smart cities—an experience-based approach	2020	Energy and urban planning	Other	Theoretical	Europe
Kumar (2020)	Kumar, D.	Urban energy system management for enhanced energy potential for upcoming smart cities	2020	Renewable energy	Hybrid energy system	Theoretical	Asia
Parks (2020)	Parks, D.	Promises and techno-politics: renewable energy and Malmö's vision of a climate-smart city	2020	Energy efficiency	Energy planning	Theoretical	Europe
Gaska and Generowicz (**2020**)	Gaska, K., Generowicz, A.	Smart computational solutions for the optimization of selected technology processes as an innovation and progress in improving energy efficiency of smart cities—a case study	2020	Energy efficiency	Smart grid	Empirical	Europe
Palanca et al. (2020)	Palanca, J., Jordán, J., Bajo, J., Botti, V.	An energy-aware algorithm for electric vehicle infrastructures in smart cities	2020	Energy efficiency	Electric vehicle	Empirical	Europe
Li et al. (2020)	Li, L., Zheng, Y., Zheng, S., Ke, H.	The new smart city programme: evaluating the effect of the internet of energy on air quality in China	2020	Energy and urban planning	Internet of energy	Empirical	Asia
Pei et al. (2020)	Pei, P., Huo, Z., Martínez, O.S., Crespo, R.G.	Minimal green energy consumption and workload management for data centres on smart city platforms	2020	Renewable energy	Hybrid energy system	Empirical	Asia

(Continued)

TABLE 6.A1 ***(Continued)***

Publications Selected for Systematic Literature Review

Ref.	Authors	Title	Year	Categories	Aim	Approach	Continent
Jettanasen et al. (2020)	Jettanasen, C., Songsukthawan, P., Ngaopitakkul, A.	Energy harvesting for smart city applications	2020	Renewable energy	Hybrid energy system	Empirical	Asia
Ye et al. (2020)	Ye, H., Li, F.M., Liu, Z.X., Deng, X.D.	A green energy consumption policy of Bluetooth mobile devices for smart cities	2020	Energy and urban planning	Green computing mode	Empirical	Asia
Anthony Jnr et al. (2020)	Anthony Jnr, B., Abbas Petersen, S., Ahlers, D., Krogstie, J.	API deployment for big data management towards sustainable energy presumption in smart cities-a layered architecture perspective	2020	Energy efficiency	Smart grid	Empirical	Europe
Gitelman et al. (2020)	Gitelman, L.D., Kozhevnikov, M.V., Starikov, E.M., Gamburg, A.V.	Sustainable energy in smart cities	2020	Energy efficiency	Smart grid	Theoretical	Asia
Deakin et al. (2020)	Deakin, M., Reid, A., Mora, L.	Smart cities: the metrics of future Internet-based developments and renewable energies of urban and regional innovation	2020	Renewable energy	Hybrid energy system	Theoretical	Europe
Serban and Lytras (**2020**)	Serban, A.C., Lytras, M.D.	Artificial intelligence for smart renewable energy sector in Europe—smart energy infrastructures for next generation smart cities	2020	Renewable energy	Hybrid energy system	Empirical	Europe

Mekhum (2020)	Mekhum, W.	Smart cities: impact of renewable energy consumption, information and communication technologies and e-governance on CO_2 emission	2020	Renewable energy	Hybrid energy system	Theoretical	Asia
Sayah et al. (2020)	Sayah, Z., Kazar, O., Lejdel, B., Laouid, A., Ghenabzia, A.	An intelligent system for energy management in smart cities based on big data and ontology	2020	Energy efficiency	Energy saving	Theoretical	Africa
Petrović and Kocić (**2020**)	Petrović, N., Kocić, D.	Data-driven framework for energy-efficient smart cities	2020	Energy efficiency	Electric vehicle	Theoretical	Europe
Fadeyi et al. (2020)	Fadeyi, O., Krejcar, O., Maresova, P., Kuca, K., Brida, K., Selamat, A.	Opinions on sustainability of smart cities in the context of energy challenges posed by cryptocurrency mining	2020	Energy efficiency	Energy saving	Theoretical	Europe
Strielkowski et al. (2020)	Strielkowski, W., Veinbender, T., Tvaronavičienė, M., Lace, N.	Economic efficiency and energy security of smart cities	2020	Energy efficiency	LED lighting	Theoretical	Europe
Abbas et al. (2020)	Abbas, S., Khan, M.A., Falcon-Morales, L.E., Rehman, A., Saeed, Y., Zareei, M., Zeb, A., Mohamed, E.M.	Modeling, simulation and optimization of power plant energy sustainability for IoT enabled smart cities empowered with deep extreme learning machine	2020	Energy efficiency	Energy planning	Empirical	Central America
Oldenbroek et al. (2020)	Oldenbroek, V., Smink, G., Salet, T., van Wijk, A.J.M.	Fuel cell electric vehicle as a power plant: techno-economic scenario analysis of a renewable integrated transportation and energy system for smart cities in two climates	2020	Renewable energy	Hydrogen	Empirical	Europe

(Continued)

TABLE 6.A1 ***(Continued)***

Publications Selected for Systematic Literature Review

Ref.	Authors	Title	Year	Categories	Aim	Approach	Continent
Shu et al. (2020)	Shu, M.; Wu, S.Z.; Wu, T.; Qiao, Z.L.; Wang, N.; Xu, F.; Shanthini, A.; Muthu, B.A.	Efficient energy consumption system using heuristic renewable demand energy optimization in smart city	2020	Renewable energy	Hybrid energy system	Empirical	Asia
Tanwar et al. (2021)	Tanwar, S.; Popat, A.; Bhattacharya, P.; Gupta, R.; Kumar, N.	A taxonomy of energy optimization techniques for smart cities: architecture and future directions	2020	Energy efficiency	Decentralized systems	Theoretical	Asia
Arif et al. (2020)	Arif, A.; Barrigon, F.A.; Gregoretti, F.; Iqbal, J.; Lavagno, L.; Lazarescu, M.T.; Ma, L.; Palomino, M.; Segura, J.L.L.	Performance and energy-efficient implementation of a smart city application on FPGAs	2020	Energy efficiency	Smart grid	Empirical	Europe
Biancardi et al. (2021)	Biancardi, M., Di Bari, A., Villani, G.	R&D investment decision on smart cities: energy sustainability and opportunity	2021	Energy and urban planning	Renewable energy and urban planning	Empirical	Europe
Kim et al. (**2021**)	Cho, K., Yang, J., Kim, T., Jang, W.	Influence of building characteristics and renovation techniques on the energy-saving performances of EU smart city projects	2021	Energy and urban planning	Smart buildings	Theoretical	Asia
Duan et al. (2021)	Duan, P., Askari, M., Hemat, K., Ali, Z.M.	Optimal operation and simultaneous analysis of the electric transport systems and distributed energy resources in the smart city	2021	Energy efficiency	Energy distribution	Empirical	Asia

Kuznetsov et al. (2021)	Kuznetsov, P., Rimar, M., Yakimovich, B., Kulikova, O., Lopusniak, M., Voronin, D., Evstigneev, V.	Parametric optimization of combined wind-solar energy power plants for sustainable smart city development	2021	Renewable energy	Combined wind-solar energy power plants	Theoretical	Europe
Hussain et al. (2021)	Hussain, M.M., Akram, R., Memon, Z.A., Nazir, M.H., Javed, W., Siddique, M.	Demand side management techniques for home energy management systems for smart cities	2021	Energy efficiency	Combined energy sources	Theoretical	Europe
Sabory et al. (2021)	Sabory, N.R., Senjyu, T., Danish, M.S.S., Hosham, A., Noorzada, A., Amiri, A.S., Muhammdi, Z.	Applicable smart city strategies to ensure energy efficiency and renewable energy integration in poor cities: Kabul case study	2021	Energy and urban planning	Urban energy planning	Theoretical	Asia
Ghadami et al. (2021)	Ghadami, N., Gheibi, M., Kian, Z., Faramarz, M.G., Naghedi, R., Eftekhari, M., Fathollahi-Fard, A.M., Dulebenets, M.A., Tian, G.	Implementation of solar energy in smart cities using an integration of artificial neural network, photovoltaic system and classical Delphi methods	2021	Renewable energy	Solar energy	Empirical	Asia
Boeri et al. (2021)	Boeri, A., Boulanger, S.O.M., Turci, G., Pagliula, S.	Enabling strategies for mixed-used PEDs: energy efficiency between smart cities and Industry 4.0 (strategie e tecnologie abilitanti per ped misti: efficienza tra smart cities e industria 4.0)	2021	Energy and urban planning	Urban energy planning	Empirical	Europe

(Continued)

TABLE 6.A1 ***(Continued)***

Publications Selected for Systematic Literature Review

Ref.	Authors	Title	Year	Categories	Aim	Approach	Continent
Xia et al. (2021)	Xia, X., Wu, X., BalaMurugan, S., Marimuthu, K.	Effect of environmental and social responsibility in energy-efficient management models for smart cities infrastructure	2021	Energy efficiency	LED lighting	Theoretical	Asia
Shen et al. (2021)	Shen, X., Yu, H., Liu, X., Bin, Q., Luhach, A.K., Saravanan, V.	The optimized energy-efficient sensible edge processing model for the internet of vehicles in smart cities	2021	Energy efficiency	Electric vehicle	Empirical	Asia
Lu et al. (2021)	Lu, C.-W., Huang, J.-C., Chen, C., Shu, M.-H., Hsu, C.-W., Tapas Bapu, B.R.	An energy-efficient smart city for sustainable green tourism industry	2021	Energy and urban planning	Tourism	Theoretical	Asia
Wang et al. (2021a)	Wang, C., Gu, J., Sanjuán Martínez, O., González Crespo, R.	Economic and environmental impacts of energy efficiency over smart cities and regulatory measures using a smart technological solution	2021	Energy efficiency	Conceptualization of energy efficiency	Theoretical	Asia
Sami et al. (2021)	Sami, M.S., Abrar, M., Akram, R., Hussain, M.M., Nazir, M.H., Khan, M.S., Raza, S.	Energy management of microgrids for smart cities: a review	2021	Energy efficiency	Micro-grids	Theoretical	Europe
Govindarajan and Ganesh (**2021**)	Govindarajan, H.K., L.S. G.	Renewable energy for electricity use in India: evidence from India's smart cities mission	2021	Renewable energy	Hybrid energy system	Theoretical	Asia

Carrera et al. (2021)	Carrera, B., Peyrard, S., Kim, K.	Meta-regression framework for energy consumption prediction in a smart city: a case study of Songdo in South Korea	2021	Energy efficiency	Energy planning	Theoretical	Asia
Zhang et al. (2021)	Zhang, X., Manogaran, G., Muthu, B.	IoT enabled integrated system for green energy into smart cities	2021	Energy efficiency	Energy saving	Empirical	Asia
Ashwin et al. (2021)	Ashwin, M., Alqahtani, A.S., Mubarakali, A.	IoT based intelligent route selection of wastage segregation for smart cities using solar energy	2021	Energy efficiency	Solar energy	Empirical	Asia
Abu-Rayash and Dincer (**2021**)	Abu-Rayash, A., Dincer, I.	Development and analysis of an integrated solar energy system for smart cities	2021	Renewable energy	Solar energy	Empirical	North America
Gorla and Chamola (**2021**)	Gorla, P., Chamola, V.	Battery lifetime estimation for energy efficient telecommunication networks in smart cities	2021	Energy efficiency	Solar energy	Empirical	Asia
Khalil et al. (2021)	Khalil, M.I., Jhanjhi, N.Z., Humayun, M., Sivanesan, S., Masud, M., Hossain, M.S.	Hybrid smart grid with sustainable energy efficient resources for smart cities	2021	Renewable energy	Hybrid energy system	Empirical	Asia
Abdel-Basset et al. (2021)	Abdel-Basset, M., Hawash, H., Chakrabortty, R.K., Ryan, M.	Energy-net: a deep learning approach for smart energy management in IoT-based smart cities	2021	Energy and urban planning	Internet-of-things	Empirical	Africa
Zhang and Yue (**2021**)	Zhang, W., Yue, M.	The application of building energy management system based on IoT technology in smart city	2021	Energy efficiency	Hybrid energy system	Empirical	Asia

(Continued)

TABLE 6.A1 ***(Continued)***

Publications Selected for Systematic Literature Review

Ref.	Authors	Title	Year	Categories	Aim	Approach	Continent
Hoang and Nguyen (**2021**)	Hoang, A.T., Pham, V.V., Nguyen, X.P.	Integrating renewable sources into energy system for smart city as a sagacious strategy towards clean and sustainable process	2021	Renewable energy	Hybrid energy system	Theoretical	Asia
Mutule et al. (2021)	Mutule, A., Domingues, M., Ulloa-Vásquez, F., Carrizo, D., García-Santander, L., Dumitrescu, A.-M., Issicaba, D., Melo, L.	Implementing smart city technologies to inspire change in consumer energy behaviour	2021	Energy efficiency	Energy consumer	Empirical	Europe
Miyasawa et al. (2021)	Miyasawa, A., Akira, S., Fujimoto, Y., Hayashi, Y.	Spatial demand forecasting based on smart meter data for improving local energy self-sufficiency in smart cities	2021	Energy and urban planning	Urban energy planning	Theoretical	Asia
Bachanek et al. (2021)	Bachanek, K.H., Tundys, B., Wiśniewski, T., Puzio, E., Maroušková, A.	Intelligent street lighting in a smart city concepts: a direction to energy saving in cities: an overview and case study	2021	Renewable energy	Hybrid energy system	Empirical	Europe
Mahmood et al. (2021)	Mahmood, D., Javaid, N., Ahmed, G., Khan, S., Monteiro, V.	A review on optimization strategies integrating renewable energy sources focusing uncertainty factor—paving path to eco-friendly smart cities	2021	Renewable energy	Hybrid energy system	Theoretical	Asia
Zekić-Sušac et al. (2021)	Zekić-Sušac, M., Mitrović, S., Has, A.	Machine learning based system for managing energy efficiency of public sector as an approach towards smart cities	2021	Energy efficiency	Energy planning	Empirical	Europe

Hajduk and Jelonek (**2021**)	Hajduk, S., Jelonek, D.	A decision-making approach based on TOPSIS method for ranking smart cities in the context of urban energy	2021	Energy and urban planning	Energy planning	Theoretical	Europe
Vrabie (2021)	Vrabie, C.	Converting municipal waste to energy through the biomass chain, a key technology for environmental issues in (smart) cities	2021	Renewable energy	Biomass	Theoretical	Europe
Xiaoyi et al. (2021)	Xiaoyi, Z., Dongling, W., Yuming, Z., Manokaran, K.B., Benny Antony, A.	IoT driven framework based efficient green energy management in smart cities using multi-objective distributed dispatching algorithm	2021	Renewable energy	Hybrid energy system	Empirical	Asia
Babar et al. (2021)	Babar, M., Khattak, A.S., Jan, M.A., Tariq, M.U.	Energy aware smart city management system using data analytics and internet of things	2021	Energy efficiency	Energy planning	Empirical	Asia
Tantau and Şanta (**2021**)	Tantau, A., Şanta, A.-M.I.	New energy policy directions in the European Union developing the concept of smart cities	2021	Energy and urban planning	Other	Theoretical	Europe
Swain et al. (2021)	Swain, A., Salkuti, S.R., Swain, K.	An optimized and decentralized energy provision system for smart cities	2021	Energy efficiency	Hybrid energy system	Empirical	Asia
Carrera et al. (**2021**)	Kim, D., Kwon, D., Park, L., Kim, J., Cho, S.	Multiscale LSTM-based deep learning for very-short-term photovoltaic power generation forecasting in smart city energy management	2021	Renewable energy	Solar energy	Empirical	Asia

(*Continued*)

TABLE 6.A1 ***(Continued)***

Publications Selected for Systematic Literature Review

Ref.	Authors	Title	Year	Categories	Aim	Approach	Continent
Uspenskaia et al. (2021)	Uspenskaia, D., Specht, K., Kondziella, H., Bruckner, T.	Challenges and barriers for net-zero/positive energy buildings and districts—empirical evidence from the smart city project Sparcs	2021	Energy efficiency	Hybrid energy system	Empirical	Europe
Elkamel et al. (2021)	Elkamel, M., Ahmadian, A., Diabat, A., Zheng, Q.P.	Stochastic optimization for price-based unit commitment in renewable energy-based personal rapid transit systems in sustainable smart cities	2021	Energy efficiency	Hybrid energy system	Empirical	North America
Avotins et al. (2021)	Avotins, A., Adrian, L.R., Porins, R., Apse-Apsitis, P., Ribickis, L.	Smart city street lighting system quality and control issues to increase energy efficiency and safety	2021	Energy and urban planning	Street lighting	Empirical	Europe
Vukovic et al. (2021)	Vukovic, N., Koriugina, U., Illarionova, D., Pankratova, D., Kiseleva, P., Gontareva, A.	Towards smart green cities: analysis of integrated renewable energy use in smart cities	2021	Energy and urban planning	Energy planning	Theoretical	Europe
Nesmachnow et al. (2020)	Nesmachnow, S., Colacurcio, G., Rossit, D.G., Toutouh, J., Luna, F.	Optimizing household energy planning in smart cities: a multiobjective approach (optimización de la planificación energética en hogares inteligentes: un enfoque multi-objetivo)	2021	Energy efficiency	Energy planning	Empirical	Latin America

Martín et al. (2021)	Martín, C., Castillo-Calzadilla, T., Zabala, K., Arrizabalaga, E., Hernández, P., Mabe, L., López, J.R., Casado, J.M., Santos, M.N., Guardo, J., Molinete, B.	The opportunity for smart city projects at municipal scale: implementing a positive energy district in Zorrozaurre	2021	Energy and urban planning	Urban energy planning	Empirical	Europe
Chen and Han (**2021**)	Chen, L., Han, P.	The construction of a smart city energy efficiency management system oriented to the mobile data aggregation of the internet of things	2021	Energy efficiency	Energy saving	Theoretical	Asia
Meenakshi et al. (2021)	Meenakshi, Mothi Kumar, K.E., Kumari, N.	A geo-spatial approach for quantifying rooftop photovoltaic energy potential in Karnal smart city, Haryana-a case study	2021	Renewable energy	Solar energy	Empirical	Asia
Wang et al. (2021)	Wang, X., Chen, Q., Wang, J.	Fuzzy rough set based sustainable methods for energy efficient smart city development	2021	Energy and urban planning	Solar energy	Empirical	Asia
Farid et al. (2021)	Farid, A.M., Alshareef, M., Badhesha, P.S., Boccaletti, C., Cacho, N.A.A., Carlier, C.-I., Corriveau, A., Khayal, I., Liner, B., Martins, J.S.B., Rahimi, F., Rossetti, R., Schoonenberg, W.C.H., Stillwell, A., Wang, Y.	Smart city drivers and challenges in energy and water systems	2021	Energy efficiency	Electric vehicle	Theoretical	North America

(Continued)

TABLE 6.A1 ***(Continued)***

Publications Selected for Systematic Literature Review

Ref.	Authors	Title	Year	Categories	Aim	Approach	Continent
Shin and Cho (**2021**)	V E S., Shin, C., Cho, Y.	Efficient energy consumption prediction model for a data analytic-enabled industry building in a smart city	2021	Energy efficiency	Energy consumer	Empirical	Asia
Martins et al. (2021)	Martins, F.; Patrao, C.; Moura, P.; de Almeida, A.T.	A review of energy modeling tools for energy efficiency in smart cities	2021	Energy efficiency	Energy efficiency planning	Theoretical	Europe
Verma (2021)	Verma, S.	Energy-efficient routing paradigm for resource-constrained internet of things-based cognitive smart city	2021	Energy efficiency	Clusters	Empirical	Asia
Hadidi et al. (2021)	Hadidi, L.A.; Rahman, S.M.; Maghrabi, A.T.	Smart city: a sustainable solution for enhancing energy efficiency and climate change mitigation in Saudi Arabia	2021	Energy efficiency	Savings in power	Empirical	Asia
Nuvvula et al. (2021)	Nuvvula, R.; Devaraj, E.; Srinivasa, K.T.	A comprehensive assessment of large-scale battery integrated hybrid renewable energy system to improve sustainability of a smart city	2021	Renewable energy	Solar energy	Empirical	Asia

7

Trends in Heat Vulnerability Research

7.1 Introduction and Background

Amid the rapid escalation of climate change and increasing global urbanisation, global warming and the urban heat island (UHI) effect have worsened notably in recent decades (Chapman et al., 2017; Kamruzzaman et al., 2018; Nguyen et al., 2022), consequently heightening urban heat vulnerability (UHV). The Intergovernmental Panel on Climate Change (IPCC) defines vulnerability as the extent to which a system, community, or society is at risk from the adverse effects of climate change. This concept includes three key components: Exposure—the nature and degree to which a system or community is exposed to major climatic changes like extreme weather and long-term climate shifts; sensitivity—a degree to which a system or community is affected by climate change in terms of its structure and functioning; and adaptive capacity, which is the ability of systems, communities, or individuals to adapt to potential damage of or cope with the consequences of climate change (Parry, 2007). This study uses this conceptual framework to explore the development and evolution of UHV.

UHI refers to a phenomenon where urban temperatures are noticeably higher than those in the surrounding suburban regions (Degirmenci et al., 2021; Deilami et al., 2018). This phenomenon is primarily attributed to increasingly dense urban structures but dwindling natural landscapes, which directly alter surface albedo and thermal properties (Mohajerani et al., 2017; Mortoja et al., 2020). Furthermore, it is also significantly influenced by (a) the type and characteristics of the building materials, which store and reradiate heat; (b) the influence of urban form, including the layout and height of buildings, that affects wind patterns and shade; and (c) other anthropogenic activities such as energy consumption and vehicular emissions that generate additional heat.

Additionally, local climate conditions and geographical location also play crucial roles in modulating the UHI effect (Deilami et al., 2018; Gago et al., 2013). Intensified UHIs directly contribute to UHV in multiple areas, where UHV is increasingly recognised as a critical issue in urban studies. For example, from the climate perspective, UHI contributes to more frequent and intense extreme heat events, like heatwaves, by extending the duration of high temperatures and narrowing the time window for relief from high-heat exposure (Li et al., 2020; Schatz & Kucharik, 2015; Tan et al., 2010; Zhao

DOI: 10.1201/9781003737049-10

et al., 2018). UHI can also disrupt atmospheric circulation patterns, leading to altered precipitation patterns such as reduced rainfall or increased intensity of storms (Li et al., 2020). Additionally, the heat trapped within urban areas can exacerbate air pollution by promoting the formation of ground-level ozone and other pollutants, thereby contributing to deteriorating air quality (Mika et al., 2018). United Nations. (2018).

From the urban infrastructure perspective, high temperatures negatively impact buildings, roads, and public transportation systems, and affect energy consumption, material durability, and overall functionality (Ghiasi et al., 2021; Sen et al., 2022; Wu et al., 2014). For instance, it is revealed that UHI could result in a median increase of 19% in energy consumption for cooling buildings according to reviewed studies (Li et al., 2019). A global mean surface temperature rise of 1.5 and 2.0°C will increase permanent asphalt pavement deformation by 18.63 and 36.71%, respectively, compared to the baseline year of 2019 (Miao et al., 2022). Ecologically, increased temperatures stress urban vegetation and wildlife, leading to changes in species composition and ecosystem services (Cai et al., 2023; Tuczek et al., 2022). Tomlinson, C. J., Chapman, L., Thornes, J. E., & Baker, C. J. (2011).

IPCC Sixth Assessment Report indicates that at 1.5 and 2°C of warming, 3–14% and 3–18% of terrestrial species face a high risk of extinction, respectively (Parmesan et al., 2022). The study by Arunab and Mathew (2024) reveals significant UHI-pollution links that high-risk areas show 53.5% more NO_2, 21.7% more aerosols, 7.6% more HCHO, 1.2% more CO, and 2.6°C warmer than non-risk areas. The deteriorating urban thermal environment also causes an increase in heat-related hospitalisations and fatalities in the summer months, especially during heatwaves (Ho et al., 2023; Huang et al., 2023; Yadav et al., 2023). For instance, during the 2003 European heatwave, over 70,000 excess deaths were recorded, with urban areas being the most affected (Robine et al., 2008). By the 2050s, urban heat stress in 231 Chinese cities is projected to increase annual labour losses by over 0.20% of GDP compared to the 2010s (He et al., 2022a–c). Therefore, understanding and addressing UHV is crucial for promoting sustainable and resilient urban environments, in alignment with the United Nations' Sustainable Development Goals (SDGs), specifically SDG 3 (Good Health and Wellbeing), SDG 11 (Sustainable Cities and Communities), and SDG 13 (Climate Action) (UN-DESA, 2016).

The research on UHV has evolved significantly over the years due to the dynamic interplay between climate change and global urbanisation. Initially, research in this field primarily focused on exposure risk to urban heat. Early efforts were devoted to investigating the UHI effect in terms of temperature discrepancies, UHI intensity, and spatio-temporal dynamics using climatological and meteorological methodologies (Atkinson, 2003; Jin et al., 2007; Lemonsu et al., 2009). These foundational studies highlighted the broader implications of UHI and thus have fostered a deeper understanding of UHV in relation to heat exposure. Such enhanced understanding has subsequently shifted the research focus from merely understanding exposure risks to exploring the sensitivity and adaptive capacity of both the built and natural environments. In terms of built-environment context, the research attempted to measure the relationship between UHI characteristics, such as distribution and intensity, and city magnitude, morphology, and locations (Hamdi & Schayes, 2008; Steeneveld et al., 2011; Zhou et al., 2017).

Upon identifying areas highly exposed to urban heat, researchers have analysed contributing factors such as land cover, building density, architectural materials, and energy consumption that enhance this sensitivity (Bozdogan Sert et al., 2021; Chen et al., 2020; Heusinger et al., 2023; Zhou & Chen, 2018). This is crucial to assist with formulating effective strategies to mitigate UHV by enhancing the adaptive capacities of these urban environments. From the natural environment perspective, researchers are increasingly concentrating on analysing the impacts of urban heat on ecosystems of vegetation, water body, and wildlife, new insights critical to enhancing the understanding of natural sensitivity (Dai et al., 2019; Kabano et al., 2021; Wang et al., 2021). Besides, the research interest tends to explore how urban greenery and water bodies adjust urban temperature and relieve UHI intensity and frequency (Gunawardena et al., 2017). All the natural environmental research efforts facilitate urban planners to utilise ecological design and environmental planning to reduce the natural sensitivity and enhance the adaptive capacities of urban heat, thereby mitigating UHV issues (Feng et al., 2021; Su et al., 2022).

In recent years, given the occurrence of more frequent extreme heat events or heatwaves and increased heat-related morbidity and mortality, research in the field of UHV has significantly turned towards public health (Habeeb et al., 2015; Ho et al., 2023; Yadav et al., 2023). This research on population vulnerability to urban heat aims to identify and assess the impacts of urban heat, with a particular focus on human and social aspects. Previous studies have revealed that increased urban temperatures influence urban residents in different ways, as various socioeconomic and health conditions interplay. For instance, individuals suffering from conditions such as respiratory and cardiovascular illnesses, as well as those facing socioeconomic challenges like the elderly and low-income groups, are more susceptible to worsening thermal environments (He et al., 2022; Lanza et al., 2023; Zhu & Yuan, 2023). Research related to public health focuses on locating vulnerable areas and identifying populations at risk by integrating multiple factors such as age, income, access to healthcare, and housing conditions that influence the sensitivity and adaptive capacity of individuals (Karanja & Kiage, 2021; Yang et al., 2022). Centers for Disease Control and Prevention (2017).

The multi-factor integration methodology combines epidemiological and social science perspectives with traditional meteorological and urban planning approaches for a comprehensive and integral understanding of UHV. Furthermore, the research explores mitigation strategies and preventive interventions to reduce urban residents' vulnerability to heightened urban temperatures. These solutions incorporate the provision of cooling services, campaign arrangements to enhance public awareness of heatwaves, and the integration of heat health concerns into policymaking (Hatvani-Kovacs et al., 2018; He et al., 2022; Sanchez & Reames, 2019).

It is worth noting that the UHV research field has significantly benefited from advancements in geospatial analysis, remote sensing, and artificial intelligence (AI). These techniques assist in an array of aspects, from data collection to spatio-temporal analysis to more detailed and predictive analysis of the dynamics of UHV, designed to yield more accurate results, locally relevant methodologies, and efficient on-ground measures (Diem et al., 2023; Fahy et al., 2019; Lakshmi, 2023). The historical development and current state of research in UHV indicate that this field is apparently very complex and multidisciplinary and thus requires a multifaceted and interdisciplinary approach, in alignment with broader SDGs.

Numerous review studies on UHI have been published so far, with such topics as definitions and mechanisms, measurement data and techniques, impact analysis, influencing factors, mitigation, and adaptation strategies, as well as future trends and challenges (Coffel et al., 2018; Gago et al., 2013; Li et al., 2019; Stewart, 2011). Nonetheless, limited review studies specifically focus on UHV. These studies mainly concentrate on measuring vulnerability to heat by selecting spatial representative indicators and building models, mostly UHV indices (Bao et al., 2015; Szagri et al., 2023). Their focus involves understanding the interplay between environmental, social, and infrastructural factors that contribute to urban populations' vulnerability to heat hazards. Recent studies highlight the importance of comprehensive frameworks that integrate social, economic, and environmental factors for spatial representation and analysis of UHV (Karanja & Kiage, 2021; Li et al., 2022a, b). Key determinants of UHV include changes in socioeconomic disparities, inadequate infrastructure, and environmental challenges, all of which exacerbate heat stress (Bao et al., 2015; Niu et al., 2021; Soomar & Soomar, 2023).

Developing effective heat vulnerability indices and reliable predictive models is crucial for policymakers to ensure that mitigation strategies are both reliable and equitable (Kumar & Mishra, 2024; Niu et al., 2021; Sahu & Srivastava, 2020; Szagri et al., 2023). An equity-centric approach is particularly emphasised, as vulnerable populations, such as low-income communities and the elderly, often bear the brunt of extreme heat events (Li et al., 2022; Macharia & Kiage, 2024). Overall, the literature underscores the need for multifaceted and inclusive strategies to mitigate the adverse effects of urban heat and to protect vulnerable populations (Macharia & Kiage, 2024; Niu et al., 2021; Wang & Chang, 2020). There is however a lack of systematic reviews and meta-analyses exploring the historical development, evolution, and future directions of UHV research, which may impede the enhancement of understanding and further development in this field. To address this gap, this review aims to provide a comprehensive understanding of the historical development, emerging and prospective trends, and existing research clusters in UHV through a scientometric analysis of the last five decades of academic publications.

The remainder of this chapter is structured as follows: Section 7.2 details the research methodology and covers database selection, search strategy, data collection and analysis, software introduction, and validation. Section 7.3 outlines the main findings, including general observations, academic influence analysis, research clusters in UHV literature, historical development, and emerging trends in UHV research. Section 7.4 delves into the discussion of the results and goes into outlining the key findings, research directions, and our plans for future work. Section 7.5 provides a summary of the key findings and concludes the chapter.

7.2 Research Design

This chapter utilises a scientometric approach to explore UHV literature, focusing on key questions about research clusters, historical evolution, and emerging and future trends in the field. The scientometric technique aids in creating a detailed map of knowledge connections and facilitates a comprehensive understanding of research clusters for scholars (Faisal et al., 2021; Li et al., 2022).

Database Selection: A database of UHV literature was compiled using keyword searches from bibliographic sources. For this purpose, Elsevier's Scopus database was selected because of its extensive coverage. Scopus is also renowned for its sophisticated search capabilities and analytical features, which streamline data retrieval and examination processes. Additionally, its ability to export data in various formats is highly advantageous, and it is compatible with popular scientometric analysis software.

Keyword Selection and Search Strategy: This research aims to provide a comprehensive understanding of the historical development, emerging and prospective trends, and existing research clusters in UHV. Therefore, keywords related to 'urban heat' and 'vulnerability' were chosen for the search task. Keywords such as 'vulnerability', 'risk', 'exposure', 'sensitivity', 'adaptive capacity', and 'hazard' were selected based on the climate change vulnerability and risk triangle frameworks for disaster management (Crichton, 1999; Parry, 2007). To accommodate the perspective of epidemiology, 'extreme heat', 'stress', and 'susceptibility' were also chosen as the keywords. Moreover, to obtain a broader range of literature data, a fuzzy format symbol '*' was incorporated into the search query. The finalised query string used for the search was TITLE-ABS-KEY ('urban heat*' AND ('vulnerability' OR 'risk*' OR 'exposure*' OR 'sensitivit*' OR 'adaptive capacity' OR 'hazard*' OR 'stress*' OR 'susceptibility')).

Data Collection: The search for relevant literature was carried out in December 2023 and included works published up to January 2024. The first scholarly publication on the topic was in 1972; hence, that date was used as the beginning of the search period. A small subset of publications was excluded due to incomplete data, such as missing document types or author details, resulting in a final selection of 2340 publications. This collection comprised articles, conference articles, and other types of literature. The complete details of these works, such as citations, bibliographic information, abstracts, keywords, and funding information, were exported in CSV format.

Data Analysis: The CSV format was chosen for its compatibility with VOSviewer, an analysis tool developed by the Centre for Science and Technology Studies at Leiden University. VOSviewer is highly regarded for creating and visualising bibliometric networks and is used across various disciplines, including sustainable delivery in the e-commerce market (Kiba-Janiak et al., 2021), environmental engineering (Movahed & Sarmah, 2021), and smart home technologies (Li et al., 2022). In the context of this study, VOSviewer played a crucial role in examining the UHV literature database. VOSviewer is particularly advantageous for handling large datasets; it provides detailed mapping and offers intuitive graphical representations that facilitate the interpretation of complex bibliometric data. It is important to select appropriate analytical tools because of the effectiveness of tailored approaches in complex data analysis and decision-making processes (Sahu & Srivastava, 2021; Sahu et al., 2023).

Compared to other tools like CiteSpace and Gephi, VOSviewer stands out for its user-friendly interface and specialised functionalities tailored for bibliometric analysis. While CiteSpace excels in detecting and visualising emerging trends and pivotal points in the literature, and Gephi is strong in general network analysis and visualisation, VOSviewer provides a balanced combination of these features with a focus on ease of use and accessibility. Additionally, VOSviewer's ability to integrate data from multiple sources and generate clear, interpretable maps makes it an ideal choice for our study on UHV research clusters. Here is how it works:

Data Import and Preparation: VOSviewer starts by importing data from bibliographic databases like Scopus. The data typically include details such as titles, abstracts, keywords, citations, and authorship information. In this study, the data were exported in CSV format from Scopus and imported into VOSviewer.

Network Creation: VOSviewer creates various types of bibliometric networks, such as citation networks (showing relationships between publications based on citations), and co-occurrence networks (showing relationships between keywords or terms used in the literature). These networks help in understanding the structure and connections within the research field.

Visualisation: The software visualises these networks using graphical representations. Nodes in the network represent entities such as authors, papers, or keywords, and edges (lines) represent the relationships or connections between them. The node size often indicates the importance or frequency of the entity, while the thickness of the edges indicates the strength of the connection.

Clustering and Analysis: VOSviewer applies clustering techniques to group related items together. For example, in a co-occurrence network of keywords, clusters of keywords that frequently appear together in the literature are identified. This helps in recognising research themes or topics within the field.

Interpretation: The visual maps and clusters generated by VOSviewer allow researchers to interpret and analyse the data effectively. They can identify key research areas, emerging trends, influential publications, and collaborative networks among researchers.

By utilising VOSviewer, this study was able to create visual representations of UHV literature, facilitating a comprehensive understanding of the research clusters, historical development, and emerging trends within the field. The detailed maps and networks provided by VOSviewer enabled a nuanced analysis of the vast amount of bibliometric data collected.

Validation and Reliability: To ensure the analysis was both reliable and valid, multiple rounds of validation were conducted. These steps included screening for duplicates in the initial data, re-evaluating the software, and performing random checks on selected results. The potential limitations of the chosen methodology are worth noting and acknowledging. (a) The Scopus database, while extensive, may not encompass all relevant publications, potentially leading to a biased representation of the field. (b) The use of a fuzzy search could have resulted in the inclusion of non-relevant publications, impacting the precision of the analysis. (c) There may be an unconscious bias from the authors in interpreting the results, which could influence the outcomes of the analysis. (d) The analysis relied on bibliometric data, which, while informative, might not capture the full depth of qualitative insights from the individual studies.

7.3 Analysis and Results

7.3.1 General Observation

As outlined earlier, the curated database on UHV encompasses 2,340 publications published over the last 50 years (see Table 7.1 for further details).

TABLE 7.1
Data Statistics Overview

Data Source	Scopus Bibliographic Repositories
Covered period	January 1972 to January 2024
Included countries	107
Number of publications	2340
Number of authors (including corresponding authors and co-authors)	7463
Number of organisations	5784
Number of publishing sources	779

The statistical data show that the volume of publications related to UHV remained relatively low until 2001, as depicted in Figure 7.1. Post-2016, an exponential increase is evident and expected to continue, with the count rising from 111 in 2016 to 242 in 2020, and further tripling to 314 in 2021. Furthermore, publications from 2016 to 2024 constitute nearly 80% of the entire database. This surge in publication numbers reflects a growing interest in UHV in both industrial and academic circles (Cheng et al., 2021; Li et al., 2022). The predominant publication types include articles (78%), followed by conference papers (10%), reviews (5%), and book chapters (5%),

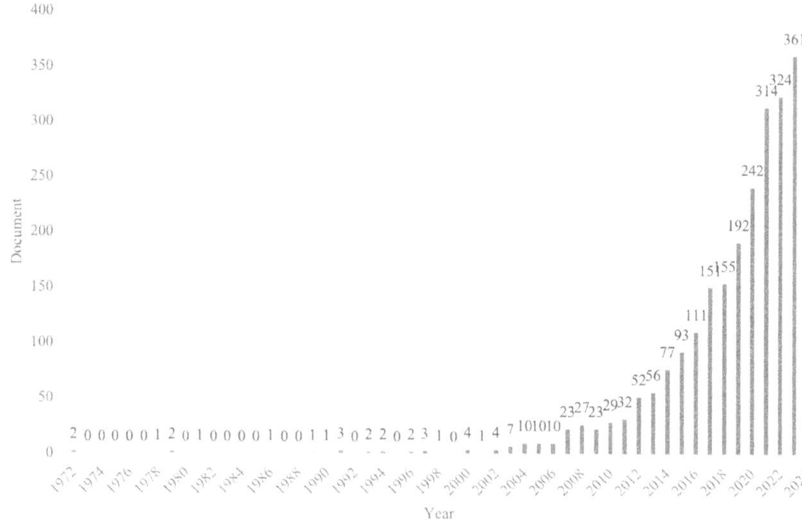

FIGURE 7.1
Trends in UHV literature.

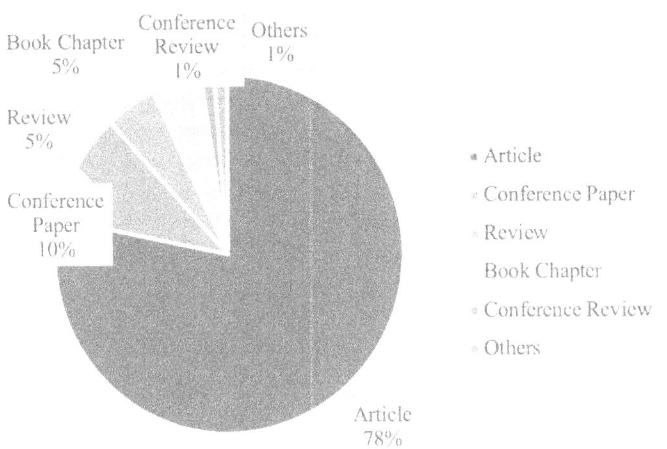

FIGURE 7.2
Document types of the UHV literature.

with other types making up a mere 2%, as shown in Figure 7.2. The primary areas covered by these publications are environmental science (26%), Earth and planetary sciences (16%), social sciences (14%), and engineering (12%) (see Figure 7.3). As illustrated in Figure 7.4, the US (26%, n = 602) and China (19%, n = 447), along with the UK (9%, n = 213), lead in terms of productivity, jointly contributing to over half of the publications.

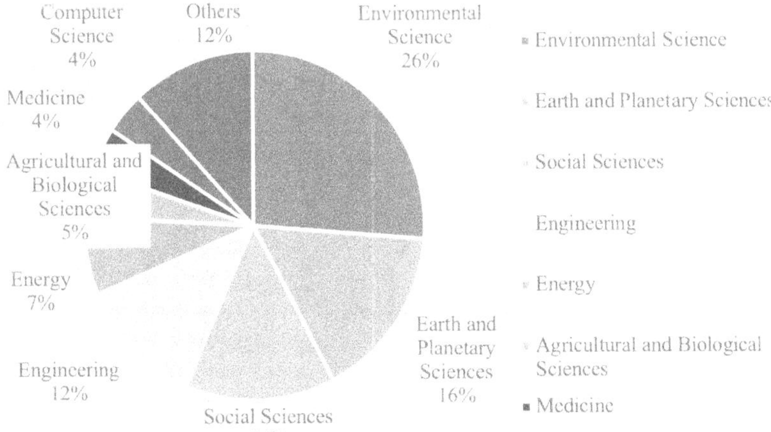

FIGURE 7.3
Subject areas of the UHV literature.

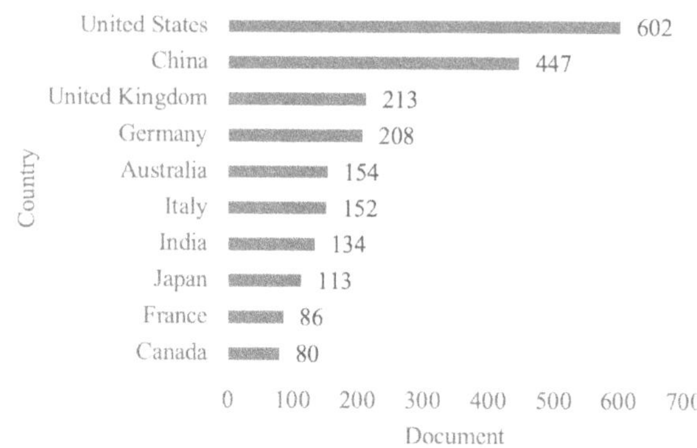

FIGURE 7.4
Top ten most prolific countries in UHV literature.

Table 7.2 lists the leading first authors by research output, along with their associated organisations. Leading the list are Santamouris, M., Pisello, A.L., and Middel, A., who are the most active first authors, contributing 1.28, 0.77, and 0.68%, respectively, to the body of selected literature. The authors from the second to the tenth position show no significant difference from their immediate successors in the number of publications, at times even sharing the same rank. Notably, authors from the US, China, and the UK dominate this list, with significant contributions also from Belgium and Germany. Belgium and Germany have a comparable number of authors to China and the UK, albeit with slightly fewer publications. Altogether, the ten most prolific authors in this field are associated with 27 distinct organisations, encompassing a total of 32 authors. Only Arizona State University in the US, University College London in the UK, the Flemish Institute for Technological Research in Belgium, and the Technical University of Berlin in Germany boast between two and three authors on this list. Interestingly, the top two

TABLE 7.2
Top Ten Most Prolific First Authors in UHV Research

Rank	Author	Affiliated Organisation	Country	Documents	%
1	Santamouris, M.	University of New South Wales	Australia	30	1.3
2	Pisello, A.L.	University of Perugia	Italy	18	0.8
3	Middel, A.	Arizona State University	The US	16	0.7
4	Bou-Zeid, E.	Princeton University	The US	15	0.6
5	Chen, F.	Hong Kong University of Science and Technology	China	14	0.6

(Continued)

TABLE 7.2 *(Continued)*

Top Ten Most Prolific First Authors in UHV Research

Rank	Author	Affiliated Organisation	Country	Documents	%
6	Mavrogianni, A.	University College London	The UK	13	0.6
6	Ren, C.	University of Hong Kong	China	13	0.6
7	Davies, M.	University College London	The UK	12	0.5
7	Hamdi, R.	Royal Meteorological Institute of Belgium	Belgium	12	0.5
7	He, B.J.	Chongqing University	China	12	0.5
7	Kusaka, H.	University of Tsukuba	Japan	12	0.5
7	Lauwaet, D.	Flemish Institute for Technological Research	Belgium	12	0.5
7	Martilli, A.	Centre for Energy, Environmental and Technological Research	Spain	12	0.5
8	De Ridder, K.	Flemish Institute for Technological Research	Belgium	11	0.5
8	Heaviside, C.	University College London	The UK	11	0.5
8	Hondula, D.M.	Arizona State University	The US	11	0.5
8	Matzarakis, A.	University of Freiburg	Germany	11	0.5
8	Meier, F.	Technical University of Berlin	Germany	11	0.5
9	Lee, X.	Yale University	The US	10	0.4
9	Masson, V.	University of Toulouse	France	10	0.4
9	Sharifi, E.	University of Adelaide	Australia	10	0.4
9	Termonia, P.	Royal Meteorological Institute	Belgium	10	##
		Ghent University	Belgium		
9	Yang, X.	Zhejiang University	China	10	0.4
10	Cartalis, C.	National and Kapodistrian University of Athens	Greece	9	0.4
10	Miao, S.	China Meteorological Administration	China	9	0.4
10	Palme, M.	Federico Santa María Technical University	Chile	9	0.4
10	Sailor, D.J.	Arizona State University	The US	9	0.4
10	Scherer, D.	Technical University of Berlin	Germany	9	0.4
10	Shandas, V.	Portland State University	The US	9	0.4
10	Stone, B.	Georgia Institute of Technology	The US	9	0.4
10	Tapper, N.J.	Monash University	Australia	9	0.4
10	Taylor, J.	Tampere University	Finland	9	0.4

Notes: Affiliation information is based on the Scopus profile and verified by both Google Scholar and ORCID profiles.

authors are not linked with institutions from the three countries with the highest productivity. Instead, they are affiliated with the University of New South Wales in Australia and the Università degli Studi di Perugia in Italy, respectively.

Table 7.3 identifies the top three institutions in UHV research: Arizona State University (the US), the Chinese Academy of Sciences, and the Ministry of Education of the People's Republic of China. China, with six institutions, and the US, with two, are leading in this area. The organisations listed predominantly include educational institutions (n = 6), national research institutions (n = 3), and national government institutions (n = 2). This distribution highlights that not only academic entities but also governmental bodies significantly invested in research on UHV.

Table 7.4 reveals that the four leading sources of publication in this field are all Elsevier journals, namely, *Urban Climate*, *Science of the Total Environment*, *Sustainable Cities and Society*, and *Building and Environment*, contributing 4.57,

TABLE 7.3

Top Ten Most Prolific Organisations

Rank	Organisation	Type	Country	Documents	%
1	Arizona State University	Educational institution	The US	86	3.7
2	Chinese Academy of Sciences	National research institution	China	83	3.5
3	Ministry of Education of the People's Republic of China	National government department	China	66	2.8
4	The University of New South Wales	Educational institution	Australia	49	2.1
5	University of Chinese Academy of Sciences	Educational institution	China	46	2.0
6	Nanjing University of Information Science & Technology	Educational institution	China	42	1.8
7	National Centre for Atmospheric Research	National research institution	The US	39	1.7
8	The University of Hong Kong	Educational institution	China	38	1.6
8	China Meteorological Administration	National government department	China	38	1.6
9	University College London	Educational institution	The UK	36	1.5
10	French National Centre for Scientific Research (CNRS)	National research institution	France	35	1.5

TABLE 7.4

Top Ten Most Prolific Publication Sources

Rank	Source	Subject Area	Type	Documents	%
1	*Urban Climate*	Environmental sciences; meteorology and atmospheric sciences	Journal	107	4.6
2	*Science of the Total Environment*	Environmental sciences	Journal	89	3.8
3	*Sustainable Cities and Society*	Green and sustainable science and technology; construction and building technology; energy and fuels	Journal	80	3.4
4	*Building and Environment*	Engineering, environmental; engineering, civil; construction and building technology	Journal	65	2.8
5	*Sustainability*	Environmental sciences; green and sustainable science and Technology	Journal	64	2.7
6	*Atmosphere*	Environmental sciences; meteorology and atmospheric sciences	Journal	40	1.7
7	*International Journal of Environmental Research and Public Health*	Environmental sciences; public, environmental and occupational health	Journal	39	1.7
8	*Environmental Research Letters*	Environmental sciences; meteorology and atmospheric sciences	Journal	38	1.6
9	*IOP Conference Series: Earth and Environmental Science*	Environmental sciences; geosciences, multidisciplinary	Conference proceeding	37	1.6
10	*Energy and Buildings*	Construction and building technology; energy and fuels; engineering, civil	Journal	35	1.5

3.80, 3.42, and 2.78%, respectively, to the total literature selected. The fifth to seventh spots are occupied by journals from the Multidisciplinary Digital Publishing Institute (MDPI), while the eighth and ninth positions are held by journals from the Institute of Physics (IOP). Elsevier also claims the tenth spot, making it responsible for half of the top publishing sources. These sources are all dedicated to reporting the latest developments in environmental science and sustainable development. In conjunction with subject area analysis (Figure 7.3), Table 7.4 indicates a primary focus on the environmental and sustainable aspects of UHV in current research.

7.3.2 Academic Influence Analysis

7.3.2.1 Citation Analysis by Publication

Impact of regional climate change on human health, by Patz et al. (2005), stands out as the most influential work in this field, receiving a notably high number of citations. It reviews empirical studies and model simulations to understand climate change impacts on human health. It focuses on direct temperature effects and acknowledges other risks like altered storm patterns and sea-level rise. Findings from the study indicate that the recent global warming trend has increased morbidity and mortality globally. It highlights the vulnerability of temperate regions, areas affected by El Niño/Southern Oscillation, sub-Saharan Africa, and UHI. The review emphasises the escalating health risks due to climate-health relationships under future climate change projections.

Safeguarding human health in the Anthropocene epoch: report of The Rockefeller Foundation–Lancet Commission on planetary health by Whitmee et al. (2015) ranked second on the ten most-cited publications. This review explores the significant threats to human health and development from environmental changes in the Anthropocene epoch. It underscores the lack of knowledge regarding the relationship between environmental changes and health, as well as the success of various interventions and policies. The review calls for interdisciplinary research and better governance to integrate social, economic, and environmental policies for planetary health. It emphasises redefining prosperity, balancing social progress with environmental sustainability, and addressing consumption patterns, population growth, and technological development to support a growing world population sustainably.

The integrated WRF/urban modelling system: development, evaluation, and applications to urban environmental problems, by Chen et al. (2011), the third most-cited paper, delves into the creation of an integrated urban modelling system that works in tandem with meteorological research and prediction. This system incorporates a range of methods for urban surface process parameterisation, varying from basic to advanced models, for tackling urban environmental challenges. It utilises detailed data on urban land use, building structures, and human-generated heat. The study assesses the model's sensitivity to various urban parameters and its ability to accurately represent UHI and intricate boundary-layer formations and thus showcases its viability as a tool for regional climate modelling. This is of great value to understanding how different urban forms and materials contribute to heat exposure and distribution, which is a key component of UHV. The WRF model has been widely used to simulate urban climate and model urban heat patterns as it provides critical data needed for UHV assessments (Macintyre et al., 2018; Sharma et al., 2018).

Table 7.5 presents the ten most-cited research publications in the field along with their details. It can be inferred from Table 7.5 that the leading publications in UHV primarily concentrate on exploring the relationships between climate change, UHI, heatwaves, and human health, using a mix of empirical

TABLE 7.5

Top Ten Most-Cited Publications

Rank	Title	Authors and Year	Research Focus	Citation
1	'Impact of regional climate change on human health'	Patz et al. (2005)	Climate change, public health, climate-related mortality, disease transmission, heatwaves, UHI, socioeconomic factors, vulnerable region.	2117
2	'Safeguarding human health in the Anthropocene epoch: Report of the Rockefeller Foundation-Lancet Commission on planetary health'	Whitmee et al. (2015)	Planetary health, environmental degradation, urban health challenges, governance and policy, resilience and adaptation.	1418
3	'The integrated WRF/urban modelling system: Development, evaluation, and applications to urban environmental problems'	Chen et al. (2011)	Urban modelling, climate modelling, UHI, urban environment issues, mesoscale modelling	856
4	'Strong contributions of local background climate to urban heat islands'	Zhao et al. (2014)	UHI, climate modelling, evaporative cooling reduction, heat convection efficiency, heat stress.	838
5	'Climate Change and Extreme Heat Events'	Luber and McGeehin (2008)	Climate change, heatwaves, extreme heat events, UHI, public health impact, heat response plans, remote sensing.	763
6	'The urban heat island and its impact on heat waves and human health in Shanghai'	Tan et al. (2010)	Global warming, UHI, meteorological records, heatwave, urbanisation, mortality rates, human health, thermal conditions.	728
7	'Neighborhood microclimates and vulnerability to heat stress'	Harlan et al. (2006)	Neighbourhood microclimates, heat stress, human thermal comfort index, socioeconomic and ethnic factors, urban planning and vegetation, vulnerability and coping resources.	714
8	'Utilising green and bluespace to mitigate urban heat island intensity'	Gunawardena et al. (2017)	UHI, urban green and blue infrastructure, evapotranspiration, microclimate, urban planning.	583
9	'Synergistic interactions between urban heat islands and heat waves: The impact in cities is larger than the sum of its parts'	Li and Bou-Zeid (2013)	UHI, heatwaves, synergistic interactions, temperature difference, heat stress, health risks, climate change, adaptation and mitigation strategies.	559
10	'The cooling effect of green spaces as a contribution to the mitigation of urban heat: A case study in Lisbon'	Oliveira et al. (2011)	UHI, climate change, urban green spaces, thermal performance, atmospheric environment, weather parameters, mean radiant temperature, mitigation strategies.	517

Notes: Min 5 citations/document; 1455 of 2340 documents meet the threshold.

research and literature reviews. Some of the publications also explore the threats of environmental issues to human health, as well as how to utilise the mitigation effects to relieve urban heat. Thereinto, more than half (n = 6) of the top ten publications focus on human health and urban heat and other environmental issues, e.g., human health protection, public health response, health vulnerability/risk assessment, and heat-related health inequity analysis. This reflects that human health and wellbeing is one of the most influential research themes of UHV research.

Three of the top ten publications focus on the climatological and meteorological aspects in this field by investigating the relationship among climate change, UHI, and extreme heat events (heatwaves). Furthermore, urban heat management is another key focus within UHV research. This includes areas such as the development of urban modelling systems and examining and applying green and blue infrastructure's cooling effects within urban settings.

7.3.2.2 Citation Analysis by Research Organisation/Institutions

The three leading institutions in terms of citations within the UHV field are all based in the US, specifically the National Centre for Atmospheric Research, Princeton University, and Arizona State University. As indicated in Table 7.6, which lists the top ten most-cited organisations, half of these are located in the US (n = 5), with China nearly matching this representation

TABLE 7.6

Top Ten Most-Cited Organisations

Rank	Organisation	Type	Country	Citation
1	National Centre for Atmospheric Research	National research institution	The US	1936
2	Princeton University	Educational institution	The US	1747
3	Arizona State University	Educational institution	The US	1296
4	University of Tsukuba	Educational institution	Japan	1184
5	Institute of Urban Meteorology, CMA, Beijing	National research institution	China	1058
6	Pacific Northwest National Laboratory	National research institution	The US	380
7	University of Chinese Academy of Sciences	Educational institution	China	354
8	Yale University	Educational institution	The US	340
9	Chinese Academy of Meteorological Sciences	National research institution	China	287
10	Chongqing University	Educational institution	China	249

Notes: Max 25 organisations/document; min 5 documents and 5 citations/organisation. Overall, 23 of 5784 organisations meet thresholds.

(n = 4). The University of Tsukuba is the only organisation from Japan, ranking fourth. Furthermore, all organisations are classified as educational institutions (n = 6) or national research institutions (n = 4), roughly split equally between the two. These results reveal that UHV research has been receiving almost equal attention from both academia and government, particularly in the US and China. These two countries have had a considerable impact on the evolution of research in this area.

7.3.2.3 ***Citation Analysis by Country***

Table 7.7 highlights the top ten countries by citation count in UHV research, along with their total link strengths and average publication years (APYs). The US emerges as the leading nation in the UHV domain, boasting the highest number of citations. China and the UK rank second and third, respectively, with similar citation figures. The 'total link strength' metric indicates the degree of each country's scholarly connections with other countries based on citations. For example, the US' total link strength is 3523, meaning a broader range of academic interactions in citations than Australia, which has a total link strength of 921. The order of the total link strengths is consistent with one of the citations among the top five countries. However, it is worth noting that it is somewhat different from the 6th to 9th. For instance, Italy (6th, n = 525) has stronger total link strength with other countries than Switzerland (7th, 376), while Switzerland (n = 3883) and Italy (n = 3758) have similar citations. The same situation exists between Canada and Japan. It is noteworthy that Japan has the most total link strengths of the four countries even if it has the least number of citations.

TABLE 7.7

Top Ten Most-Cited Countries

Rank	Country/Region	Citations	Total Link Strength	APY
1	The US	32,809	3523	2018.1
2	China	13,923	2292	2019.9
3	The UK	11,006	1619	2017.8
4	Germany	5923	1150	2018.7
5	Australia	5367	921	2019.2
6	Switzerland	3883	376	2018.7
7	Italy	3758	525	2019.4
8	Canada	3452	401	2018.7
9	Japan	3312	553	2018
10	Malaysia	2764	220	2018.9

Notes: Max 25 countries/regions per document; min 5 documents and 5 citations per country/region. Overall, 59 of 107 countries/regions meet these thresholds. APY: Average publication year.

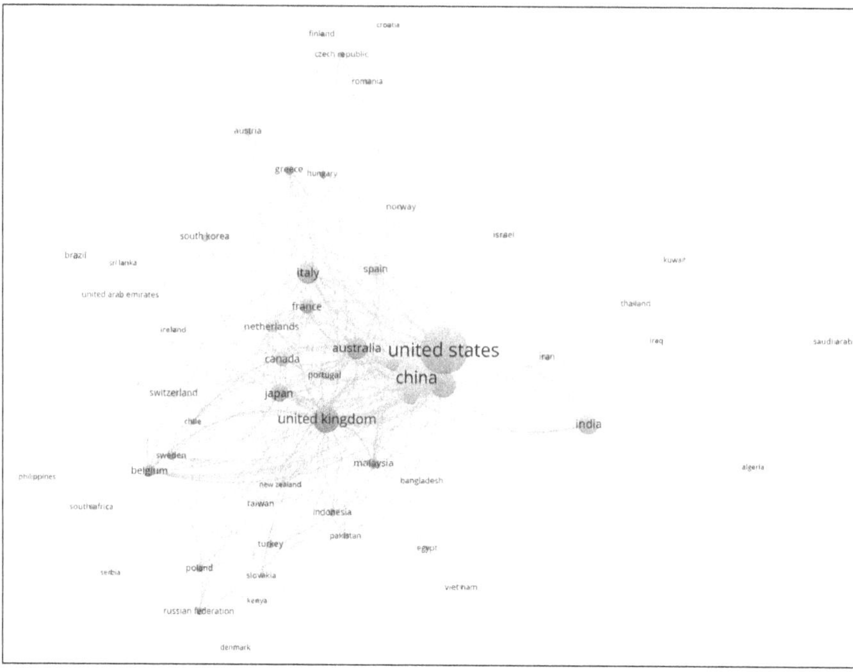

FIGURE 7.5
Citation network map by countries.

Figure 7.5 presents a citation network map categorised by country or region. In this visualisation, each circle represents a country or region, with the size of the circle indicating the citation volume of publications from that area. The connecting lines between circles represent scholarly ties, with shorter lines indicating stronger academic relationships. The circle colours group them into clusters based on their citation connections, identifying them as scientific communities. This implies that countries/regions within the same community are closely interconnected in terms of citations (Van Eck & Waltman, 2017).

As shown in Figure 7.5, even though the five most-cited countries/regions are part of five distinct citation communities, they exhibit very strong citation connections with one another, particularly noticeable among the US (first), China (second), and Canada (fourth). Furthermore, the composition of citation communities shows that the countries/regions are mostly from North America, East and South Asia, Europe, and Oceania. This suggests that research on UHV is not dominated by any single country or continent. Instead, academic communities worldwide maintain a consistent level of interest and engage in ongoing scholarly communication.

Figure 7.6 presents the visualisation of APYs in the citation network map, categorised by countries/regions. In this visualisation, the colour of each circle denotes the timeline of APYs, where darker hues signify earlier periods

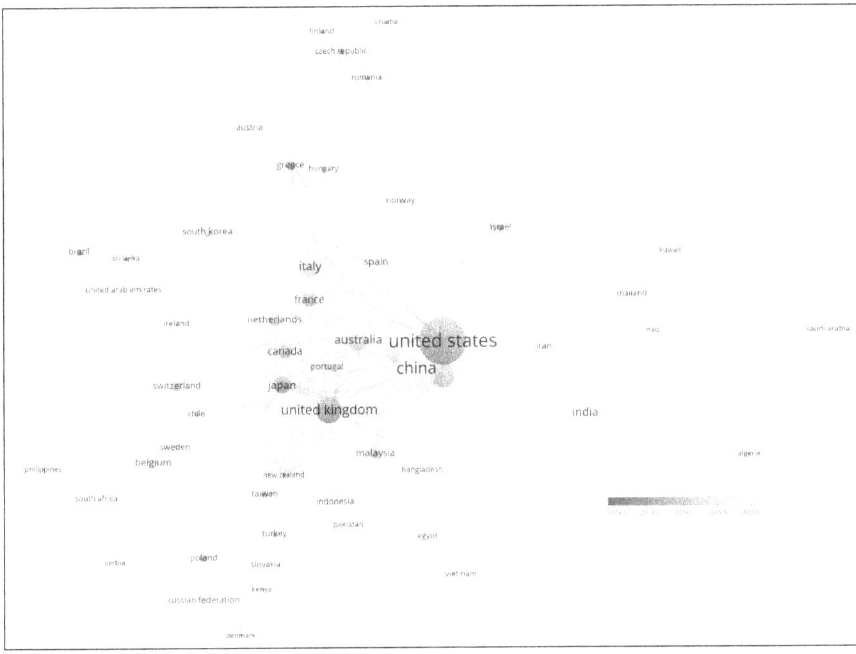

FIGURE 7.6
Citation network map by countries and publication year.

and lighter hues indicate more recent publications. The map shows that UHV research was intensively conducted from 2018 to 2020 which is an indication that it is an emerging research area. There is no notable difference among the countries/regions regarding the APY. UHV research originating from the US, the UK, Japan, and Greece emerged relatively earlier than other countries/regions such as China, Spain, and Austria. Interestingly, China and India eventually have caught up in the race with the other most-cited countries/regions.

7.3.3 Research Clusters of the UHV Literature

This study identified research clusters by examining the occurrence and connection strength of 1166 keywords extracted from the UHV literature database. Out of these, 1166 keywords with a minimum of five occurrences each were chosen for further examination. The outcome of this analysis revealed ten distinct clusters within the current UHV research. These primary research clusters have been organised into five groups as follows:

- *Cluster 1: Urban Planning and Architectural Design*—this cluster of research focuses on evaluating and reducing urban heat. It examines how urban structures contribute to the heat island effect, particularly in terms of atmospheric temperature. This includes studying the impact of land cover planning, architectural design, transportation infrastructure optimisation, and building energy management.

- *Cluster 2: Urban Environment and Ecology*—this cluster of research contributes to UHV mainly by analysing the relieving effects of green and blue infrastructure on the urban thermal environment, along with a focus on ecosystem services and environmental protection.
- *Cluster 3: Urban Public Health and Wellbeing*—this cluster of research focuses on the vulnerability of urban residents to urban heat, from public health, population vulnerability, risk assessment, and epidemiology perspectives.
- *Cluster 4: Urban Climate and Meteorology*—this cluster of research delves into the climatic and meteorological aspects of the heat island through weather monitoring and forecasting, climate modelling, meteorological modelling, and air pollution monitoring.
- *Cluster 5: Urban Remote Sensing*—this cluster of research focuses on the applications of remote sensing technology in the field of UHV, particularly in terms of satellite imagery and related products. Examples include imagery from satellites such as Landsat and MODIS, as well as products related to land use, landform, and land surface temperature.

Clusters 1–4 are extended research clusters, each representing a distinct extended research orientation within the field of UHV. Cluster 5, urban remote sensing, is a technical cluster that offers crucial support to the other clusters in this area. This support includes providing data, methodologies, and platforms. Figure 7.7 illustrates the network map of research clusters. In this map, every circle corresponds to a keyword, and its size denotes how frequently the keyword appears. The colour of each circle represents the keyword's research cluster, while the spacing between circles indicates the level of correlation and resemblance among them. The map uncovers the following insights: (a) The clusters are interconnected to some degree, with no cluster being isolated. For example, the distribution of Cluster 1 significantly overlaps with Clusters 2 and 4, and Cluster 4 also overlaps considerably with Cluster 5. Clusters 2 and 3 overlap as well; (b) the key occurrence circles of Clusters 1, 2, and 4 (heat island, climate change, and urbanisation) are positioned close to the centre of the cluster network map, acting as connectors among the clusters; (c) in UHV research, the key circles within each cluster are interconnected with lines, demonstrating strong connections between the clusters.

Table 7.8 lists the ten most common keywords, as well as information on their APY, occurrence count, number of links, total link strength, and associated clusters. This table consolidates similar keywords in singular and plural forms within the same cluster, such as 'city' and 'cities', or 'human'

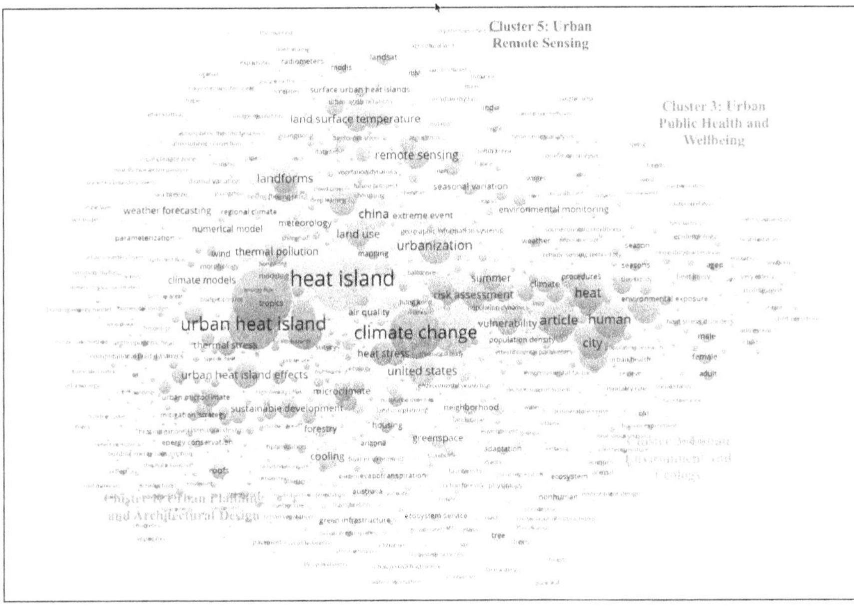

FIGURE 7.7
UHV research clusters network map (by keyword occurrences).

TABLE 7.8
Top Ten Occurrence Keywords in UHV Literature

Rank	Keyword	APY	Occurrences	Links	Total Link Strength	Cluster
1	Heat island	2018	772	1131	13,385	4
2	Atmospheric temperature	2019	707	1117	12,106	1
3	Climate change	2018	538	1075	9218	1
4	Urban heat island	2018	536	1077	9097	1
5	Human[a]	2018	450	1,557	11,573	3
6	City[a]	2020	440	1636	11,333	3
7	Urban area	2018	406	1043	8050	2
8	Urban planning	2018	273	880	4513	1
9	Urbanisation	2018	268	885	4989	2
10	Air temperature	2017	251	891	4,810	4

Notes: Min 5 occurrences/keyword; 1166 of 8580 keywords meet threshold.
[a] Merge for singular/plural in same cluster; Exclusionary for irrelevant keywords.

and 'humans', and excludes irrelevant terms like 'article'. Employing a set of exclusion criteria, this approach minimises repetition in the data and accurately highlights the trends in UHV research, clearly delineating research clusters or broader research directions. The keyword 'heat island' in Cluster 4, which has the highest total link strength, signifies that studies on heat islands within urban climate and meteorology have been the most impactful area in UHV research up to early 2024. It also maintains the strongest connection with keywords from all other clusters. Another key term from Cluster 4, 'air temperature', ranked 10th, most attracted attention from researchers earlier than other keywords, indicating an early research focus on capturing and analysing air temperature.

Cluster 1 includes the highest-ranking keywords in the top ten, specifically the second, third, fourth, and eighth positions. These keywords collectively hold the greatest total link strength, far exceeding that of other clusters, suggesting that urban planning and architectural design have been the most influential and popular trends in UHV research. Keywords 'human' and 'city' from Cluster 3 occupy the middle-ranking positions and achieve the second-highest total link strength, indicating significant attention to urban public health and wellbeing dimensions. Cluster 2, with keywords in the seventh and ninth positions, shows relatively weak connections with other clusters in this field. Furthermore, the absence of any keywords from Cluster 5 in the list indicates that urban remote sensing has not been a popular research orientation in UHV research so far, being considered more of an auxiliary technical tool for data collection, processing, and analysis.

7.3.4 Historical Development of UHV Research

In this subsection, the historical development of UHV research is traced through the analysis of the ranking and density map of keyword occurrence. The keyword selection process merges singular and plural forms and excludes irrelevant terms. The intention behind this methodology is to diminish repetition and offer a clearer insight into the distinct research clusters and broader research directions within the field of UHV.

From 1972 to 2005, the top ten occurrence list contains only eight eligible keywords, with the majority originating from Cluster 4, urban climate and meteorology, as shown in Table 7.9. Due to their limited number, these keywords are distributed discretely, as illustrated in Figure 7.8a, indicating a lack of strong connections among them. Additionally, Clusters 3, urban public health and wellbeing, and 5, urban remote sensing, have no representation in the list, while Cluster 1, urban planning and architectural design, and Cluster 2, urban environment and ecology, contribute a minority of keywords. However, this scenario changed between 2006 and 2010. Although Cluster 4 continues to dominate the list, Clusters 1, 2, and 3 experience an increase in occurrences and keyword numbers, particularly Cluster 1.

TABLE 7.9

Top Ten Occurrence Keywords in UHV Literature (Ranks Per Five-Year Periods)

Cluster	Keyword	Occurrence Per Period					Rank Per Period				
		1972–2005	2006–2010	2011–2015	2016–2020	2021–2024	1972–2005	2006–2010	2011–2015	2016–2020	2021–2024
1	Climate change	9	28	71	216	214	2	3	4	3	4
1	Urban planning	6	13	41	100	113	3	10	7	10	8
1	Urban heat island[a]	–	42	97	198	207	–	2	2	4	5
1	Atmospheric temperature	–	26	88	289	303	–	4	3	1	2
1	Sensitivity analysis	–	13	–	–		–	10	–	–	–
1	Surface temperature	–	–	28	–		–	–	10	–	–
2	Environmental impact	5	–	–	–		4	–	–	–	–
2	Urban area	–	21	50	157	174	–	6	6	6	7
2	Urbanisation	–	15	40	106	105	–	9	8	9	10
3	Human[a]	–	25	53	179	189	–	5	5	5	6
3	Risk assessment	–	–	28	–		–	–	10	–	–
3	City[a]	–	–	–	154	249	–	–	–	7	3
3	Heat	–	–	–	111		–	–	–	8	–
3	Hot temperature	–	–	–	–	110	–	–	–	–	9
4	Heat island	11	52	128	273	308	1	1	1	2	1
4	Urban atmosphere	6	–	–	–		3		–	–	–
4	Air temperature	5	20	40	–		4	7	8	–	–
4	Computer simulation	5	15	–	–		4	9	–	–	-
4	Meteorology	5	–	–	–		4	–	–	–	–
4	Heat flux	–	21	–	–		–	6	–	–	–
4	Urban climate	–	16	–	–		–	8	–	–	–
5	Land use	–	–	32	–		–	–	9	–	–
5	Remote sensing	–	–	32	–		–	–	9	–	–
5	Landforms	–	–	28	–		–	–	10	–	–

Notes: From 1972 to 2024, the minimum occurrence threshold for keywords was 5. Over the periods 1972–2005, 2006–2010, 2011–2015, 2016–2020, and 2021–2024, out of 296, 1297, 2441, 4050, and 4761 keywords, 9, 76, 199, 485, and 512 keywords met this threshold, respectively.

[a] Keywords were merged for singular and plural forms in the same cluster, excluding irrelevant keywords.

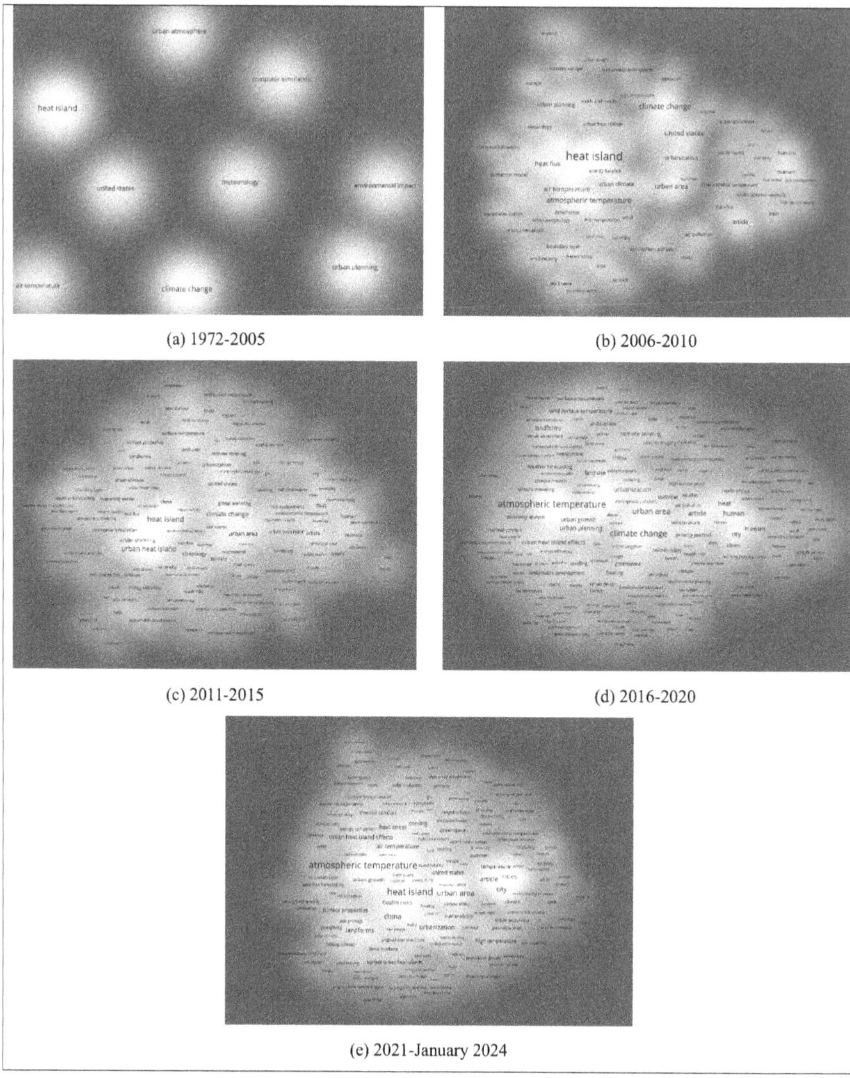

FIGURE 7.8
UHV keywords occurrence density map by periods.

This cluster now matches Cluster 4 with five keywords in the list and occupies the second to fourth positions, achieving total occurrences nearly equal to those of Cluster 4. Cluster 3 makes its first appearance in the middle of the list, while Cluster 5 remains absent. Figure 7.8b, illustrating the density map for the period 2006–2010, shows the distribution of keywords from Clusters 1 and 4 in adjacent regions. These keywords form a tight network, resulting in the map's largest concentration. Cluster 1's keywords also exhibit close proximity to those in Cluster 2, leading to the formation of the second-largest dense aggregation. In contrast, Cluster 3's keywords are situated in a more distant region, signifying a less robust link with the keywords of Clusters 1 and 4.

From 2011 to 2015, Cluster 1 dominated the list with the highest number of keywords and occurrences, as shown in Table 7.9. Meanwhile, Cluster 4 maintained the top position. However, Cluster 4 had only two keywords on the list, the same number as Clusters 2 and 3, and even fewer than Cluster 5, which appeared with three keywords at the ninth and tenth positions. In the density map covering the years 2011–2015, the most prominent aggregation continued to be formed by the keywords from Clusters 1 and 4, as observed in Figure 7.8c. The connection between Clusters 1 and 2 was strengthened, as was the distribution aggregation of these two clusters. The third-largest aggregation was formed by Cluster 3. It is worth noting that the distribution of the entire research field was becoming more integrated due to the increase in keywords and publications, leading to stronger connections among clusters and aggregations.

From 2016 to 2020, apart from 'heat island', no other keywords from Cluster 4 appear in the list (see Table 7.9). Cluster 1 continues to dominate, occupying three of the top four positions. Notably, 'atmospheric temperature' takes over the first position. Cluster 3 becomes the second largest group, replacing Cluster 2. Cluster 5 disappears from the ranking again. On the density map for 2016–2020, Clusters 1 and 4 separately establish individual aggregations (Figure 7.8d). The aggregation formed by Clusters 1 and 2 remains, bridging the separate aggregations of Clusters 3 and 4. The aggregation of Cluster 3 becomes more concentrated and moves closer to the distribution of the other aggregations.

From 2021 to January 2024, the keyword counts of each cluster, excluding Cluster 5, remain the same, and the ranking positions undergo minor changes. An exception is the keyword 'Cities' from Cluster 3, which moves up to the third position from the seventh, and 'hot temperature' replaces 'heat'. On the density map from 2021 to January 2024, there are five aggregations in total, four of which are mainly contributed by Clusters 1, 2, and 4. A new aggregation emerges, connecting these three clusters into a close group at the bottom (Figure 7.8e), indicating strong connections and collaboration among them. Cluster 3 contributes to the largest aggregation during this period, remaining relatively independent from the other aggregations.

Overall, the key findings revealed that

- Urban climate and meteorology (Cluster 4) is one of the earliest clusters. It forms the foundation of UHV research by understanding the UHI effect from climatological and meteorological perspectives. It gradually lost its dominant position in this field, being overtaken by Cluster 1.
- Urban planning and architectural design (Cluster 1) has replaced the leading position of urban climate and meteorology (Cluster 4),

evolving into the most popular research cluster in UHV over the past two decades. It has consistently received considerable attention in UHV research and maintains wide and strong connections with Clusters 2 and 4.

- Urban environment and ecology (Cluster 2) has received stable research attention throughout all periods as one of the earliest research clusters. This cluster has strong connections with Cluster 1.
- Urban remote sensing (Cluster 5) has attracted less attention than other clusters, only receiving notable focus from 2011 to 2015 in UHV research.
- Urban public health and wellbeing (Cluster 3) has been garnering a certain degree of research interest since 2006, as an earlier cluster in UHV research. It is evident that more research attention has been directed to this cluster during this period, making it the second most prominent cluster in terms of keyword occurrence and total link strength since 2016, and after 2021, it has contributed to the largest research aggregation.

7.3.5 Emerging Trends of UHV Research

Analysing the keywords leading in occurrences from 1972 to January 2024 (Table 7.9) and their rankings over each five-year interval, it becomes apparent that UHV is a relatively recent field of study. Prior to 2006, it attracted minimal research endeavours, and significant interest developed only after 2016. The focus on urban public health and wellbeing has steadily increased in keyword occurrences and total link strength, making it the second largest cluster since 2016. After 2021, it forms the largest research aggregation, as illustrated in Figure 7.8e. This suggests that the emphasis on studying UHV within the scope of public health and wellbeing has gained greater significance and attracted more research interest over time.

In the visualisation depicting research clusters networked by APYs, every circle symbolises a keyword, with its size corresponding to the frequency of occurrence (Figure 7.9). The connecting lines between circles illustrate the scholarly connection between keyword pairs, where shorter lines suggest a stronger relevance. The colour gradient of the circles reflects the timeline of the keywords' APYs, transitioning from earlier (darker colour) to more recent (lighter colour) timeframes. According to the map, the research primarily concentrates between 2016 and 2021, indicating that UHV is a relatively new research area.

This result aligns with the conclusions drawn earlier. Urban public health and wellbeing is denoted by the lightest colour overall in all clusters, which means that this field has experienced significant development in recent years. It has many noticeable light-coloured circles, e.g., heat, human, city, risk assessment, and vulnerability. It is noteworthy that this research cluster is

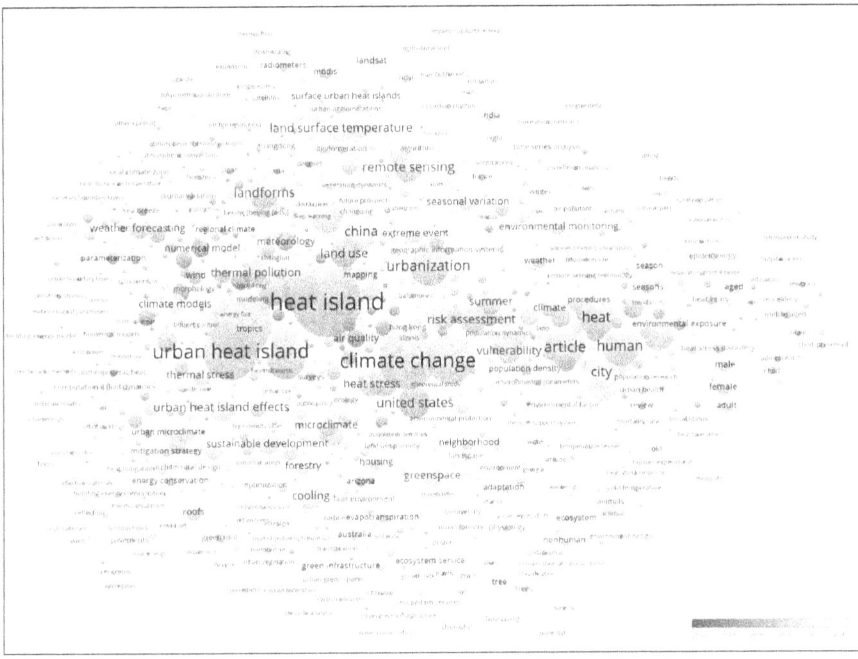

FIGURE 7.9
UHV research clusters network map (by keyword occurrences and average publication per year).

more focused on the humans' vulnerability/risk assessment to heat. Besides, the keyword 'city' is the lightest coloured one, indicating the research attention has been turned to residents in cities. The connection of the urban public health and wellbeing cluster with all other clusters does indicate that this area of research is interdisciplinary.

The colours of Cluster 4 circles, focused on urban climate and meteorology, are notably darker compared to other clusters, featuring the fewest light-coloured circles. This suggests that this research cluster is the earliest research orientation in this field, and fewer new research topics and areas of interest have emerged in this cluster in recent years. The cluster of urban public health and wellbeing (Cluster 3) also displays some small-sized dark circles on the right. These keywords pertain to epidemiology and public health, for example, hospitalisation, newborns, and heat stress disorder, indicating that this research cluster initially attracted attention from epidemiology. Besides the emerging keywords mentioned, other emerging keywords within this cluster include environmental monitoring, environmental factors, weather, climate, and socioeconomic status. Building on the foundations of epidemiology, recent research in this cluster has incorporated concepts from multiple disciplines, such as environmental science, climatology, sociology, and economics. This interdisciplinary approach explains why this cluster has research connections with all other clusters. The interdisciplinary collaborations have significantly contributed to recent developments in UHV research, and urban public health and wellbeing has evolved into an influential research cluster in the urban studies field.

7.4 Findings and Discussion

This chapter has analysed 2340 academic publications from 1972 to January 2024, with the goal of elucidating the historical development, emerging trends, prospects, and key research areas in the study of UHV. The notable growth trend in publications on this subject since 2016 suggests that it has attracted considerable attention from academic and industry sectors. This surge in attention may be attributed to the advancement of remote sensing technologies and the significant development and collaboration of multiple disciplines (Diem et al., 2023; Zhou & Chen, 2018).

Between 2010 and 2015, significant breakthroughs in remote sensing included the launch of high-resolution satellites (e.g., Landsat 8 and the Sentinel series) (Lu et al., 2023; Varnakovida & Ko, 2023), open data policies (Zhu et al., 2019), cloud computing for big data analysis (Gorelick et al., 2017), advances in deep learning and AI for image interpretation (Maxwell et al., 2018), the expansion of hyperspectral sensing (Bioucas-Dias et al., 2013), and improved integration with precise mapping and GIS (Gupta, 2012). These developments have significantly contributed to UHV research by enhancing Earth Observation capabilities, enabling accurate monitoring of surface temperature variations in urban areas, and aiding the study of UHI (Keramitsoglou, 2012).

The dynamic evolution of remote sensing technology has significantly enhanced its application in assessing heat-related risks, aiding in urban planning, and bolstering disaster management efforts, thereby deepening our understanding of UHV (Diem et al., 2023; Jedlovec et al., 2017; Xue et al., 2022). Research focusing on population vulnerability to urban heat, such as extreme heat events, saw a notable increase after 2016, as highlighted by Cheng et al. (2021) and Li et al. (2022). The focus on integrating thermal infrared remote sensing with multiple indicators like socio-demographic data, health statuses, and environmental elements to assess urban populations' vulnerability to increasing urban heat has grown. It is a rising trend in examining how UHI affect human health, especially considering aspects like exposure risk, sensitivity, and the ability to adapt (Sagris & Sepp, 2017).

The US, China, and the UK are the top three contributors, accounting for nearly half of the total selected publications on UHV. These countries also dominate the top ten lists of productive authors, along with Belgium and

Germany. Additionally, China and the US lead the list of the top ten most productive organisations. This indicates that as an emerging research area, UHV has seen significant contributions from North America, East Asia, and Europe. Nevertheless, there is a conspicuous lack of representation from African and South American nations in various productive categories, such as country, author, and organisation contributions, with only a single prolific author hailing from Chile. This implies that UHV research or its practical applications in these two continents might still be at an early stage or that the urgency of this research area is not widely recognised.

As an emerging research topic, UHV requires research attention from all over the world, which is significant for global climate change adaptation and equity (Hsu et al., 2021). An analysis of academic impact, determined by citation frequency, reveals that the US holds a leading position in UHV research, with China and the UK also showing significant contributions in this area. The US and China cover almost all the top ten most productive organisations, presumably due to the urgency to climate change and rapid urbanisation (Chen et al., 2023; Hoehne et al., 2018). These regions experience intensified UHI effects and face increased health risks from extreme temperatures. High population densities and advanced research capabilities drive these regions to prioritise studying and mitigating urban heat impacts, both for local welfare and global climate response efforts (Chen et al., 2023; Lin et al., 2017).

Based on the subject areas classified by Scopus, approximately 70% of the existing publications on UHV fall within the fields of environmental science, earth and planetary sciences, social sciences, and engineering. These fields host the most influential publications, which often focus on multiple domains and the interdisciplinary applications of UHV. Moreover, the ten leading sources of publications in this domain exhibit expertise across diverse disciplines, including environmental and social sciences, sustainable urban development, and building engineering. These findings indicate that the current knowledge structure of UHV research leans towards contributions from and collaborations across different research/knowledge domains. However, research incorporating advanced technical perspectives from computer science, such as artificial intelligence, machine learning, and big data Analysis, needs to be extensively explored.

The main research clusters of UHV are identified as (a) urban planning and architectural design; (b) urban environment and ecology; (c) urban public health and wellbeing; (d) urban climate and meteorology; (e) urban remote sensing. This finding underscores that research on UHV is not confined to a single cluster but rather encompasses multiple perspectives, including planning, environmental, social, and climatological perspectives. Urban remote sensing stands out as a distinct technical perspective, providing technological and practical support to the other clusters. The collaboration between theoretical and technical perspectives contributes to the comprehensive development of the field of UHV. Furthermore, this chapter develops a framework illustrating the evolution and implications of UHV research based on

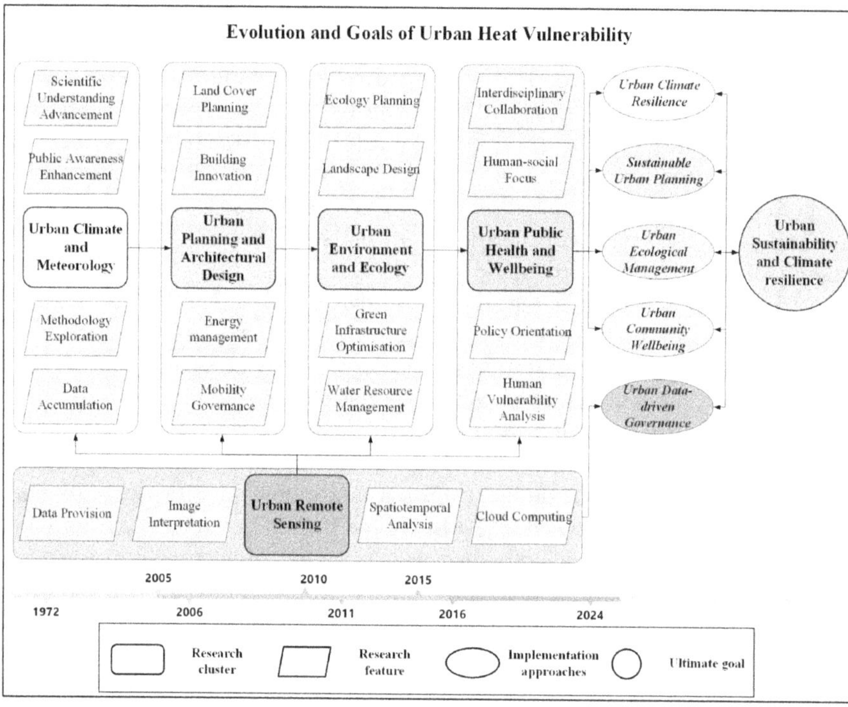

FIGURE 7.10
Evolution and goals of UHV research.

the clusters (Figure 7.10). It highlights the characteristics, practical implications, and research interests of each cluster within a time-framed structure. The framework illustrates how UHV research supports urban sustainability and climate resilience by outlining five key implementation strategies or sub-objectives. These approaches include urban climate resilience, sustainable urban planning, urban ecological management, data-driven urban governance, and enhancing urban community wellbeing. Each research cluster in the field of UHV primarily contributes to one of these sub-objectives, which are indicated by corresponding colours.

In the earliest times, research on UHV received limited attention. However, as climate change accelerated and global urbanisation increased, extreme events such as heatwaves began occurring more frequently, drawing more research focus to this area (Luber & McGeehin, 2008; Tan et al., 2010; Zhang et al., 2009). Research on UHV initially focused predominantly on urban climate and meteorology, with these topics forming the core clusters of study in the field until 2010. This indicates that the initial research focus was on understanding UHV from climatological and meteorological perspectives, such as building climate and meteorology models to simulate and capture the UHI effect (Atkinson, 2003; Jin et al., 2007; Miao et al., 2009; Oleson et al.,

2008). This might be because early studies aimed to grasp fundamental atmospheric principles, forming the scientific groundwork to enhance policymaking and public awareness (Lemonsu et al., 2009; Stone Jr, 2007).

During this period, concepts like climate change and global warming were very new to the public and governments and lacked scientific and systematic understanding. Meanwhile, the scientific community attempted to explore methodologies and collect fundamental data for better analysis and understanding of UHV. As public awareness and policies evolved, along with advancements in interdisciplinary research, the focus shifted from basic climatic and meteorological studies to incorporating broader disciplines like social sciences, urban planning, and public health. This cluster improves the conceptual understanding of UHV and establishes a solid foundation in climatological and meteorological data and methods. This groundwork supports further research collaboration and ultimately contributes to enhancing urban climate resilience.

As understanding and public awareness of the UHI effect improved, both research and political focus shifted towards the implications of urban planning and architectural design on managing UHV. As one of the earliest extended research clusters, the cluster of urban planning and architectural design rapidly developed between 2006 and 2010 and became a major cluster since 2011. During this period, rapid global urbanisation and climate change, especially global warming, interacted with and exacerbated the UHI effect. While global warming and urbanisation increased baseline temperatures and heat events, the UHI effect also contributed to climate change through increased energy consumption and greenhouse gas emissions (Luber & McGeehin, 2008; Tan et al., 2010; Zhang et al., 2009). This worsening of urban thermal environments shifted research attention from climatological aspects to urban planning, fostering more interdisciplinary collaboration. It involved investigating how urban planning and architectural design, such as in aspects of land cover, building materials, energy consumption, and transport infrastructure, affect the UHI and utilising sustainable urban planning to mitigate this issue (Che-Ani et al., 2009; Huang et al., 2011; Okeil, 2010; Yigitcanlar et al., 2008). For instance, reducing building density to facilitate heat dissipation through natural ventilation became a strategy (Allegrini & Carmeliet, 2017; He et al., 2021).

Innovations in building design also gained attention, including the use of reflective materials, green roofs, and improved ventilation designs (Qin, 2015; Sanchez & Reames, 2019). Energy management played a role by enhancing building energy efficiency, promoting renewable energy use, and reducing unnecessary energy consumption (Tian et al., 2021; Zhou et al., 2020). Furthermore, for mobility governance, optimising the locations of roads and pavements, reducing transportation emissions, and promoting public transportation were also recognised as effective strategies (Chapman et al., 2013; Coutts et al., 2010; Hendel et al., 2016; Kamruzzaman et al., 2015). It is worth noting that technological advancements, particularly in remote sensing, GIS, and urban modelling, have made it feasible to study the complex interplay between urban structures and heat vulnerability (Banzhaf & Netzband, 2011; Luber & McGeehin, 2008; Miao et al., 2009).

Urban environment and ecology have been key research clusters in the study of UHV and have attracted consistent scholarly interest since 2006, with peaks appearing in 2011 and 2015. As global urbanisation accelerates, the interaction between urban environments and ecosystems is garnering increasing research and political attention (Chapman et al., 2013; Jenerette et al., 2011; Livesley et al., 2016). This includes examining how urban ecosystems influence the UHI effect and the impact of urban planning and management on ecological balance. With the exacerbation of the UHI effect, research into the ecological environment of cities has become more crucial for understanding and adapting to these changes. Moreover, during this period, green and blue infrastructure's cooling effect on mitigating temperatures has been widely recognised (Coutts et al., 2013; Whitmee et al., 2015). This recognition has led to a surge in research into urban environments and ecosystems, exploring more urban ecological management approaches to reducing UHV.

Additionally, growing public interest in environmental protection and sustainable lifestyles has further propelled research in this field (Topay, 2012; Whitmee et al., 2015). Overall, research in this cluster has concentrated on ecological planning and protection, landscape design, green infrastructure optimisation, and water resource management (Chessa & Susca, 2023; Topay, 2012; Yao et al., 2023). Ecological planning emphasises maintaining ecological balance and biodiversity during urbanisation. Its goal is to reduce UHV by enhancing ecological services and building ecological networks (Livesley et al., 2016; Park et al., 2021). Landscape design aims to strategically plan and design green spaces, providing more shaded areas for the public and leveraging vegetation transpiration to reduce temperatures (Meerow & Newell, 2017; Yang et al., 2018). Optimising both natural and built green infrastructures improves urban environmental services, such as drainage systems and microclimates (Gunawardena et al., 2017; Liu et al., 2021). These improvements contribute to creating a more thermally comfortable urban environment by reducing heat accumulation and offering natural cooling effects.

Urban remote sensing is unique among research clusters as it primarily focuses on technical perspectives, offers vital data and methodological support to other clusters, and contributes to data-driven governance. Unlike the other four main clusters, it is not an extended research cluster, which may contribute to its relatively lower academic attention than other clusters. This specificity leads to less thematic overlap with other clusters, yet urban remote sensing, particularly supported by machine learning, maintains strong interconnections with them due to the essential role it plays in data acquisition and analysis (Estoque et al., 2020; Kimothi et al., 2023; Li et al., 2020; Venter et al., 2020). It has received the most attention between 2011 and 2015 compared to other periods.

The deep application of remote sensing technology in UHV research has benefited from significant developments in various aspects of remote sensing technologies during this period. Post-2015, remote sensing has been further applied in the UHV field. Although its role, as a novel introduction to the field, may have diminished compared to the 2011–2015 period, it has transitioned into a fundamental and indispensable technical approach. This evolution reflects not just the maturity of remote sensing technologies but also their integration into interdisciplinary research. The expansion and deepening of remote sensing applications underscore its evolving significance in comprehensively understanding and addressing UHV challenges. The advancements in remote sensing for UHV, from 2010 onwards, also go hand in hand with the developments of AI techniques, particularly machine learning and its subset, deep learning (Li et al., 2023).

During 2011–2015, the historical image accumulation and the launch of high-resolution multispectral satellites, such as Landsat 8 and the Sentinel series, have been particularly beneficial for accurately measuring surface temperatures and identifying hotspots (Mitraka & Berger, 2012; Shi et al., 2021; Zhu et al., 2021; Zoran & Dida, 2012). This contributes to a better understanding of the UHI effect on a fine-grained scale. Open data policies and developments are on the rise. They offer greater data access to the public, leading to more collaborative, multidisciplinary research (Gorelick et al., 2017; Wulder et al., 2022). Advances have been made not only in capturing rich urban datasets but also in the speed and computational capacity, including cloud computing and integration with AI and deep learning. The increasing proliferation of computational technology enables large-scale data analysis (Chen et al., 2022; Sagan et al., 2020). Cloud computing platforms, such as Google Earth Engine, offer user functions for faster data processing, larger data storage, and efficient data sharing (Gomes et al., 2020; Gorelick et al., 2017).

Breakthroughs in AI have improved the interpretation and analysis of information extracted from remote sensing images, for example, to identify the spatio-temporal patterns and changes of urban heat characteristics (Gorelick et al., 2017; Ramani et al., 2023; Wulder et al., 2022). As the integration of precise mapping with powerful spatio-temporal analysis opens new capabilities for incorporating urban planning information, environmental conditions, and socioeconomic characteristics in urban studies analysis, the field of UHV is flourishing. In recent years, the popularity of unmanned aerial vehicles has further enabled researchers to collect customised fine-grained images in targeted areas with low economic costs and high efficiency (Lyu et al., 2023; Rodríguez et al., 2022). Remote sensing, with its provision and sharing of high-resolution, long-term data such as surface temperatures, vegetation cover, and urban layouts, enables researchers to analyse spatio-temporal patterns of UHV worldwide over time.

In recent years, a shift from environmental UHV to public health and wellbeing vulnerabilities has been recognised. Urban public health and wellbeing have emerged as significant research clusters, developing rapidly since 2016 and contributing to the largest research aggregation, especially after

2021. Initially, this cluster received attention from a medical perspective. Researchers aimed to establish a medical understanding of the urban heat impact on human health, focusing particularly on young and elderly populations (Lai & Cheng, 2010; Lissner et al., 2012). This was done by exploring the qualitative and quantitative relationship between epidemiological data, such as hospitalisation and mortality rates, and atmospheric temperature (Hondula & Barnett, 2014; O'Neill & Ebi, 2009; Urban et al., 2014).

Nevertheless, recent research has shifted focus, utilising multidisciplinary approaches to measure population vulnerability to extreme heat, especially in urban areas. It has become clear that people living in urban areas are more vulnerable to increased occurrences of extreme heat and higher health risks. This situation is worsened by the growing density of urban populations and the intensification of the UHI effect (Mika et al., 2018; Estoque et al., 2020; Zander et al., 2018). The focus shift has benefited from an enhanced understanding of UHV, as well as theoretical and technological advancements in fields like climatology, meteorology, urban planning, and environmental science.

Socioeconomic conditions also affect the adaptive capacity of people to urban heat. For instance, low-income groups and marginalised communities often find it more difficult to adapt to and mitigate the impacts of UHI effects (Hsu et al., 2021; Voelkel et al., 2018). Consequently, research and policy are now increasingly focusing on identifying vulnerable groups and mitigating their vulnerability to urban heat. With the advancements in theory and technology in other research clusters, it is now possible to measure population vulnerability in a multidisciplinary way, not limited to the medical angle. For example, research in this cluster utilises multiple indicators, including sociodemographic, health, and environmental characteristics, to measure population heat vulnerability in terms of exposure risk, sensitivity, and adaptive capacity (Karanja & Kiage, 2021; Räsänen et al., 2019; Sun et al., 2022). The results, in turn, guide the formulation of mitigation strategies in urban planning, environmental improvement, and climate adaptation, contributing to the creation of liveable environments and the enhancement of human wellbeing (Deng & Wong, 2020; Hatvani-Kovacs et al., 2016).

The findings from this research are instrumental in guiding the development of targeted and effective policies, thereby contributing to the achievement of SDG 3 (Good Health and Wellbeing), SDG 11 (Sustainable Cities and Communities), and SDG 13 (Climate Action) (Li et al., 2022; Mushore et al., 2018; Weber et al., 2015). Ultimately, UHV research will help achieve the goal of urban sustainability and climate resilience. Overall, while UHV research is a relatively new field, it is expected to rapidly expand in the future with increased interdisciplinary collaboration and a stronger focus on human–social dimensions. There are several key directions and gaps that future research could explore.

- One promising direction involves capturing and managing the urban thermal environment from various perspectives to develop innovative urban climate adaptation strategies. Future research in urban climate and meteorology will likely focus on developing advanced urban climate models, conducting microclimate studies, and exploring how global climate change affects urban areas. These investigations will offer new insights into the intricate relationships between urban environments and climate dynamics.

- In terms of urban planning and architectural design, future studies should continue to emphasise sustainable urban planning strategies, green building designs, and energy-efficient management practices. These initiatives are vital for alleviating UHV by decreasing heat accumulation and enhancing thermal comfort in urban areas. Future studies could also examine the impact of different urban design strategies, including green roofs, reflective materials, and optimised city layouts.
- Studies of urban environments and ecosystems will need to concentrate on optimising green infrastructures, ecological planning, and landscape design to balance urban heat. Research in this area should explore how different types of vegetation and water features can alleviate UHI effects, enhance biodiversity, and provide ecosystem services that contribute to urban resilience.
- The focus on public health and wellbeing will become increasingly central as our understanding deepens regarding how urban heat affects diverse populations, including vulnerable groups like the elderly and low-income communities. Future research should aim to devise mitigation strategies that enhance urban residents' health and wellbeing. This could involve integrating epidemiological data with socioeconomic and environmental factors to identify at-risk populations and develop targeted interventions.
- A deeper application of remote sensing technologies is expected, featuring innovations in high-resolution imaging and sophisticated data analytics, to enhance the precision in monitoring and analysing urban thermal environment and their impacts on urban environments and human health. The integration of advanced computational sciences like artificial intelligence, machine learning, and big data analysis with traditional urban heat studies will provide powerful tools for data-driven urban governance. These advancements will enable more accurate and timely decision-making to address UHV challenges.

By exploring the above-mentioned future research areas and addressing the identified gaps, the field of UHV can significantly contribute to the development of sustainable, resilient, and liveable urban environments. Although this review primarily focuses on the scientometric analysis of UHV research, it is important to acknowledge the challenges associated with real-time applications of UHV research (Kumar & Mishra, 2024). Real-time data collection and integration pose significant hurdles, including the deployment of sufficient sensors, ensuring data quality, and managing large volumes of data. Technical infrastructure investments, such as high-quality sensors, data transmission networks, and data storage solutions, are essential yet costly, particularly for resource-constrained urban areas. Interdisciplinary coordination is critical but can be complex, requiring effective collaboration between urban planners, public health officials, meteorologists, and data scientists. Additionally, scalability issues may arise when transitioning from small-scale pilot projects to broader urban implementations. Policymakers must navigate regulatory, political, and social barriers to translate UHV research into actionable policies. Furthermore, public engagement and awareness are vital for the success of real-time applications, necessitating robust communication strategies to ensure community support and participation. By addressing these challenges, future UHV research can more effectively bridge the gap between theoretical insights and practical and real-time applications.

Building on the findings of our comprehensive scientometric analysis, our future work will focus on several strategic areas to advance the understanding and mitigation of UHV. We plan to initiate interdisciplinary research projects that bring together experts from urban planning, environmental science, public health, and advanced technologies, with the aim of developing integrated approaches to UHV. One primary objective is to refine and enhance urban climate models by incorporating detailed microclimate data and high-resolution remote sensing imagery, and by leveraging machine learning algorithms and big data analytics to create predictive models for real-time urban heat management. We plan to conduct in-depth studies to address the needs and vulnerabilities of at-risk populations, such as the elderly, children, and low-income communities, and thus could inform better design and pilot community–based adaptation programmes. Additionally, policy development is another key focus, where we intend to collaborate with policymakers to translate our research findings into actionable policies and urban planning guidelines. Finally, we will expand our public engagement efforts through educational programmes and outreach initiatives to build community resilience, raise public awareness about UHV, and promote community involvement in mitigation efforts. These endeavours will help to address critical gaps identified in the field and contribute to the development of sustainable, resilient, and health-promoting urban environments.

7.5 Conclusion

UHV poses a significant challenge worldwide. The research in this field has evolved from measuring vulnerabilities to developing adaptation and mitigation strategies. With the rise of anthropogenic climate change in recent decades, UHV has become a heightened source of risk to flora, fauna, and people. This is reflected in the noticeable increase in UHV research over the last two decades. While there has been significant growth in UHV research, a systematic review of the literature analysing its development and evolution over a significant period is notably absent. Therefore, this study employed a scientometric approach to map the literature from 1972 to January 2024, with the purpose of providing a consolidated comprehension of the historical development, emerging and prospective trends, and existing research clusters in UHV.

This chapter conducted a systematic scientometric analysis of 2340 academic works focused on UHV, covering publications over the last five decades. The analysis provided a concise and up-to-date overview of the field; it encapsulated the systematic knowledge in the field and the developmental trajectory of research in this area. The key findings from this study can be summarised as follows:

- *Emerging and Rapidly Expanding Research Area:* UHV is identified as an emerging research area with a significant increase in publications since 2016. The field is expected to continue expanding rapidly, in response to the growing recognition of the impacts of UHI and climate change on urban populations.
- *Five Interconnected Research Clusters:* UHV research is clustered around five main areas: Urban planning and architectural design, urban environment and ecology, urban public health and wellbeing, urban climate and meteorology, and urban remote sensing. These clusters highlight the interdisciplinary nature of UHV research and the importance of integrating different perspectives to address urban heat challenges effectively.
- *Shift Towards Public Health and Wellbeing:* Recent research trends indicate a shift in focus towards public health and wellbeing, with an apparent emphasis on the human–social dimensions of UHV. This highlights the need for interdisciplinary collaboration by integrating epidemiological data with socioeconomic and environmental factors to develop targeted interventions for vulnerable populations.
- *Significant Role of Remote Sensing and Advanced Technologies:* Remote sensing, supported by advancements in machine learning and data analytics, plays a critical role in UHV research. These technologies provide detailed and high-resolution data, which is critical for precise monitoring and analysis of urban heat and its effects on urban environments and human health.
- *Alignment with SDGs:* UHV research aligns with several SDGs, particularly SDG 3 (Good Health and Wellbeing), SDG 11 (Sustainable Cities and Communities), and SDG 13 (Climate Action). The research contributes to global efforts to achieve urban sustainability and climate resilience.

There are many implications that can be drawn from this study. Some of them are outlined below:

- *Interdisciplinary Collaboration:* The findings underscore the importance of interdisciplinary collaboration in advancing UHV research. Integrating urban planning, environmental science, public health, and advanced technologies is essential for developing comprehensive strategies to mitigate urban heat impacts.
- *Focus on Vulnerable Populations:* As UHV research continues to evolve, there is a critical need to focus on vulnerable populations, such as the elderly, low-income groups, and individuals with existing health issues. Tailored interventions and policies are necessary to enhance their resilience to urban heat.
- *Technological Advancements:* Leveraging technological advancements, particularly in remote sensing and data analytics, will be key to improving real-time monitoring and decision-making processes. These technologies will enable more effective urban heat management and policy implementation.
- *Policy Implications:* Policymakers should prioritise integrating UHV considerations into urban planning and public health strategies. Addressing the regulatory, political, and social barriers will be crucial for translating research findings into actionable policies.

Acknowledgements

This chapter, with permission from the copyright holder, is a reproduced version of the following journal article: Li, F., Yigitcanlar, T., Li, W., Nguyen, K., Nepal, M., & Dur, F. (2024). Understanding urban heat vulnerability: scientometric analysis of five decades of research. *Urban Climate*, 56(1) 102035.

References

Allegrini, J., & Carmeliet, J. (2017). Coupled CFD and building energy simulations for studying the impacts of building height topology and buoyancy on local urban microclimates. *Urban Climate, 21,* 278–305. https://doi.org/10.1016/j.uclim.2017.07.005

Arunab, K. S., & Mathew, A. (2024). Quantifying urban heat island and pollutant nexus: A novel geospatial approach. *Sustainable Cities and Society, 101,* 105117. https://doi.org/10.1016/j.scs.2023.105117

Atkinson, B. W. (2003). Numerical modelling of urban heat-island intensity. *Boundary-Layer Meteorology, 109*, 285–310. https://doi.org/10.1023/A:1025820326672

Banzhaf, E., & Netzband, M. (2011). Monitoring urban land use changes with remote sensing techniques. *Applied Urban Ecology: A Global Framework*, 18–32. https://doi.org/10.1002/9781444345025.ch3

Bao, J., Li, X., & Yu, C. (2015). The construction and validation of the heat vulnerability index, a review. *International Journal of Environmental Research and Public Health, 12*(7), 7220–7234. https://doi.org/10.3390/ijerph120707220

Bioucas-Dias, J. M., Plaza, A., Camps-Valls, G., Scheunders, P., Nasrabadi, N., & Chanussot, J. (2013). Hyperspectral remote sensing data analysis and future challenges. *IEEE Geoscience and Remote Sensing Magazine, 1*(2), 6–36. https://doi.org/10.1109/MGRS.2013.2244672

Bozdogan Sert, E., Kaya, E., Adiguzel, F., Cetin, M., Gungor, S., Zeren Cetin, I., & Dinc, Y. (2021). Effect of the surface temperature of surface materials on thermal comfort: A case study of Iskenderun (Hatay, Turkey). *Theoretical and Applied Climatology, 144*(1–2), 103–113. https://doi.org/10.1007/s00704-021-03524-0

Cai, Z., La Sorte, F. A., Chen, Y., & Wu, J. (2023). The surface urban heat island effect decreases bird diversity in Chinese cities. *Science of the Total Environment, 902*, 166200. https://doi.org/10.1016/j.scitotenv.2023.166200

Centers for Disease Control and Prevention (2017). Heat-related deaths—United States, 2004–2014. *MMWR Morbidity and Mortality Weekly Report, 66*(35), 935–939.

Chapman, L., Azevedo, J. A., & Prieto-Lopez, T. (2013). Urban heat & critical infrastructure networks: A viewpoint. *Urban Climate, 3*, 7–12. https://doi.org/10.1016/j.uclim.2013.04.001

Chapman, S., Watson, J. E., Salazar, A., Thatcher, M., & McAlpine, C. A. (2017). The impact of urbanization and climate change on urban temperatures: A systematic review. *Landscape Ecology, 32*, 1921–1935. https://doi.org/10.1007/s10980-017-0561-4

Che-Ani, A. I., Shahmohamadi, P., Sairi, A., Mohd-Nor, M. F. I., Zain, M. F. M., & Surat, M. (2009). Mitigating the urban heat island effect: Some points without altering existing city planning. *European Journal of Scientific Research, 35*(2), 204–216.

Chen, F., Kusaka, H., Bornstein, R., Ching, J., Grimmond, C. S. B., Grossman-Clarke, S., …, & Zhang, C. (2011). The integrated WRF/urban modelling system: Development, evaluation, and applications to urban environmental problems. *International Journal of Climatology, 31*(2), 273–288. https://doi.org/10.1002/joc.2158

Chen, J., Chen, S., Fu, R., Li, D., Jiang, H., Wang, C., …, & Hicks, B. J. (2022). Remote sensing big data for water environment monitoring: Current status, challenges, and future prospects. *Earth's Future, 10*(2), e2021EF002289. https://doi.org/10.1029/2021EF002289

Chen, K., Boomsma, J., & Holmes, H. A. (2023). A multiscale analysis of heatwaves and urban heat islands in the western US during the summer of 2021. *Scientific Reports, 13*(1), 9570. https://doi.org/10.1038/s41598-023-35621-7

Chen, Y., Wu, J., Yu, K., & Wang, D. (2020). Evaluating the impact of the building density and height on the block surface temperature. *Building and Environment, 168*, 106493. https://doi.org/10.1016/j.buildenv.2019.106493

Cheng, W., Li, D., Liu, Z., & Brown, R. D. (2021). Approaches for identifying heat-vulnerable populations and locations: A systematic review. *Science of The Total Environment, 799*, 149417. https://doi.org/10.1016/j.scitotenv.2021.149417

Chessa, C., & Susca, T. (2023). Development of an LCA characterization factor to account UHI local effect on terrestrial ecosystems damage category: Evaluation of European Bombus and Onthophagus genera heat-stress mortality. *Science of The Total Environment, 897,* 165183. https://doi.org/10.1016/j.scitotenv.2023.165183

Coffel, E. D., de Sherbinin, A., Horton, R. M., Lane, K., Kienberger, S., & Wilhelmi, O. (2018). The science of adaptation to extreme heat. In: *Resilience* (pp. 89–103). Elsevier. https://doi.org/10.1016/B978-0-12-811891-7.00007-4

Coutts, A., Beringer, J., & Tapper, N. (2010). Changing urban climate and CO2 emissions: Implications for the development of policies for sustainable cities. *Urban Policy and Research,* 28(1), 27–47. https://doi.org/10.1080/08111140903437716

Coutts, A. M., Tapper, N. J., Beringer, J., Loughnan, M., & Demuzere, M. (2013). watering our cities: The capacity for water sensitive urban design to support urban cooling and improve human thermal comfort in the Australian context. *Progress in Physical Geography, 37*(1), 2–28. https://doi.org/10.1177/0309133312461032

Crichton, D. (1999). The risk triangle. *Natural Disaster Management, 102*(3), 102–103.

Dai, Z., Guldmann, J. M., & Hu, Y. (2019). Thermal impacts of greenery, water, and impervious structures in Beijing's Olympic area: A spatial regression approach. *Ecological Indicators, 97,* 77–88. https://doi.org/10.1016/j.ecolind.2018.09.041

Degirmenci, K., Desouza, K. C., Fieuw, W., Watson, R. T., & Yigitcanlar, T. (2021). Understanding policy and technology responses in mitigating urban heat islands: A literature review and directions for future research. *Sustainable Cities and Society, 70,* 102873. https://doi.org/10.1016/j.scs.2021.102873

Deilami, K., Kamruzzaman, M., & Liu, Y. (2018). Urban heat island effect: A systematic review of spatio-temporal factors, data, methods, and mitigation measures. *International Journal of Applied Earth Observation and Geoinformation, 67,* 30–42. https://doi.org/10.1016/j.jag.2017.12.009

Deng, J. Y., & Wong, N. H. (2020). Impact of urban canyon geometries on outdoor thermal comfort in central business districts. *Sustainable Cities and Society, 53,* 101966. https://doi.org/10.1016/j.scs.2019.101966

Diem, P. K., Nguyen, C. T., Diem, N. K., Diep, N. T. H., Thao, P. T. B., Hong, T. G., & Phan, T. N. (2023). Remote sensing for urban heat island research: Progress, current issues, and perspectives. *Remote Sensing Applications: Society and Environment,* 101081. https://doi.org/10.1016/j.rsase.2023.101081

Estoque, R. C., Ooba, M., Seposo, X. T., Togawa, T., Hijioka, Y., Takahashi, K., & Nakamura, S. (2020). Heat health risk assessment in Philippine cities using remotely sensed data and social-ecological indicators. *Nature Communications, 11*(1), 1581. https://doi.org/10.1038/s41467-020-15218-8

Fahy, B., Brenneman, E., Chang, H., & Shandas, V. (2019). Spatial analysis of urban flooding and extreme heat hazard potential in Portland, OR. *International Journal of Disaster Risk Reduction, 39,* 101117. https://doi.org/10.1016/j.ijdrr.2019.101117

Faisal, A., Yigitcanlar, T., Kamruzzaman, M., & Paz, A. (2021). Mapping two decades of autonomous vehicle research: A systematic scientometric analysis. *Journal of Urban Technology, 28*(3–4), 45–74. https://doi.org/10.1080/10630732.2020.1780868

Feng, R., Wang, F., Wang, K., Wang, H., & Li, L. (2021). Urban ecological land and natural-anthropogenic environment interactively drive surface urban heat island: An urban agglomeration-level study in China. *Environment International, 157,* 106857. https://doi.org/10.1016/j.envint.2021.106857

Gago, E. J., Roldan, J., Pacheco-Torres, R., & Ordóñez, J. (2013). The city and urban heat islands: A review of strategies to mitigate adverse effects. *Renewable and Sustainable Energy Reviews, 25*, 749–758. https://doi.org/10.1016/j.rser.2013.05.057

Ghiasi, S., Passe, U., Zhou, Y., & Thompson, J. (2021, September). A sensitivity analysis to investigate urban heat island impact on building energy consumption. In Building Simulation Conference Proceedings (Vol. 17). https://doi.org/10.26868/25222708.2021.30873

Gomes, V. C., Queiroz, G. R., & Ferreira, K. R. (2020). An overview of platforms for big earth observation data management and analysis. *Remote Sensing, 12*(8), 1253. https://doi.org/10.3390/rs12081253

Gorelick, N., Hancher, M., Dixon, M., Ilyushchenko, S., Thau, D., & Moore, R. (2017). Google earth engine: Planetary-scale geospatial analysis for everyone. *Remote Sensing of Environment, 202*, 18–27. https://doi.org/10.1016/j.rse.2017.06.031

Gunawardena, K. R., Wells, M. J., & Kershaw, T. (2017). Utilising green and bluespace to mitigate urban heat island intensity. *Science of the Total Environment, 584*, 1040–1055. https://doi.org/10.1016/j.scitotenv.2017.01.158

Gupta, R. (2012). Temporal and spatial variations of urban heat island effect in Jaipur city using satellite data. *Environment and Urbanization Asia, 3*(2), 359–374. https://doi.org/10.1177/0975425312473232

Habeeb, D., Vargo, J., & Stone, B. (2015). Rising heat wave trends in large US cities. *Natural Hazards, 76*, 1651–1665. https://doi.org/10.1007/s11069-014-1563-z

Hamdi, R., & Schayes, G. (2008). Sensitivity study of the urban heat island intensity to urban characteristics. *International Journal of Climatology: A Journal of the Royal Meteorological Society, 28*(7), 973–982. https://doi.org/10.1002/joc.1598

Harlan, S. L., Brazel, A. J., Prashad, L., Stefanov, W. L., & Larsen, L. (2006). Neighborhood microclimates and vulnerability to heat stress. *Social Science & Medicine, 63*(11), 2847–2863. https://doi.org/10.1016/j.socscimed.2006.07.030

Hatvani-Kovacs, G., Belusko, M., Skinner, N., Pockett, J., & Boland, J. (2016). Drivers and barriers to heat stress resilience. *Science of the Total Environment, 571*, 603–614. https://doi.org/10.1016/j.scitotenv.2016.07.028

Hatvani-Kovacs, G., Bush, J., Sharifi, E., & Boland, J. (2018). Policy recommendations to increase urban heat stress resilience. *Urban Climate, 25*, 51–63. https://doi.org/10.1016/j.uclim.2018.05.001

He, B. J., Zhao, D., Dong, X., Zhao, Z., Li, L., Duo, L., & Li, J. (2022a). Will individuals visit hospitals when suffering heat-related illnesses? Yes, but…. *Building and Environment, 208*, 108587. https://doi.org/10.1016/j.buildenv.2021.108587

He, B. J., Zhao, D., Dong, X., Xiong, K., Feng, C., Qi, Q., …, & Pathak, M. (2022b). Perception, physiological and psychological impacts, adaptive awareness and knowledge, and climate justice under urban heat: A study in extremely hot-humid Chongqing, China. *Sustainable Cities and Society, 79*, 103685. https://doi.org/10.1016/j.scs.2022.103685

He, B. J., Zhao, D., Xiong, K., Qi, J., Ulpiani, G., Pignatta, G., …, & Jones, P. (2021). A framework for addressing urban heat challenges and associated adaptive behavior by the public and the issue of willingness to pay for heat resilient infrastructure in Chongqing, China. *Sustainable Cities and Society, 75*, 103361. https://doi.org/10.1016/j.scs.2021.103361

He, C., Zhang, Y., Schneider, A., Chen, R., Zhang, Y., Ma, W., …, & Kan, H. (2022c). The inequality labor loss risk from future urban warming and adaptation strategies. *Nature Communications, 13*(1), 3847. https://doi.org/10.1038/s41467-022-31145-2

Hendel, M., Gutierrez, P., Colombert, M., Diab, Y., & Royon, L. (2016). Measuring the effects of urban heat island mitigation techniques in the field: Application to the case of pavement-watering in Paris. *Urban Climate, 16*, 43–58. https://doi.org/10.1016/j.uclim.2016.02.003

Heusinger, J., Bruchmann, N., & Weber, S. (2023). Modeling the impacts of building energy efficiency on the thermal microclimate in a midsize German city. *Urban Climate, 52*, 101678. https://doi.org/10.1016/j.uclim.2023.101678

Ho, J. Y., Shi, Y., Lau, K. K., Ng, E. Y., Ren, C., & Goggins, W. B. (2023). Urban heat island effect-related mortality under extreme heat and non-extreme heat scenarios: A 2010–2019 case study in Hong Kong. *Science of the Total Environment, 858*, 159791. https://doi.org/10.1016/j.scitotenv.2022.159791

Hoehne, C. G., Hondula, D. M., Chester, M. V., Eisenman, D. P., Middel, A., Fraser, A. M., …, & Gerster, K. (2018). Heat exposure during outdoor activities in the US varies significantly by city, demography, and activity. *Health & Place, 54*, 1–10. https://doi.org/10.1016/j.healthplace.2018.08.014

Hondula, D. M., & Barnett, A. G. (2014). Heat-related morbidity in Brisbane, Australia: Spatial variation and area-level predictors. *Environmental Health Perspectives, 122*(8), 831–836. https://doi.org/10.1289/ehp.1307496

Hsu, A., Sheriff, G., Chakraborty, T., & Manya, D. (2021). Disproportionate exposure to urban heat island intensity across major US cities. *Nature Communications, 12*(1), 2721. https://doi.org/10.1038/s41467-021-22799-5

Huang, G., Zhou, W., & Cadenasso, M. L. (2011). Is everyone hot in the city? Spatial pattern of land surface temperatures, land cover and neighborhood socioeconomic characteristics in Baltimore, MD. *Journal of Environmental Management, 92*(7), 1753–1759. https://doi.org/10.1016/j.jenvman.2011.02.006

Huang, W. T. K., Masselot, P., Bou-Zeid, E., Fatichi, S., Paschalis, A., Sun, T., …, & Manoli, G. (2023). Economic valuation of temperature-related mortality attributed to urban heat islands in European cities. *Nature Communications, 14*(1), 7438. https://doi.org/10.1038/s41467-023-43135-z

Jedlovec, G., Crane, D., & Quattrochi, D. (2017). Urban heat wave hazard and risk assessment. *Results in Physics, 7*, 4294–4295. https://doi.org/10.1016/j.rinp.2017.10.056

Jenerette, G. D., Harlan, S. L., Stefanov, W. L., & Martin, C. A. (2011). Ecosystem services and urban heat riskscape moderation: Water, green spaces, and social inequality in Phoenix, USA. *Ecological Applications, 21*(7), 2637–2651. https://doi.org/10.1890/10-1493.1

Jin, M., Shepherd, J. M., & Peters-Lidard, C. (2007). Development of a parameterization for simulating the urban temperature hazard using satellite observations in climate model. *Natural Hazards, 43*, 257–271. https://doi.org/10.1007/s11069-007-9117-2

Kabano, P., Lindley, S., & Harris, A. (2021). Evidence of urban heat island impacts on the vegetation growing season length in a tropical city. *Landscape and Urban Planning, 206*, 103989. https://doi.org/10.1016/j.landurbplan.2020.103989

Kamruzzaman, M., Deilami, K., & Yigitcanlar, T. (2018). Investigating the urban heat island effect of transit oriented development in Brisbane. *Journal of Transport Geography, 66*, 116–124. https://doi.org/10.1016/j.jtrangeo.2017.11.016

Kamruzzaman, M., Hine, J., & Yigitcanlar, T. (2015). Investigating the link between carbon dioxide emissions and transport-related social exclusion in rural Northern Ireland. *International Journal of Environmental Science and Technology, 12*(11), 3463–3478. https://doi.org/10.1007/s13762-015-0771-8

Karanja, J., & Kiage, L. (2021). Perspectives on spatial representation of urban heat vulnerability. *Science of The Total Environment, 774*, 145634. https://doi.org/10.1016/j.scitotenv.2021.145634

Keramitsoglou, I. (2012, June). Advanced earth observation methodologies for the study of the thermal environment of cities. In *2012 Second International Workshop on Earth Observation and Remote Sensing Applications* (pp. 11–15). IEEE. https://doi.org/10.1109/EORSA.2012.6261125

Kiba-Janiak, M., Marcinkowski, J., Jagoda, A., & Skowrońska, A. (2021). Sustainable last mile delivery on e-commerce market in cities from the perspective of various stakeholders. Literature review. *Sustainable Cities and Society, 71*, 102984. https://doi.org/10.1016/j.scs.2021.102984

Kimothi, S., Thapliyal, A., Gehlot, A., Aledaily, A. N., Bilandi, N., Singh, R., …, & Akram, S. V. (2023). Spatio-temporal fluctuations analysis of land surface temperature (LST) using remote sensing data (LANDSAT TM5/8) and multifractal technique to characterize the urban heat islands (UHIs). *Sustainable Energy Technologies and Assessments, 55*, 102956. https://doi.org/10.1016/j.seta.2022.102956

Kumar, R., & Mishra, S. K. (2024). Assessing the impact of heat vulnerability on urban public spaces using a fuzzy-based unified computational technique. *AI & SOCIETY*, 1–18. https://doi.org/10.1007/s00146-024-01904-4

Lai, L. W., & Cheng, W. L. (2010). Urban heat island and air pollution—An emerging role for hospital respiratory admissions in an urban area. *Journal of Environmental Health, 72*(6), 32–36.

Lakshmi, R. P. (2023). Urban heat island: A review of effects and predictions with deep and machine learning technologies. *Geoinformatics for Sustainable Urban Development*, 71–84. https://doi.org/10.1201/9781003331001-5

Lanza, K., Jones, J., Acuña, F., Coudert, M., Bixler, R. P., Kamath, H., & Niyogi, D. (2023). Heat vulnerability of Latino and Black residents in a low-income community and their recommended adaptation strategies: A qualitative study. *Urban Climate, 51*, 101656. https://doi.org/10.1016/j.uclim.2023.101656

Lemonsu, A., Belair, S., & Mailhot, J. (2009). The new Canadian urban modelling system: Evaluation for two cases from the Joint Urban 2003 Oklahoma City experiment. *Boundary-Layer Meteorology, 133*, 47–70. https://doi.org/10.1007/s10546-009-9414-2

Li, D., & Bou-Zeid, E. (2013). Synergistic interactions between urban heat islands and heat waves: The impact in cities is larger than the sum of its parts. *Journal of Applied Meteorology and Climatology, 52*(9), 2051–2064. https://doi.org/10.1175/JAMC-D-13-02.1

Li, F., Yigitcanlar, T., Nepal, M., Thanh, K. N., & Dur, F. (2022a). Understanding urban heat vulnerability assessment methods: A PRISMA review. *Energies, 15*(19), 6998. https://doi.org/10.3390/en15196998

Li, F., Yigitcanlar, T., Nepal, M., Nguyen, K., & Dur, F. (2023). Machine learning and remote sensing integration for leveraging urban sustainability: A review and framework. *Sustainable Cities and Society, 96*, 104653. https://doi.org/10.1016/j.scs.2023.104653

Li, L., Zha, Y., & Wang, R. (2020). Relationship of surface urban heat island with air temperature and precipitation in global large cities. *Ecological Indicators, 117*, 106683. https://doi.org/10.1016/j.ecolind.2020.106683

Li, W., Yigitcanlar, T., Liu, A., & Erol, I. (2022b). Mapping two decades of urban heat vulnerability research: A systematic scientometric analysis. *Technological Forecasting and Social Change, 179*, 121676. https://doi.org/10.1016/j.techfore.2022.121676

Li, X., Zhou, Y., Yu, S., Jia, G., Li, H., & Li, W. (2019). Urban heat island impacts on building energy consumption: A review of approaches and findings. *Energy, 174*, 407–419. https://doi.org/10.1016/j.energy.2019.02.183

Li, Y., Schubert, S., Kropp, J. P., & Rybski, D. (2020). On the influence of density and morphology on the urban heat island intensity. *Nature Communications, 11*(1), 2647. https://doi.org/10.1038/s41467-020-16461-9.

Lin, P., Lau, S. S. Y., Qin, H., & Gou, Z. (2017). Effects of urban planning indicators on urban heat island: A case study of pocket parks in high-rise high-density environment. *Landscape and Urban Planning, 168*, 48–60. https://doi.org/10.1016/j.landurbplan.2017.09.024

Lissner, T. K., Holsten, A., Walther, C., & Kropp, J. P. (2012). Towards sectoral and standardised vulnerability assessments: The example of heatwave impacts on human health. *Climatic Change, 112*, 687–708. https://doi.org/10.1007/s10584-011-0231-5

Liu, Z., Cheng, W., Jim, C. Y., Morakinyo, T. E., Shi, Y., & Ng, E. (2021). Heat mitigation benefits of urban green and blue infrastructures: A systematic review of modeling techniques, validation and scenario simulation in ENVI-met V4. *Building and Environment, 200*, 107939. https://doi.org/10.1016/j.buildenv.2021.107939

Livesley, S. J., McPherson, E. G., & Calfapietra, C. (2016). The urban forest and ecosystem services: Impacts on urban water, heat, and pollution cycles at the tree, street, and city scale. *Journal of Environmental Quality, 45*(1), 119–124. https://doi.org/10.2134/jeq2015.11.0567

Lu, L., Xu, Y., Tang, J., & Hu, G. (2024). High-resolution satellite imagery analysis of coseismic landslides and liquefaction induced by the 2024 MW 7.4 Hualien earthquake, Taiwan, China. *Earthquake Research Advances*, 100356.

Luber, G., & McGeehin, M. (2008). Climate change and extreme heat events. *American Journal of Preventive Medicine, 35*(5), 429–435. https://doi.org/10.1016/j.amepre.2008.08.021

Lyu, R., Pang, J., Tian, X., Zhao, W., & Zhang, J. (2023). How to optimize the 2D/3D urban thermal environment: Insights derived from UAV LiDAR/multispectral data and multi-source remote sensing data. *Sustainable Cities and Society, 88*, 104287. https://doi.org/10.1016/j.scs.2022.104287

Macharia, C. W., & Kiage, L. M. (2024). Conceptualizing heat vulnerability: Equity-centered approaches for comprehensive resilience in a changing climate. *Natural Hazards*, 1–19. https://doi.org/10.1007/s11069-024-06440-4

Macintyre, H. L., Heaviside, C., Taylor, J., Picetti, R., Symonds, P., Cai, X. M., & Vardoulakis, S. (2018). Assessing urban population vulnerability and environmental risks across an urban area during heatwaves–Implications for health protection. *Science of the Total Environment, 610*, 678–690. https://doi.org/10.1016/j.scitotenv.2017.08.062

Maxwell, A. E., Warner, T. A., & Fang, F. (2018). Implementation of machine-learning classification in remote sensing: An applied review. *International Journal of Remote Sensing, 39*(9), 2784–2817. https://doi.org/10.1080/01431161.2018.1433343

Meerow, S., & Newell, J. P. (2017). Spatial planning for multifunctional green infrastructure: Growing resilience in Detroit. *Landscape and Urban Planning, 159*, 62–75. https://doi.org/10.1016/j.landurbplan.2016.10.005

Miao, S., Chen, F., LeMone, M. A., Tewari, M., Li, Q., & Wang, Y. (2009). An observational and modeling study of characteristics of urban heat island and boundary layer structures in Beijing. *Journal of Applied Meteorology and Climatology, 48*(3), 484–501. https://doi.org/10.1175/2008JAMC1909.1

Miao, Y., Sheng, J., & Ye, J. (2022). An assessment of the impact of temperature rise due to climate change on asphalt pavement in China. *Sustainability, 14*(15), 9044. https://doi.org/10.3390/su14159044

Mika, J., Forgo, P., Lakatos, L., Olah, A. B., Rapi, S., & Utasi, Z. (2018). Impact of 1.5 K global warming on urban air pollution and heat island with outlook on human health effects. *Current Opinion in Environmental Sustainability, 30,* 151–159. https://doi.org/10.1016/j.cosust.2018.05.013

Mitraka, Z., & Berger, M. (2012). *Land surface emissivity estimation from synergistic use of Sentinel-2 and Sentinel-3 data based on spatial-spectral unmixing techniques.*

Mohajerani, A., Bakaric, J., & Jeffrey-Bailey, T. (2017). The urban heat island effect, its causes, and mitigation, with reference to the thermal properties of asphalt concrete. *Journal of Environmental Management, 197,* 522–538. https://doi.org/10.1016/j.jenvman.2017.03.095

Mortoja, M., Yigitcanlar, T., & Mayere, S. (2020). What is the most suitable methodological approach to demarcate peri-urban areas? A systematic review of the literature. *Land Use Policy, 95,* 104601. https://doi.org/10.1016/j.landusepol.2020.104601

Movahed, S. M. A., & Sarmah, A. K. (2021). Global trends and characteristics of nano- and micro-bubbles research in environmental engineering over the past two decades: A scientometric analysis. *Science of the Total Environment, 785,* 147362. https://doi.org/10.1016/j.scitotenv.2021.147362

Mushore, T. D., Mutanga, O., Odindi, J., & Dube, T. (2018). Determining extreme heat vulnerability of Harare Metropolitan City using multispectral remote sensing and socio-economic data. *Journal of Spatial Science, 63*(1), 173–191. https://doi.org/10.1080/14498596.2017.1290558

Nguyen, C. T., Chidthaisong, A., Limsakul, A., Varnakovida, P., Ekkawatpanit, C., Diem, P. K., & Diep, N. T. H. (2022). How do disparate urbanization and climate change imprint on urban thermal variations? A comparison between two dynamic cities in Southeast Asia. *Sustainable Cities and Society, 82,* 103882. https://doi.org/10.1016/j.scs.2022.103882

Niu, Y., Li, Z., Gao, Y., Liu, X., Xu, L., Vardoulakis, S., …, & Liu, Q. (2021). A systematic review of the development and validation of the heat vulnerability index: Major factors, methods, and spatial units. *Current Climate Change Reports, 7*(3), 87–97. https://doi.org/10.1007/s40641-021-00173-3

O'Neill, M. S., & Ebi, K. L. (2009). Temperature extremes and health: Impacts of climate variability and change in the United States. *Journal of Occupational and Environmental Medicine, 51*(1), 13–25. https://doi.org/10.1097/JOM.0b013e318173e122

Okeil, A. (2010). A holistic approach to energy efficient building forms. *Energy and Buildings, 42*(9), 1437–1444. https://doi.org/10.1016/j.enbuild.2010.03.013

Oleson, K. W., Bonan, G. B., Feddema, J., & Vertenstein, M. (2008). An urban parameterization for a global climate model. Part II: Sensitivity to input parameters and the simulated urban heat island in offline simulations. *Journal of Applied Meteorology and Climatology, 47*(4), 1061–1076. https://doi.org/10.1175/2007JAMC1598.1

Oliveira, S., Andrade, H., & Vaz, T. (2011). The cooling effect of green spaces as a contribution to the mitigation of urban heat: A case study in Lisbon. *Building and Environment, 46*(11), 2186–2194. https://doi.org/10.1016/j.buildenv.2011.04.034

Park, C. Y., Park, Y. S., Kim, H. G., Yun, S. H., & Kim, C. K. (2021). Quantifying and mapping cooling services of multiple ecosystems. *Sustainable Cities and Society, 73,* 103123. https://doi.org/10.1016/j.scs.2021.103123

Parmesan, C., Morecroft, M. D., & Trisurat, Y. (2022). Climate change 2022: Impacts, adaptation and vulnerability (Doctoral dissertation, GIEC).

Parry, M. L. (Ed.). (2007). *Climate change 2007-impacts, adaptation and vulnerability: Working group II contribution to the fourth assessment report of the IPCC* (Vol. 4). Cambridge University Press.

Patz, J. A., Campbell-Lendrum, D., Holloway, T., & Foley, J. A. (2005). Impact of regional climate change on human health. *Nature, 438*(7066), 310–317. https://doi.org/10.1038/nature04188

Qin, Y. (2015). A review on the development of cool pavements to mitigate urban heat island effect. *Renewable and Sustainable Energy Reviews, 52*, 445–459. https://doi.org/10.1016/j.rser.2015.07.177

Ramani, V., Arjunan, P., Poolla, K., & Miller, C. (2023). Semantic segmentation of longitudinal thermal images for identification of hot and cool spots in urban areas. *Building and Environment*, 111112. https://doi.org/10.1016/j.buildenv.2023.111112

Räsänen, A., Heikkinen, K., Piila, N., & Juhola, S. (2019). Zoning and weighting in urban heat island vulnerability and risk mapping in Helsinki, Finland. *Regional Environmental Change, 19*, 1481–1493. https://doi.org/10.1007/s10113-019-01491-x

Robine, J. M. et al (2008). Death toll exceeded 70,000 in Europe during the summer of 2003. *Comptes Rendus Biologies, 331*(2), 171–178. https://doi.org/10.1016/j.crvi.2007.12.001

Rodríguez, M. V., Melgar, S. G., & Márquez, J. M. A. (2022). Assessment of aerial thermography as a method of in situ measurement of radiant heat transfer in urban public spaces. *Sustainable Cities and Society, 87*, 104228. https://doi.org/10.1016/j.scs.2022.104228

Sagan, V., Peterson, K. T., Maimaitijiang, M., Sidike, P., Sloan, J., Greeling, B. A., ..., & Adams, C. (2020). Monitoring inland water quality using remote sensing: Potential and limitations of spectral indices, bio-optical simulations, machine learning, and cloud computing. *Earth-Science Reviews, 205*, 103187. https://doi.org/10.1016/j.earscirev.2020.103187

Sagris, V., & Sepp, M. (2017). Landsat-8 TIRS data for assessing urban heat island effect and its impact on human health. *IEEE Geoscience and Remote Sensing Letters, 14*(12), 2385–2389. https://doi.org/10.1109/LGRS.2017.2765703

Sahu, K., & Srivastava, R. K. (2020). Needs and importance of reliability prediction: An industrial perspective. *Information Sciences Letters, 9*(1), 33–37. http://dx.doi.org/10.18576/isl/090105

Sahu, K., & Srivastava, R. K. (2021). Predicting software bugs of newly and large datasets through a unified neuro-fuzzy approach: Reliability perspective. *Advances in Mathematics: Scientific Journal, 10*(1), 543–555. https://doi.org/10.37418/amsj.10.1.54

Sahu, K., Srivastava, R. K., Kumar, S., Saxena, M., Gupta, B. K., & Verma, R. P. (2023). Integrated hesitant fuzzy-based decision-making framework for evaluating sustainable and renewable energy. *International Journal of Data Science and Analytics, 16*(3), 371–390. https://doi.org/10.1007/s41060-023-00426-4

Sanchez, L., & Reames, T. G. (2019). Cooling Detroit: A socio-spatial analysis of equity in green roofs as an urban heat island mitigation strategy. *Urban Forestry & Urban Greening, 44*, 126331. https://doi.org/10.1016/j.ufug.2019.04.014

Schatz, J., & Kucharik, C. J. (2015). Urban climate effects on extreme temperatures in Madison, Wisconsin, USA. *Environmental Research Letters, 10*(9), 094024. https://doi.org/10.1088/1748-9326/10/9/094024.

Sen, S., Li, H., & Khazanovich, L. (2022). Effect of climate change and urban heat islands on the deterioration of concrete roads. *Results in Engineering, 16*, 100736. https://doi.org/10.1016/j.rineng.2022.100736

Sharma, A., Woodruff, S., Budhathoki, M., Hamlet, A. F., Chen, F., & Fernando, H. J. S. (2018). Role of green roofs in reducing heat stress in vulnerable urban communities—A multidisciplinary approach. *Environmental Research Letters, 13*(9), 094011. https://doi.org/10.1088/1748-9326/aad93c

Shi, H., Xian, G., Auch, R., Gallo, K., & Zhou, Q. (2021). Urban heat island and its regional impacts using remotely sensed thermal data—A review of recent developments and methodology. *Land, 10*(8), 867. https://doi.org/10.3390/land10080867

Soomar, S. M., & Soomar, S. M. (2023). Identifying factors to develop and validate a heat vulnerability tool for Pakistan—A review. *Clinical Epidemiology and Global Health, 19*, 101214. https://doi.org/10.1016/j.cegh.2023.101214

Steeneveld, G. J., Koopmans, S., Heusinkveld, B. G., Van Hove, L. W. A., & Holtslag, A. A. M. (2011). Quantifying urban heat island effects and human comfort for cities of variable size and urban morphology in the Netherlands. *Journal of Geophysical Research: Atmospheres, 116*(D20). https://doi.org/10.1029/2011JD015988

Stewart, I. D. (2011). A systematic review and scientific critique of methodology in modern urban heat island literature. *International Journal of Climatology, 31*(2), 200–217. https://doi.org/10.1002/joc.2141

Stone Jr, B. (2007). Urban and rural temperature trends in proximity to large US cities: 1951–2000. *International Journal of Climatology: A Journal of the Royal Meteorological Society, 27*(13), 1801–1807. https://doi.org/10.1002/joc.1555

Su, W., Zhang, L., & Chang, Q. (2022). Nature-based solutions for urban heat mitigation in historical and cultural block: The case of Beijing Old City. *Building and Environment, 225*, 109600. https://doi.org/10.1016/j.buildenv.2022.109600

Sun, Y., Li, Y., Ma, R., Gao, C., & Wu, Y. (2022). Mapping urban socio-economic vulnerability related to heat risk: A grid-based assessment framework by combing the geospatial big data. *Urban Climate, 43*, 101169. https://doi.org/10.1016/j.uclim.2022.101169

Szagri, D., Nagy, B., & Szalay, Z. (2023). How can we predict where heatwaves will have an impact? – A literature review on heat vulnerability indexes. *Urban Climate, 52*, 101711. https://doi.org/10.1016/j.uclim.2023.101711

Tan, J., Zheng, Y., Tang, X., Guo, C., Li, L., Song, G., …, & Chen, H. (2010). The urban heat island and its impact on heat waves and human health in Shanghai. *International Journal of Biometeorology, 54*, 75–84. https://doi.org/10.1007/s00484-009-0256-x

Tian, L., Li, Y., Lu, J., & Wang, J. (2021). Review on urban heat island in China: Methods, its impact on buildings energy demand and mitigation strategies. *Sustainability, 13*(2), 762. https://doi.org/10.3390/su13020762

Tomlinson, C. J., Chapman, L., Thornes, J. E., & Baker, C. J. (2011). Including the urban heat island in spatial heat health risk assessment strategies: A case study for Birmingham, UK. *International Journal of Health Geographics, 10*(1), 1–14. https://doi.org/10.1186/1476-072X-10-42

Topay, M. (2012). Importance of thermal comfort in the sustainable landscape planning. *Journal of Environmental Protection and Ecology, 13*(3), 1480–1487.

Tuczek, M., Degirmenci, K., Desouza, K. C., Watson, R. T., Yigitcanlar, T., & Breitner, M. H. (2022). Mitigating urban heat with optimal distribution of vegetation and buildings. *Urban Climate, 44*, 101208. https://doi.org/10.1016/j.uclim.2022.101208

UN-DESA. (2016). *Transforming our world: The 2030 agenda for sustainable development.*

United Nations. (2018). *World Urbanization Prospects: The 2018 Revision*. Retrieved from https://population.un.org/wup/Publications/

Urban, A., Davídkovová, H., & Kyselý, J. (2014). Heat-and cold-stress effects on cardiovascular mortality and morbidity among urban and rural populations in the Czech Republic. *International Journal of Biometeorology, 58*, 1057–1068. https://doi.org/10.1007/s00484-013-0693-4

Van Eck, N. J., & Waltman, L. (2017). Citation-based clustering of publications using CitNetExplorer and VOSviewer. *Scientometrics, 111*, 1053–1070. https://doi.org/10.1007/s11192-017-2300-7

Varnakovida, P., & Ko, H. Y. K. (2023). Urban expansion and urban heat island effects on Bangkok metropolitan area in the context of eastern economic corridor. *Urban Climate, 52*, 101712. https://doi.org/10.1016/j.uclim.2023.101712

Venter, Z. S., Krog, N. H., & Barton, D. N. (2020). Linking green infrastructure to urban heat and human health risk mitigation in Oslo, Norway. *Science of the Total Environment, 709*, 136193. https://doi.org/10.1016/j.scitotenv.2019.136193

Voelkel, J., Hellman, D., Sakuma, R., & Shandas, V. (2018). Assessing vulnerability to urban heat: A study of disproportionate heat exposure and access to refuge by socio-demographic status in Portland, Oregon. *International Journal of Environmental Research and Public Health, 15*(4), 640. https://doi.org/10.3390/ijerph15040640

Wang, C., & Chang, H. T. (2020). Hotspots, heat vulnerability and urban heat islands: An interdisciplinary review of research methodologies. *Canadian Journal of Remote Sensing, 46*(5), 532–551. https://doi.org/10.1080/07038992.2020.1816815

Wang, X., Dallimer, M., Scott, C. E., Shi, W., & Gao, J. (2021). Tree species richness and diversity predicts the magnitude of urban heat island mitigation effects of greenspaces. *Science of The Total Environment, 770*, 145211. https://doi.org/10.1016/j.scitotenv.2021.145211

Weber, S., Sadoff, N., Zell, E., & de Sherbinin, A. (2015). Policy-relevant indicators for mapping the vulnerability of urban populations to extreme heat events: A case study of Philadelphia. *Applied Geography, 63*, 231–243. https://doi.org/10.1016/j.apgeog.2015.07.006

Whitmee, S., Haines, A., Beyrer, C., Boltz, F., Capon, A. G., de Souza Dias, B. F., …, & Yach, D. (2015). Safeguarding human health in the Anthropocene epoch: Report of the Rockefeller Foundation–Lancet commission on planetary health. *The Lancet, 386*(10007), 1973–2028. http://dx.doi.org/10.1016/S0140-6736(15)60901-1

Wu, P. S., Hsieh, C. M., & Hsu, M. F. (2014). Using heritage risk maps as an approach to estimating the threat to materials of traditional buildings in Tainan (Taiwan). *Journal of Cultural Heritage, 15*(4), 441–447. https://doi.org/10.1016/j.culher.2013.10.005

Wulder, M. A., Roy, D. P., Radeloff, V. C., Loveland, T. R., Anderson, M. C., Johnson, D. M., …, & Cook, B. D. (2022). Fifty years of Landsat science and impacts. *Remote Sensing of Environment, 280*, 113195. https://doi.org/10.1016/j.rse.2022.113195

Xue, X., He, T., Xu, L., Tong, C., Ye, Y., Liu, H., …, & Zheng, X. (2022). Quantifying the spatial pattern of urban heat islands and the associated cooling effect of blue–green landscapes using multisource remote sensing data. *Science of the Total Environment, 843*, 156829. https://doi.org/10.1016/j.scitotenv.2022.156829

Yadav, N., Rajendra, K., Awasthi, A., Singh, C., & Bhushan, B. (2023). Systematic exploration of heat wave impact on mortality and urban heat island: A review from 2000 to 2022. *Urban Climate, 51*, 101622. https://doi.org/10.1016/j.uclim.2023.101622

Yang, S., Ding, L., & Prasad, D. (2022). A multi-sector causal network of urban heat vulnerability coupling with mitigation. *Building and Environment, 226*, 109746. https://doi.org/10.1016/j.buildenv.2022.109746
Yang, W., Lin, Y., & Li, C. Q. (2018). Effects of landscape design on urban microclimate and thermal comfort in tropical climate. *Advances in Meteorology, 2018*. https://doi.org/10.1155/2018/2809649
Yao, L., Sailor, D. J., Yang, X., Xu, G., & Zhao, L. (2023). Are water bodies effective for urban heat mitigation? Evidence from field studies of urban lakes in two humid subtropical cities. *Building and Environment, 245*, 110860. https://doi.org/10.1016/j.buildenv.2023.110860
Yigitcanlar, T., Fabian, L., & Coiacetto, E. (2008). Challenges to urban transport sustainability and smart transport in a tourist city: The Gold Coast, Australia. *Open Transportation Journal, 2*, 29–46. https://doi.org/10.2174/1874447800802010029
Zander, K. K., Cadag, J. R., Escarcha, J., & Garnett, S. T. (2018). Perceived heat stress increases with population density in urban Philippines. *Environmental Research Letters, 13*(8), 084009. https://doi.org/10.1088/1748-9326/aad2e5
Zhang, D. L., Shou, Y. X., & Dickerson, R. R. (2009). Upstream urbanization exacerbates urban heat island effects. *Geophysical Research Letters, 36*(24). https://doi.org/10.1029/2009GL041082
Zhao, L., Lee, X., Smith, R. B., & Oleson, K. (2014). Strong contributions of local background climate to urban heat islands. *Nature, 511*(7508), 216–219. https://doi.org/10.1038/nature13462
Zhao, L., Oppenheimer, M., Zhu, Q., Baldwin, J. W., Ebi, K. L., Bou-Zeid, E., …, & Liu, X. (2018). Interactions between urban heat islands and heat waves. *Environmental Research Letters, 13*(3), 034003. https://doi.org/10.1088/1748-9326/aa9f73
Zhou, B., Rybski, D., & Kropp, J. P. (2017). The role of city size and urban form in the surface urban heat island. *Scientific Reports, 7*(1), 4791. https://doi.org/10.1038/s41598-017-04242-2
Zhou, X., Carmeliet, J., Sulzer, M., & Derome, D. (2020). Energy-efficient mitigation measures for improving indoor thermal comfort during heat waves. *Applied Energy, 278*, 115620. https://doi.org/10.1016/j.apenergy.2020.115620
Zhou, X., & Chen, H. (2018). Impact of urbanization-related land use land cover changes and urban morphology changes on the urban heat island phenomenon. *Science of the Total Environment, 635*, 1467–1476. https://doi.org/10.1016/j.scitotenv.2018.04.091
Zhu, J. S., Ren, H. Z., Ye, X., Zeng, H., Nie, J., Jiang, C. C., & Guo, J. X. (2021). Ground validation of land surface temperature and surface emissivity from thermal infrared remote sensing data: A review. *National Remote Sensing Bulletin, 25*(8), 1538–1566.
Zhu, W., & Yuan, C. (2023). Urban heat health risk assessment in Singapore to support resilient urban design—By integrating urban heat and the distribution of the elderly population. *Cities, 132*, 104103. https://doi.org/10.1016/j.cities.2022.104103
Zhu, Z., Wulder, M. A., Roy, D. P., Woodcock, C. E., Hansen, M. C., Radeloff, V. C., …, & Scambos, T. A. (2019). Benefits of the free and open Landsat data policy. *Remote Sensing of Environment, 224*, 382–385. https://doi.org/10.1016/j.rse.2019.02.016
Zoran, M. A., & Dida, M. R. (2012, October). Satellite remote sensing data for urban heat waves assessment and human health impacts. In: *Earth resources and environmental remote sensing/GIS applications III* (Vol. 8538; pp. 340–347). SPIE. https://doi.org/10.1117/12.974411

8

Machine Learning and Remote Sensing Approaches

8.1 Introduction and Background

8.1.1 Introduction

The profound impacts of climate change are increasingly apparent. In the face of rapid urbanisation, cities are grappling with an intensified urban heat island effect, a climate change phenomenon that elevates temperatures in urban areas relative to their rural counterparts (Degirmenci et al., 2021; Kamruzzaman et al., 2018). Such temperature spikes exacerbate the risks of extreme heat, contributing to a surge in heat-related illnesses and fatalities. In Australia, for instance, extreme heat has emerged as the leading cause of weather-related injuries and deaths, with a startling 7104 hospitalisations and 293 fatalities over the span of a decade, as reported by the Australian Institute of Health and Welfare (Australian Institute of Health and Welfare, 2023).

Historically, the nuanced aspects of urban heat vulnerability received limited research attention (Harlan et al., 2006; Vescovi et al., 2005). The introduction of the Heat Vulnerability Index, by Reid et al. (2009), marked a turning point, catalysing a wave of studies that utilise demographics, socio-economic status, health conditions, and environmental data to assess heat vulnerability (Boumans et al., 2014; Cheng et al., 2021; Zhang et al., 2018). While index models incorporating these indicators have gained popularity, they often suffer from subjective judgments and potential inaccuracies. This is largely due to the absence of universal criteria and references for selecting indicators and developing methodologies, especially in the diverse contexts of various urban environments (Adger, 2006; Chen et al., 2018; Niu et al., 2021; Tate, 2012). Moreover, the reliance on static data in these models poses a limitation, as they may not fully capture the dynamic and evolving nature of urban vulnerabilities to heat (Li et al., 2022).

8.1.1.1 The Current State of Urban Heat Vulnerability Assessment

The increasing urbanisation under changing global climate conditions has made the study of urban heat vulnerability critically important (He et al., 2019; Kazak, 2018). As cities expand and weather patterns change, the

DOI: 10.1201/9781003737049-11

well-being of urban residents is increasingly at risk. It is essential to understand the complexities of vulnerability to mitigate the issue (Boumans et al., 2014; Holec et al., 2021; Zhang et al., 2018). Despite concerted efforts, achieving methodological standardisation in this area remains a challenge (Li et al., 2022). Researchers are grappling with multiple factors contributing to urban heat vulnerability in their pursuit of a comprehensive understanding (Weber et al., 2015; Wilson & Chakraborty, 2019).

In the realm of urban heat vulnerability indicator selection, the absence of universal standards poses significant challenges. The indicators are often influenced by local contexts and researcher preferences (Chen et al., 2018; Heaton et al., 2014; Tate, 2012). While this diversity reflects the complexity of the issue, it complicates the comparison of studies and the establishment of best practices (Adger, 2006; Niu et al., 2021). Furthermore, finding fitting data proxies for these indicators is challenging. Many studies build their own indicators and data systems, but the reliability varies, leading to inconsistencies in study outcomes (Barron et al., 2018; Bu et al., 2024; Prosdocimi & Klima, 2020; Yin et al., 2021).

Spatial and temporal considerations add further complexity to these assessments. Many studies use datasets from census tracts and administrative areas (Cheng et al., 2021; Dong et al., 2020; Oh et al., 2017), but this approach has been criticised for potential biases. The Modifiable Areal Unit Problem (MAUP) is a significant issue, leading some researchers to adopt grid-scale spatial analyses for improved accuracy (Ho et al., 2015; Krstic et al., 2017). On the temporal front, there is a predominance of short-term, post-1990s assessments, highlighting a gap in longitudinal research that could provide insights into evolving vulnerabilities (Li et al., 2022).

The development of heat vulnerability indexing models has been a popular approach among researchers. These models often focus on specific aspects like biophysical or social vulnerabilities. However, they overlook the complex interactions between the natural/built environment, individual biophysical and psychological conditions, and socio-economic status, underscoring the need for further exploration. Integrative methods, including composite heat vulnerability indices that combine biophysical and socio-demographic indicators, have been proposed. However, the integration of these indicators brings up issues related to potential cumulative effects (Kazak, 2018; Macintyre et al., 2017; Tomlinson et al., 2011). In these models, the method of assigning weights is vital, with popular approaches including Equal Weighting (EW) and Principal Component Analysis (PCA). Using PCA, however, presents unique difficulties, particularly in terms of naming components and maintaining stringent data accuracy (Johnson et al., 2012; Kamal et al., 2021).

Result validation in urban heat vulnerability research is paramount. Unfortunately, only a few studies incorporate validation procedures, raising questions about their authenticity and credibility. Studies that perform validations often use heat-related health outcomes for validation data, including emergency hospital admissions, as well as morbidity and mortality rates. However, the use of these outcomes as validation variables presents its own challenges, for example, the unavailability of direct data and reliance on proxies, which may introduce inaccuracies (Conlon et al., 2020; Estoque et al., 2020). Additionally, there is an emerging call to consider the mental health impacts of extreme heat, an aspect often neglected in current studies (Chen et al., 2019; Yoo et al., 2021).

In summary, while the research on urban heat vulnerability has considerably advanced, numerous challenges remain. Firstly, a lack of a rigorous approach to specifying relevant indicators for heat vulnerability assessment leads to inconsistencies in the study outcomes and their comparisons, as well as the reliability of the results. Secondly, an adaptive, dynamic modelling approach designed to sufficiently capture both rich historical data and real-time urban planning/environment-related data for predictive urban heat vulnerability analysis to address the rapidly changing urban environment for urban resilience and sustainability is needed. Lastly, multifaceted validation methods should be the essential element of urban heat vulnerability modelling to ensure its reliability and accuracy. The integrative framework proposed in this study is designed to address these challenges.

Machine learning (ML) and remote sensing (RS) have shown great promise in urban studies for applications such as land cover/use classification and disaster risk management (Huang et al., 2019; Wang et al., 2022). Their potential in urban heat vulnerability assessment, however, remains underutilised. RS, with its unparalleled spatial and spectral data from platforms like Landsat and MODIS (moderate resolution imaging spectroradiometer), is invaluable for land use classification and heat mapping (Kafy et al., 2022; Osaku et al., 2015). Complementing this, ML's prowess in data analytics can provide deeper insights into the urban environments (Uniyal et al., 2022; Wu et al., 2022). Although there has been some incorporation of ML in urban heat vulnerability research (Kim & Kim, 2024; Yin et al., 2021), the focus has predominantly been on data processing and preparation, with less emphasis on urban heat vulnerability identification and its in-depth analysis.

8.1.2 The Emerging Trend of Machine Learning and Remote Sensing Integration

Urban studies are witnessing a remarkable transformation, marked by the fusion of ML and RS. This integration is largely powered by ML algorithms like Support Vector Machines (SVMs), Random Forests (RFs), and Convolutional Neural Networks (CNNs) (Li et al., 2023). Each of these methods has its unique strengths: SVMs are adept at classification tasks; RFs excel in decision-making processes; and CNNs are unparalleled in processing visual data. These algorithms are particularly effective in urban applications. CNNs, for example, have shown outstanding performance in analysing satellite imagery for land use classification (Huang et al., 2019; Zhu et al., 2022).

The versatility of RFs and SVMs extends across diverse ML scenarios (Frantz et al., 2021; Guo et al., 2021; Jamali, 2019; Jato-Espino et al., 2022), with their adoption driven by reliability and tool accessibility (Abunnasr & Mhawej, 2022).

In the realm of urban studies, supervised learning techniques are prominent, primarily due to their effectiveness with labelled datasets. These supervised learning approaches iteratively train models on input–expected output (also known as labels) pairs to learn the function that maps the input to the output, and adapt to new, unseen data—a crucial advantage in urban contexts with abundant labelled data (Casali et al., 2022). Conversely, the laborious nature of data labelling has instead steered researchers towards unsupervised learning. These unsupervised learning techniques, which discern the hidden patterns and insights from the given data without pre-set labels, offer alternative methodologies for urban data analysis (Demarchi et al., 2014; Oh et al., 2017). RS has revolutionised urban studies by offering dynamic, high-resolution images that reveal a range of spatial characteristics. When coupled with ML, these tools surpass traditional/static models. Such integration is essential for precise urban land use classification, monitoring urban expansion, and evaluating environmental influences on urban well-being (Deng & Wu, 2013; Priem et al., 2019; Yigitcanlar, 2006).

Most recently, Li et al. introduced a pivotal framework to harmonise ML and RS in urban studies (Li et al., 2023). This framework (Figure 8.1) promotes a more structured approach for comprehensive urban analyses. It emphasises increased automation in image processing and data analytics, a critical move away from conventional manual methods (Oh et al., 2017; Patino & Duque, 2013), while acknowledging some inherent challenges related to algorithmic biases and the need for precise modelling of parameters (Galaz et al., 2021). Despite its transformative potential, the integration of ML and RS in urban studies faces some challenges, with the major ones being the biases

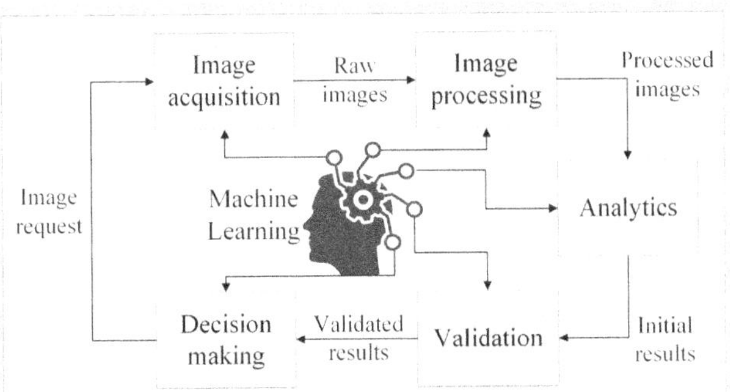

FIGURE 8.1
ML and RS integration framework for urban sustainability (Frantz et al., 2021).

inherent in algorithms and an ongoing need for clear, well-defined parameters to guide the use of these advanced tools.

In line with the United Nations Sustainable Development Goals (SDGs) established by the United Nations (United Nations, 2023), the integration of ML and RS is poised to significantly enhance urban sustainability due to its potential to produce increased accuracy (Chi et al., 2022; Touzani & Granderson, 2021), real-time monitoring capability (Abunnasr & Mhawej, 2022; Peng et al., 2019), and adaptability to a dynamic urban environments (Ghaffarian et al., 2021; Zhang et al., 2023). As urban settings rapidly evolve and regenerate, the adaptive and precise capabilities of these integrated technologies will be invaluable for sustainable and resilient urban development, including urban heat vulnerability studies.

Urban heat vulnerability assessment is critical due to rising urbanisation and climate change impacts, yet current approaches face significant challenges and limitations. These include the lack of standardised criteria for selecting indicators, leading to inconsistencies and inaccuracies; insufficient spatial resolution and temporal coverage that fail to capture the dynamic nature of urban heat vulnerability; and subjective judgments in indicator selection, which introduce biases. Additionally, while ML and RS have potential, they are underutilised, focusing more on data processing than comprehensive analysis. The validation of results also remains a challenge, with few studies incorporating robust procedures, often relying on proxies. Furthermore, many models do not fully integrate the complex spatial and temporal dynamics of urban environments, including interactions between natural and built environments and socio-economic factors. This study aims to bridge the gaps by integrating ML techniques with RS data.

Building upon our research findings, this chapter proposes a novel framework called the *'Urban Heat vulnErability Analysis Tool (U-HEAT)'*. U-HEAT is designed to assess urban heat vulnerability through an innovative conceptual framework that leverages the strengths of both ML and RS. It develops universal criteria for indicator selection, reducing subjectivity and enhancing reliability. By leveraging high-resolution RS data and advanced ML techniques, U-HEAT provides detailed, grid-level evaluations, capturing the spatial and temporal dynamics of urban heat vulnerability. This approach allows for the precise identification of vulnerable areas and improves temporal coverage through historical and predictive mapping. U-HEAT's active learning capabilities ensure continuous improvement in predictive accuracy, and its robust validation framework using heat-related health outcomes enhances result reliability. This innovative framework equips urban planners and policymakers with actionable insights for identifying vulnerable areas, developing mitigation strategies, and enhancing urban resilience and sustainability.

The primary objective of this study is to outline the theoretical underpinnings and potential applications of U-HEAT, emphasising its conceptual nature. The authors intend for future research to empirically validate and analyse the framework's performance. This framework aims to provide actionable tools for policymakers and urban planners to identify vulnerable localities and communities. By focusing on the creation of well-informed, precise, and effective mitigation strategies, U-HEAT aspires to play a pivotal role in enhancing the well-being and future of urban populations. The integration of ML's advanced data analytic capabilities with RS's high-resolution spatial data promises to improve urban heat vulnerability assessments, setting a solid foundation for future empirical studies and practical implementations (Koumetio Tekouabou et al., 2022; Li et al., 2024; Yigitcanlar, 2010a, b).

8.1.3 The Focus of This Study

This study presents significant contributions to addressing urban heat vulnerability through the establishment of the U-HEAT framework, characterised by four major advancements. First, it contributes by developing a comprehensive set of universal criteria, creating a benchmark for the selection and categorisation of indicators in this field. Second, this study demonstrates the effective application of ML and RS in conducting comprehensive and precise long-term urban heat assessments. This integration allows for detailed evaluations at a granular, grid-level scale, ensuring that the spatial and temporal dynamics of urban heat vulnerability are accurately captured and analysed. Third, it introduces an innovative predictive approach, filling a crucial gap in existing methodologies, and enabling the forecasting of future trends in urban heat vulnerability. Finally, this study offers a robust, enduring, and sustainable framework, specifically designed for the continuous, precise, and focused monitoring and management of the challenges associated with urban heat vulnerability.

Following the introduction in Section 8.1, Section 8.2 outlines the U-HEAT methodology, covering its components, features, and contributions. Section 8.3 details the approach for mapping and managing urban heat vulnerability. Section 8.4 discusses challenges, future applications, limitations, and broader implications. Finally, Section 8.5 concludes by emphasising the necessity and innovation of U-HEAT in addressing climate issues and advancing urban sustainability and resilience.

8.2 Research Design

An overarching methodology proposed in this study consists of a structured, three-step process. First and foremost, this study synthesises results from two systematic reviews conducted in accordance with PRISMA guidelines. The first review critically evaluates methodologies used in prior research on urban heat vulnerability, identifying the key areas needing refinement (Li et al., 2022). The second review extends this analysis to the combined application of ML and RS in urban studies (Li et al., 2023). These reviews have laid the foundation for a comprehensive understanding of the current practices and emerging trends in the field.

Building upon these reviews, the second part of this study introduces U-HEAT for assessing urban heat vulnerability. U-HEAT integrates ML with RS techniques to achieve the precise, grid-scale mapping of urban heat impacts. The framework is unique due to its loop structure, guiding users from the selection of indicators to the continuous monitoring of urban heat vulnerability evolutions, a critical effort towards standardising urban heat vulnerability assessments. Detailed descriptions of the U-HEAT components, their interactions, and the specific ML techniques employed are described below, including U-HEAT's advancements over the existing methodologies.

Lastly, a structured workshop was conducted with domain experts to validate the coherence and alignment of the proposed U-HEAT with established theories and best practices. This process was essential in guiding and designing the presentations and discussions for the subsequent workshops. Expert reviews involve professionals in urban planning, environmental sustainability, climate change, and ML, offering practical insights and diverse perspectives on the relevance and applicability of U-HEAT in real-world scenarios and its refinement.

8.2.1 The Urban Heat Vulnerability Analysis Framework

The primary goal of U-HEAT is to mitigate urban heat vulnerability by providing a structured approach and guidance for future research in the area, as well as a tool for heat vulnerability prediction, intervention, and policy making. The proposed framework is presented in Figure 8.2.

A key aspect of U-HEAT is its structured, component-based approach, which is crucial for mapping urban heat vulnerability effectively. This approach can be broken down into four main components:

Component 1—Indicator and Data Selection: The first component of U-HEAT involves a meticulous process of selecting relevant indicators and gathering proxy data. This phase is critical as it establishes the foundation for accurate vulnerability assessment. By analysing various socio-economic, environmental, and health-related indicators, U-HEAT ensures a comprehensive understanding of the factors contributing to urban heat vulnerability. The integration of RS data enhances this process by providing high-resolution spatial and temporal data that captures the dynamic nature of urban environments. This combination of indicators and RS data allows for a nuanced analysis, identifying vulnerable populations and areas with precision. The rigorous selection and collection of data in this phase are essential for creating a robust and reliable vulnerability assessment framework, which is crucial for informing urban planning and policy decisions.

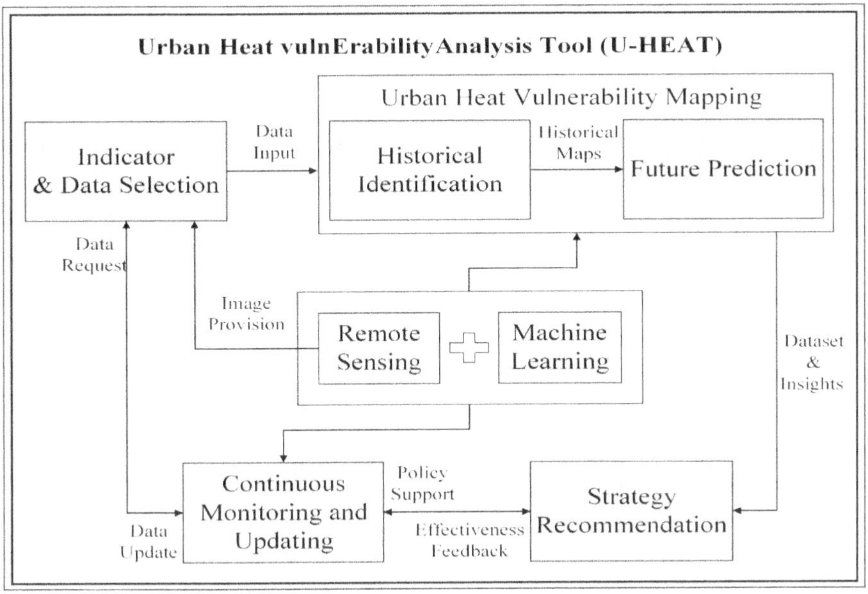

FIGURE 8.2
Framework of urban heat vulnerability analysis with ML and RS.

Component 2—Urban Heat Vulnerability Mapping: Mapping is a core component of U-HEAT, divided into two crucial phases:

- *Historical Mapping:* Historical mapping in U-HEAT utilises advanced ML and RS techniques to reconstruct detailed heat maps from diverse data sources. This phase is pivotal as it provides a comprehensive understanding of past and present heat vulnerabilities, allowing for the identification of long-term trends and spatial patterns in urban heat exposure. By fusing various socio-economic, environmental, and health-related data with high-resolution RS imagery, U-HEAT generates a nuanced portrayal of how urban heat vulnerability has evolved over time. This retrospective analysis is essential for establishing a baseline, understanding the historical context of current vulnerabilities, and identifying persistent hotspots that require targeted interventions. The detailed historical maps produced in this phase serve as a foundational reference for subsequent predictive mapping efforts, ensuring that future projections are grounded in a robust empirical understanding of past conditions.
- *Predictive Mapping:* Building on the insights gained from historical mapping, predictive mapping in U-HEAT integrates urban planning data to forecast future trends and distributions of urban heat

vulnerability. This phase leverages the predictive power of ML models to simulate how urban heat patterns might evolve under various scenarios, such as climate change, population growth, and urban development. By incorporating forward-looking data, such as planned infrastructure projects and anticipated demographic shifts, U-HEAT can generate projections that inform proactive urban planning and policymaking. The predictive mapping capability is crucial for identifying emerging areas of concern and guiding the implementation of preventative measures. This forward-thinking approach ensures that cities can anticipate and mitigate future heat risks, enhancing their resilience and adaptability to climate change. Predictive mapping transforms U-HEAT from a reactive tool into a proactive planning resource, enabling urban planners to design cities that are better equipped to handle the challenges of rising temperatures.

Component 3—Strategy Recommendation: The strategy recommendation component leverages U-HEAT's ability to identify vulnerable areas and populations across different time frames. By simulating various scenarios, such as climate change impacts and potential interventions, U-HEAT can predict how urban heat vulnerability will evolve. This feature allows for a comparative analysis of different strategies, helping urban planners and policymakers to identify the most effective heat mitigation and adaptation measures. The ability to adjust data inputs and simulate hypothetical scenarios provides a dynamic tool for exploring various intervention strategies, making U-HEAT an invaluable resource for informed decision-making. This component ensures that strategies are evidence-based and tailored to address the specific needs of vulnerable populations, promoting effective urban heat management.

Component 4—Continuous Monitoring and Updating: To maintain its relevance and accuracy, U-HEAT incorporates a dynamic model that continuously adapts to new data. This ongoing monitoring and updating process is critical for ensuring that the framework remains responsive to evolving urban heat challenges. By integrating fresh data, U-HEAT can refine its models and improve its predictive accuracy over time. This adaptability makes U-HEAT a sustainable tool for long-term use, capable of providing up-to-date insights into urban heat vulnerability. Continuous monitoring ensures that urban planners and policymakers have access to the latest information, enabling them to respond proactively to emerging heat risks. This component underscores the importance of a flexible and iterative approach to urban heat management, ensuring that strategies remain effective in the face of changing environmental conditions.

The U-HEAT framework's active learning capabilities are integral to its effectiveness and adaptability in managing urban heat challenges. These capabilities are designed to operate within an iterative loop, where continuous monitoring and feedback play crucial roles. As new data becomes available, U-HEAT actively learns from it, refining its indicator selection, enhancing mapping accuracy, and updating strategies accordingly. This means that U-HEAT does not require complete system re-training each time new data are introduced. Instead, the model incrementally improves its predictive accuracy, becoming more precise and reliable over time.

Active learning within U-HEAT ensures that the framework remains dynamic and responsive to changing conditions. This approach allows the model to adapt to new patterns and emerging trends in urban heat vulnerability, making it a continually evolving tool. By incorporating fresh data, U-HEAT can adjust its assessments and recommendations in real-time, providing urban planners and policymakers with the most up-to-date insights. This capability is vital for addressing the unpredictable and often rapid changes associated with urban heat variations, driven by factors like climate change and urban development.

The cyclical process facilitated by U-HEAT's active learning involves constant refinement and improvement. As the model processes new data, it fine-tunes its parameters and recalibrates its predictions, ensuring that the framework remains accurate and relevant. This ongoing adaptation enhances U-HEAT's ability to identify and mitigate urban heat risks effectively. By maintaining an up-to-date and accurate understanding of urban heat vulnerabilities, U-HEAT supports the development of proactive and informed strategies, helping cities to better prepare for and respond to the challenges posed by rising temperatures. In summary, the active learning capabilities of U-HEAT provide a robust mechanism for continuous improvement and adaptability. This ensures that U-HEAT is not just a static tool for current analysis, but a forward-thinking solution that evolves to meet future urban heat challenges, making it an essential resource for sustainable urban planning and resilience-building.

The U-HEAT framework is innovative in its granular, grid-scale approach to evaluating long-term urban heat vulnerability. The U-HEAT framework achieves granular, grid-scale urban heat vulnerability mapping by meticulously aggregating and analysing data using advanced ML techniques in conjunction with high-resolution RS data. This integration allows U-HEAT to segment urban areas into fine, grid-scale segments, incorporating RS data such as Land Surface Temperature (LST), land use patterns, population density, the Normalised Difference Vegetation Index (NDVI), and the Normalised Difference Built-up Index (NDBI). Each grid cell is analysed individually for a detailed assessment of urban heat vulnerability, helping solve the MAUP problem (Figure 8.3). ML enhances this process by intelligently processing and interpreting the RS data, identifying complex patterns such as the correlation between built-up areas and increased heat vulnerability. This combination of ML and RS not only facilitates the processing of

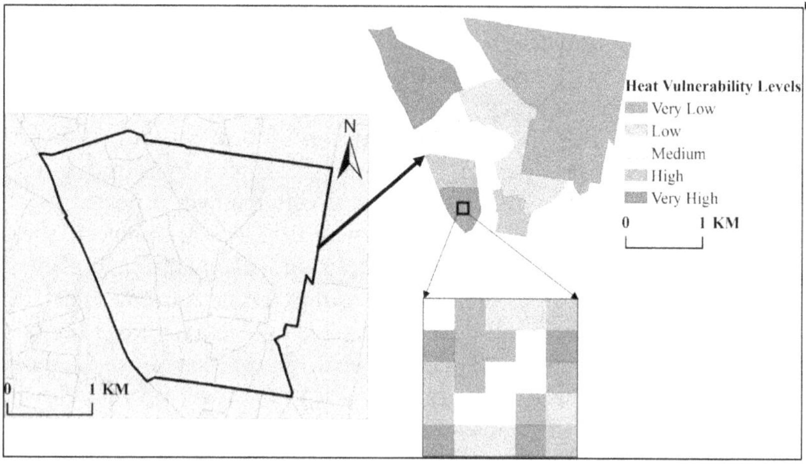

FIGURE 8.3
Likely differences of mapping urban heat vulnerability in census tracts and grid scale.

vast datasets but also enables nuanced analysis at a granular level, crucial for pinpointing specific hotspots of heat vulnerability and facilitating targeted intervention strategies. It harnesses the power of ML and RS to offer the following advantages:

- *Relevance:* Rooted in established frameworks and empirical research, U-HEAT's approach to selecting indicators and gathering data is both relevant and representative.
- *Precision:* By transitioning from broad statistical areas to a more detailed grid-scale, U-HEAT provides a finer-grained and accurate depiction of urban heat vulnerability, benefiting from the integration of ML and RS.
- *Comprehensiveness:* U-HEAT not only maps historical data, but also predicts future urban heat trends, resulting in spatially detailed and temporally extensive outcomes.
- *Sustainability:* The U-HEAT framework's ability to recommend mitigation strategies, adapt to new data, and provide ongoing monitoring highlights its sustainability.
- In sum, this study aims to establish U-HEAT as a benchmark framework for addressing urban heat vulnerability, with the following four distinct objectives and potential contributions.
- *Criteria Development:* To formulate a universal set of criteria for the selection and categorisation of indicators, establishing a reference framework.

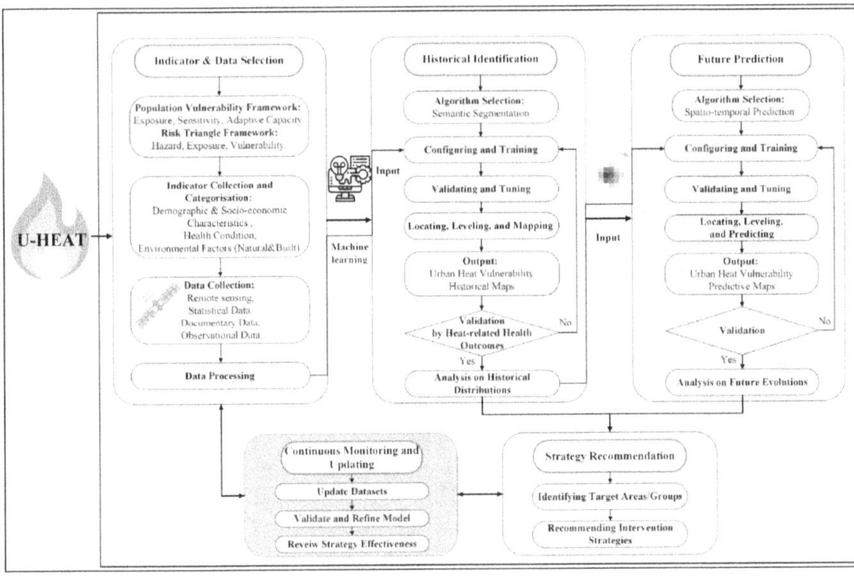

FIGURE 8.4
Flowchart of U-HEAT implementation.

- *Feasibility Demonstration:* To showcase the practicality of conducting long-term, grid-scale, and precise assessments by integrating ML and RS technologies.
- *Predictive Methodology:* To bridge the existing gap in predictive methods by introducing an innovative approach for forecasting urban heat vulnerability trends in future decades.
- *Framework Proposal:* To offer a robust, enduring, and sustainable framework for the continuous, accurate, and focused monitoring and management of urban heat vulnerability challenges.

Figure 8.4 provides a detailed workflow of U-HEAT's implementation, with various steps involved within each component. Section 8.4 delves into each of these components.

8.3 Integrated Urban Heat Vulnerability Analysis with Machine Learning and Remote Sensing

As mentioned before, this study has developed a systematic framework for urban heat analysis by leveraging ML and RS. This framework has five key components, each of which is discussed below.

8.3.1 Indicators and Data Selection

8.3.1.1 Popular Reference Frameworks

Based on the findings from our recent study (Li et al., 2022) on current urban heat vulnerability assessment research, the selection of heat-related indicators and the determination of data proxies are crucial for the quantitative assessment of urban heat vulnerability (Karanja & Kiage, 2021; Niu et al., 2021). Typically, indicators are chosen based on two popular disaster evaluation frameworks, the Population Vulnerability Framework and the Risk Triangle Framework.

The Population Vulnerability Framework was proposed by the Intergovernmental Panel on Climate Change (IPCC) (Intergovernmental Panel on Climate Change, 2001). The IPCC emphasises three key elements when addressing vulnerability: Exposure, sensitivity, and adaptive capacity. Exposure to extreme heat, whether in daily life or at work, is a primary cause of heat vulnerability (Rinner et al., 2010; Wu et al., 2022). Differences in individual sensitivities to extreme heat—often determined by demographics and socio-economic status—mean that people experience varying levels of heat vulnerability (Depietri et al., 2013; Guo et al., 2021). As a result, individuals pursue mitigation solutions to reduce the risk of exposure to extreme heat, such as seeking green spaces, using air conditioning, and accessing medical relief. The availability of these heat-relief services and facilities constitutes adaptive capacity (Harlan et al., 2006; Hulley et al., 2019).

The other widely used theoretical framework is the Risk Triangle Framework, proposed by Crichton (Crichton, 1999). This framework also encompasses three elements: Hazard, exposure, and vulnerability, with vulnerability viewed as an integral component (Buscail et al., 2012; Dugord et al., 2014). While 'exposure' bears some similarities in definition across both frameworks, in the Risk Triangle Framework, it is often represented by population density. Conversely, 'hazard' in this framework closely aligns with the concept of 'exposure' in the Population Vulnerability Framework, representing the spatio-temporal distribution of harmful heat (Dong et al., 2020; Paranunzio et al., 2021).

As depicted in Figure 8.5, despite the different descriptions of their underlying elements, both indicator selection frameworks exhibit some overlap. This similarity often results in confusion during the indicator selection process. Cheng et al. (2021) and Li et al. (2022) noted that the same indicators are frequently interpreted as different elements due to the lack of generic criteria. Additionally, even within studies adopting the same theoretical framework, interpretations of identical indicators can differ. To eliminate this confusion, this chapter has organised the indicators used in current studies and recategorised them more distinctly. These categories include demographic and socio-economic characteristics, health conditions, and environmental factors (see Table 8.1). Table 8.1 also shows the descriptions and data sources of the indicators.

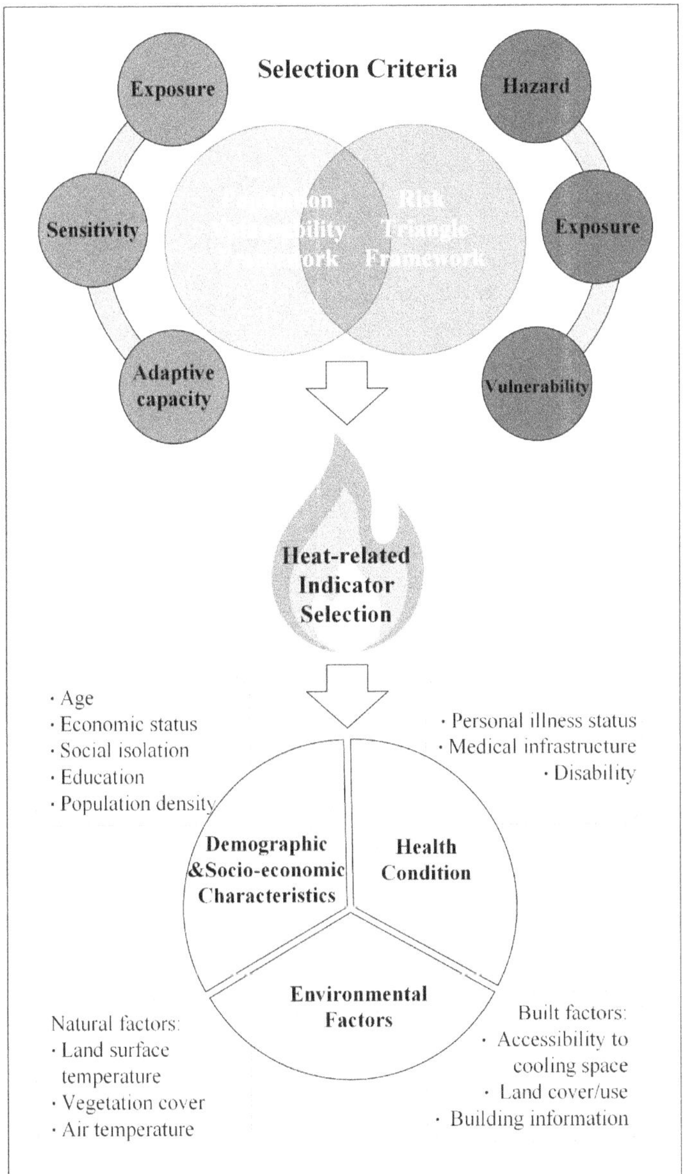

FIGURE 8.5
Integrated vulnerability and risk indicator framework.

TABLE 8.1

Frequently Used Indicators and Categories in Urban Heat Vulnerability (Li et al., 2022)

Categories	Indicators	Descriptions	Data Sources
Socio-demographic characteristics	Age	% of population over 65, below 5 or in a specific range	Census and demographic data
	Economic status	% of population with high/low incomes; local financial status	Census and demographic data
	Social isolation	% of elderly population living alone or living in a group	Census and demographic data
	Education	% of population with a low education level	Census and demographic data
	Population density	Number of population/households per study unit	Census and demographic data, satellite imagery data
Health conditions	Personal illness status	% population with pre-existing physical/mental illness	Census and demographic data, health and medical data
	Medical infrastructure	Number of medical workers/facilities/institutions; or distance to medical institutions	Health and medical data, Google Maps
	Disability	% population with a disability	Census and demographic data, health and medical data
Environmental factors (natural)	Land surface temperature	Daytime/night-time land surface temperature	Satellite imagery data
	Vegetation cover	%/area of vegetation	Satellite imagery data
	Air temperature	Daytime/night-time mean/maximum/minimum air temperature	Meteorological data
Environmental factors (built)	Accessibility to cooling space	Area of or distance to green space/open space/water body/cooling facilities	Satellite imagery data and Google Maps
	Land cover/use	Area of developed urban land cover	Satellite imagery data
	Building information	Building density/height/type	Satellite imagery data

8.3.1.2 Indicator Collection and Categorisation

This study gathered all heat-related indicators from the findings of our research (Li et al., 2022). To ensure a broad acceptance and clear understanding of the chosen indicators, this study classified them into three primary categories, focusing on those that are employed most in the field. These categories were formulated with a specific focus on human health and well-being in response to harmful thermal conditions. This approach to categorisation marks a significant improvement over previous methods, which were often less explicit. The framework for selecting indicators enables us to pinpoint specific areas of daily life that are more likely to be affected by heat-related issues, thereby guiding the implementation of preventive strategies against hazardous heat conditions. Each category of indicators reflects, either directly or indirectly, the vulnerability of urban populations to heat exposure. This vulnerability is assessed through various lenses, including physical characteristics, physiological responses, psychological factors, and environmental conditions. For instance, while individuals over 65 years old are generally at a higher risk in extreme heat situations (Cheng & Brown, 2020; Luber & McGeehin, 2008), access to public health services and cooling facilities can significantly reduce this risk (Alonso & Renard, 2020; Wang et al., 2022). Such external support systems can effectively offset physiological constraints.

Validation indicators, which are as crucial as assessment indicators, ensure the accuracy and reliability of methodologies in research. The systematic review found that less than 30% of studies included a validation step, often hindered by data unavailability. This lack of validation can significantly undermine the credibility of the results (Frantz et al., 2021; Hulley et al., 2019). To mitigate this issue, this chapter compiled a list of validation indicators frequently used in previous studies, specifically focusing on urban heat vulnerability assessments. The analysis revealed that data on heat-related morbidity and mortality are regularly employed in these validation processes (Hu et al., 2017; Kwon et al., 2020). Importantly, Reid et al. established a clear positive correlation between chronic diseases, such as cardiovascular, respiratory conditions, and diabetes, and heat-related mortality rates (Reid et al., 2012). This insight led the following research to focus on using heat-related mortality data for validation purposes. This chapter further categorised diseases contributing to mortality as either directly or indirectly related to heat exposure (see Table 8.2). This categorisation allows for a more nuanced understanding of the impacts of heat on health. The morbidity and mortality data from these diseases will be integral in the ML model evaluation, providing robust assessment results of urban heat vulnerability.

The indicator selection process in the U-HEAT framework is carefully detailed to ensure a thorough understanding of urban heat vulnerability. Indicators are prioritised based on their relevance, representativeness, data quality, and alignment with established theoretical frameworks. Demographic factors like age and economic status are critical, as they affect individuals'

TABLE 8.2

Heat-Related Health Conditions for Validation Indicators

Type of Condition	Diseases	ICD-10 Codes
Direct heat-related conditions	Heat stroke	X30
	Dehydration	E86
	Hyperpyrexia	R50.9
Indirect heat-related conditions (the impact of heat on pre-existing conditions)	Cardiovascular diseases	I00-I99
	Respiratory diseases	J00-J99
	Diabetes	E10-E14
	Renal disease	N00-N29
	Nervous disorders	G00-G99
	Cerebrovascular disease	I60-I69
	Mental health conditions	F00-F99

sensitivity and adaptive capacity to heat stress. Health conditions and access to medical resources are also key, influencing an individual's ability to cope with extreme heat. Indicators are chosen based on the availability of reliable data, with socio-demographic data sourced from census bureaus and environmental data from high-resolution satellite imagery.

The selected indicators align with frameworks like the Population Vulnerability and Risk Triangle Frameworks, ensuring consistency with other studies. This approach allows for a nuanced analysis of urban heat vulnerability, incorporating socio-demographic, health, and environmental factors. High-resolution RS data provide precise spatial information, while longitudinal data enable the analysis of temporal trends. The rigorous selection and data collection process is crucial for creating a reliable vulnerability assessment framework, essential for informing urban planning and policy decisions. This careful process ensures the U-HEAT framework offers a robust assessment of urban heat vulnerability, supporting informed decision-making and effective policy development.

8.3.1.3 Data Collection and Preprocessing

Determining the right indicators is the first step of U-HEAT, followed by collecting corresponding data proxies from reliable sources. Data quality, encompassing integrity, accuracy, consistency, and more, is critical for the framework's operation. Table 8.1 shows the authentic data sources utilised in the reviewed articles. All socio-demographic data are freely available from census datasets provided by national/local census bureaus. Additionally, for population density data, it is not only available from statistical datasets, but also from some satellite imagery products, such as WorldPop, which offers detailed data at a 100 m resolution, invaluable for scaling urban heat vulnerability from census tracts to grids. Health-related data, crucial for assessment as well as validation, including individual health conditions, public medical infrastructures, and morbidity and mortality statistics, come from public datasets provided by health departments or public well-being organisations and programmes. Census datasets also include some information on pre-existing illnesses and disabilities distributions in terms of ages, genders, and regions. To assess accessibility to medical infrastructure, Google Maps offers a promising way to measure the distance to the nearest available medical infrastructure through Point of Interest data. It is also a surprisingly valuable resource in the context of measuring accessibility to cooling spaces.

For environmental factors, temperature (LST), vegetation cover (NDVI), and building density (NDBI) are considered. These metrics are often derived from satellite images, such as Landsat, MODIS, and Sentinel, which offer resolutions up to 30, 250, and 10 m, respectively. Landsat images provide high-resolution thermal data crucial for studying surface temperature dynamics and identifying urban heat islands. The long historical record of Landsat enhances its capability to analyse trends over extended periods. MODIS offers frequent, high-resolution thermal images essential for capturing diurnal variations in urban heat patterns, helping to understand how heat accumulates and dissipates in urban areas. Sentinel-2 provides data on the NDVI and the NDBI, which are crucial for assessing vegetation cover and built-up areas. These indicators help understand the factors contributing to urban heat vulnerability and validate the model's predictions. Additionally, Geoscience Australia provides a public 25 m land cover/use dataset for free download, saving time on classification. Digital Earth Australia offers detailed land cover/use information, vital for analysing how different land uses contribute to urban heat vulnerability. OpenStreetMap provides data on the availability of cooling spaces and medical resources, which are integrated into the analysis to evaluate adaptive capacities. These diverse RS data sources collectively enhance the understanding of urban heat vulnerability and support the comprehensive assessment of environmental factors.

One key environmental data in text format is air temperature, observed by various sites and released by meteorological bureaus and related institutions and organisations. The collected data come in diverse formats, including text, shape, and raster. Standardising this into a raster grid format is crucial for grid-scale assessment. One potential approach involves inputting statistical and record data into corresponding administrative unit shapes as properties, and then interpolating these into raster grids. Zhang et al. utilised raster land cover data to evenly interpolate census data into urban area grids (Zhang et al., 2018). This is a good reference based on the assumption of the same distribution of demographic features, but it ignores the spatial heterogeneity of population distribution. Therefore, this chapter proposes to involve population density data as another reference for interpolation, which enables the generation of more reasonable raster socio-demographic data.

While the U-HEAT framework comprehensively lists its data sources, it is important to acknowledge potential limitations such as availability, quality, and representativeness, which can impact the accuracy of urban heat vulnerability assessments. A significant issue is the variability in data quality and availability across different regions and time periods. Socio-demographic data, though generally reliable, may suffer from being outdated or inconsistently collected, as census data are typically gathered every five to ten years, potentially missing recent changes. To address this, the framework incorporates supplementary data from more frequently updated sources like WorldPop, which provides higher resolution and current estimates of population density.

Environmental data from satellite imagery, while precise, may face challenges like atmospheric conditions, sensor malfunctions, and temporal resolution inconsistencies. For instance, Landsat offers high spatial resolution, but is limited by a 16-day temporal resolution, which might miss short-term temperature variations. To mitigate this, the framework uses multiple sources, including MODIS for frequent thermal imaging and Sentinel-2 for detailed vegetation and built-up area assessments, ensuring comprehensive environmental data representation.

Health-related data, crucial for validating vulnerability assessments, may encounter issues such as data privacy, limited access, and inconsistencies in records, with reliance on proxies like hospital admissions potentially introducing inaccuracies. The framework addresses this by utilising well-established correlations, such as those between chronic diseases and heat-related mortality, and integrating health infrastructure data from Google Maps to assess medical service accessibility.

The framework also recognises the challenge of integrating diverse data formats and employs interpolation methods and ML algorithms to standardise and accurately transform data for grid-scale assessment. By using multiple data sources, advanced processing techniques, and continuous updates, the U-HEAT framework effectively mitigates these limitations, ensuring robust and reliable urban heat vulnerability assessments.

8.3.2 Historical Mapping of Urban Heat Vulnerability

8.3.2.1 Two Scenarios of Historical Mapping

In the U-HEAT framework, historical mapping involves using past data to analyse and map urban heat vulnerability patterns.

Scenario 1—Spatial and Temporal Extrapolation: The first scenario is training an ML model using data from a selected area spanning over time to identify correlations between urban heat vulnerability measures and inputs from collected data areas. These inputs include multiple factors such as socio-demographic characteristics, health conditions, and environmental factors. Then, this trained model can be applied: (a) Spatially to new similar geographical areas that lack direct urban heat vulnerability measures; and (b) temporally to historical time periods when such measures are absent. This approach allows the model to extrapolate and estimate urban heat vulnerability for different areas and times, providing insights into periods and locations not directly observed or measured. This offers a comprehensive historical view of urban heat vulnerability.

Scenario 2—Fine-tuning for a Tailored Area with Limited Data: Training an ML model usually requires a large amount of data, which may not always be available for a specific area. In this scenario of limited data availability for a specific area, an ML model can be first trained using a general dataset (e.g., from other areas). Once trained, the ML model can be further fine-tuned or adapted with a small amount of localised data available in the specific area of interest. This fine-tuning process enables the model to adapt its predictions to the unique conditions of each new area and time frame. For instance, if detailed urban heat vulnerability measures from specific urban sectors for recent years are available, these are integrated into the model, refining its predictive capabilities. This scenario highlights the dynamic and adaptive nature of the U-HEAT framework, enabling it to continually learn, adapt, and improve its accuracy with each new dataset. This provides a more precise and context-specific mapping of urban heat vulnerability across different times and locations.

8.3.2.2 Challenges and Algorithm Selection for Historical Mapping

Mapping urban heat vulnerability accurately across large areas and extensive historical periods is a multifaceted challenge. This task requires the navigation of the complexities of data granularity, the completeness of historical records, and the dynamic nature of urban environments (Guo et al., 2021; Yin et al., 2021). For fine-scale mapping, high-quality, detailed data is essential to pinpoint minor vulnerability variations within cities. However, longitudinal datasets often suffer from temporal inconsistencies and spatial inhomogeneities due to evolving data collection methodologies, leading to potential uncertainties in historical trend analysis (Xiao et al., 2022). Additionally, the dynamic nature of cities, shaped by ongoing developments, infrastructural changes, and socio-economic shifts, complicates historical comparisons. As urban landscapes and populations evolve, vulnerability assessments must account for historical shifts in population density, land use, and socio-economic conditions, which influence how people experience and adapt to heat (Dugord et al., 2014; Li et al., 2024).

To ensure data accuracy and reliability, several mitigation strategies are employed. First, data preprocessing techniques like normalisation and interpolation address temporal inconsistencies and spatial inhomogeneities. Normalisation standardises data from different sources, making them comparable, while interpolation estimates missing data points, enhancing dataset completeness. Additionally, integrating supplementary data sources, such as high-resolution RS imagery and detailed socio-economic data, fills gaps and provides a comprehensive view of urban heat vulnerability. RS data offers insights into LST and vegetation cover, while socio-economic data highlights population vulnerability factors like income, age, and access to medical facilities. Advanced ML algorithms, such as CNNs and long short-term memories (LSTMs), further enhance analysis granularity and accuracy. CNNs effectively extract spatial features from high-resolution RS images, enabling precise urban heat island mapping, while LSTMs excel at capturing temporal patterns, making them ideal for analysing changes in urban heat vulnerability over time. These algorithms help identify complex patterns and interactions within the data, leading to more accurate predictions and assessments.

In mapping historical urban heat vulnerability, selecting appropriate RS images and ML algorithms is vital. High-resolution satellite imagery, such as from Landsat's Thermal Infrared Sensor (TIRS) and MODIS, is critical for providing historical thermal data to study surface temperature dynamics and urban heat (Hammer et al., 2020; Loughnan et al., 2012). Using a series of Landsat images can illuminate changes in land use and urban expansion, offering insights into shifting heat vulnerability. Meanwhile, CNNs stand out for their capabilities in managing spatial complexities in RS data and maintaining the hierarchical structure of images (Guo et al., 2021; Huang et al., 2019). For studying temporal patterns, LSTM networks, a type of recurrent neural network (RNN), are effective, capturing the progression of changes over time with precision (Lammers et al., 2021; Li & Demir, 2023; Ronneberger et al., 2015; Wei et al., 2021; Yan et al., 2022).

In the U-HEAT framework, spatial mapping is innovatively approached as a semantic segmentation problem in ML. Unlike traditional methods like linear regression, which estimate urban heat vulnerability based on individual locations in satellite images, semantic segmentation considers both individual locations and their neighbouring regions. This method enhances the accuracy and consistency of mapping by accounting for the interconnected nature of urban environments and the influence of adjacent areas on heat vulnerability. U-HEAT's flexible design allows for the integration of various ML algorithms suited to semantic segmentation in historical urban heat vulnerability mapping.

Modern semantic segmentation approaches fall into two main categories: CNN-based and Transformer-based. CNNs, like U-Net, excel at processing spatial hierarchies in complex urban landscapes, making them ideal for high-precision segmentation in dense urban environments with fine-grained details. U-Net's encoder–decoder architecture captures both context and spatial information, making it particularly effective for urban heat vulnerability mapping where precise boundaries are essential. On the other hand, Transformers, such as UNetFormer, are superior in modelling global information and context, making segmentation both accurate and efficient. Their self-attention mechanism captures long-range dependencies within data, enhancing the understanding of complex urban landscapes and their thermal characteristics.

The U-HEAT framework leverages these algorithms to ensure robustness and versatility. U-Net can be used for detailed segmentation, providing high-resolution mapping of urban heat islands, while Transformers can model broader contextual information, identifying patterns and trends over larger spatial extents. By incorporating both historical and real-time RS data, U-HEAT allows for the continuous monitoring and dynamic updating of urban heat vulnerability assessments. For instance, historical Landsat data can be used to train a U-Net to recognise long-term trends in LST changes, while current Sentinel-2 data can be processed using UNetFormer for the up-to-date segmentation of urban green spaces and built-up areas. This adaptability makes U-HEAT a comprehensive tool for urban heat vulnerability assessment, enabling urban planners to select the most suitable methods based on the specific characteristics of the urban area and the nature of the historical data, ensuring precise and actionable insights for mitigating urban heat risks.

8.3.2.3 Model Development, Validation, and Effective Communication of Results

Developing an effective training and validation process for ML models like U-Net, used in mapping urban heat vulnerability, is crucial for their reliability and practicality in sustainable urban planning and policymaking (Yigitcanlar, 2010). A comprehensive training phase requires a meticulously organised dataset encompassing a variety of data types such as socio-demographic information, health statistics, and environmental factors (Cheng et al., 2021; Li et al., 2022). The objective is to capture the complex and detailed aspects of urban heat vulnerability, which vary across different urban layouts and environmental conditions. A stratified sampling method is key to ensuring the dataset accurately represents diverse urban environments (Stehman et al., 2011), from densely populated residential areas to industrial zones, green spaces, and water bodies, reflecting their unique thermal profiles and seasonal variations.

Additionally, it is essential to combine the strengths of RS and ground-based data. Integrating high-resolution satellite imagery with detailed on-the-ground measurements creates a robust training set, offering both extensive coverage and precision (Amani & Shafizadeh-Moghadam, 2023). This approach lays a solid foundation for the algorithms to learn and interpret the complex patterns of urban heat disparities. A crucial component of the training set is labelled data, particularly from the validation dataset, which includes statistics on heat-related health issues. This strategic inclusion of real-world health outcomes enables the model to associate thermal anomalies with potential health risks, enhancing its predictive accuracy and making its insights highly relevant to public health and urban policymaking (Wang et al., 2023a, b).

The validation phase in urban heat vulnerability mapping is essential for ensuring the model's rigour and accuracy. We use k-fold cross-validation as a key technique to enhance the model's robustness (Vu et al., 2022). In k-fold cross-validation, the dataset is split into k equally sized subsets (folds). The model is trained on k–1 of these subsets and tested on the remaining one. This process is repeated k times, with each subset serving as the test data once. The results from each fold are averaged to produce a single performance metric. This technique maximises data use and helps detect and reduce overfitting and underfitting by exposing the model to different data subsets. The evaluation metrics include accuracy, precision, recall, and the F1 score, offering a balanced measure of performance. Accuracy measures the proportion of correctly predicted instances, providing an overall performance overview. Precision indicates the proportion of true positive predictions among all positive predictions, reflecting the model's exactness. Recall measures the proportion of true positive predictions among all actual positives, highlighting the model's ability to identify all relevant cases. The F1 score, the harmonic mean of precision and recall, balances these two metrics, providing a single measure that considers both false positives and false negatives.

The primary validation indicators are heat-related morbidity and mortality rates, chosen for their relevance. These rates are direct measures of the health impacts of extreme heat, making them highly suitable for validating urban heat vulnerability models. They reflect the real-world consequences of heat exposure on human health, with a well-established positive correlation between chronic diseases (like cardiovascular, respiratory conditions, and diabetes) and heat-related mortality rates (Hu et al., 2017; Kwon et al., 2020). This correlation makes these indicators especially useful for assessing the model's accuracy and reliability. Using heat-related morbidity and mortality rates allows for a direct assessment of how well the model predicts real-world health outcomes, linking model predictions to actual impacts.

These indicators provide a solid benchmark for evaluating the model's performance, enabling comparisons between predictions and observed health outcomes to assess accuracy and generalisability. Validating the model against independent datasets with real-world health outcomes avoids bias and overfitting, ensuring the model's results are generalisable to new, unseen data (Wang et al., 2023). Advanced validation techniques like confusion matrices, receiver operating characteristic (ROC) curves, and area under the curve (AUC) metrics also provide detailed evaluations of the model's predictive capabilities, measuring the trade-offs between true and false positive rates. This comprehensive approach ensures that the U-HEAT framework offers reliable and actionable insights for urban heat vulnerability assessment, equipping planners and policymakers with accurate, up-to-date information for mitigating heat-related risks (Wang et al., 2023).

The expected grid-scale map of urban heat vulnerability is presented in Figure 8.6. This map categorises vulnerability into five levels, as established by prior studies: Very low, low, medium, high, and very high (Azhar et al., 2017;

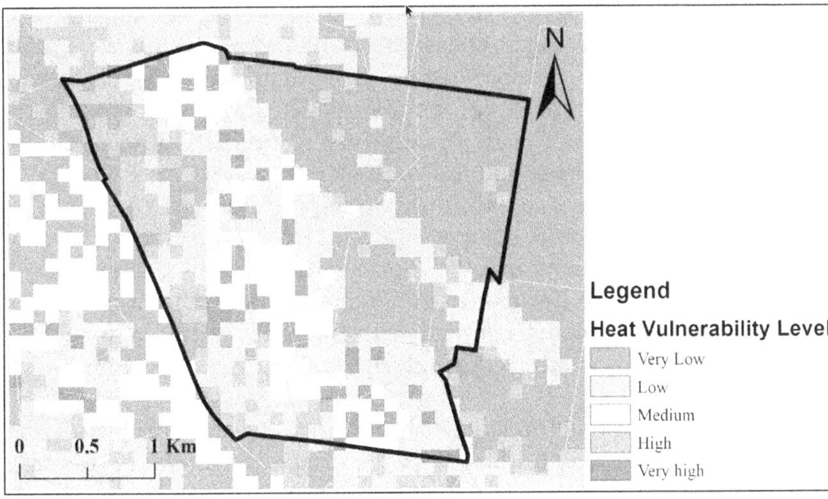

FIGURE 8.6
Visualisation of a hypothetical urban heat vulnerability mapping.

Guo et al., 2019). Unlike earlier mappings, this grid-scale approach offers a more detailed view of heat vulnerabilities within administrative boundaries. This approach aids urban planners in pinpointing vulnerabilities more precisely. The grid-based maps enable us not only to identify areas of high vulnerability through anomalies, but also to gain a deeper understanding of the spatial coherence in the distribution of these vulnerabilities.

For visualising long-term and grid-scale historical urban heat trends, geographic information systems (GIS) platforms are indispensable. They enable the integration and analysis of multi-temporal, multi-source data, applying spatial statistics to identify trends, patterns, and hotspots. Enhanced with visualisation techniques like heat maps, change detection overlays, and temporal sliders, these maps vividly illustrate the dynamics of urban heat vulnerability over time. The insights gained are actionable for urban planners and policymakers, aiding in identifying areas vulnerable to heat that need mitigation efforts, such as increased greenery or reflective surfaces. Temporal analysis helps evaluate the effectiveness of past policies and guides future initiatives (Shang et al., 2023). By analysing the progression of heat vulnerability, policies can be tailored to specific phases of urban development, promoting dynamic and sustainable urban planning.

8.3.3 Future Prediction of Urban Heat Vulnerability

8.3.3.1 Lack of Future Prediction

A research gap in current studies is the lack of effective methods for predicting future urban heat vulnerability (Li et al., 2022). This shortfall in predictive methods limits the ability to anticipate and respond to future urban challenges, affecting preparedness and urban resilience. The difficulty in creating forward-looking assessment methods for urban heat vulnerability stems from several factors. Firstly, increasing urbanisation and climate change bring more uncertainty to prediction research compared to historical identification. Urban heat vulnerability is influenced by a complex mix of biophysical, socio-demographic, and environmental factors, each with its own set of complexities and interconnections. Capturing these in future predictions may be more challenging because these phenomena have not fully occurred yet. In this context, the ever-changing nature of urban ecosystems, coupled with the multifaceted nature of vulnerability, makes forecasting particularly challenging (Paranunzio et al., 2021).

Secondly, the lack of methodologies leads to a noticeable absence of methodological reference and uniformity in forecasting research. This might be due to path dependence (Matthews et al., 2015), which describes how researchers or policymakers are inclined to make decisions based on previous studies and policies. As a result, forecasting urban heat vulnerability receives little attention. However, the importance of forecasting in managing urban heat vulnerability issues cannot be denied. Thirdly, the lack of a consistent historical understanding hinders the forecasting process. Many studies focus on data post-1990, offering a limited view and reducing the potential for a long-term understanding and prediction of evolving vulnerabilities (Li et al., 2022). A comprehensive understanding of historical patterns and changes is missing, which is essential for predicting future evolution.

The combination of ML and RS presents a promising solution to these challenges in forecasting urban heat vulnerabilities. RS provides detailed insights into urban landscape dynamics, crucial for identifying changes in urban microclimates (Ferreira et al., 2021; Mitchell & Chakraborty, 2015). ML, with its ability to handle large datasets, helps identify patterns in these insights (Das et al., 2021; Wieland & Pittore, 2016). Together, they might effectively identify evolving urban heat trends, pinpoint vulnerable areas, and predict emerging concerns. The importance of this approach is further heightened when combined with historical urban heat vulnerability maps, which act as empirical references, depicting past trends in urban heat vulnerability. Using these maps to train ML models improves their predictive accuracy, ensuring predictions are grounded in historical data. Moreover, ML's flexibility in integrating varied data from multiple sources in a more comprehensive understanding of urban heat vulnerability enhances the accuracy of predictions (Tao et al., 2023). In summary, the integration of ML and RS may aptly fill the gaps in predicting urban heat vulnerability. This approach shifts the focus from just understanding current vulnerabilities to actively preparing for future challenges, equipping urban areas with the tools to handle upcoming complexities.

8.3.3.2 Challenges and Algorithm Selection for Future Prediction

Predicting future urban heat vulnerability at a grid scale, even with historical maps and urban planning data, faces significant challenges. The main issue is the ever-changing nature of urban environments (Paranunzio et al., 2021). Factors like urban expansion, land use changes, and different building materials can significantly alter heat absorption and release, making it difficult to rely on past patterns to forecast future conditions. Moreover, urban planning efforts like increasing green spaces and upgrading building standards can greatly reduce heat effects, reducing the reliability of historical trends for future predictions.

Climate change exacerbates these challenges by unpredictably altering baseline temperatures and the frequency of extreme heat events (Tomlinson et al., 2011). Socio-economic changes, such as shifts in demographics and economic growth, add further complexity (Alonso & Renard, 2020). These changes can impact population density and resource distribution, and intensify the urban heat island effect. Technological advancements in construction and environmental policies can also change established patterns of heat vulnerability (Degirmenci et al., 2021). Therefore, while historical and planning data are valuable, their application to future scenarios must be approached cautiously, considering the various evolving factors that impact urban heat dynamics.

Selecting the right ML algorithms for long-term urban heat vulnerability prediction requires a strategic approach. Techniques like RF and Gradient Boosting Machines effectively handle nonlinear patterns and interactions among socio-environmental factors (Al-Aizari et al., 2022; Tella et al., 2021), accommodating diverse urban data and climate change impacts. Conversely, deep learning algorithms like CNNs, which excel in analysing complex spatial data, are better suited for examining RS imagery and detecting urban landscape changes (Singleton et al., 2022). We propose framing future urban heat vulnerability prediction as a spatio-temporal ML challenge, integrating geographical and time-based dimensions. This approach is particularly suited for dynamic urban environments, where spatial layout and temporal changes are crucial. Spatio-temporal models synthesise spatial dimensions (like city layouts) and temporal dimensions (such as seasonal variations) to predict patterns shaped by urban expansion, land use changes, and socio-economic shifts. ConvLSTM, which combines the spatial processing of CNNs with the temporal sequencing of LSTM networks, is ideal for scenarios demanding high precision in handling temporal dynamics (Papadomanolaki et al., 2021; Zhu et al., 2021).

ConvLSTM captures spatial features from RS data, like satellite images, and processes them over time to predict changes in urban heat patterns. For instance, it can analyse historical LST data to forecast future urban heat vulnerability, considering seasonal and long-term trends. Another example is STCNN (Spatio-Temporal CNN), which excels in analysing complex urban landscapes and processing large-scale geospatial data (He et al., 2019; Xiao et al., 2023). STCNNs leverage convolutional layers to extract spatial features from high-resolution RS imagery, such as vegetation cover and urban density, while temporal layers track changes over time. This makes STCNNs effective in diverse cities, allowing detailed analysis of how various urban features contribute to heat vulnerability over time.

Additionally, STGCNs (Spatio-Temporal Graph Convolutional Networks) are particularly useful for structured spatial data analysis, such as urban networks, assessing the impact of urban development and infrastructural changes on heat vulnerability (Yu et al., 2017; Xu et al., 2021). STGCNs extend CNNs and LSTMs by incorporating graph structures to model complex relationships between urban entities. For instance, STGCNs can evaluate how changes in transportation networks, building layouts, and green spaces influence heat distribution and vulnerability, enabling a more comprehensive analysis of urban environments by considering both spatial connections and temporal evolution.

Integrating these advanced ML algorithms within the U-HEAT framework ensures robust and versatile heat vulnerability assessments. ConvLSTM, with its spatio-temporal data handling, is particularly valuable for predicting future urban heat trends based on historical RS data. STCNNs offer detailed spatial analysis of current urban conditions, while STGCNs provide insights into the broader impact of urban planning decisions. By leveraging these algorithms, U-HEAT can dynamically adapt to new data inputs, continuously refining its predictions and offering actionable insights for urban planners and policymakers. This comprehensive approach enhances the framework's ability to address the complexity of urban heat vulnerability, making it a powerful tool for sustainable urban development and climate resilience planning.

8.3.3.3 ***Model Development, Validation, and Presentation of Results***

To effectively train models for predicting future urban heat vulnerability, it is crucial to consider several key factors for accuracy and effectiveness. Initially, a deep understanding of historical urban heat vulnerability patterns is necessary (Loughnan et al., 2012). This involves examining past maps to identify trends and establish a baseline for future comparisons. Additionally, incorporating urban planning or projected data is critical (Kim & Newman, 2020). This includes projected changes in land use, infrastructure, and vegetation cover, which heavily influence urban heat vulnerability dynamics. Combining these data sources is challenging but essential to accurately represent the complex interactions affecting urban heat vulnerability.

For instance, in training a ConvLSTM network, these elements are pivotal. ConvLSTM, adept in handling both spatial and temporal data due to its fusion of CNNs and LSTM networks (Dehghani et al., 2023; Ma et al.,

2022), is ideal for this task. Training involves using historical heat maps for spatial understanding and urban planning data to add a temporal perspective, anticipating how heat distribution might evolve. It is vital to expose the model to a diverse array of scenarios, including varying climate conditions, urban layouts, and time scales. This diversity is key to enabling the model to generalise and make reliable predictions in changing urban environments. As urban areas grow and evolve, the adaptability and predictive capabilities of models like ConvLSTM become increasingly important for effective urban planning and heat management.

Validating predictions of future urban heat vulnerability is crucial for ensuring accuracy and reliability (Maier et al., 2014). Without future real-world data, the process involves examining the model's ability to replicate historical patterns and apply them to future scenarios. This includes using recent data (e.g., the 2020 map) for validation, ensuring the model captures both short-term fluctuations and long-term trends. Validation assesses the model's skill in capturing the temporal and spatial dynamics of urban heat vulnerability. Consistency over time and spatial precision are key—the model must accurately identify areas in urban landscapes prone to heat vulnerability. Sensitivity analyses under different urban planning scenarios also provide insights into the model's robustness and flexibility.

When applying these principles to a ConvLSTM model for predicting urban heat vulnerability, the process involves complex steps. The LSTM components are tested against temporal sequences of real-world historical data to ensure accurate time-related pattern capture. The convolutional parts, responsible for spatial feature detection, are validated by comparing predictions with historical maps, confirming the model's ability to interpret spatial distribution variances. The model is also tested under hypothetical urban development scenarios, evaluating its adaptability to changes in urban planning. Its effectiveness is measured with statistical metrics, offering a thorough assessment and identifying areas for improvement. This ongoing validation and refinement cycle is crucial for enhancing performance, ensuring it serves as a reliable resource for urban planners and policymakers.

Predictive urban heat vulnerability maps are essential for advancing urban planning and policymaking. These maps offer more than historical data, providing a forward-thinking view crucial for adapting to rapid urban changes (Loughnan et al., 2012). They project future heat vulnerabilities, enabling planners and policymakers to prepare for upcoming challenges. This proactive approach is vital for identifying future high-risk areas, guiding interventions, and shaping resilient infrastructure. Predictive maps offer insights for developing urban policies, from heat-resilient building regulations to public health plans for heatwaves (Paranunzio et al., 2021). Adopting predictive heat maps marks a significant shift in urban planning, emphasising long-term sustainability and resilience against increasing heat-related challenges.

8.3.4 Strategy Recommendation

The impact of certain research results on the strategy design is often not fully acknowledged. Many studies highlight their potential to influence policy decisions, yet their influence is sometimes limited by the constraints of specific space and time frames (Li et al., 2022). A considerable number of these studies focus on evaluations within a single year, constrained by administrative boundaries. While these studies provide foundational insights, they may not offer the comprehensive analysis needed for definitive policymaking decisions (Weber et al., 2015; Yigitcanlar, 2010). It is important to recognise that assessments based on administrative scales might suit broad management strategies, but the diverse needs of individuals within these areas call for more personalised policy approaches. Therefore, a one-size-fits-all policy might not be the most effective approach.

To develop truly thorough and influential policies, ongoing monitoring and detailed data analysis are essential (Khan & Sudheer, 2022; Mushore et al., 2017), features that U-HEAT embodies. Combining ML's analytical power with RS's ability to capture LST in high resolutions, U-HEAT is an innovative and promising approach. RS allows for long-term and accurate temperature data collection through thermal bands, while ML excels in fusing various data types, offering deep insights into urban heat vulnerability with sophisticated computational analysis. U-HEAT sets a high bar for urban policy development and is also a ground-breaking framework for assessing strategy effectiveness. Its framework supports ongoing self-validation and refinement of parameters as data evolves. Strategies are reassessed by comparing outcomes before and after implementation, ensuring continuous improvement and relevance.

8.3.5 Continuous Monitoring and Updating

Remarkably, many existing studies have not highlighted the critical importance of continuous monitoring for effective disaster prevention and management in urban planning (Koumetio Tekouabou et al., 2022). This gap is partly due to the lack of long-term focus in these studies, as well as their methodological constraints, which are often tied to specific datasets, time frames, and contexts. Current methods typically lack the flexibility needed to remain relevant across different scenarios. The notable absence of robust predictive techniques hinders the establishment of continuous monitoring, a key element for informed policy development (Paranunzio et al., 2021). Both retrospective analysis and forward-looking insights are crucial for crafting wise policies. Achieving ongoing monitoring requires a combination of sustained research, methodological diversity, and iterative processes.

U-HEAT stands out as a sophisticated and efficient tool for the continuous assessment of urban heat vulnerability and its changes over time.

Constructed as an iterative system, U-HEAT is designed to be adaptable. As new data become available, U-HEAT actively incorporates this information, allowing for model refinements based on updated datasets and facilitating comparisons between earlier predictions and recent results. This adaptability ensures that U-HEAT remains a dynamic tool, capable of responding to evolving urban heat challenges. By integrating fresh data, U-HEAT continuously improves its predictive accuracy, thereby providing urban planners and policymakers with the most up-to-date insights.

The continuous monitoring and updating process in U-HEAT involves several key elements. Firstly, it emphasises the importance of an iterative approach to data integration. By regularly updating its models with new socio-economic, environmental, and health-related data, U-HEAT ensures that its assessments reflect the latest conditions and trends. This real-time data integration is critical for maintaining the relevance and accuracy of the framework, enabling it to detect and adapt to new patterns of urban heat vulnerability.

Secondly, U-HEAT's ability to refine its models based on new data inputs allows for the ongoing enhancement of mapping accuracy. This means that as new information is fed into the system, U-HEAT recalibrates its predictions, improving the precision of its vulnerability assessments. This continuous refinement process ensures that urban heat maps remain current and reliable, providing a solid foundation for decision-making.

Moreover, U-HEAT's active learning capabilities enable the model to learn from new data without the need for complete system re-training. This incremental learning approach allows U-HEAT to adjust its parameters and improve its performance efficiently, making it a sustainable tool for long-term use. By incorporating the latest data and refining its models accordingly, U-HEAT can offer proactive insights that help urban planners anticipate and mitigate future heat risks.

The continuous monitoring and updating feature of U-HEAT also extends to strategy evaluation and policy adaptation. By comparing expected trends with actual developments, U-HEAT provides a robust mechanism for evaluating the effectiveness of implemented policies. This comparative analysis ensures that policies are not only thoroughly evaluated but also appropriately modified when necessary, enhancing their impact and efficacy. This feedback loop is essential for creating resilient urban environments that can adapt to changing heat conditions over time.

In summary, U-HEAT's continuous monitoring and updating process is fundamental to its ability to provide ongoing, accurate assessments of urban heat vulnerability. By incorporating new data, refining its models, and enabling proactive policy adjustments, U-HEAT ensures that urban planners and policymakers have access to the most current information. This dynamic and iterative approach makes U-HEAT a vital tool for long-term urban heat management, helping cities to effectively address and adapt to the challenges posed by rising temperatures.

8.4 Findings and Discussion

8.4.1 Key Challenges and Limitations in Existing Approaches

This study highlighted the key challenges and limitations of existing urban heat vulnerability assessment approaches as follows:

Lack of Standardised Criteria and References: Many existing studies on urban heat vulnerability assessment suffer from the absence of standardised criteria and references for selecting indicators and developing methodologies. This lack of standardisation leads to inconsistencies and potential inaccuracies in the assessment results, making it challenging to compare findings across different studies and contexts.

Insufficient Spatial Resolution and Temporal Coverage: Traditional approaches often do not meet the needs of urban planning due to their insufficient spatial resolution and temporal coverage. Many studies rely on static data from census tracts or administrative areas, which can introduce biases and fail to capture the dynamic and evolving nature of urban heat vulnerability. Additionally, short-term assessments predominate, leaving a gap in longitudinal research that could provide insights into evolving vulnerabilities over time.

Subjective Judgments in Indicator Selection: The selection of indicators in many urban heat vulnerability studies is often based on subjective judgments. The absence of universal criteria for selecting and categorising indicators complicates the comparison of studies and the establishment of best practices. This subjectivity can lead to potential biases and variations in the reliability of the assessment outcomes.

Limited Integration of ML and RS: While ML and RS have shown great promise in urban studies, their potential in urban heat vulnerability assessment remains underutilised. Existing approaches predominantly focus on data processing and preparation, with less emphasis on in-depth analysis and identification of urban heat vulnerability. This limitation hinders the development of comprehensive and precise long-term urban heat assessments.

Challenges in Validating Results: Result validation is paramount in urban heat vulnerability research, yet only a few studies incorporate robust validation procedures. The use of heat-related health outcomes, such as emergency hospital admissions and mortality rates, as validation variables presents challenges due to the unavailability of direct data and reliance on proxies, which may introduce inaccuracies. Additionally, the mental health impacts of extreme heat are often neglected in current studies.

Complexity of Spatial and Temporal Dynamics: Urban heat vulnerability assessments must account for the complex interactions between natural and built environments, individual biophysical and psychological conditions, and socio-economic status. Many existing models focus on specific aspects, such as biophysical or social vulnerabilities, but fail to integrate these dimensions comprehensively. The MAUP and the need for grid-scale spatial analyses for improved accuracy further complicate these assessments.

U-HEAT aims to overcome these challenges by integrating ML with RS to establish a standardised and sustainable approach for urban heat vulnerability assessment. U-HEAT develops universal criteria for selecting and categorising indicators, reducing subjectivity, and enhancing the comparability and reliability of assessments. By leveraging high-resolution RS data and advanced ML techniques, U-HEAT provides detailed evaluations at a granular grid-level scale, accurately capturing the spatial and temporal dynamics of urban heat vulnerability. The integration of ML and RS significantly enhances urban heat vulnerability analysis by improving spatial resolution, temporal coverage, and accuracy. ML algorithms such as CNNs and LSTM networks process high-resolution RS data, including LST and NDVI, to create detailed grid-scale maps.

This approach allows for the precise identification of vulnerable areas by considering the influence of neighbouring regions. Temporal coverage is improved through historical and predictive mapping, enabling the assessment of long-term trends and future scenarios. The framework's active learning capabilities ensure continuous improvement in predictive accuracy by integrating new data. Additionally, U-HEAT incorporates a robust validation framework using heat-related health outcomes and other relevant data to ensure the accuracy and reliability of results. Its innovative predictive approach enables forecasting future trends in urban heat vulnerability, providing a structured component-based framework that integrates multiple dimensions of vulnerability for detailed mapping and management. This comprehensive tool aims to equip urban planners and policymakers with actionable insights for identifying vulnerable localities, developing mitigation strategies, and enhancing urban resilience and sustainability.

8.4.2 Prospective Applications

The framework proposed is at the conceptual stage and will need to be applied in pilot testing for recalibration once all datasets are ready. This step will provide invaluable insights and possible improvements, guiding further actions. Then, further testing and extensive applications on relevant case studies are needed to understand U-HEAT's effectiveness and adaptability in different local contextual settings. The results at this initial case testbed stage also require feedback from local authorities and expert professionals.

This feedback is critical for validating the results and making them accurate, locally relevant, and widely applicable. Data consistency and integration are also significant challenges, given the variety of data sources. This needs to be tackled by using data from reliable sources, standardising it through proven processing techniques, and applying ML for multi-modal data fusion. Additionally, there are questions about U-HEAT's scalability and its effectiveness in larger, more complex urban areas. The likely approach is to start with smaller, representative regions in the pilot study and, upon its success, expand it to a broader case study. The step-by-step approach aims to validate and refine U-HEAT and explore its impact on urban research. By explaining the methods, addressing concerns, and proposing solutions, it is important that we work towards creating a robust, scalable tool for analysing urban heat vulnerability in various contexts. Ultimately, collaboration with the academic and urban planning communities to keep improving and broadening U-HEAT's utility is a necessity.

8.4.3 Contributions to Sustainable Development

The development of U-HEAT, aimed at assessing and mitigating urban heat vulnerability, is closely tied to several SDGs. U-HEAT supports these goals by providing critical insights and tools for urban planning and policymaking. It primarily aligns with SDG 11, which focuses on creating sustainable cities and communities. This goal emphasises the need for urban areas to be inclusive, safe, resilient, and sustainable. Urban heat vulnerability affects these aspects, as increased temperatures can compromise urban livability and safety. U-HEAT aids in designing resilient and sustainable urban spaces that better cope with rising temperatures, integrating green spaces, sustainable materials, and heat-mitigating urban designs. Additionally, U-HEAT is closely related to SDG 13, centred on climate action. The increasing instances of urban heat islands, driven by climate change, make their mitigation vital to climate action strategies (Liu et al., 2024a, b; Tuczek et al., 2022). U-HEAT helps cities adapt to and mitigate the significant impact of climate change. It also ties into SDG 3, which focuses on good health and well-being, as heatwaves pose major health risks, particularly to vulnerable populations (Guo et al., 2024; Kitchley, 2024; Wu et al., 2024). By reducing heat-related health risks, U-HEAT promotes public health and well-being, ensuring safer urban environments for all, especially the vulnerable. Moreover, U-HEAT addresses SDG 10, which seeks to reduce inequalities. Urban heat often disproportionately impacts lower-income communities, making its assessment and mitigation crucial for reducing social disparities. U-HEAT helps identify and protect those most at risk from urban heat, contributing to reducing inequalities. By offering a comprehensive approach to urban heat vulnerability, U-HEAT not only improves immediate urban living conditions but also plays a crucial role in sustainable development, ensuring a more equitable and resilient future for urban populations.

8.4.4 Implications in Policy and Public Engagement

The U-HEAT framework has significant implications for sustainable urban development and climate change adaptation, as well as for planning, policymaking, and public engagement.

Enhanced Urban Planning: U-HEAT enables urban planners to design cities that are more resilient to heatwaves by providing high-resolution, grid-scale data on urban heat vulnerability. This detailed information facilitates the precise identification of heat-vulnerable areas, allowing for targeted interventions such as increasing green spaces and optimising building materials to reduce heat absorption. The framework supports comprehensive methodological guidance throughout the disaster management process, aiding in the preparation, mapping, strategic recommendations, and ongoing monitoring of urban heat issues.

Informed Policymaking: The predictive capabilities of U-HEAT are crucial for proactive policymaking. By forecasting future urban heat vulnerability trends, policymakers can implement strategies to mitigate these risks before they become critical. This includes zoning regulations, urban greening initiatives, and infrastructure improvements to enhance urban resilience against rising temperatures. The high-resolution, long-term datasets provided by U-HEAT support evidence-based governance and planning, pinpointing highly vulnerable areas that require preventive interventions or targeted resource allocations.

Public Health and Well-being: U-HEAT focuses on socio-economic and health-related indicators to prioritise vulnerable populations, such as the elderly and those with pre-existing conditions, in heat mitigation strategies. This approach supports sustainable development by promoting equitable urban environments where all residents have access to necessary resources and protections against extreme heat. By integrating datasets like green infrastructure, housing conditions, and cultural backgrounds, U-HEAT can analyse how these indicators affect inequity, leading to more targeted strategy development and resource allocation (Mitchell & Chakraborty, 2015). For example, planting vegetation in areas with green inequity and distributing heat-relief resources to disadvantaged groups can help create a livable, health-friendly environment for all communities.

Long-term Sustainability: U-HEAT's continuous monitoring and updating process ensures that urban heat vulnerability assessments remain current and relevant. This dynamic approach allows cities to adapt their strategies over time, maintaining resilience in the face of evolving climate conditions. The framework's ability to refine its models based on new data inputs allows for the ongoing enhancement of mapping accuracy, ensuring that strategies remain effective and responsive to changing urban heat challenges.

Public Engagement: U-HEAT promotes public engagement by offering online map resources that help residents understand the heat vulnerability of their areas. These resources, implemented by various national and state governments, raise awareness and provide tools to mitigate heat exposure risks in daily life. The framework also advocates for more public engagement opportunities, such as training modules and discussion workshops, to effectively integrate policy interventions and public involvement (Wang et al., 2023). This collective effort aids in managing urban heat vulnerability more efficiently and sustainably.

The U-HEAT framework is adaptable to various global urban contexts, but it requires further testing in diverse local settings to gauge its effectiveness. Implementing U-HEAT in cities with different climates, socio-economic conditions, and urban layouts is essential to recalibrate the model, ensuring it accurately reflects the unique heat vulnerability factors of each region. Collaboration with local planners, scientists, and health officials is crucial for refining the framework with region-specific data, making U-HEAT's recommendations more relevant and effective. Comparative studies across regions can also identify global patterns and regional nuances, enhancing our understanding of urban heat risks and informing tailored intervention strategies. By addressing these broader implications, U-HEAT aims to be a versatile tool for global urban resilience and sustainable development.

Implementing U-HEAT involves addressing several critical considerations. First, data accuracy and consistency are essential; high-quality datasets are necessary for reliable assessments, but integrating them from various sources can be challenging. Continuous updates and robust algorithms are required to maintain predictive accuracy, demanding long-term commitment. Second, technical expertise and training for urban planners are crucial to effectively use U-HEAT outputs; without proper training, its potential may not be fully realised. Third, ethical data handling and equitable resource access are vital; protecting socio-economic and health data privacy while prioritising vulnerable populations aligns with sustainable development goals. Finally, effective public engagement is necessary for U-HEAT's success; raising awareness and ensuring inclusive access to digital tools fosters a collective effort to manage urban heat vulnerability. Addressing these key considerations will enable U-HEAT to significantly enhance urban resilience to heatwaves and support sustainable urban development.

8.4.5 Assumptions and Limitations

The development of U-HEAT is based on several assumptions and has potential limitations. The selection of indicators and data assumes that they accurately reflect urban heat factors and that the data sources are reliable and representative (Cheng et al., 2021; Li et al., 2022). This chapter compiles frequently used indicators and data sources as a standard reference, but supplementing suitable indicators for different urban environments may encounter biases and subjective preferences. The process is also limited by data availability and quality. Urban environments are complex, and chosen indicators may not capture all nuances, especially in areas with limited data collection infrastructure. Moreover, data quality can vary significantly due to socio-economic and technological factors, potentially leading to skewed assessments.

In integrating ML and RS, RS provides high-resolution images of urban environments, essential for data on LST and infrastructure. The assumption is that these images are consistent and representative. However, data quality can be affected by atmospheric conditions and technological constraints. For example, Landsat provides 30 m spatial resolution images but only captures data every 16 days at 10 AM, possibly missing peak temperatures and heat variations during heatwaves. MODIS offers daily LST data but with a 1 km spatial resolution (Hulley et al., 2019). Interpreting RS data within complex urban landscapes can introduce errors or misinterpretations.

ML algorithms are expected to uncover patterns and correlations within large datasets, providing accurate predictions based on success in other urban domains. However, their effectiveness depends on the quality of input data. Issues like overfitting or underfitting can limit the models' ability to generalise, leading to inaccuracies in new or evolving scenarios. The 'black box' nature of complex ML models poses challenges in transparency and interpretability (Wang et al., 2023), crucial for justifying assessments. Additionally, the computational demands of processing large datasets require significant resources, which may not always be available.

Continuous monitoring is assumed to be feasible and integral for U-HEAT's adaptability. This involves regularly collecting and integrating new data to refine models and adjust strategies. However, it requires consistent funding and manpower, which may not always be available. Challenges in timeliness and accuracy of data updates could impact the framework's responsiveness to evolving urban heat challenges (Paranunzio et al., 2021). Despite these potential limitations, U-HEAT remains a promising tool for assessing and mitigating urban heat vulnerability. By limiting bias and enhancing the feasibility of each component—selecting suitable indicators, RS images, and ML algorithms—we can improve the framework's integrity and build a more consolidated tool.

8.5 Conclusion

The growing challenge of urbanisation and climate change has intensified the focus on urban heat, particularly its disproportionate impact on vulnerable groups like the elderly and those with pre-existing health conditions. While previous studies have examined the spatial distribution and key drivers of urban heat vulnerability, they often lacked rigour in indicator selection, spatio-temporal coverage, and validation techniques.

The primary objective of this study is to outline the theoretical underpinnings and potential applications of U-HEAT, emphasising its conceptual nature. U-HEAT introduces a novel approach by integrating ML with RS, enabling the precise mapping and forecasting of urban heat vulnerability at a detailed resolution. This approach is crucial for informed urban planning and disaster management. U-HEAT also combines historical data with predictive mapping, offering valuable insights for ongoing monitoring and mitigation, thereby aiding urban planners and policymakers in addressing urban heat island effects and protecting communities.

U-HEAT marks a significant step forward in understanding urban heat vulnerabilities, with important implications for urban sustainability and resilience. By melding advanced technologies and data, U-HEAT enriches academic research and serves as a crucial resource for cities looking to enhance their ability to adapt to climate-related challenges. This framework underscores the importance of interdisciplinary research in strengthening urban resilience and highlights the need for sustainable urban ecosystems capable of facing the complex challenges posed by climate change and increased urban heat.

Lastly, we conclude by highlighting the following key contributions of this study: (a) The introduction of an innovative predictive method for forecasting urban heat vulnerability; (b) the development of universal criteria and a reference framework for selecting urban heat vulnerability indicators; (c) the demonstration of the feasibility of long-term, grid-scale, precise urban heat vulnerability assessments; (d) the proposal of a robust, enduring, and sustainable framework for monitoring and managing urban heat vulnerability; and (e) support for urban sustainability through the integration of ML and RS in the urban heat vulnerability framework.

Future research should explore additional indicators, such as psychological factors and detailed socio-economic variables, to enhance vulnerability assessments. Refining the ML algorithms in the U-HEAT framework, by incorporating advanced deep learning techniques and improving data handling, could further boost predictive accuracy and model robustness. Empirical validation and broad application across various urban contexts are also necessary to ensure the framework's adaptability and effectiveness. By refining and expanding U-HEAT, future studies can contribute to more resilient and sustainable urban environments, better equipped to protect vulnerable populations and ensure urban well-being amid rising temperatures and climate change.

Acknowledgements

This chapter, with permission from the copyright holder, is a reproduced version of the following journal article: Li, F., Yigitcanlar, T., Nepal, M., Nguyen,

K., & Dur, F., (2024). A novel urban heat vulnerability analysis: Integrating machine learning and remote sensing for enhanced insights. *Remote Sensing, 16*(16), 3032.

References

Abunnasr, Y., & Mhawej, M. (2022). Towards a combined Landsat-8 and Sentinel-2 for 10-m land surface temperature products: The Google Earth Engine monthly Ten-ST-GEE system. *Environmental Modelling & Software, 155*, 105456. https://doi.org/10.1016/j.envsoft.2022.105456

Adger, W. N. (2006). Vulnerability. *Global Environmental Change, 16*, 268–281. https://doi.org/10.1016/j.gloenvcha.2006.02.006

Al-Aizari, A. R., Al-Masnay, Y. A., Aydda, A., Zhang, J., Ullah, K., Islam, A. R. M. T., Habib, T., Kaku, D. U., Nizeyimana, J. C., Al-Shaibah, B., & Khalil, Y. M. (2022). Assessment analysis of flood susceptibility in tropical desert area: A case study of Yemen. *Remote Sensing, 14*, 4050. https://doi.org/10.3390/rs14164050

Alonso, L., & Renard, F. (2020). A comparative study of the physiological and socio-economic vulnerabilities to heat waves of the population of the metropolis of Lyon (France) in a climate change context. *International Journal of Environmental Research and Public Health, 17*, 1004. https://doi.org/10.3390/ijerph17031004

Amani, S., & Shafizadeh-Moghadam, H. (2023). A review of machine learning models and influential factors for estimating evapotranspiration using remote sensing and ground-based data. *Agricultural Water Management, 284*, 108324. https://doi.org/10.1016/j.agwat.2023.108324

Australian Institute of Health and Welfare. (2023). *Let's talk about the weather: Injuries related to extreme weather.* Available online: https://www.aihw.gov.au/reports/injury/extreme-weather-injuries/contents/about (accessed on 19 June 2024).

Azhar, G., Saha, S., Ganguly, P., Mavalankar, D., & Madrigano, J. (2017). Heat wave vulnerability mapping for India. *International Journal of Environmental Research and Public Health, 14*, 357. https://doi.org/10.3390/ijerph14040357

Barron, L., Ruggieri, D., & Branas, C. (2018). Assessing vulnerability to heat: A geo-spatial analysis for the city of Philadelphia. *Urban Science, 2*, 38. https://doi.org/10.3390/urbansci2020038

Boumans, R., Phillips, D., Victery, W., & Fontaine, T. (2014). Developing a model for effects of climate change on human health and health–environment interactions: Heat stress in Austin, Texas. *Urban Climate, 8*, 78–99. https://doi.org/10.1016/j.uclim.2014.03.001

Bu, S., Smith, K. L., Masoud, F., & Sheinbaum, A. (2024). Spatial distribution of heat vulnerability in Toronto, Canada. *Urban Climate, 54*, 101838. https://doi.org/10.1016/j.uclim.2024.101838

Buscail, C., Upegui, E., & Viel, J. (2012). Mapping heatwave health risk at the community level for public health action. *International Journal of Health Geographics, 11*, 38. https://doi.org/10.1186/1476-072X-11-38

Casali, Y., Aydin, N. Y., & Comes, T. (2022). Machine learning for spatial analyses in urban areas: A scoping review. *Sustainable Cities and Society, 85*, 104050. https://doi.org/10.1016/j.scs.2022.104050

Chen, N., Lin, P., & Guo, Y. (2019). Long-term exposure to high temperature associated with the incidence of major depressive disorder. *Science of the Total Environment, 659*, 1016–1020. https://doi.org/10.1016/j.scitotenv.2018.12.434

Chen, Q., Ding, M., Yang, X., Hu, K., & Qi, J. (2018). Spatially explicit assessment of heat health risk by using multi-sensor remote sensing images and socioeconomic data in Yangtze River Delta, China. *International Journal of Health Geographics, 17*, 13. https://doi.org/10.1186/s12942-018-0135-y

Cheng, W., & Brown, R. (2020). An energy budget model for estimating the thermal comfort of children. *International Journal of Biometeorology, 64*, 1355–1366. https://doi.org/10.1007/s00484-020-01916-x

Cheng, W., Li, D., Liu, Z., & Brown, R. (2021). Approaches for identifying heat-vulnerable populations and locations: A systematic review. *Science of the Total Environment, 799*, 149417. https://doi.org/10.1016/j.scitotenv.2021.149417

Chi, Y., Fan, M., Zhao, C., Yang, Y., Fan, H., Yang, X., Yang, J., & Tao, J. (2022). Machine learning-based estimation of ground-level NO_2 concentrations over China. *Science of the Total Environment, 807*, 150721. https://doi.org/10.1016/j.scitotenv.2021.150721

Conlon, K., Mallen, E., Gronlund, C., Berrocal, V., Larsen, L., & O'Neill, M. (2020). Mapping human vulnerability to extreme heat: A critical assessment of heat vulnerability indices created using principal components analysis. *Environmental Health Perspectives, 128*, 97001. https://doi.org/10.1289/EHP4030

Crichton, D. (1999). The risk triangle. In J. Ingleton (Ed.), *Natural disaster management* (pp. 102–103). Tudor Rose. ISBN 0-9536140-1-8.

Das, T., Jana, A., Mandal, B., & Sutradhar, A. (2021). Spatio-temporal pattern of land use and land cover and its effects on land surface temperature using remote sensing and GIS techniques: A case study of Bhubaneswar City, Eastern India (1991–2021). *GeoJournal, 87*, 765–795. https://doi.org/10.1007/s10708-021-10541-z

Degirmenci, K., Desouza, K., Fieuw, W., Watson, R., & Yigitcanlar, T. (2021). Understanding policy and technology responses in mitigating urban heat islands: A literature review and directions for future research. *Sustainable Cities and Society, 70*, 102873. https://doi.org/10.1016/j.scs.2021.102873

Dehghani, A., Moazam, H. M. Z. H., Mortazavizadeh, F., Ranjbar, V., Mirzaei, M., Mortezavi, S., Ng, J. L., & Dehghani, A. (2023). Comparative evaluation of LSTM, CNN, and ConvLSTM for hourly short-term streamflow forecasting using deep learning approaches. *Ecological Informatics, 75*, 102119. https://doi.org/10.1016/j.ecoinf.2023.102119

Demarchi, L., Canters, F., Cariou, C., Licciardi, G., & Chan, J. C.-W. (2014). Assessing the performance of two unsupervised dimensionality reduction techniques on hyperspectral APEX data for high resolution urban land-cover mapping. *ISPRS Journal of Photogrammetry and Remote Sensing, 87*, 166–179. https://doi.org/10.1016/j.isprsjprs.2013.10.012

Deng, C., & Wu, C. (2013). The use of single-date MODIS imagery for estimating large-scale urban impervious surface fraction with spectral mixture analysis and machine learning techniques. *ISPRS Journal of Photogrammetry and Remote Sensing, 86*, 100–110. https://doi.org/10.1016/j.isprsjprs.2013.09.010

Depietri, Y., Welle, T., & Renaud, F. (2013). Social vulnerability assessment of the Cologne urban area (Germany) to heat waves: Links to ecosystem services. *International Journal of Disaster Risk Reduction, 6*, 98–117. https://doi.org/10.1016/j.ijdrr.2013.10.001

Dong, J., Peng, J., He, X., Corcoran, J., Qiu, S., & Wang, X. (2020). Heatwave-induced human health risk assessment in megacities based on heat stress-social vulnerability-human exposure framework. *Landscape and Urban Planning, 203,* 103907. https://doi.org/10.1016/j.landurbplan.2020.103907

Dugord, P., Lauf, S., Schuster, C., & Kleinschmit, B. (2014). Land use patterns, temperature distribution, and potential heat stress risk: The case study Berlin, Germany. *Computers, Environment and Urban Systems, 48,* 86–98. https://doi.org/10.1016/j.compenvurbsys.2014.07.005

Estoque, R., Ooba, M., Seposo, X., Togawa, T., Hijioka, Y., Takahashi, K., & Nakamura, S. (2020). Heat health risk assessment in Philippine cities using remotely sensed data and social-ecological indicators. *Nature Communications, 11,* 1581. https://doi.org/10.1038/s41467-020-15218-8

Ferreira, F., Pereira, E., Gonçalves, A., Costa, R., & Bezerra, F. (2021). An explicitly spatial approach to identify heat vulnerable urban areas and landscape patterns. *Urban Climate, 40,* 101021. https://doi.org/10.1016/j.uclim.2021.101021

Frantz, D., Schug, F., Okujeni, A., Navacchi, C., Wagner, W., van der Linden, S., & Hostert, P. (2021). National-scale mapping of building height using Sentinel-1 and Sentinel-2 time series. *Remote Sensing of Environment, 252,* 112128. https://doi.org/10.1016/j.rse.2020.112128

Galaz, V., Centeno, M. A., Callahan, P. W., Causevic, A., Patterson, T., Brass, I., Baum, S., Farber, D., Fischer, J., Garcia, D., & McPhearson, T. (2021). Artificial intelligence, systemic risks, and sustainability. *Technology in Society, 67,* 101741. https://doi.org/10.1016/j.techsoc.2021.101741

Ghaffarian, S., Roy, D., Filatova, T., & Kerle, N. (2021). Agent-based modelling of post-disaster recovery with remote sensing data. *International Journal of Disaster Risk Reduction, 60,* 102285. https://doi.org/10.1016/j.ijdrr.2021.102285

Guo, F., Zheng, R., Zhao, J., Zhang, H., & Dong, J. (2024). Framework of street grid-based urban heat vulnerability assessment: Integrating entropy weight method and BPNN model. *Urban Climate, 56,* 102067. https://doi.org/10.1016/j.uclim.2024.102067

Guo, H., Huang, J. J., Chen, B., Guo, X., & Singh, V. P. (2021). A machine learning-based strategy for estimating non-optically active water quality parameters using Sentinel-2 imagery. *International Journal of Remote Sensing, 42,* 1841–1866. https://doi.org/10.1080/01431161.2020.1846222

Guo, X., Huang, G., Jia, P., & Wu, J. (2019). Estimating fine-scale heat vulnerability in Beijing through two approaches: Spatial patterns, similarities, and divergence. *Remote Sensing, 11,* 2358. https://doi.org/10.3390/rs11202358

Hammer, J., Ruggieri, D., Thomas, C., & Caum, J. (2020). Local extreme heat planning: An interactive tool to examine a heat vulnerability index for Philadelphia, Pennsylvania. *Journal of Urban Health, 97,* 519–528. https://doi.org/10.1007/s11524-020-00443-9

Harlan, S., Brazel, A. J., Prashad, L., Stefanov, W., & Larsen, L. (2006). Neighborhood microclimates and vulnerability to heat stress. *Social Science & Medicine, 63,* 2847–2863. https://doi.org/10.1016/j.socscimed.2006.07.030

He, C., Ma, L., Zhou, L., Kan, H., Zhang, Y., Ma, W., & Chen, B. (2019). Exploring the mechanisms of heat wave vulnerability at the urban scale based on the application of big data and artificial societies. *Environment International, 127,* 573–583. https://doi.org/10.1016/j.envint.2019.01.057

He, Z., Chow, C. Y., & Zhang, J. D. (2019). STCNN: A spatio-temporal convolutional neural network for long-term traffic prediction. In *Proceedings of the 2019 20th IEEE International Conference on Mobile Data Management (MDM)*, Hong Kong, China, 10–13 June 2019, pp. 226–233. https://doi.org/10.1109/MDM.2019.00-53

Heaton, M., Sain, S., Greasby, T., Uejio, C., Hayden, M., Monaghan, A., Boehnert, J., Sampson, K., Banerjee, D., Nepal, V., & Wilhelmi, O.V. (2014). Characterizing urban vulnerability to heat stress using a spatially varying coefficient model. *Spatial and Spatio-temporal Epidemiology, 8*, 23–33. https://doi.org/10.1016/j.sste.2014.01.002

Ho, H., Knudby, A., & Huang, W. (2015). A spatial framework to map heat health risks at multiple scales. *International Journal of Environmental Research and Public Health, 12*, 16110–16123. https://doi.org/10.3390/ijerph121215046

Holec, J., Šveda, M., Szatmári, D., Feranec, J., Bobáľová, H., Kopecká, M., & Šťastný, P. (2021). Heat risk assessment based on mobile phone data: Case study of Bratislava, Slovakia. *Natural Hazards, 108*, 3099–3120. https://doi.org/10.1007/s11069-021-04816-4

Hu, K., Yang, X., Zhong, J., Fei, F., & Qi, J. (2017). Spatially explicit mapping of heat health risk utilizing environmental and socioeconomic data. *Environmental Science & Technology, 51*, 1498–1507. https://doi.org/10.1021/acs.est.6b04355

Huang, F., Yu, Y., & Feng, T. (2019). Automatic extraction of impervious surfaces from high Resolution remote sensing images based on deep learning. *Journal of Visual Communication and Image Representation, 58*, 453–461. https://doi.org/10.1016/j.jvcir.2018.11.041

Hulley, G., Shivers, S., Wetherley, E., & Cudd, R. (2019). New ECOSTRESS and MODIS land surface temperature data reveal fine-scale heat vulnerability in cities: A case study for Los Angeles county, California. *Remote Sensing, 11*, 2136. https://doi.org/10.3390/rs11182136

Intergovernmental Panel on Climate Change. (2001). *Climate change 2007: Impacts, adaptation and vulnerability*. Intergovernmental Panel on Climate Change: Geneva, Switzerland.

Jamali, A. (2019). Evaluation and comparison of eight machine learning models in land use/land cover mapping using Landsat 8 OLI: A case study of the Northern region of Iran. *SN Applied Sciences, 1*, 1527. https://doi.org/10.1007/s42452-019-1527-8

Jato-Espino, D., Manchado, C., Roldán-Valcarce, A., & Moscardó, V. (2022). ArcUHI: A GIS add-in for automated modelling of the urban heat island effect through machine learning. *Urban Climate, 44*, 101203. https://doi.org/10.1016/j.uclim.2022.101203

Johnson, D., Stanforth, A., Lulla, V., & Luber, G. (2012). Developing an applied extreme heat vulnerability index utilizing socioeconomic and environmental data. *Applied Geography, 35*, 23–31. https://doi.org/10.1016/j.apgeog.2012.04.006

Kafy, A. A., Saha, M., Faisal, A.-A., Rahaman, Z. A., Rahman, M. T., Liu, D., Fattah, M. A., Al Rakib, A., AlDousari, A. E., Rahaman, S. N., Hasan, M. Z. & Ahasan, M. A. K. (2022). Predicting the impacts of land use/land cover changes on seasonal urban thermal characteristics using machine learning algorithms. *Building and Environment, 217*, 109066. https://doi.org/10.1016/j.buildenv.2022.109066

Kamal, N., Ash'aari, Z., Abdullah, A., Kusin, F., Mohamat Yusuff, F., Sharaai, A., & Mohd Ariffin, N. (2021). Extreme heat vulnerability assessment in tropical region: A case study in Malaysia. *Climate and Development, 4*, 472–486. https://doi.org/10.1080/17565529.2021.1937030

Kamruzzaman, M., Deilami, K., & Yigitcanlar, T. (2018). Investigating the urban heat island effect of transit oriented development in Brisbane. *Journal of Transport Geography, 66*, 116–124. https://doi.org/10.1016/j.jtrangeo.2017.11.016

Karanja, J., & Kiage, L. (2021). Perspectives on spatial representation of urban heat vulnerability. *Science of the Total Environment, 774*, 145634. https://doi.org/10.1016/j.scitotenv.2021.145634

Kazak, J. (2018). The use of a decision support system for sustainable urbanization and thermal comfort in adaptation to climate change actions: The case of the Wrocław Larger Urban Zone (Poland). *Sustainability, 10*, 1083. https://doi.org/10.3390/su10041083

Khan, A., & Sudheer, M. (2022). Machine learning-based monitoring and modeling for spatio-temporal urban growth of Islamabad. *The Egyptian Journal of Remote Sensing and Space Sciences, 25*, 541–550. https://doi.org/10.1016/j.ejrs.2022.03.012

Kim, Y., & Kim, Y. (2024). Deep learning-based analytic framework using comprehensive urbanization index for heat vulnerability assessment in urban areas. *Expert Systems with Applications, 235*, 121140. https://doi.org/10.1016/j.eswa.2023.121140

Kim, Y., & Newman, G. (2020). Advancing scenario planning through integrating urban growth prediction with future flood risk models. *Computers, Environment and Urban Systems, 82*, 101498. https://doi.org/10.1016/j.compenvurbsys.2020.101498

Kitchley, J. L. (2024). A framework to assess the contextual composite heat vulnerability index for a heritage city in India – A case study of Madurai. *Sustainable Cities and Society, 101*, 105119. https://doi.org/10.1016/j.scs.2023.105119

Koumetio Tekouabou, S. C., Diop, E. B., Azmi, R., Jaligot, R., & Chenal, J. (2022). Reviewing the application of machine learning methods to model urban form indicators in planning decision support systems: Potential, issues and challenges. *The Journal of King Saud University Computer and Information Sciences, 34*, 5943–5967. https://doi.org/10.1016/j.jksuci.2021.08.007

Krstic, N., Yuchi, W., Ho, H., Walker, B., Knudby, A., & Henderson, S. (2017). The heat exposure integrated deprivation index (HEIDI): A data-driven approach to quantifying neighborhood risk during extreme hot weather. *Environment International, 109*, 42–52. https://doi.org/10.1016/j.envint.2017.09.011

Kwon, Y., Lee, D., & Kwon, Y. (2020. Is sensible heat flux useful for the assessment of thermal vulnerability in Seoul (Korea)? *International Journal of Environmental Research and Public Health, 17*, 963. https://doi.org/10.3390/ijerph17030963

Lammers, R., Li, A., Nag, S., & Ravindra, V. (2021). Prediction models for urban flood evolution for satellite remote sensing. *Journal of Hydrology, 603*, 127175. https://doi.org/10.1016/j.jhydrol.2021.127175

Li, F., Yigitcanlar, T., Li, W., Nepal, M., Nguyen, K., & Dur, F. (2024). Understanding urban heat vulnerability: Scientometric analysis of five decades of research. *Urban Climate, 56*, 102035. https://doi.org/10.1016/j.uclim.2024.102035

Li, F., Yigitcanlar, T., Nepal, M., Nguyen, K., & Dur, F. (2022). Understanding urban heat vulnerability assessment methods: A PRISMA review. *Energies, 15*, 6998. https://doi.org/10.3390/en15196998

Li, F., Yigitcanlar, T., Nepal, M., Nguyen, K., & Dur, F. (2023). Machine learning and remote sensing integration for leveraging urban sustainability: A review and framework. *Sustainable Cities and Society, 96*, 104653. https://doi.org/10.1016/j.scs.2023.104653

Li, Z., & Demir, I. (2023). U-net-based semantic classification for flood extent extraction using SAR imagery and GEE platform: A case study for 2019 Central US flooding. *Science of the Total Environment, 869,* 161757. https://doi.org/10.1016/j.scitotenv.2023.161757

Liu, Y., Guaralda, M., Yigitcanlar, T., Limb, M., & Garcia-Hansen, V. (2024a). Navigating urban climate design implementation challenges: Insights from Brisbane's built environment experts. *Journal of Urban Design,* 1–20. https://doi.org/10.1080/13574809.2024.2376680

Liu, Y., Yigitcanlar, T., Guaralda, M., Degirmenci, K., & Liu, A. (2024b). Spatial modelling of urban wind characteristics: Review of contributions to sustainable urban development. *Buildings, 14,* 737. https://doi.org/10.3390/buildings14030737

Loughnan, M., Nicholls, N., & Tapper, N. (2012). Mapping heat health risks in urban areas. *International Journal of Population Research, 2012,* 518687. https://doi.org/10.1155/2012/518687

Luber, G., & McGeehin, M. (2008). Climate change and extreme heat events. *American Journal of Preventive Medicine, 35,* 429–435. https://doi.org/10.1016/j.amepre.2008.08.021

Ma, Y., Hu, Y., Moncrieff, G. R., Slingsby, J. A., Wilson, A. M., Maitner, B., & Zhou, R. Z. (2022). Forecasting vegetation dynamics in an open ecosystem by integrating deep learning and environmental variables. *International Journal of Applied Earth Observation and Geoinformation, 114,* 103060. https://doi.org/10.1016/j.jag.2022.103060

Macintyre, H., Heaviside, C., Taylor, J., Picetti, R., Symonds, P., Cai, X., & Vardoulakis, S. (2017). Assessing urban population vulnerability and environmental risks across an urban area during heatwaves: Implications for health protection. *Science of the Total Environment, 610–611,* 678–690. https://doi.org/10.1016/j.scitotenv.2017.08.062

Maier, G., Grundstein, A., Jang, W., Li, C., Naeher, L., & Shepherd, M. (2014). Assessing the performance of a vulnerability index during oppressive heat across Georgia, United States. *Weather, Climate, and Society, 6,* 253–263. https://doi.org/10.1175/WCAS-D-13-00037.1

Matthews, T., Lo, A. Y., & Byrne, J. A. (2015). Reconceptualizing green infrastructure for climate change adaptation: Barriers to adoption and drivers for uptake by spatial planners. *Landscape and Urban Planning, 138,* 155–163. https://doi.org/10.1016/j.landurbplan.2015.02.010

Mitchell, B., & Chakraborty, J. (2015). Landscapes of thermal inequity: Disproportionate exposure to urban heat in the three largest US cities. *Environmental Research Letters, 10,* 115005. https://doi.org/10.1088/1748-9326/10/11/115005

Mushore, T., Mutanga, O., Odindi, J., & Dube, T. (2017). Determining extreme heat vulnerability of Harare Metropolitan City using multispectral remote sensing and socio-economic data. *Journal of Spatial Science, 63,* 173–191. https://doi.org/10.1080/14498596.2017.1290558

Niu, Y., Li, Z., Gao, Y., Liu, X., Xu, L., Vardoulakis, S., & Liu, Q. (2021). A systematic review of the development and validation of the heat vulnerability index: Major factors, methods, and spatial units. *Current Climate Change Reports, 7,* 87–97. https://doi.org/10.1007/s40641-021-00173-3

Oh, K., Lee, M., & Jeon, S. (2017). Development of the Korean climate change vulnerability assessment tool (VESTAP): Centered on health vulnerability to heat waves. *Sustainability, 9,* 1103. https://doi.org/10.3390/su9071103

Osaku, D., Nakamura, R. Y. M., Pereira, L. A. M., Pisani, R. J., Levada, A. L. M., Cappabianco, F. A. M., Falcão, A. X., & Papa, J. P. (2015). Improving land cover classification through contextual-based optimum-path forest. *Information Sciences, 324*, 60–87. https://doi.org/10.1016/j.ins.2015.06.020

Papadomanolaki, M., Vakalopoulou, M., & Karantzalos, K. (2021). A deep multitask learning framework coupling semantic segmentation and fully convolutional LSTM networks for urban change detection. *IEEE Transactions on Geoscience & Remote Sensing, 59*, 7651–7668. https://doi.org/10.1109/TGRS.2021.3055584

Paranunzio, R., Dwyer, E., Fitton, J., Alexander, P., & O'Dwyer, B. (2021). Assessing current and future heat risk in Dublin City, Ireland. *Urban Climate, 40*, 100983. https://doi.org/10.1016/j.uclim.2021.100983

Patino, J. E., & Duque, J. C. (2013). A review of regional science applications of satellite remote sensing in urban settings. *Computers, Environment and Urban Systems, 37*, 1–17. https://doi.org/10.1016/j.compenvurbsys.2012.06.003

Peng, B., Meng, Z., Huang, Q., & Wang, C. (2019). Patch similarity convolutional neural network for urban flood extent mapping using bi-temporal satellite multispectral imagery. *Remote Sensing, 11*, 2492. https://doi.org/10.3390/rs11212492

Priem, F., Okujeni, A., van der Linden, S., & Canters, F. (2019). Comparing map-based and library-based training approaches for urban land-cover fraction mapping from Sentinel-2 imagery. *International Journal of Applied Earth Observation and Geoinformation, 78*, 295–305. https://doi.org/10.1016/j.jag.2019.02.003

Prosdocimi, D., & Klima, K. (2020). Health effects of heat vulnerability in Rio de Janeiro: A validation model for policy applications. *SN Applied Sciences, 2*, 1948. https://doi.org/10.1007/s42452-020-03750-7

Reid, C., Mann, J., Alfasso, R., English, P., King, G., Lincoln, R., & Balmes, J. (2012). Evaluation of a heat vulnerability index on abnormally hot days: An environmental public health tracking study. *Environmental Health Perspectives, 120*, 715–720. https://doi.org/10.1289/ehp.1103766

Reid, C. E., O'Neill, M. S., Gronlund, C. J., Brines, S. J., Brown, D. G., Diez-Roux, A. V., & Schwartz, J. (2009). Mapping community determinants of heat vulnerability. *Environmental Health Perspectives, 117*, 1730–1736. https://doi.org/10.1289/ehp.0900683

Rinner, C., Patychuk, D., Bassil, K., Nasr, S., Gower, S., & Campbell, M. (2010). The role of maps in neighborhood-level heat vulnerability assessment for the City of Toronto. *Cartography and Geographic Information Science, 37*, 31–44. https://doi.org/10.1559/152304010790588089

Ronneberger, O., Fischer, P., & Brox, T. (2015). U-net: Convolutional networks for biomedical image segmentation. *Lecture Notes in Computer Science, 9351*, 234–241. https://doi.org/10.1007/978-3-319-24574-4_28

Shang, W. L., Chen, Y., Yu, Q., Song, X., Chen, Y., Ma, X., Chen, X., Tan, Z., Huang, J., & Ochieng, W. (2023). Spatio-temporal analysis of carbon footprints for urban public transport systems based on smart card data. *Applied Energy, 352*, 121859. https://doi.org/10.1016/j.apenergy.2023.121859

Singleton, A., Arribas-Bel, D., Murray, J., & Fleischmann, M. (2022). Estimating generalized measures of local neighbourhood context from multispectral satellite images using a convolutional neural network. *Computers, Environment and Urban Systems, 95*, 101802. https://doi.org/10.1016/j.compenvurbsys.2022.101802

Stehman, S. V., Hansen, M. C., Broich, M., & Potapov, P. V. (2011). Adapting a global stratified random sample for regional estimation of forest cover change derived from satellite imagery. *Remote Sensing of Environment, 115,* 650–658. https://doi.org/10.1016/j.rse.2010.10.009

Tao, C., Jia, M., Wang, G., Zhang, Y., Zhang, Q., Wang, X., Wang, Q., & Wang, W. (2023). Time-sensitive prediction of NO_2 concentration in China using an ensemble machine learning model from multi-source data. *Journal of Environmental Sciences, 137,* 30–40. https://doi.org/10.1016/j.jes.2023.02.026

Tate, E. (2012). Social vulnerability indices: A comparative assessment using uncertainty and sensitivity analysis. *Natural Hazards, 63,* 325–347. https://doi.org/10.1007/s11069-012-0152-2

Tella, A., Balogun, A.-L., Adebisi, N., & Abdullah, S. (2021). Spatial assessment of PM10 hotspots using random forest, k-nearest neighbour and Naïve Bayes. *Atmospheric Pollution Research, 12,* 101202. https://doi.org/10.1016/j.apr.2021.101202

Tomlinson, C., Chapman, L., Thornes, J., & Baker, C. (2011). Including the urban heat island in spatial heat health risk assessment strategies: A case study for Birmingham, UK. *International Journal of Health Geographics, 10,* 42. https://doi.org/10.1186/1476-072X-10-42

Touzani, S., & Granderson, J. (2021). Open data and deep semantic segmentation for automated extraction of building footprints. *Remote Sensing, 13,* 2578. https://doi.org/10.3390/rs13132578

Tuczek, M., Degirmenci, K., Desouza, K., Watson, R., Yigitcanlar, T., & Breitner, M. (2022). Mitigating urban heat with optimal distribution of vegetation and buildings. *Urban Climate, 44,* 101208. https://doi.org/10.1016/j.uclim.2022.101208

United Nations. (2023). *Global indicator framework for the sustainable development goals and targets of the 2030 agenda for sustainable development.* United Nations: San Francisco, CA.

Uniyal, S., Purohit, S., Chaurasia, K., Rao, S. S., & Amminedu, E. (2022). Quantification of carbon sequestration by urban forest using Landsat 8 OLI and machine learning algorithms in Jodhpur, India. *Urban Forestry & Urban Greening, 67,* 127445. https://doi.org/10.1016/j.ufug.2021.127445

Vescovi, L., Rebetez, M., & Rong, F. (2005). Assessing public health risk due to extremely high temperature events: Climate and social parameters. *Climate Research, 30,* 71–78. https://doi.org/10.3354/cr030071

Vu, H. L., Ng, K. T. W., Richter, A., & An, C. (2022). Analysis of input set characteristics and variances on k-fold cross validation for a recurrent neural network model on waste disposal rate estimation. *Journal of Environmental Management, 311,* 114869. https://doi.org/10.1016/j.jenvman.2022.114869

Wang, H., Yang, J., Chen, G., Ren, C., & Zhang, J. (2023a). Machine learning applications on air temperature prediction in the urban canopy layer: A critical review of 2011–2022. *Urban Climate, 49,* 101499. https://doi.org/10.1016/j.uclim.2023.101499

Wang, P., Bayram, B., & Sertel, E. (2022). A comprehensive review on deep learning based remote sensing image super-resolution methods. *Earth-Science Reviews, 234,* 104110. https://doi.org/10.1016/j.earscirev.2022.104110

Wang, S., Sun, Q. C., Huang, X., Tao, Y., Dong, C., Das, S., & Liu, Y. (2023b). Health-integrated heat risk assessment in Australian cities. *Environmental Impact Assessment Review, 102,* 107176. https://doi.org/10.1016/j.eiar.2023.107176

Weber, S., Sadoff, N., Zell, E., & de Sherbinin, A. (2015). Policy-relevant indicators for mapping the vulnerability of urban populations to extreme heat events: A case study of Philadelphia. *Applied Geography, 63*, 231–243. https://doi.org/10.1016/j.apgeog.2015.07.006

Wei, G., Zhang, Z., Ouyang, X., Shen, Y., Jiang, S., Liu, B., & He, B.-J. (2021). Delineating the spatial-temporal variation of air pollution with urbanization in the belt and road initiative area. *Environmental Impact Assessment Review, 91*, 106646. https://doi.org/10.1016/j.eiar.2021.106646

Wieland, M., & Pittore, M. (2016). Large-area settlement pattern recognition from Landsat-8 data. *ISPRS Journal of Photogrammetry and Remote Sensing, 119*, 294–308. https://doi.org/10.1016/j.isprsjprs.2016.06.010

Wilson, B., & Chakraborty, A. (2019). Mapping vulnerability to extreme heat events: Lessons from metropolitan Chicago. *Journal of Environmental Planning and Management, 62*, 1065–1088. https://doi.org/10.1080/09640568.2018.1462475

Wu, H., Xu, Y., Zhang, M., Su, L., Wang, Y., & Zhu, S. (2024). Spatially explicit assessment of the heat-related health risk in the Yangtze River Delta, China, using multisource remote sensing and socioeconomic data. *Sustainable Cities and Society, 104*, 105300. https://doi.org/10.1016/j.scs.2024.105300

Wu, X., Ou, J., Wen, Y., Liu, X., He, J., & Zhang, J. (2022). Developing a data-fusing method for mapping fine-scale urban three-dimensional building structure. *Sustainable Cities and Society, 80*, 103716. https://doi.org/10.1016/j.scs.2022.103716

Xiao, Y., Wang, Y., Yuan, Q., He, J., & Zhang, L. (2022). Generating a long-term (2003–2020) hourly 0.25° global PM2.5 dataset via spatiotemporal downscaling of CAMS with deep learning (DeepCAMS). *Science of the Total Environment, 848*, 157747. https://doi.org/10.1016/j.scitotenv.2022.157747

Xiao, Z., Li, H., Jiang, H., Li, Y., Alazab, M., Zhu, Y., & Dustdar, S. (2023). Predicting urban region heat via learning arrive-stay-leave behaviors of private cars. *IEEE Transactions on Intelligent Transportation Systems, 24*, 10843–10856. https://doi.org/10.1109/TITS.2023.3276704

Xu, Z., Li, J., Lv, Z., Wang, Y., & Li, H. (2021). A prediction method for population density in Key areas. In *Proceedings of the International Symposium on Artificial Intelligence and Robotics 2021*, Fukuoka, Japan, 28 October 2021; SPIE: Bellingham, WA, USA; Volume 11884, pp. 13–28. https://doi.org/10.1117/12.2600837

Yan, C., Fan, X., Fan, J., & Wang, N. (2022). Improved U-Net remote sensing classification algorithm based on multi-feature fusion perception. *Remote Sensing, 14*, 1118. https://doi.org/10.3390/rs14051118

Yigitcanlar, T. (2006). Australian Local Governments' practice and prospects with online planning. *URISA Journal, 18*, 7–17.

Yigitcanlar, T. (2010a). *Rethinking sustainable development: Urban management, engineering, and design*. IGI Global.

Yigitcanlar, T. (2010b). *Sustainable urban and regional infrastructure development: Technologies, applications and management*. IGI Global.

Yin, Y., Grundstein, A., Mishra, D. R., Ramaswamy, L., Tonekaboni, N., & Dowd, J. (2021). DTEx: A dynamic urban thermal exposure index based on human mobility patterns. *Environment International, 155*, 106573. https://doi.org/10.1016/j.envint.2021.106573

Yoo, E., Eum, Y., Roberts, J., Gao, Q., & Chen, K. (2021). Association between extreme temperatures and emergency room visits related to mental disorders: A multi-region time-series study in new York, USA. *Science of the Total Environment, 792,* 148246. https://doi.org/10.1016/j.scitotenv.2021.148246

Yu, B., Yin, H., & Zhu, Z. (2017). Spatio-temporal graph convolutional networks: A deep learning framework for traffic forecasting. *Arxiv Prepr,* arXiv:1709.04875. https://doi.org/10.48550/arXiv.1709.04875

Zhang, P., Liu, L., Yang, L., Zhao, J., Li, Y., Qi, Y., Ma, X., & Cao, L. (2023). Exploring the response of ecosystem service value to land use changes under multiple scenarios coupling a mixed-cell cellular automata model and system dynamics model in Xi'an, China. *Ecological Indicators, 147,* 110009. https://doi.org/10.1016/j.ecolind.2023.110009

Zhang, W., McManus, P., & Duncan, E. (2018). A raster-based subdividing indicator to map urban heat vulnerability: A case study in Sydney, Australia. *International Journal of Environmental Research and Public Health, 15,* 2516. https://doi.org/10.3390/ijerph15112516

Zhu, Q., Sun, Y., Guan, Q., Wang, L., & Lin, W. (2022). A weakly pseudo-supervised decorrelated subdomain adaptation framework for cross-domain land-use classification. *IEEE Transactions on Geoscience & Remote Sensing, 60,* 1–13. https://doi.org/10.1109/TGRS.2022.3170335

Zhu, Y., Geiß, C., So, E., & Jin, Y. (2021). Multitemporal relearning with convolutional LSTM models for land use classification. *IEEE Journal of Selected Topics in Applied Earth Observations and Remote Sensing, 14,* 3251–3265. https://doi.org/10.1109/JSTARS.2021.3055784

9

Urban Heat Vulnerability and Spatial Inequity

9.1 Introduction and Background

Climate change, with its hallmark of global warming, has led to an increase in the frequency and intensity of extreme weather events, including heatwaves that pose a significant threat to the stability and survival of human societies and natural systems (Intergovernmental Panel on Climate Change (IPCC), 2021; Mortoja & Yigitcanlar, 2020). Rapid urbanisation exacerbates this problem through the phenomenon known as the 'urban heat island effect', where urban areas experience higher temperatures than their suburban counterparts (Degirmenci et al., 2021; Ellena et al., 2023). This effect significantly increases the heat vulnerability of city residents, particularly those who live in areas lacking green spaces and adequate ventilation facilities (Sharma et al., 2018; Zha et al., 2024). Additionally, socioeconomic disparities play a critical role; individuals and communities with lower economic status and those marginalised socially are disproportionately affected by extreme heat events (Cheng et al., 2021; Forceville et al., 2024).

Urban heat vulnerability is a multifaceted phenomenon. Many factors such as vulnerable age groups (under 5 and above 65), poverty, low education, poor health conditions, and limited access to medical infrastructures further heighten the vulnerability of city residents to the adverse effects of heat. Tackling this grave challenge requires comprehensive research toolkits and action (Ellena et al., 2020; Kim et al., 2023). The hazards associated with heat vulnerability manifest in both health risks and economic losses, significantly impacting societies at various levels. Extreme heat increases the likelihood of heat stroke, cardiovascular diseases, respiratory problems, as well as other health issues, especially among the elderly, children, and individuals with chronic conditions (Basu, 2009; Campbell et al., 2018; Ho et al., 2023; Kovats & Hajat, 2008). Economically, such conditions lead to heightened energy demands, decreased work productivity, and infrastructure damage, thereby exerting widespread effects on economic activities (Sun et al., 2014, 2024). Therefore, understanding the implications and distribution of heat vulnerability is crucial for developing effective adaptation and mitigation measures that enhance the resilience of humans to future extreme heat events.

DOI: 10.1201/9781003737049-12

An increasing amount of heat vulnerability research has been conducted since Reid et al. (2009) proposed an innovative heat vulnerability index (HVI). This index recognises potential areas where mitigation interventions are urgently needed (Christenson et al., 2017; Chuang & Gober, 2015; Kumari & Kitchley, 2024; Wolf & McGregor, 2013). Heat vulnerability research aims to inform policymakers in crafting evidence-based policies and interventions aimed at protecting the most vulnerable populations and mitigating the adverse effects of extreme heat. It also helps to raise public awareness about heat vulnerability and its potential impacts. These studies integrate different heat-related indicators to spatially and temporally identify vulnerable populations and areas that may be exposed to increased extreme heat. For instance, the research by Chen et al. (2021) integrated 18 exposure, sensitivity, and adaptive capacity indicators to investigate the heat vulnerability of different community types in heat-exposed areas. Song et al. (2020) used 14 socio-demographic and environmental indicators and combined them to explore the spatial distribution of heat vulnerability at the community level of a high-density city.

The selection of appropriate indicators is a critical aspect of heat vulnerability research, as it directly influences the accuracy and relevance of vulnerability assessments (Cheng et al., 2021; Li et al., 2022). Over the years, researchers have employed various indicators to capture the multifaceted nature of heat vulnerability, often categorising them into demographic, socioeconomic, health, environmental, and infrastructural factors (Alonso & Renard, 2020; Li et al., 2024a; Oh et al., 2017). For instance, studies such as those by Reid et al. (2009) and Harlan et al. (2006) have underscored the importance of including demographic indicators like age and socioeconomic status, which are pivotal in determining the sensitivity of populations to heat. These studies argue that elderly individuals, young children, and low-income populations are particularly vulnerable to heat stress due to their limited adaptive capacity (Estoque et al., 2020; Rinner et al., 2010; Zhang et al., 2019). Moreover, environmental indicators, such as land surface temperature (LST) and vegetation cover, have been widely recognised as essential components of heat vulnerability assessments. Studies by Mitchell and Chakraborty (2015) and Hulley et al. (2019) highlight the significance of incorporating LST as an exposure indicator, given its direct relationship with heat intensity. Vegetation cover, often measured using the normalised difference vegetation index (NDVI), serves as an adaptive capacity indicator, reflecting the cooling effect of green spaces in urban environments (Johnson et al., 2012). Additionally, infrastructural indicators, such as the availability of cooling centres and the density of healthcare facilities, are also crucial, as demonstrated by the work of Uejio et al. (2011) and He et al. (2019), which linked the accessibility of such infrastructure to reduced heat-related mortality.

Nonetheless, significant gaps remain in the current body of research. Firstly, indicator selection is often impacted by over-reliance on local contexts and researchers' subjective judgement (Cheng et al., 2021; Li et al., 2022).

There is a lack of generic criteria and a universal reference for selecting heat-related indicators for heat vulnerability assessment research. Consequently, significant disparities exist in the literature for indicator selection. The number of indicators used in the existing research ranges from 2 to 30, and for categories they vary from 2 to 7. Such disparities can inevitably lead to grossly inaccurate heat vulnerability assessment results, thereby causing misguidance for policymaking. Furthermore, most studies have not rigorously validated the heat vulnerability assessment results (Kamruzzaman et al., 2018; Li et al., 2022). Although a few studies have attempted to validate the indices they proposed, some of these studies have not yielded convincing results. For example, the validation results from Estoque et al. (2020) show that the Pearson correlation coefficient (PCC) between the HVI and heat-related deaths is 0.436. Additionally, in the study by Wang et al. (2023), the PCCs relating HVI values to deaths, death rates, and hospital admissions during summer are all below 0.1. Although these results confirm a positive relationship between the proposed HVI and heat-related deaths, the relationship is not strong. This highlights the need for more rigorous and convincing validation. Addressing these gaps, this study introduces the comprehensive HVI (CHVI), which aims to standardise indicator selection and incorporate a more rigorous validation process.

Thermal inequity refers to the uneven distribution of heat exposure among different socioeconomic and demographic groups, increasingly highlighted in urban studies. Research demonstrates that lower socioeconomic neighbourhoods are more vulnerable to extreme heat due to limited access to cooling resources and green spaces. Studies by Dialesandro et al. (2021) and Johnson (2022) emphasise that socioeconomic disparities and residential segregation exacerbate these vulnerabilities, with marginalised communities bearing the brunt of heat exposure. Urban design plays a crucial role in mitigating thermal inequity. Agathangelidis et al. (2019) and Saverino et al. (2021) argue for climate-responsive architecture and urban planning that can create more equitable thermal environments. Additionally, studies by Wei et al. (2024) and Gabbe et al. (2023) highlight the importance of district cooling systems and energy-efficient technologies in reducing urban heat, particularly in cities with hot climates. Urban greening emerges as a vital strategy for addressing thermal inequity. Research by Byrne et al. (2016) and Burbidge et al. (2022) shows that enhancing access to green spaces can significantly reduce heat exposure in disadvantaged areas, highlighting the need for optimised green infrastructure. The health impacts of thermal inequity are well-documented. Colucci et al. (2023) emphasise the need for targeted public health interventions to protect vulnerable populations from heat-related illnesses. The integration of public health strategies with urban planning, as discussed by Chakraborty (2024) and Thomas and Butters (2018), is critical to mitigating heat-related risks. As urbanisation accelerates and climate change intensifies, addressing thermal inequity becomes increasingly crucial for fostering sustainable and equitable urban environments. However, limited research has focused on assessing thermal inequity across different heat vulnerability groups using diverse indicators. To bridge this gap, this study investigates thermal inequity by analysing the relationship between heat vulnerability and a range of indicators, including demographic groups, socioeconomic status, health conditions, living conditions, and transportation.

The *State of the Climate 2022* report (CSIRO & Bureau of Meteorology, 2022) indicates that since 1910, Australia has warmed by an average of 1.47°C, with the most rapid increases occurring since 1950. Every decade since then has been warmer than the previous one, with 2019 being the warmest year on record. The years 2013–2020 are among the ten warmest in Australian history. This warming trend is evident across all months, with both daytime and night-time temperatures rising. Notably, 2019 experienced 41 extremely warm days—approximately three times more than any year before 2000—and 33 days when national daily maximum temperatures exceeded 39°C, surpassing the total from the 59 years between 1960 and 2018. Over the last 50 years, the number of days breaking heat records has doubled (Steffen et al., 2015). Additionally, heatwave events are becoming more frequent and intense in Australia, especially in capital cities. For example, Sydney, Melbourne, and Brisbane are projected to face substantial increases in extreme heat by mid-century (Climate Council, 2024). Sydney could endure twice as many days above 35°C, potentially resulting in three weeks of extreme heat each summer. Melbourne, where heatwaves already start 17 days earlier than in the past, is expected to see double the number of days above 35°C by 2050. Brisbane may experience a threefold increase in days exceeding 35°C and a significant rise in warm nights where temperatures remain above 25°C, emphasising the urgent need for climate adaptation in these urban areas.

Australia, especially its capital cities where about 70% of the population resides (Australian Bureau of Statistics (ABS), 2021), is facing a series of economic losses and human health hazards due to rising temperatures and extreme weather conditions (Adams et al., 2022; Climate Council, 2014). Rising temperatures are projected to cost Australia about $19 billion by 2030 and could escalate to $211 billion by 2050 in terms of lost agricultural and labour productivity (including human health effects) (Steffen et al., 2019). As the impacts of urbanisation and climate change intensify, the country is experiencing higher temperatures, more extreme droughts, fiercer fire seasons, and more frequent extreme heat events (Australian Museum, 2023). Extreme heat can be attributed to significant economic losses across various sectors, notably in agriculture, healthcare, and infrastructure, affecting labour productivity and increasing energy and insurance costs (Steffen et al., 2019; Tuczek et al., 2022). Human Rights Watch emphasises that extreme heat has caused more deaths in Australia in the past 200 years than any other natural disaster, further highlighting the urgent need for comprehensive mitigation and adaptation strategies. Australian Bureau of Statistics (ABS). (2023), Wolf et al. (2014) and Yang et al. (2014)

Despite this, limited research has been dedicated to assessing heat vulnerability in Australia. Presently, only two studies have examined heat vulnerability across all capital cities. For example, Wang et al.'s (2023) study incorporated health-related factors into the assessment framework to investigate heat vulnerability in Australia's capital cities. Elizabeth Loughnan et al. (2014), on the other hand, proposed a spatial index of heat vulnerability to predict emergency service demand during extreme heat events. Nevertheless, Elizabeth Loughnan et al.'s (2014) research lacked comprehensiveness and consistency in indicator selection and thus might reduce the reliability of the research when applied to the Australian capital cities. Although Wang et al. (2023) innovatively introduced heat-related diseases into the heat vulnerability assessment framework, the results may have overestimated the effect of each disease and overlooked the impact of environmental and other socioeconomic factors.

This chapter proposes a CHVI to address the research gap in assessing heat vulnerability for Australian capital cities. By leveraging the vulnerability evaluation framework for climate disasters developed by the Intergovernmental Panel on Climate Change (IPCC) (Parry, 2007), the CHVI assesses heat vulnerability—including exposure, sensitivity, and adaptive capacity—of the smallest geographical units used by the Australian Bureau of Statistics (ABS) to report census data, Statistical Area 1 (SA1). Assessing heat vulnerability at the SA1 level is advantageous because it reveals finer spatial distinctions in vulnerability, enabling policymakers and agencies to more accurately target small communities or blocks that need special attention. To the best of our knowledge, the CHVI is the first comprehensive, validated index to integrate demographic, socioeconomic, health, and environmental indicators within the context of Australia's capital cities. This chapter contributes to knowledge in many ways. It provides a methodological reference for global heat vulnerability assessments, particularly in the selection of indicators. Focusing on Australia's capital cities, it enhances public understanding of heat vulnerability at the most detailed statistical level. This chapter explores thermal inequity by examining the relationship between heat vulnerability and various indicators, including population groups, socioeconomic status, health conditions, living conditions, and transportation. Spatially visualised outcomes of heat vulnerability and thermal inequity from this study could aid in the development of effective policies and strategies. This study also contributes to addressing several United Nations Sustainable Development Goals (SDGs) (United Nations, 2023), particularly SDG 11 (Sustainable Cities and Communities), SDG 13 (Climate Action), and SDG 10 (Reduced Inequalities).

The remainder of the chapter is organised into four sections. Section 9.2 introduces the study area and the methodological approach used to build the CHVI and conduct an inequity analysis. This includes the theoretical framework, assessment indicators, establishment methods, and validation methods employed for the study. Section 9.3 offers a comprehensive presentation of the heat vulnerability results from the perspectives of exposure, sensitivity, adaptive capacity, and CHVI, alongside thermal inequity in Australia's capital cities. Section 9.4 analyses and discusses the key findings from each perspective along with the interpretations of the results. In doing so, it examines underlying causality and the broader implications for the field, as well as study limitations. Finally, Section 9.5 concludes the chapter by summarising the key findings and their implications for theory and practice and outlines avenues for future research.

9.2 Research Design

The rationale and necessity for proposing the CHVI stem from the significant shortcomings in current studies on heat vulnerability indices, which often lack comprehensiveness and standardisation in indicator selection and validation. Existing studies rely heavily on local contexts and subjective judgement, leading to inconsistencies and inaccuracies in vulnerability assessments. To address these gaps, the CHVI standardises indicator selection by incorporating a broad range of socio-demographic, health, and environmental factors, ensuring a holistic assessment (see Figure 9.1). The CHVI is constructed using the IPCC vulnerability evaluation framework, encompassing exposure, sensitivity, and adaptive capacity. Indicators are chosen based on a systematic literature review (Li et al., 2022), which systematically analysed various methods for assessing urban heat vulnerability worldwide, with a particular focus on the selection and categorisation of indicators. It provides a comprehensive framework covering key indicators that are widely used and validated in urban heat vulnerability assessments. These indicators are categorised into three main groups: Demographic and socioeconomic characteristics, health conditions, and environmental factors. Based on this literature, the selection of indicators is scientifically grounded and widely applicable, thereby enhancing the validity and credibility of our study's assessment of urban heat vulnerability. The indicators are processed using normalisation and principal component analysis (PCA) to extract key features. Validation is rigorously conducted using heat-related mortality data from Australian capital cities, with PCC demonstrating strong positive correlations between CHVI scores and heat-related deaths, thereby confirming the index's robustness and reliability. Finally, this study conducts a thermal inequity analysis by examining the existence of spatial inequity between selected indicators and mapped heat vulnerability. This comprehensive approach enhances the accuracy of heat vulnerability assessments and thermal equity, providing valuable insights for informing climate resilience and adaptation strategies.

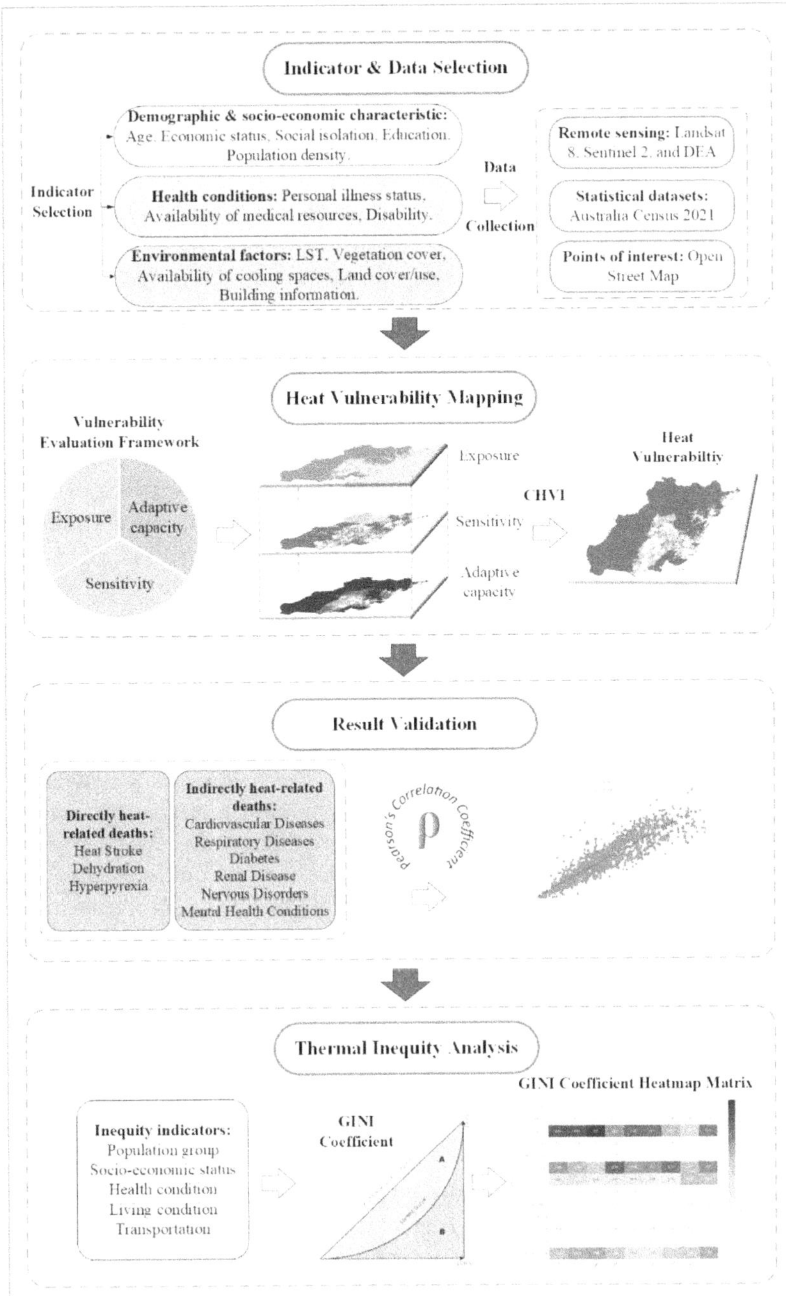

FIGURE 9.1
Research framework.

The objectives of this study are the following:

- **Develop** and **validate** a CHVI by integrating socio-demographic, health, and environmental indicators at the SA1 scale and validating it with heat-related mortality data.
- **Analyse** thermal inequity using the Gini coefficient, focusing on factors such as the Indigenous population, unemployment rates, English proficiency, and public transportation usage.
- **Investigate** the spatial distribution of heat vulnerability, exposure, sensitivity, and adaptive capacity in Australian capital cities, highlighting regional variations and urban–rural gradients.
- **Provide** insights for climate resilience and adaptation strategies to aid policymakers and urban planners in developing targeted interventions for vulnerable populations.
- **Contribute** to SDG 11 (Sustainable Cities and Communities), SDG 13 (Climate Action), and SDG 10 (Reduced Inequalities) through the study's findings and recommendations.

9.2.1 Study Area

The study areas selected in this study include the Greater Capital City Statistical Areas (GCCSAs), which cover the eight capital cities in Australia, namely, Greater Sydney (GSYD), Greater Melbourne (GMEL), Greater Brisbane (GBNE), Greater Adelaide (GADL), Greater Perth (GPER), Greater Hobart (GHBA), Greater Darwin (GDRW), and the Australian Capital Territory (ACT). As Figure 9.2 shows, the light-coloured regions on the map delineate the urban areas of Australia's capital cities, encompassing densely populated areas with substantial residential, commercial, and infrastructural development. These urban areas are encircled by suburban areas, which function as transitional areas featuring lower-density residential neighbourhoods, along with various commercial and public services. Beyond these suburban belts, the rural areas are positioned at the periphery of the cities, distinguished by their low population density, predominance of agricultural lands, and natural landscapes. Any reference to a capital city in this chapter refers to the GCCSAs. GCCSAs are defined to represent the functional areas of Australia's state/territory capital cities. These areas are crucial for understanding the dynamics of urbanisation and demographic changes throughout the country. Australia, being one of the most highly urbanised countries, had an urbanisation rate of 86.6% in 2023, with an estimated annual growth rate of 1.27% leading up to 2025 (Central Intelligence Agency, 2024). It should be noted that a significant portion of the Australian population lives in urban areas. The Australian Census 2021 shows that the capital cities account for about 66.95% of the total population in all 39,862 SA1s (ABS, 2022). As urbanisation accelerates, Australia, particularly capital cities, is facing higher temperatures and

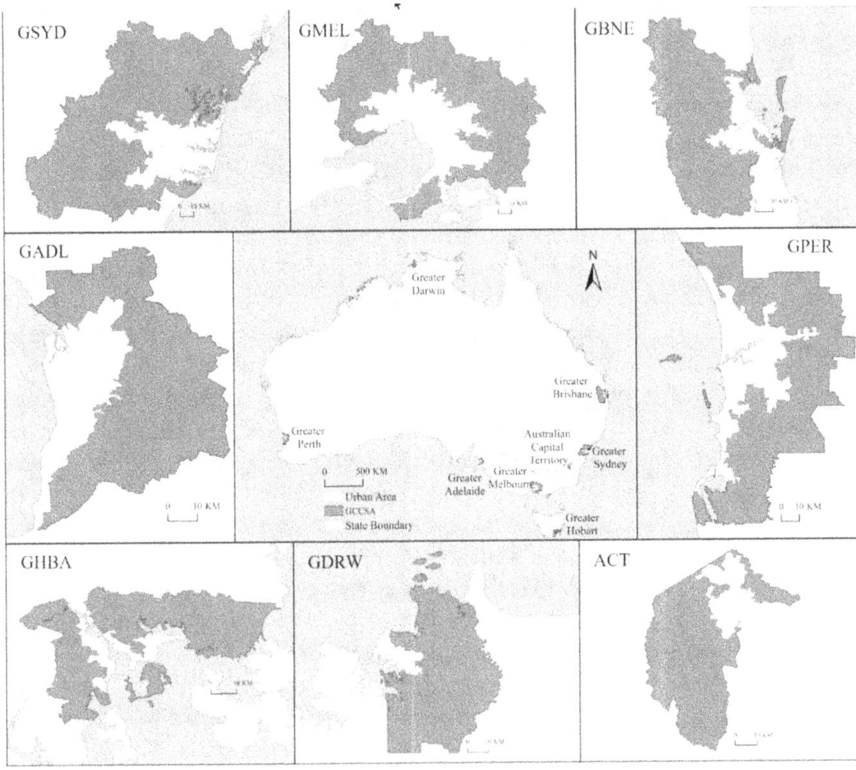

FIGURE 9.2
Study areas: Australia's capital cities with urban boundaries.

more frequent and intense extreme heat events, such as heatwaves, droughts, and bushfires (Steffen, 2015). These heat disasters have caused physical and mental health impacts on Australians. Despite this, there are still limited studies focused on investigating heat vulnerability in Australia, not to mention fine-scale assessments. Therefore, it is essential to assess heat vulnerability in Australia's capital cities and provide adaptation strategies and mitigation solutions to reduce this vulnerability.

9.2.2 Comprehensive Heat Vulnerability Index

9.2.2.1 IPCC Framework for Heat Vulnerability Assessment

This study adopts the IPCC's vulnerability evaluation framework for assessing heat vulnerability. As one of the most popular frameworks in this field, this framework has been widely used in recent studies (Rinner et al., 2010; Wilson & Chakraborty, 2019; Wu et al., 2022). Exposure, sensitivity, and adaptive capacity are three fundamental components of the framework used to assess and understand the impacts of increased high temperatures and heatwaves on populations, communities, or systems (Parry, 2007). Exposure refers to the degree to which a population, system, or area is subject to potentially harmful heat conditions. It considers factors like temperature, urban heat island effects, and the frequency, duration, and intensity of heatwaves, essentially measuring the extent of heat hazards an area or population directly faces (Kamal et al., 2022; Kim et al., 2017). Sensitivity refers to the susceptibility of a population or system to harm when exposed to heat hazards. It takes into account how factors such as age, health status, built environment, and social determinants of health might influence the ability of individuals or communities to withstand high temperatures (Dizdaroglu et al., 2012; Dong et al., 2020; Heaton et al., 2014). Sensitivity emphasises the internal factors that make certain populations more likely to suffer adverse outcomes when exposed to extreme heat. Adaptive capacity includes factors such as the availability and effectiveness of public health interventions, cooling centres, social support networks, and infrastructure resilience (Chen et al., 2021; He et al., 2019). It concerns the resources and strategies available to mitigate the impact of heat exposure and reduce sensitivity. In summary, while exposure and sensitivity increase heat vulnerability, adaptive capacity offers a mitigating effect. This framework helps to identify vulnerable populations and, subsequently, develop strategies to reduce heat-related risks.

Based on a rigorous and systematic literature review of measures assessing heat vulnerability, this study meticulously selects indicators that represent the vulnerability components (see Figure 9.1) (Li et al., 2022). This thorough review ensures the representativeness and relevance of the indicators while minimising subjective judgement. Unlike prior studies, which often categorically divided heat vulnerability indicators based on either demographic, socioeconomic, or environmental factors in isolation (Barron et al., 2018; Kershaw & Millward, 2012; Macnee & Tokai, 2016), this study introduces a novel approach by systematically re-categorising these indicators into three balanced categories: Demographic and socioeconomic characteristics, health conditions, and natural/built-environment factors (see Table 9.1). This re-categorisation is innovative because it achieves a more balanced and integrated assessment, which addresses the common issue of overemphasis on particular types of indicators in earlier research. By ensuring that these categories are balanced, the method enhances the comprehensiveness and clarity of the heat vulnerability assessment, allowing for a more holistic understanding of how different factors interact to influence heat vulnerability.

Additionally, this methodology integrates widely accepted indicators with new insights from recent studies, reflecting both the broad validation across the field and the latest advancements in heat vulnerability research. The new insights specifically focus on the integration of direct and indirect heat-related health conditions into the assessment framework. While traditional studies often emphasised environmental and socioeconomic factors, our approach uniquely includes personal illness status as an indicator, covering a

TABLE 9.1

Information of Heat Vulnerability Indicators

Categories	Indicators	Descriptions	Effects	Categories	Data Source
Demographic and socioeconomic characteristics	Age	Population aged below 5 or over 65	(+)	Sensitivity	Census 2021
	Economic status	Population with low incomes less than $800 weekly	(+)	Sensitivity	Census 2021
	Social isolation	Population living alone	(+)	Sensitivity	Census 2021
	Education	Population with a low education level below Grade 12	(+)	Sensitivity	Census 2021
	Population density	Population density of each SA1	(+)	Sensitivity	WorldPop
Health conditions	Personal illness status	Population with heat-related long-term physical/mental illness	(+)	Sensitivity	Census 2021
	Availability of medical resources	Availability of medical resources such as hospital/pharmacy/doctor/chemist	(−)	Adaptive capacity	OpenStreetMap
	Disability	Population with disability	(+)	Sensitivity	Census 2021
Environmental factors (natural)	Land surface temperature	Daytime land surface temperature	(+)	Exposure	Landsat 8
	Vegetation cover	Mean value of NDVI	(−)	Adaptive capacity	Sentinel 2
Environmental factors (built)	Availability of cooling spaces	Availability to public places with free cooling services	(−)	Adaptive capacity	OpenStreetMap
	Land cover/use	Area of developed urban land cover	(+)	Sensitivity	Digital Earth Australia
	Building information	Mean value of NDBI	(+)	Sensitivity	Sentinel 2

range of heat-related diseases such as heat stroke, cardiovascular, and respiratory conditions. This integration is informed by recent studies, such as Reid et al. (2012), Conlon et al. (2020), and Estoque et al. (2020), which highlight the significant impact of these health conditions on heat vulnerability. These insights are crucial for providing a more nuanced and comprehensive assessment, ensuring that the model accurately reflects the health-related dimensions of vulnerability. This unique combination of systematic selection and novel categorisation sets this study apart, providing a robust framework for accurately assessing heat vulnerability. This re-categorisation further explicitly explains the vulnerability indicators from the perspectives of socioeconomic attributes, health, and environment, maintaining a balance among them.

For exposure, this chapter utilises LST images from Landsat 8, which offer the finest spatial resolution of 30 m among free public temperature datasets to date (Zhang et al., 2018, 2023). These LST images are retrieved and processed in Google Earth Engine (GEE) with assistance from Dr. Qian's team at the Royal Melbourne Institute of Technology and A. Prof. Siqin Wang at the University of Southern California. The same applies to the NDVI and normalised difference built-up index (NDBI) data used in this study. Another commonly used data type for capturing exposure is air temperature, known for its high temporal resolution. However, it is measured by atmospheric monitoring stations which often have low spatial resolutions and a discontinuous distribution, making it less suitable for capturing detailed exposure characteristics at fine scales (Cheng et al., 2021; Naserikia et al., 2023). This chapter has used summer LST images from 2016 to 2020 during December, January, and February in Australia's capital cities, a period typically vulnerable to high temperatures and heatwaves. This process aims to minimise the potential impact of clouds and missing data on capturing exposure. The mean values of LST are calculated and normalised from 0 to 1 at the SA1 scale using the min–max normalisation method with ArcGIS Pro 2.9 as

$$X_{norm} = \frac{\left(X - X_{min}\right)}{\left(X_{max} - X_{min}\right)} \tag{9.1}$$

where X_{norm} is the normalised value, X is the original value, and X_{min} and X_{max} are the minimum and maximum values in the original data, respectively.

Sensitivity indicators include age, economic status, social isolation, education, personal illness status, and disability. It has been widely acknowledged that populations at the extremes of age, as well as those with poor socioeconomic status and health, are more sensitive to heat than other groups (Cresswell, 2023; Kwon et al., 2020). For instance, young (0–4 years old) and elderly (above 65 years old) populations, along with those who are unhealthy or disabled, may not be able to take timely and efficient actions to respond to heat due to immature, diseased, or degraded body functions (Cheng et al., 2021; Li et al., 2022). Additionally, economic status plays a crucial role, as those with lower incomes may not be able to afford cooling resources, such as air conditioning or other means of cooling, thereby increasing their sensitivity to heat (Paranunzio et al., 2021). Socially isolated individuals are more sensitive because they lack the social support networks that can provide assistance during extreme heat events (Li et al., 2022). Similarly, lower education levels can hinder individuals' ability to understand and respond effectively to heat warnings, reducing their capacity to protect themselves from heat-related risks (Cheng et al., 2021). All these data are retrieved and collected from the ABS Census 2021 (ABS, 2022). Population density, land cover/use, and building information contribute to sensitivity from the perspective of the living environment. Population density places a burden on public infrastructure and services to reduce heat impacts, and artificial land cover/use along with dense building aggregations are principal contributors to the urban heat island effect (Buscail et al., 2012). Population density data are provided by WorldPop at a resolution of 100 m, land cover/use data are collected from the Australian Urban Research Infrastructure Network (AURIN) at 25 m, and NDBI data for buildings are retrieved and calculated from Sentinel 2 images (10 m) using GEE. The NDBI values are derived from the Shortwave Infrared (SWIR, Band 11) and Near-Infrared (NIR, Band 8) bands. In this analysis of nine indicators to build sensitivity, this chapter employs PCA to simplify the data structure and extract important data features (see Eq. (9.2)).

PCA is adept at reducing dimensionality by converting variables into principal components (PC) that explain the majority of the variability in the data (Guo et al., 2019). It can also reduce the impact of multicollinearity among variables and enhance the stability and interpretability of the model. By eliminating unnecessary variables and noise, PCA aids in building more streamlined models that may offer better generalisability, as they rely mainly on the most significant features of the data (Wold et al., 1987). All sensitivity data are processed and normalised in ArcGIS Pro 2.9 before applying PCA on SPSS 29.0.0.0.

$$Sensitivity = \sum_{i=1}^{n} \frac{E(\sigma^2)_i}{C(\sigma^2)} \sum_{j=1}^{m} \frac{\alpha_j}{\sqrt{\lambda_i}} X_j \tag{9.2}$$

where $E(\sigma^2)_i$ represents the proportion of variance explained of PC i, $C(\sigma^2)$ represents the cumulative proportion of variance explained, α_j represents the loading of indicator j, λ_i represents the eigenvalue of PC i, X_j represents the indicator j, and the right part of the equation represents the value of PC i.

Adaptive capacity incorporates indicators of the availability of medical resources, vegetation cover, and cooling spaces. These indicators reflect the availability of services, resources, and infrastructure, which can provide comfortable cooling spaces for the public to avoid the risk of heat exposure and offer necessary medical support for sensitive groups (Chen et al., 2021; He et al., 2019). This chapter calculates the values of adaptive capacity by summing the three indicators with equal weights and normalising them. This method has been commonly adopted in similar studies to ensure a balanced and comprehensive assessment of adaptive capacity across different regions (Chuang & Gober, 2015; Zhang et al., 2023). The equal weighting is applied to prevent any single indicator from disproportionately influencing the overall adaptive capacity score, thereby providing a more accurate reflection of the true adaptive potential of a region. Moreover, no studies to date have found significant differences in the significance of these three indicators, further supporting the rationale and effectiveness of using an equal weighting approach. Vegetation cover, sourced from Sentinel 2 images with a 10 m spatial resolution and a five-day temporal resolution, involves retrieving summer images to capture peak vegetation growth. The mean values of NDVI were calculated for each SA1 using GEE and ArcGIS Pro 2.9. The NDVI values were derived from the Red (Band 4) and NIR (Band 8) bands of the Sentinel 2 imagery, ensuring high-quality and cloud-free images. To assess the availability of medical resources and cooling spaces, this chapter collects points of interest (POI) data from OpenStreetMap (OSM) and screens them to identify typical types based on previous studies (Li et al., 2021; Wang et al., 2023).

POIs related to medical resources include hospitals, pharmacies, doctors, and chemists. Cooling spaces POIs encompass public places providing free cooling services, such as supermarkets, libraries, cinemas, and swimming pools. To quantify the availability of medical resources and cooling spaces, this chapter employs kernel density analysis (KDA) using ArcGIS Pro 2.9 (see Eq. (9.3)), which is a widely recognised method for assessing spatial distribution patterns of point-based data (Li et al., 2020; Yu et al., 2017). KDA is particularly useful in this context because it transforms discrete point data into a continuous surface, providing a probabilistic estimate of resource density across space. This approach not only smooths out the data's inherent discreteness but also facilitates a more intuitive and visually interpretable understanding of spatial patterns (Yu et al., 2017). Furthermore, KDA generates specific density values for each location, which allows for a quantitative comparison of resource availability across different areas. This capability is crucial for evaluating accessibility and informing policy decisions aimed at optimising the distribution of medical and cooling resources (Delso et al., 2018; Spencer & Angeles, 2007).

$$P_i = \frac{1}{n\pi R^2} \times \sum_{j=1}^{n} K_j \left(1 - \frac{D_{ij}^2}{R^2}\right)^2 \tag{9.3}$$

where P_i represents the kernel density value at point i, K_j represents the weight assigned to point j, D_{ij} denotes the Euclidean distance between points i and j, R signifies the bandwidth ($D_{ij} < R$) of the regular area, and n represents the quantity of point j in the regular area.

Based on the vulnerability evaluation framework of the IPCC, which includes components of exposure, sensitivity, and adaptive capacity, along with a comprehensive selection of indicators covering demographic and socioeconomic characteristics, health conditions, and environmental factors, this chapter proposes the CHVI. This study uses the CHVI to investigate heat vulnerability in Australia's capital cities at the SA1 scale. Considering the contributing effects of exposure and sensitivity and the mitigating effect of adaptive capacity on heat vulnerability and referring to previous heat vulnerability research and IPCC's report (Cai et al., 2019; Kamal et al., 2022; Oh et al., 2017; Parry, 2007), the CHVI is calculated using Eq. (9.4) and is normalised by using ArcGIS Pro 2.9. Heat vulnerability is categorised into five levels according to values and visualised on ArcGIS Pro 2.9: Very low (0%–20%), low (20%–40%), medium (40%–60%), high (60%–80%), and very high (80%–100%) (Wang et al., 2023). This categorisation criterion is also adopted in the visualisations of exposure, sensitivity, and adaptive capacity.

$$Heat\ Vulnerability = Exposure + Sensitivity - Adaptive\ Capacity \tag{9.4}$$

9.2.2.2 Validation with Heat-Related Deaths Datasets

Previous literature reviews (Cheng et al., 2021; Li et al., 2022; Niu et al., 2021) reveal that data on heat-related deaths are typically employed to validate the ability to capture heat vulnerability. This approach is widely recognised and justified because the significant correlation between heat vulnerability and heat-related deaths has been proven and acknowledged in this field (Conlon et al., 2020; Estoque et al., 2020). For instance, Reid et al. (2012) established a clear positive correlation between chronic diseases, such as cardiovascular and respiratory conditions, and diabetes, with heat-related mortality rates. This study further enhances this validation method by categorising diseases with codes from the International Classification of Disease, 10th Version (ICD-10), as either directly or indirectly related to heat vulnerability (see Table 9.2). This categorisation provides a more nuanced understanding of the impacts of heat on human health. This chapter collected death data from these heat-related diseases in Australia's capital cities in 2021.

Using death data at the Statistical Area Level 2 (SA2) scale—the most detailed death data currently available from the ABS heat-related death data were calculated based on the total population (Death_P) and total deaths (Death_D) at the SA2 level. The SA2 is a geographical area defined by the ABS as part of the Australian Statistical Geography Standard (ASGS). These areas are designed to represent communities that interact socially and economically. They are typically used to present detailed regional data, such as health statistics and economic information. SA1s are nested within SA2s, meaning that an SA2 is composed of multiple SA1s. In this study, although the SA1 scale offers finer spatial resolution, the availability of death data at this scale is limited. The ABS currently only provides reliable death data at the SA2 scale. Therefore, we used SA2-scale data to calculate heat-related deaths. This choice ensures the completeness and reliability of the data used in our analysis while still providing sufficient spatial resolution to explore regional variations in heat vulnerability.

After data collection for validation, this study utilised the PCC to validate the ability of CHVI to capture heat vulnerabilities (Estoque et al., 2020; Zhang et al., 2019). This method is acknowledged as a robust and authoritative

TABLE 9.2

Heat-Related Death Data of Australia's Capital Cities for Validation

		NSW	VIC	QLD	SA	WA	TAS	NT	ACT
Direct heat-related illnesses	Heat stroke (X30)	2	6	6	6	6	0	6	0
	Dehydration (E86)	27	34	18	11	5	1	1	5
	Hyperpyrexia (R50.9)	0	0	2	0	0	0	0	0
Indirect heat-related illnesses	Cardiovascular diseases (R50.9)	13,659	10,198	7773	3399	3628	1200	217	548
	Respiratory diseases (J00–J99)	4287	3103	2356	1049	1091	381	101	154
	Diabetes (E10–E14)	1735	1290	989	418	493	162	88	51
	Renal disease (N00–N29)	973	946	421	247	240	57	38	30
	Nervous disorders (G00–G99)	3074	2737	1815	988	1073	208	45	162
	Mental health conditions (F00–F99)	3661	2368	2119	995	868	278	53	165
Total		27,418	20,682	15,499	7113	7404	2287	549	1115
Total population		8,072,163	6,503,491	5,156,138	1,781,516	2,660,026	557,571	232,605	454,499
Total deaths		52,485	41,093	31,367	13,607	14,993	4435	1141	2162
Death rate of total population (%)		0.34	0.32	0.30	0.40	0.28	0.41	0.24	0.25
Death rate of total deaths (%)		52.24	50.33	49.41	52.27	49.38	51.57	48.12	51.57

validation technique. Additionally, this study is exploring other potential validation methods to further enhance the robustness of the assessment, such as using hospitalisation records, emergency service calls, and heat-related morbidity data. For consistency between variables, values of CHVI for SA2s were derived by summing up the values of subordinate SA1s. The PCC is an index to evaluate the linear correlation between two random variables, X and Y (see Eq. (9.5)). If the correlation coefficient is positive (PCC close to 1), the variables are directly related; if it is negative (PCC close to –1), then they are inversely related. In this chapter, the results are validated by evaluating the significance of the correlation between CHVI and heat-related mortality. All PCC results are calculated with MATLAB R2003b, a tool renowned for its precision in computational processes, ensuring consistent application of the formula across different datasets.

$$\rho = \frac{\sum_{i=1}^{n}(X_i - \bar{X})(Y_i - \bar{Y})}{\sqrt{\sum_{i=1}^{n}(X_i - \bar{X})^2} \times \sqrt{\sum_{i=1}^{n}(Y_i - \bar{Y})^2}} \tag{9.5}$$

where ρ is the PCC between variables X and Y, n represents the sample size, X_i and Yi are individual values of datasets X and Y, respectively, and $\bar{X}$ and $\bar{Y}$ are the mean values of X and Y.

9.2.3 Inequity Analysis

This chapter employs the Gini coefficient to investigate inequity in heat vulnerability across selected indicators at the SA1 scale throughout Australia's capital cities. These indicators include population groups, socioeconomic status, health conditions, living conditions, and transportation (see Table 9.3). All data, except for the distance measurements, were retrieved from the ABS Census 2021 (ABS, 2022) and subjected to calculations and statistical analyses. The distances to the city centre were calculated using administrative area maps (ABS, 2021) in ArcGIS Pro 2.9. The selection of these indicators was informed by the need to address research gaps in understanding the multidimensional aspects of inequity related to heat vulnerability, such as nationality, race, language, employment, living conditions, and transportation. Limited research has examined how these factors intersect and contribute to heat vulnerability across diverse population groups. To address this gap, this study includes indicators based on previous studies (Dialesandro et al., 2021; Johnson, 2022; Li et al., 2022; Mitchell & Chakraborty, 2015), Australian conditions, and data availability to offer a more comprehensive analysis of thermal inequity.

The inclusion of Australian citizens, migrants, and Indigenous populations enables an examination of how heat vulnerability might differentially impact these groups. For instance, Indigenous populations in Australia, in particular, may exhibit distinct vulnerabilities due to historical and ongoing

TABLE 9.3

Information of Inequity Analysis Indicators

Categories	Indicators	Descriptions
Population group	Australian	Population of Australian citizens
	Migrant	Population of residents born overseas (not in Australia)
	Indigenous	Population of Aboriginal and Torres Strait Islanders
Socioeconomic status	High income	Population with an income above $1000 weekly (AUD)
	High education	Population with non-school qualifications
	English proficiency	Population using other languages and speaking English not well or not at all
	Unemployment	Unemployment rate
Health condition	Health	Population without long-term physical/ mental illness
	Assistance	Population provided with unpaid assistance
Living condition	Private dwelling	Population owning private dwellings
	Distance	Distance to city centre
Transportation	Vehicle	Dwellings with motor vehicles
	Commuting method	Population travelling to work by using public transportation

socioeconomic disadvantages. Socioeconomic status indicators, such as high income, high education, English proficiency, and unemployment rate, were selected to capture the economic and social dimensions of inequity. These factors are critical as they influence access to resources like healthcare, cooling facilities, and information, which are essential for mitigating heat-related risks. Health condition indicators focus on populations without long-term physical or mental illness and those receiving unpaid assistance, as these factors are relevant to an individual's capacity to cope with extreme heat, making them crucial for assessing heat vulnerability.

Living and transportation conditions, including private dwelling ownership, proximity to the city centre, vehicle ownership, and commuting methods, were selected for their role in influencing an individual's exposure to heat and their ability to access cooler areas or healthcare facilities. The investigation of these indicators is driven by their potential to reveal inequities in heat vulnerability across different segments of the population. By analysing the Gini coefficient across these indicators, this study aims to quantify the degree of inequity and identify the population groups or socioeconomic conditions that contribute most to these disparities. The insights gained from this analysis are intended to inform targeted interventions and policies to reduce heat-related risks in vulnerable communities. In summary, the selection of these specific indicators was aimed not only at filling existing research gaps but also at representing key dimensions of inequity that are relevant to understanding and addressing heat vulnerability in Australian capital cities.

The Gini coefficient (see Eq. (9.6)), a statistical measure of income distribution inequality ranging from 0 (perfect equality) to 1 (perfect inequality), serves as the tool for this analysis (Deininger & Squire, 1996). It is a widely accepted measure due to its standardised calculation method and interpretation, making research findings comprehensible and comparable on an international scale. Although originally devised for measuring economic inequality, the Gini coefficient has been effectively adapted for analysing other types of inequalities, including those related to health, education, and environmental conditions, such as heat vulnerability (Dai et al., 2022; Wüstemann et al., 2017). The United Nations Development Programme (UNDP) categorises Gini coefficients into five levels: Extremely low (<0.2), low (0.2–0.29), medium (0.3–0.39), high (0.4–0.59), and extremely high (≥0.6). These levels measure different degrees of inequity in heat vulnerability among various populations and areas. All Gini coefficients are calculated with MATLAB R2003b.

$$Gini = 1 - \sum_{i=1}^{n} (X_k - X_{k-1})(Y_k + Y_{k-1}) \tag{9.6}$$

where n represents the total number of SA1s in Australia's capital cities, X_k denotes the cumulative proportion of heat vulnerability, arranged by a certain inequity indicator from the lowest to the highest, with X_0 typically set to 0, Y_k signifies the cumulative proportion of inequity indicators, and Y_0 is also typically set to 0.

The calculation of the Gini coefficient in this study specifically involves determining the disparity between the cumulative distribution of heat vulnerability and the cumulative distribution of the SA1 being analysed. n represents the total number of SA1s in Australia's capital cities. The formula used in Eq. (9.6) integrates these cumulative proportions, where X_k and Y_k represent the cumulative percentages of the heat vulnerability indicator and the inequity indicator, respectively, ordered from the lowest to the highest. The stepwise differences between these cumulative percentages are then summed across all intervals, with the result subtracted from 1 to yield the Gini coefficient. This process captures the extent of inequality, providing a quantitative measure of how heat vulnerability is distributed across different regions. By setting both X_0 and Y_0 to 0, the calculation aligns with the traditional approach to measuring inequality, ensuring that the resulting Gini coefficient reflects the actual disparities present within the data. This methodical approach underscores the robustness and reliability of the Gini coefficient as a measure of inequity in environmental conditions.

9.3 Analysis and Results

9.3.1 Heat Vulnerability

9.3.1.1 Exposure

The spatial patterns of exposure of capital cities are captured and presented in Figure 9.3. The eight GCCSA regions can be categorised into three tiers based on the proportions of higher exposure levels observed in Figure 9.4. The first tier, characterised by the highest proportions of higher exposure levels, includes GBNE, GADL, and GPER. The second tier, with medium proportions of higher exposure levels, includes GSYD, GDRW, and ACT. The

FIGURE 9.3
Spatial distribution of exposure in Australia's capital cities. Notes: (a) GSYD, (b) GMEL, (c) GBNE, (d) GADL, (e) GPER, (f) GHBA, (g) GDRW, (h) ACT. The same order in Figure 9.5, Figure 9.7, and Figure 9.9.

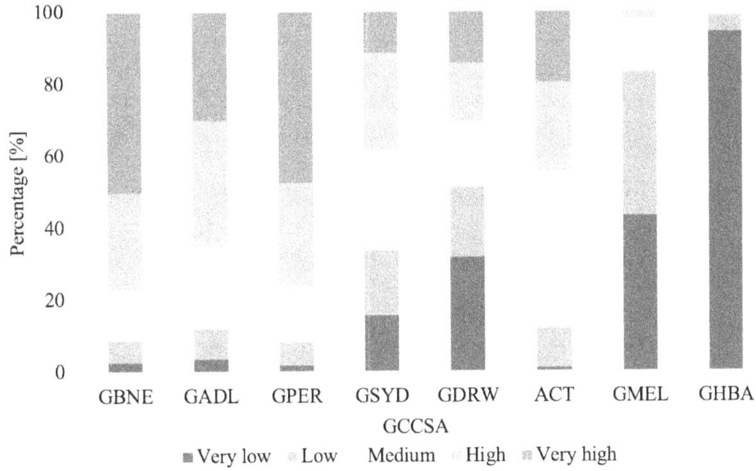

FIGURE 9.4
Proportions of exposure levels in Australia's capital cities.

third tier includes GMEL and GHBA with the lowest proportions of higher exposure levels. These tiers are visually represented by the orange and red segments in Figure 9.4. GBNE, GADL, and GPER belong to the first category, which features a noticeable portion of SA1s with very high-, high-, and medium-level exposure. These SA1s are not only centralised in Brisbane, Adelaide, and Perth cities but are also found in other parts of the GCCSAs. For instance, the west and south of GBNE, the east of Adelaide, and the southwest and northeast of Perth have clusters of higher-level (high and very high levels) exposure. It is worth noting that GPER presents a strip of very high exposure extending along the western part of the city. Low- and very low level SA1s are primarily distributed in the north of GBNE, the middle of GADL, and the east of GPER.

The second category comprises GSYD, GDRW, and ACT, which display a relatively lower yet similar proportion of SA1s with high- and very high level exposure. These SA1s are primarily concentrated in the urban areas of Sydney, Darwin, and Canberra, as well as the northeast of GSYD. Notably, the central areas (downtowns) in Canberra do not experience high exposure to heat. Sharp gradients of decreasing exposure levels extend towards peri-urban and rural areas (Mortoja et al., 2020). GMEL and GHBA belong to the third group because both have a low portion of higher-level exposure SA1s and a high portion of lower-level (low and very high levels) exposure SA1s. The second reason is that the higher-level exposure is not primarily focused on the urban areas of Melbourne and Hobart. For example, higher-level exposures are found in the northwest of GMEL, while GHBA does not have any higher-level clusters.

9.3.1.2 Sensitivity

Table 9.4 shows the results of the PCA analysis. The result of the Kaiser–Meyer–Olkin (KMO) test is 0.792, which is above the threshold of 0.70, indicating a good sampling adequacy for the analysis. Additionally, the result of Bartlett's test suggests that the variables are sufficiently correlated for PCA ($p < 0.01$). These tests indicate a moderate to high proportion of shared variance among the variables, confirming that the PCA results are meaningful. The communalities of all indicators are generally high, which indicates that the PCA effectively captures the main variance of the variables, yielding overall positive results. Three PCs were generated by the PCA, with a cumulative variance of 70.6%. This means the PCA results can explain 70.6% of the data variance. PC1 includes five contributing indicators, all with loadings above 0.80, pointing to the characteristics of socioeconomic and health profiles.

Two indicators, population density (0.799) and social isolation (0.595) contribute to PC2, which indicates the living environment and status based on the distribution of the population. PC3 incorporates indicators of land cover/use and building information, with loadings of 0.802 and 0.601, respectively. This component reflects the layout and architectural context of urban areas. The results of the PCA are used to build a sensitivity index, the results of which are shown in Figure 9.5.

Overall, the spatial distributions of sensitivity in capital cities resemble the distribution of exposure. The SA1s with higher-level sensitivity are primarily located in the urban areas of the GCCSAs. However, the distribution

TABLE 9.4

PCA Results of Sensitivity Indicators

	PC1	PC2	PC3	Communalities
Personal illness status	**0.934**	−0.054	−0.036	0.877
Age	**0.880**	−0.084	−0.074	0.786
Education	**0.826**	−0.219	0.076	0.736
Disability	**0.816**	−0.049	0.070	0.674
Economic status	**0.812**	0.065	0.043	0.665
Population density	−0.024	**0.799**	−0.318	0.739
Social isolation	0.490	**0.595**	−0.135	0.613
Land cover/use	−0.028	0.049	**0.802**	0.647
Building information	0.043	0.506	**0.601**	0.619
Eigenvalue	3.897	1.315	1.143	
% of Variance	43.299	14.615	12.702	
Cumulative (%)	43.299	57.914	70.616	
KMO test		0.792		
Bartlett's test ($p < 0.01$)		125,403.534		

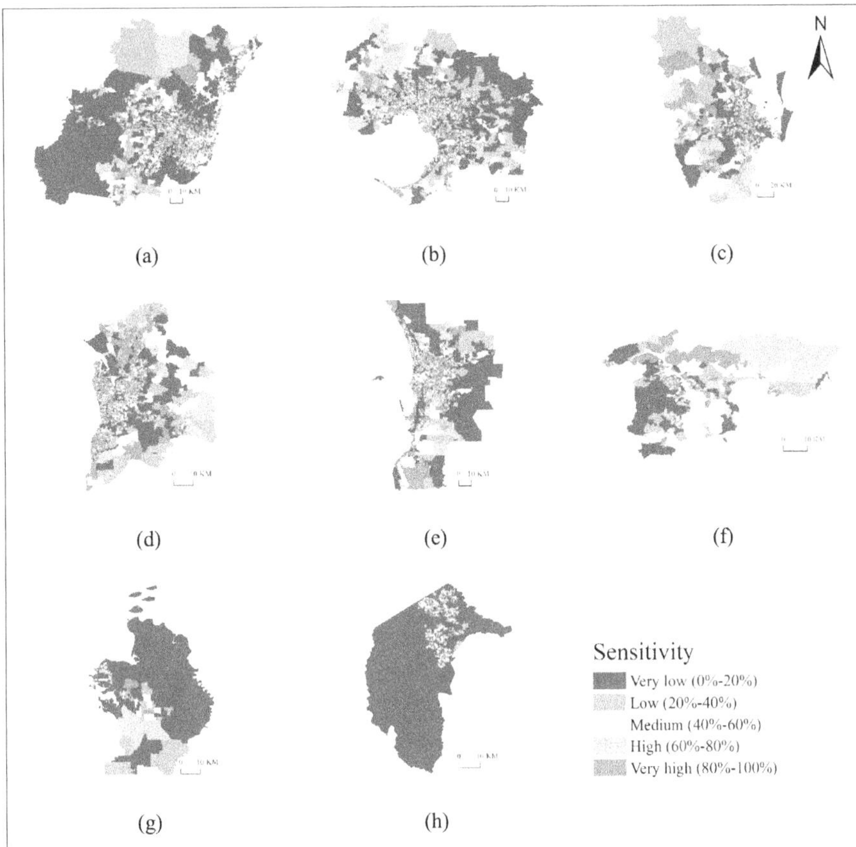

FIGURE 9.5
Spatial distribution of sensitivity in Australia's capital cities.

pattern shifts from a gradient of decreasing levels extending from urban to peri-urban and rural areas to a mixed distribution of various levels, especially within urban areas. GADL, GPER, and GHBA have a high proportion of SA1s with higher-level sensitivity (see Figure 9.6). In GADL, very high level areas have shifted from the northwest part to encompass all western parts, while the eastern part features high- and medium-level sensitivity. GPER does not present a strip of very high level sensitivity in the western part anymore; instead, higher sensitivity SA1s are located in the middle and along the western coastal areas. It is noteworthy that higher levels of sensitivity have noticeably increased in GHBA, distributed along the coastal line areas. GSYD, GMEL, and GBNE have a relatively lower rate of SA1s with higher-level sensitivity.

The distribution of higher sensitivity in GSYD extends to the peri-urban and rural areas, with urban areas showing more medium and lower sensitivity

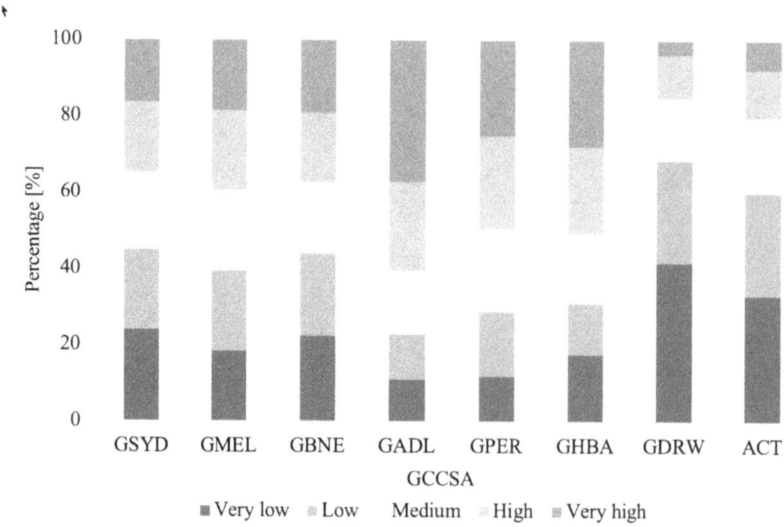

FIGURE 9.6
Proportions of sensitivity levels in Australia's capital cities.

SA1s. Compared to the exposure distribution, GMEL, similar to GHBA, has more higher-level sensitivity SA1s, mainly concentrated in the urban areas of Melbourne. Higher levels of sensitivity are still distributed in the west and east of GBNE. GDWR and ACT have the lowest rate of SA1s with higher-level sensitivity. The distribution of high-sensitivity SA1s is similar to the exposure distribution, with higher-level SA1s concentrated in the urban areas of Darwin and Canberra. Additionally, clear sensitivity boundaries still exist in the peri-urban areas.

9.3.1.3 Adaptive Capacity

The distribution of adaptive capacity generally exhibits features opposite to those of exposure distribution (see Figure 9.7). It shows a gradient of increasing adaptive capacity levels extending towards peri-urban and rural areas. SA1s with higher levels of exposure usually have lower levels of adaptive capacity. However, this differs in the central areas of capital cities. For instance, despite higher levels of exposure, the inner city and surrounding areas of Sydney, Brisbane, Adelaide, Perth, and Hobart possess very high level of adaptive capacity. It is worth noting that Parramatta in GSYD also has a very high level of adaptive capacity alongside higher-level exposure. Additionally, the southeast of Melbourne features SA1s with lower levels of adaptive capacity, while exposure is also at lower levels. SA1s with lower-level adaptive capacity are found in the coastal areas of the River Derwent and the middle area of GHBA. Figure 9.8 shows that GSYD, GMEL, GBNE,

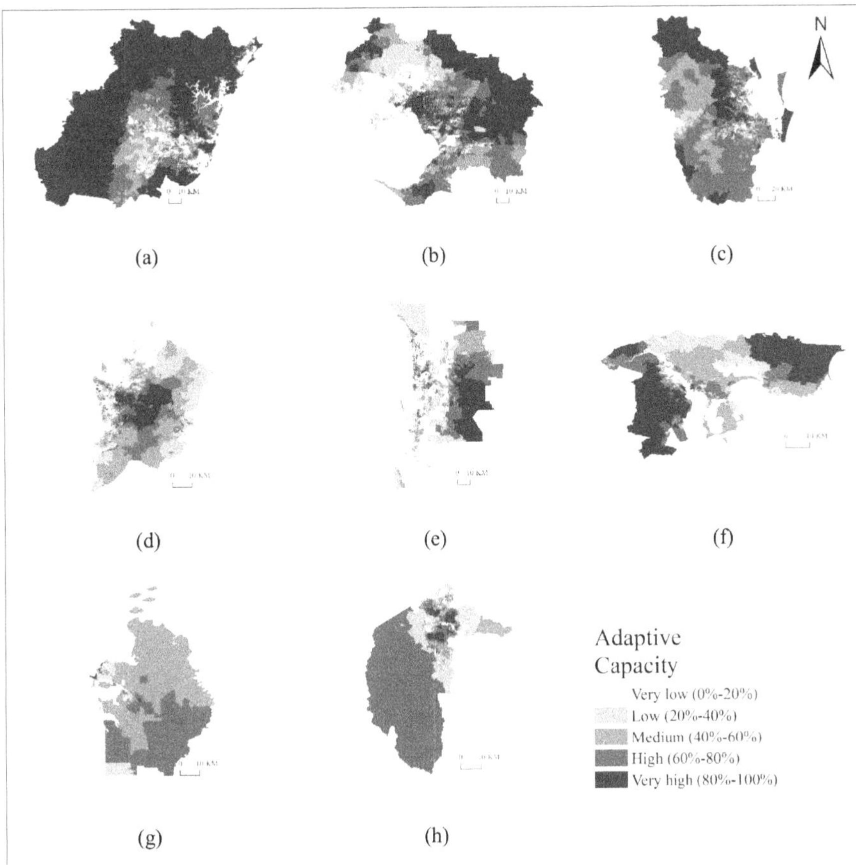

FIGURE 9.7
Spatial distribution of adaptive capacity in Australia's capital cities.

and ACT have high percentages of SA1s with higher-level adaptive capacity. Conversely, GADL, GPER, and GDRW have a significant portion of SA1s with lower-level adaptive capacity.

9.3.1.4 Comprehensive Heat Vulnerability Index

Figures 9.9 and 9.10 illustrate that the distribution of heat vulnerability is similar to the exposure distribution, with subtle changes in terms of level rates and spatial variations. Figure 9.10 reveals that GBNE, GADL, and GPER are the GCCSAs with high proportions of higher-level heat vulnerability. The rate of very high level vulnerability decreases in GBNE while it increases in GADL and GPER compared to their exposure rates, benefiting from the reduction in the number of SA1s classified as very high level. Although SA1s

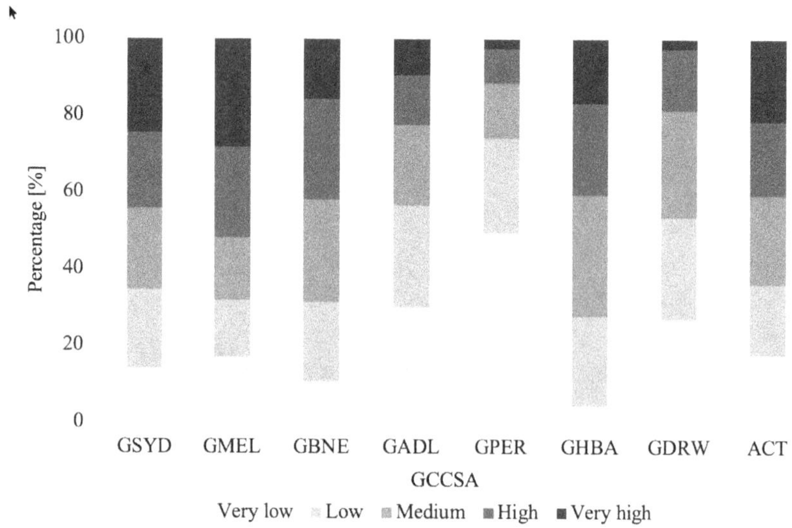

FIGURE 9.8
Proportions of adaptive levels in Australia's capital cities.

with higher-level heat vulnerability remain concentrated in the eastern and western parts of GBNE, there is a noticeable decrease in these SA1s within the central and southern areas, shifting towards lower levels. In GADL, the very high levels extend southward and are found in the western coastal areas, with an increase in lower-level SA1s within the central regions. GPER also showcases a strip of higher-level heat vulnerability, similar to its exposure pattern, extending along the city's western part but with an increased presence of very high level SA1s.

In terms of the rate of SA1s with higher-level heat vulnerability, GSYD, GDRW, and ACT rank in the second tier. Both GSYD and ACT see an increase in lower-level SA1s, primarily in the central areas of Sydney and Canberra, while the distribution of SA1s at higher levels remains only with some shifts between different levels. GMEL and GHBA have the lowest number of SA1s with higher levels of heat vulnerability, with GHBA lacking any SA1s at the very high level. Noteworthy changes in these two GCCSAs include the emergence of more higher-level SA1s in the northwest of GMEL and the dispersed distribution of very high level SA1s, including one in the southeast of Melbourne. High- and medium-level SA1s in GHBA are few and dispersed, mainly in the coastal areas of the River Derwent and central regions.

Figures 9.3, 9.5, 9.7, and 9.9 reveal distinct spatial distribution relationships between heat vulnerability and the factors of exposure, sensitivity, and adaptive capacity. Exposure typically decreases from urban centres to suburban areas, reflecting the urban heat island effect, where central areas with dense buildings and scarce green spaces experience higher temperature exposure.

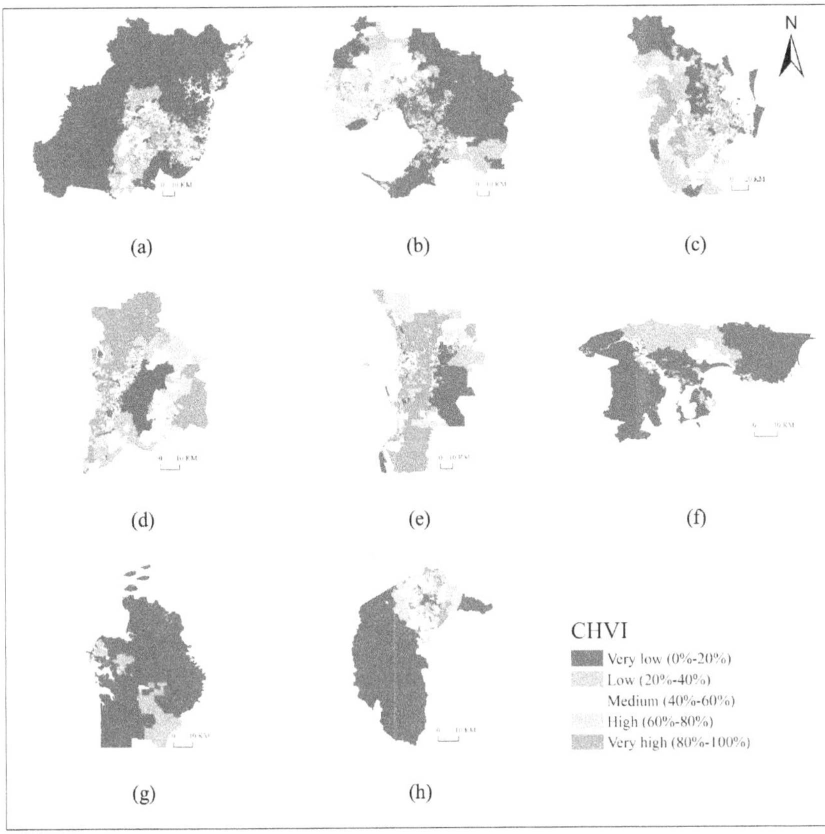

FIGURE 9.9
Spatial distribution of heat vulnerability in Australia's capital cities.

In GCCSAs like GBNE, GADL, and GPER, high exposure is observed not only in city centres but also in certain peripheral areas. Heat vulnerability closely follows this pattern, with high-exposure areas often exhibiting higher vulnerability, especially where adaptive capacity is low. Sensitivity, which is often similarly distributed to exposure, shows more pronounced variations in some areas. High-sensitivity regions are usually found in city cores with dense populations and complex built environments, such as in GADL, GPER, and GHBA, while GCCSAs like GSYD and GBNE see high sensitivity extending into surrounding areas. This spatial relationship indicates that high sensitivity in urban cores can exacerbate heat vulnerability, particularly in high-exposure zones. Adaptive capacity, on the other hand, usually increases from urban centres to suburbs, showing an inverse spatial pattern to exposure. However, some city centres, despite high exposure, demonstrate low overall heat vulnerability due to strong adaptive capacity, as seen in Sydney, Brisbane, Adelaide, Perth, and Hobart. Conversely, peripheral areas

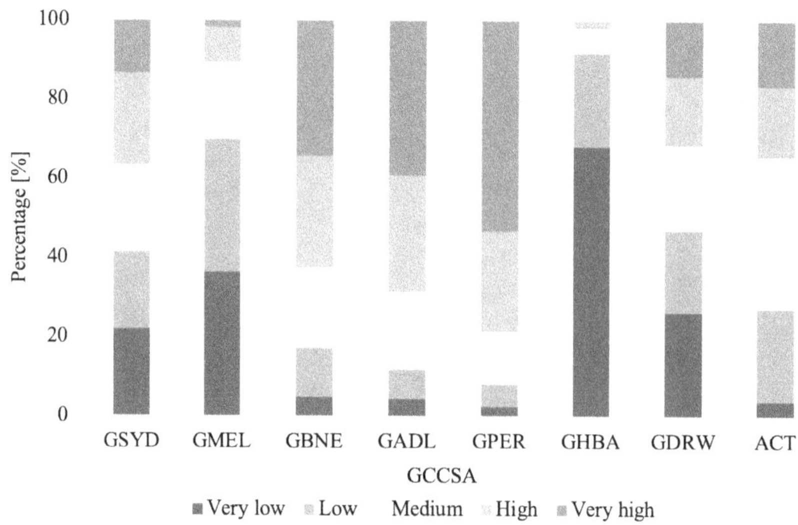

FIGURE 9.10
Proportions of heat vulnerability levels in Australia's capital cities.

in GADL, GPER, and GDRW, with low adaptive capacity, face higher heat vulnerability. These relationships underscore the need for urban planning and policy to prioritise enhancing adaptive capacity in high-exposure and high-sensitivity areas to effectively reduce overall heat vulnerability.

Figure 9.11 examines the SA1 proportions across eight GCCSAs at different levels, revealing that heat vulnerability is significantly shaped by the interaction of exposure, sensitivity, and adaptive capacity. Generally, higher exposure and sensitivity lead to a higher CHVI, while higher adaptive capacity mitigates this vulnerability. In the very low category, GHBA has the highest CHVI, driven by high exposure and moderate sensitivity, despite low adaptive capacity. Conversely, GSYD, GMEL, and GDRW show moderately high CHVI due to moderate exposure and sensitivity, with varying levels of adaptive capacity. Cities like GPER, which have higher adaptive capacity, demonstrate lower CHVI, underscoring the mitigating role of adaptive capacity. In the low category, GSYD, GMEL, and GDRW have the higher CHVI, primarily due to high exposure levels. Sensitivity peaks in ACT, resulting in a higher CHVI for ACT compared to GSYD and GDRW. In this category, exposure is the key determinant, with sensitivity contributing to vulnerability. In the medium category, GSYD and ACT have the highest CHVI due to high exposure. GHBA, which has the highest adaptive capacity, shows a lower CHVI, indicating that while exposure drives CHVI, adaptive capacity can mitigate it. In the high category, GADL and GPER have higher CHVI, driven by high exposure in GADL and high sensitivity in GPER, where low

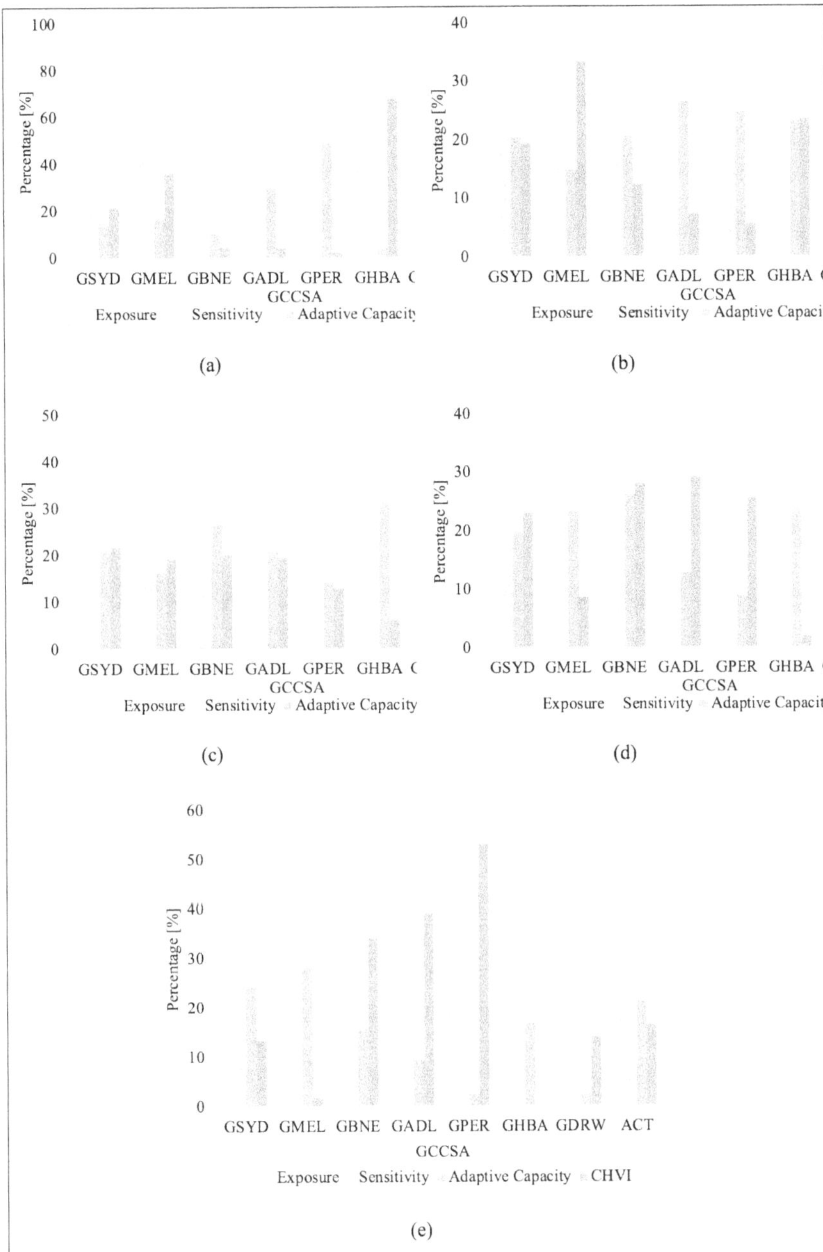

FIGURE 9.11
Level proportions of heat vulnerability and its components in Australia's capital cities. Note: (a) Very low, (b) low, (c) medium, (d) high, (e) very high.

adaptive capacity contributes to the higher CHVI. Finally, in the very high category, GPER exhibits the highest CHVI, driven by extremely high exposure and sensitivity, coupled with the lowest adaptive capacity. GBNE and GADL follow closely due to similar factors. In summary, exposure is a critical factor across all categories, with sensitivity also playing a significant role. Additionally, adaptive capacity can significantly mitigate heat vulnerability. Therefore, addressing exposure, sensitivity, and adaptive capacity is essential for managing heat vulnerability across these cities.

To validate the CHVI, two death parameters, namely, Death_P and Death_D, are calculated. They represent deaths based on the total population and total deaths, respectively, and are derived from the collected data on deaths caused by heat-related illnesses. Figure 9.12 illustrates that the PCCs between CHVI and Death_P is 0.92, and the one between CHVI and Death_D is 0.68 (both $p < 0.001$). These figures indicate strong positive relationships between CHVI and heat-related deaths. The significant correlation suggests that CHVI could be a robust tool for identifying populations at greater risk of heat vulnerability.

9.3.2 Inequity Analysis

The heatmap in Figure 9.13 displays the Gini coefficients across Australian capital cities and each of them. These coefficients assess the thermal inequity between heat vulnerability and various inequity indicators. In this case, high Gini values represent a significant inequity existing between heat vulnerability and inequity indicators. Table 9.5 lists the rates of CHVI values alongside these inequity indicators for each level of heat vulnerability. If inequity is observed, these rates can further explain it, revealing either positive or negative effects, especially if one indicator is disproportionately distributed across different levels of vulnerability. Ultimately, this analysis could reveal whether thermal inequity exists within specific groups and what kind of inequity it is. Based on UNDP Gini index categories, it is notable that Indigenous populations exhibit, on average, an extremely high inequity throughout Australia (0.82). Table 9.5 shows that the rates of Indigenous populations are much lower than CHVI's at medium and lower vulnerability levels while significantly higher at higher levels. This proves the existence of thermal inequity, with Indigenous people being particularly susceptible to a high risk of heat vulnerability. Additionally, a significant gap exists between the capital cities. For instance, GMEL has the highest Gini coefficient at 0.90, while GDRW has the lowest at 0.56, categorising it as the only capital city in the high-inequity level. GSYD, GADL, GPER, and ACT show similar Gini coefficients, fluctuating between 0.75 and 0.83. The Gini coefficients for GBNE and GHBA are both below 0.70, at 0.69 and 0.64, respectively.

English proficiency (0.69) and commuting methods (0.61) also exhibit extremely high inequity with heat vulnerability in Australia. However, the situations of the two indicators are entirely different. Table 9.5 shows that

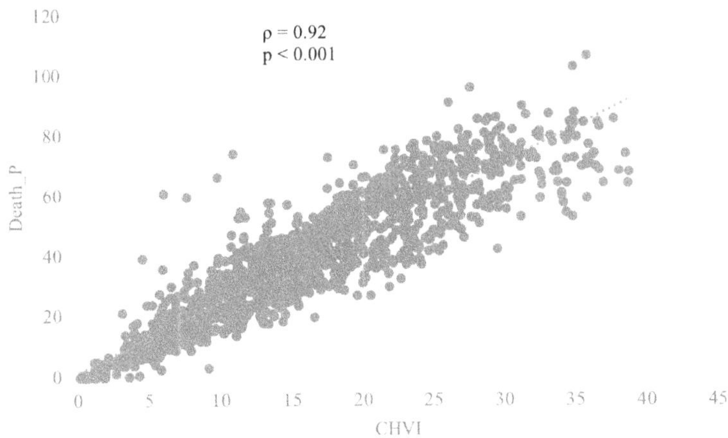

(a)

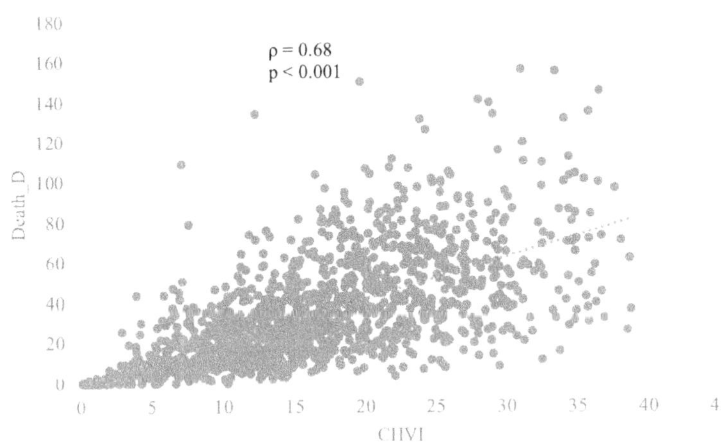

(b)

FIGURE 9.12
Validation results of PCC between heat vulnerabilities are (a) Death_P and (b) Death_D.

populations with limited English proficiency have lower rates than CHVI at the very low vulnerability level, but higher rates at medium and higher vulnerability levels, especially at the high level. These values indicate that these individuals are also suffering from thermal inequity. In contrast, people who commute by public transportation have much higher rates than

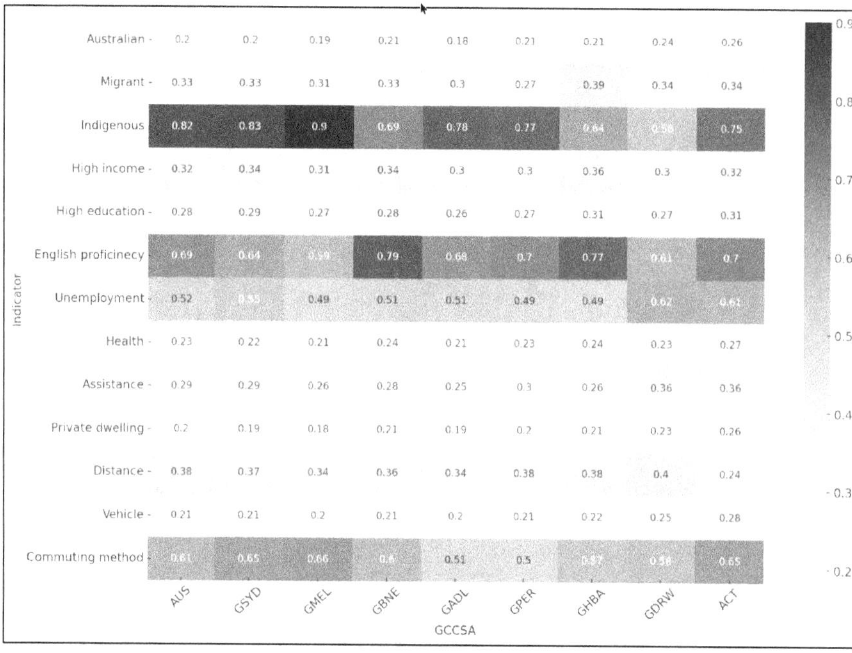

FIGURE 9.13
Results of inequity analysis with Gini coefficients for indicators in Australia's capital cities.

TABLE 9.5
Rates (%) of CHVI and Inequity Analysis Indicators in Each Heat Vulnerability Level

	Very Low	Low	Medium	High	Very High
CHVI	15.19	18.48	20.40	21.96	23.98
Australian	17.16	19.52	19.61	20.66	23.05
Migrant	15.83	18.92	19.59	21.50	24.16
Indigenous	8.09	11.78	16.68	24.62	38.82
High income	19.98	20.84	19.72	19.38	20.09
High education	20.00	20.64	19.48	19.44	20.45
Health	17.39	19.58	19.72	20.77	22.54
Private dwelling	17.07	19.49	19.61	20.78	23.05
Unemployment	14.86	16.90	19.01	22.71	26.52
Vehicle	17.22	19.74	19.48	20.40	23.16
Commuting method	22.86	18.44	19.17	19.32	20.20
English proficiency	13.14	18.86	20.54	23.45	24.01
Distance	19.04	19.80	19.08	20.78	21.30
Assistance	18.44	20.46	19.61	19.93	21.55

CHVI at the very low vulnerability level and lower rates at other levels. This reveals that this group is also experiencing thermal inequity, however, with a more positive effect. Moreover, both indicators experience notable fluctuations in the Gini coefficients across the capital cities. In terms of English proficiency, GBNE and GHBA have relatively high Gini coefficients of 0.79 and 0.77, respectively. GSYD (0.64), GMEL (0.59), and GDRW (0.61) exhibit similar, relatively low values, with GMEL being the only capital city remaining at the high-inequity level. The remaining capital cities have Gini coefficients around 0.70. Regarding the commuting methods, four capital cities—GSYD, GMEL, GBNE, and ACT—display values from 0.60 to 0.70, categorising them at the extremely high inequity level. The values for the other capitals are under 0.60, placing them in the high inequity level.

All Gini coefficients for unemployment fall in the high inequity level. According to Table 9.5, employment rates tend to be higher at the higher vulnerability levels and vary at other levels. This demonstrates a positive correlation between heat vulnerability and employment rates. Indicators such as migrant status, high income, and distance to the city centres are at the medium-inequity level across Australia. Only the Gini coefficients for migrant status in GPER (0.27) and distance to the city centre in GDRW (0.40) fall outside the medium-level range. Other Gini index analysis indicators are all at the low-inequity level, despite some capital cities' values falling outside this range. This detailed analysis underscores that while certain variables contribute to lower levels of inequality in heat vulnerability, others—especially those related to the Indigenous population, English proficiency, unemployment, and commuting habits—indicate higher levels of vulnerability.

9.4 Findings and Discussion

Current research on heat vulnerability assessment using indices is limited by incomplete indicator selection and lacks thermal environment inequity analysis (Cheng et al., 2021; Li et al., 2022; Wang et al., 2023). In an effort to bridge this research gap, this chapter has introduced the CHVI, which attempts to integrate comprehensive socio-demographic, health, and environmental indicators to evaluate heat vulnerability in Australia's capital cities at the SA1 level. Subsequently, it generates the most up-to-date and detailed heat vulnerability maps for these cities, offering valuable insights for heat vulnerability management and mitigation efforts. The CHVI is built based on the vulnerability assessment framework to climate change, which includes three components: Exposure, sensitivity, and adaptive capacity. The framework is proposed and promoted by the IPCC and United Nations Environment Programme (UNEP). The assessment is crucial for enhancing the understanding of heat vulnerability within the context of Australia, where there is a limited body of research focused on the country's vulnerability to increasingly high temperatures. Most studies concentrate on the major cities of Melbourne and Sydney (El-Zein & Tonmoy, 2015; Loughnan et al., 2012; Zhang et al., 2018), with other capital cities receiving limited research attention. Moreover, reaching the SA1 level, the finest level of analysis, is a rarity in these studies. Furthermore, Australia's capital cities, home to about 70% of the country's residents, have been experiencing more frequent and intense heatwaves and rising temperatures.

This study has further conducted an inequity analysis using Gini coefficients to compare heat vulnerability results with multidimensional indicators related to population groups, socioeconomic status, health conditions, living conditions, and transportation. The findings highlight the inequity present in Australia's capital cities and shed light on the promotion of thermal equity. This chapter's key finding is that the distribution of exposure in Australian capitals typically exhibits a gradient, decreasing from urban to rural areas, confirming the urban heat island effect pattern in the literature (Dong et al., 2020; Tomlinson et al., 2011). However, exceptions are noted in specific regions: The west and south of GBNE, the east of GADL, and the western peri-urban and rural areas of GPER, which display higher levels of exposure. This deviation is attributed to the different climatic conditions of these three capital cities. Brisbane has a humid subtropical climate, while the others feature a Mediterranean climate, both known for their hot summers. In contrast, GMEL and GHBA, characterised by climates with warm summers and higher latitudes, exhibit relatively low exposure levels, particularly GHBA. Another contributing factor is the lack of green vegetation cover, especially forests, in areas of higher exposure, which would offer cooling effects to moderate temperatures. For instance, the obvious middle boundary between higher-level and lower-level exposure is also the boundary of forests and artificial land uses.

Additionally, this chapter employs PCA to capture sensitivity to heat. For this, the indicators are divided into three PCs, comprising socioeconomic and health conditions, living environment and status, and urban layout and architectural context. Overall, higher-level sensitivity is primarily observed in the high-density SA1s, especially in the city centres, with more variations within regions. This distribution aligns with the findings of some heat vulnerability research, which presents a heterogeneous pattern across study units and a trend concentrating in city centres and specific areas (Hulley et al., 2019; Kwon et al., 2020; Wolf & McGregor, 2013). These studies all highlight that highly populated areas consistently experience high sensitivity and that housing, health, and disability conditions are key attributes linked to this heightened sensitivity. Urban areas, which are characterised by higher population densities, inherently exhibit greater diversity in socioeconomic and environmental conditions. This diversity is reflected in varied educational backgrounds, economic statuses, and characteristics of the built environment, influencing the distribution of sensitivity. High-density urban areas tend to aggregate individuals with various sensitivities, including those related to health, economic capacity, and social networks.

The distribution of adaptive capacity generally exhibits characteristics opposite to that of exposure, with a gradient of increasing adaptive capacity levels extending towards peri-urban and rural areas, except in city centres and specific areas. This chapter captures adaptive capacity from the perspective of the accessibility of natural and artificial cooling services and resources, as well as the relieving effects of medical infrastructure. The spatial pattern is attributed to the distribution of these services and resources, such as green spaces, public cooling spaces, and medical facilities (Bu et al., 2024; Zhang et al., 2023). The mentioned exceptions may be due to city centres being well-planned, with an abundance of cooling and medical services and infrastructures. This convenience contributes to a higher adaptive capacity of these areas.

The comprehensively developed CHVI in this study well integrates exposure, sensitivity, and adaptive capacity, based on a popular climate change vulnerability assessment framework from the IPCC (Parry, 2007). PCCs between heat vulnerability and heat-related deaths are being used to assess CHVI's ability to capture heat vulnerability. The results show that PCCs of 0.68 with Death_D and 0.92 with Death_P are significant ($p < 0.01$). These figures demonstrate the CHVI's capacity to measure heat vulnerability, offering a useful approach for related research and work. Heat vulnerability distribution closely aligns with exposure trends but exhibits more regional variations. Moreover, city centres and specific urban areas show lower levels of heat vulnerability despite higher exposure levels. The fundamental reason for heat vulnerability is the risk of heat exposure, which is influenced by sensitivity and can be mitigated by adaptive capacity (Parry, 2007).

Thus, the spatial pattern of heat vulnerability distribution in Australia's capital cities reflects these factors: The interaction between exposure, sensitivity, and adaptive capacity in shaping heat vulnerability. For instance, GBNE, GADL, and GPER have the highest heat vulnerability in the first tier. Compared to exposure, GBNE's high-level heat vulnerability tends to decrease due to reduced sensitivity and increased adaptive capacity. Unlike the other two capital cities, GADL's higher-level SA1s extend to the south and expand in the west, owing to higher sensitivity in these areas. In GPER, lower adaptive capacity contributes to an increase in higher-level SA1s in the west. This finding suggests that targeted strategies and policies aimed at mitigating sensitivity and enhancing adaptive capacity should be implemented based on the actual heat vulnerability in different areas (Bu et al., 2024; Hulley et al., 2019). Higher-level heat vulnerability is mainly concentrated in the SA1s of GSYD, GDRW, and ACT, indicating the need for more effort in planning and management in these areas during heat vulnerability mitigation and intervention. GMEL and GHBA have relatively low rates of higher-level heat vulnerability, suggesting that customised strategies and policies might be feasible for these areas.

The results of inequity analysis by using Gini coefficients reflect significant inequity of heat vulnerability in Australia's capital cities. SA1s with more Indigenous populations, populations with limited English-speaking proficiency, and populations who commute using public transportation are facing extremely high thermal inequity. SA1s with higher unemployment rates are facing high-level inequity when facing increasingly high temperatures. Only people commuting by public transport show a positive effect, with more individuals in this group located at the very low vulnerability level. Disparities are also evident among capital cities; for example, GMEL exhibits extremely high inequity, while GDRW has the lowest Gini coefficient, at 0.56, at the high level. Additionally, indicators like migrant status, high income, and distance to city centres are assessed at a medium-inequity level and the remaining indicators are all at the low-inequity level, suggesting a complex interplay of variables influencing heat vulnerability across Australia. This pattern of inequity might stem from a combination of socioeconomic and geographical factors (Mallen et al., 2019; Mitchell & Chakraborty, 2015).

Indigenous communities, often located in more remote or climatically harsh regions, face systemic disadvantages that exacerbate their vulnerability to heat. Populations with limited English proficiency are more vulnerable to heat hazards. This is because they may have difficulty accessing effective cooling infrastructure, such as air conditioning or public cooling centres. Additionally, due to language barriers, these groups might not receive or understand warnings and information about how to cope with high temperatures. This lack of information can lead to insufficient protective measures being taken during periods of extreme heat. In Australia, more people use public transportation in low heat vulnerability areas primarily due to the well-developed public transit infrastructure in urban centres, where comprehensive networks of buses, trains, and trams are efficiently managed to accommodate high population densities (Pitot et al., 2006; Yigitcanlar et al., 2007, 2008). This efficiency, combined with the challenges of driving such as parking and traffic, makes public transportation a more convenient and sustainable option for daily commuting in these densely populated areas. The variation among capitals suggests that urban design, density, and local climate adaptation strategies significantly influence heat vulnerability. Cities with better infrastructure, green spaces, and efficient public transport systems are likely to mitigate some effects of heat, whereas those without such amenities expose their populations to greater risk (Estoque et al., 2020). This analysis underscores the importance of targeted policies and interventions that address the specific vulnerabilities of diverse communities within Australia's urban and rural landscapes.

This study provides insights and guidance regarding indicator selection, index establishment, and validation of CHVI for heat vulnerability research. Initially, this study selects heat-related indicators based on a systematic review of heat vulnerability measurement and assessment. The selected indicators are the most frequently used ones in current studies (Kim et al., 2017;

Rinner et al., 2010; Wilson & Chakraborty, 2019), and their representativeness and reliability have already been demonstrated. This approach helps to enhance the robustness of research by avoiding subjective judgement and reducing over-reliance on local contexts. The indicator framework includes socio-demographic characteristics, health conditions, and environmental factors. These three aspects capture heat vulnerability from the perspectives of subjects' socioeconomic and demographic sensitivity, medical susceptibility due to poor health, and the adaptation and adjustment effects of built and natural environments.

The incorporation of direct and indirect heat-related illnesses in the health condition category is also unique to this study. This integration is significant for heat vulnerability assessment because illnesses such as heat stroke, cardiovascular diseases, and respiratory diseases have been proven to be highly related to populations' vulnerability to heat (Reid et al., 2012). However, these factors have often been overlooked during the indicator selection process in heat vulnerability assessment studies. In this research, the integration of these health factors is improved by including all heat-related disease metrics under the indicator of personal illness status. Considering each disease as a separate indicator could cause an overestimation of how much health issues contribute to heat vulnerability due to cumulative effects. Indeed, apart from representativeness and comprehensiveness, achieving a fine balance between these measures is another crucial factor during indicator selection.

Secondly, this chapter adopts a solid and widely recognised vulnerability evaluation framework of climate change issues, acknowledged and consolidated by international organisations such as the IPCC and UNEP, as well as by researchers to assess heat vulnerability (Parry, 2007; Wilson & Chakraborty, 2019; Wu et al., 2022). This framework suggests that a comprehensive assessment of the potential impacts of climate change must consider three components of vulnerability: Exposure, sensitivity, and adaptive capacity. It emphasises the evaluation of the risk of heat exposure and the contributing effects of sensitivity. Notably, adaptive capacity is an innovative component that highlights the mitigation contributions of both the objective environment, whether it is human-made or natural. This is the primary reason the chapter utilises this framework instead of the risk triangle framework, which includes exposure, hazard, and vulnerability but overlooks the adaptive capacity in mitigating heat vulnerability. The omission of an evaluation of adaptation and adjustment facts could lead to errors and uncertainties, particularly in the accurate assessment of heat vulnerability. For instance, without considering adaptive capacity, such as access to cooling centres, green spaces, and medical infrastructure, the assessment might overestimate the vulnerability of a population that has significant resources to cope with heat exposure. Additionally, adaptive capacity varies across different regions. Ignoring this variation can introduce uncertainties into the results, as areas with similar heat exposure and sensitivity might actually differ significantly in vulnerability due to these adaptive resources. This could result in the misallocation of resources or inadequate policy responses. The CHVI, constructed based on this vulnerability evaluation framework, can reduce these errors and uncertainties, as evidenced by the validation results using PCCs. Thirdly, the figure-supported superior measuring capability of CHVI represents another key highlight of this chapter. Literature reviews point out that the absence of validation components is quite common in heat vulnerability research (Cheng et al., 2021; Li et al., 2022). Statistics reveal that less than one-third of heat vulnerability studies incorporate essential validation sections (Li et al., 2022). Even among studies that undertake validation efforts, the chosen validation indicators often face issues of data unavailability and lack convincing validity (Estoque et al., 2020; Nayak et al., 2018; Wang et al., 2021). For instance, indicators such as all-cause hospitalisations, emergency calls, and deaths, often selected for validation, have not been definitively linked to heat vulnerability. Additionally, due to data unavailability, studies use validation data on a large scale or from nearby years as substitutes, which can affect the validity of the validation.

Moreover, the validation outcomes reported in heat vulnerability research are somewhat vague and are not convincing. For example, in the study by Wang et al. (2023), PCCs relating HVI values to deaths, death rates, and hospital admissions during summer are all below 0.1. Furthermore, in the study of Mallen et al. (2019), the R-square value from a bivariate regression between HVI values and total deaths is only 0.03. These figures suggest a lack of significant relationships between heat vulnerability and deaths, implying that the validation performance may not robustly support the research findings and conclusions. In contrast, this chapter employs metrics of direct and indirect heat-related diseases to validate the performance of CHVI (Conlon et al., 2020; Estoque et al., 2020; Reid et al., 2012). The high PCCs demonstrate significant correlations between CHVI scores and heat-related deaths, indicating an exceptional ability to measure and assess heat vulnerability. Consequently, the robustness and reliability of the CHVI proposed in this study are further strengthened. Last but not least, the standardisation and good performance of the processes for indicator selection, index creation, and validation underscore the CHVI's potential for high applicability in future research across different regions.

This study has significant implications for policymakers, urban planners, and the general public. By providing the most up-to-date and fine-scale fundamental datasets on exposure, sensitivity, adaptive capacity, and heat vulnerability, the study offers essential information for informed policymaking (Li et al., 2024b). The detailed insights into underlying factors and spatial patterns of heat vulnerability allow urban planners and policymakers to devise targeted and effective strategies and policies tailored to areas with high-heat vulnerability (Bu et al., 2024; Zhang et al., 2023). These strategies can include improving urban infrastructure, enhancing green spaces, and ensuring better access to cooling centres, particularly in areas identified as highly vulnerable. The inequity analysis further emphasises the need to address the vulnerabilities of specific groups, such as Indigenous populations, populations with limited English proficiency, and unemployed individuals. Policymakers can use this information to prioritise interventions and allocate resources to these vulnerable groups, ensuring that mitigation efforts are equitable and inclusive. This could involve community-based outreach programmes, language-specific heat risk communication, and targeted social support services to enhance adaptive capacity.

Additionally, the availability of heat vulnerability datasets to the public fosters greater awareness and understanding of heat risks in their living environments. Public access to this information empowers communities to advocate for necessary changes and adopt personal measures to mitigate heat exposure. Furthermore, the study highlights the importance of public engagement and education. Researchers advocate for more public events to raise awareness and improve understanding of the thermal environment and thermal equity. These events can take the form of workshops, training modules, and interactive educational platforms aimed at educating the public about the impacts of heat vulnerability and the importance of sustainable urban planning (Wang et al., 2023). By translating research findings into actionable strategies and engaging the community, this study contributes to building more resilient and equitable urban environments.

This study significantly aligns with and contributes to several United Nations SDGs (United Nations, 2023), particularly SDG 11 (Sustainable Cities and Communities), SDG 13 (Climate Action), and SDG 10 (Reduced Inequalities), by addressing and providing solutions to heat vulnerability and thermal inequity in Australia's capital cities (Li et al., 2022; Szagri et al., 2023). By introducing the CHVI and employing the Gini coefficient to analyse inequities in heat exposure, this research directly supports the creation of resilient and sustainable urban environments (SDG 11), emphasising the need for inclusive and equitable cities that are prepared for climate-related disasters. Furthermore, by focusing on the impacts of climate change on urban heat vulnerability, the study contributes to SDG 13 by highlighting the importance of climate mitigation and adaptation strategies. The examination of thermal inequity, particularly among vulnerable populations, aligns with SDG 10's objective of reducing inequalities within and among countries. Through its comprehensive approach to measuring heat vulnerability and addressing inequity, the research provides a foundation for informed policymaking and strategic interventions aimed at mitigating heat-related risks, promoting environmental justice, and ultimately contributing to the realisation of these interconnected SDGs (Zeng et al., 2022).

There are some limitations in this research. The majority of the heat vulnerability indicators are collected from the Australian census dataset, published every five years. The latest dataset available was published in 2021, and this chapter utilises it to measure the sensitivity component. Despite it being the most current dataset available to date, there might be a certain degree of time lag, potentially resulting in missed captures of recent changes in heat vulnerability. Therefore, updating the heat vulnerability assessment in line with the updates of census data is needed. Furthermore, investigating an innovative representative dataset that can precisely and efficiently depict heat vulnerability is another planned work for the future.

High spatial resolutions of LST images from Landsat 8 are one of the main reasons this chapter adopts them to identify heat exposure at the SA1 level. Their high availability and accuracy also contribute to their effectiveness and popularity in heat vulnerability research (Zhang et al., 2018, 2023). Nonetheless, the temporal resolution of Landsat LST images is 16 days, which may be considered coarse. This is the primary reason for using five years' worth of summer LST images in the study so as to enrich the sample volume and to avoid missing exposure characteristics in summers. The authors aim to seek high-temporal-resolution temperature data to delineate exposure, such as MODIS LST and air temperatures observed by atmospheric sites. These two types of data can provide daily temperature data, which is sufficient for heat vulnerability assessment (Rinner et al., 2010; Tomlinson et al., 2011). However, the spatial resolution of these datasets is unsatisfactory, with MODIS LST having a 1 km spatial resolution and air temperature data having an even lower spatial resolution based on sparsely distributed observation sites.

Additionally, combining surface and air temperatures can better capture people's perception of heat due to their direct exposure every day. Exploring a method to integrate the advantages of these available temperature datasets in heat vulnerability assessment could be promising. Although the finest spatial scale for heat-related deaths was at the SA2 level, the authors are still seeking potential validation datasets, such as hospitalisation records, emergency service calls, and heat-related morbidity data. The authors are also seeking collaboration with national and state health departments, which can provide finer datasets. Once the dataset is updated, the results will be further validated and consolidated. Authors' future research is expected to conduct historical heat vulnerability assessments in Australia's capital cities. It is essential to evaluate the evolution of heat vulnerability and investigate the underlying contributors to heat vulnerability. Furthermore, the CHVI will be applied in typical regions experiencing continuously high temperatures, more intense, and more frequent heat extreme events, such as Phoenix, Miami, and Los Angeles in America; Beijing, Shanghai, and Chongqing in China; and Athens, Rome, and Paris in Europe. It is beneficial for evaluating the applicability of CHVI and for the management and mitigation of heat vulnerability in local contexts. Finally, finer-scale heat vulnerability mapping is also on the authors' future agenda. Current heat vulnerability measurements are almost at the administrative level, which is useful for macro-scale management and intervention. Although studies have focused on the finest level, the results cannot be used to capture the subtle characteristics within the study unit.

9.5 Conclusion

With the overarching aim to address the existing gap for comprehensive and systematic heat vulnerability assessment and validation methods, this chapter introduced a CHVI and applied it to the Australian context. The index accommodates demographic, socioeconomic, health, and environmental indicators. The chapter also expanded thermal inequity analysis for Australia's capital cities. The study has unveiled a thorough understanding of heat vulnerability at the detailed SA1 level, along with new empirical insights and datasets to guide further research. The key findings of the study can be summarised as follows:

a. The heat vulnerability components feature different distribution characteristics in the Australian capital cities. Exposure generally represents a gradient of decreasing levels from urban to rural areas, with subtle variations inside regions. Sensitivity displays similar spatial patterns, however, with more pronounced regional differences. Adaptive capacity is inversely related to exposure, except in the city centres and some SA1s clusters.
b. CHVI demonstrates a promising capacity in measuring heat vulnerability, as indicated by PCCs of 0.68 and 0.92 with heat-related death data. The distribution of heat vulnerability closely aligns with exposure trends but exhibits more regional variations, and city centres feature lower levels of heat vulnerability.
c. Thermal inequity exists among SA1s, particularly influenced by factors such as Indigenous population, unemployment, English proficiency, and using public transportation for commuting. The situation in each capital city varies slightly in terms of these inequity factors.

This study offers methodological guidance for the assessment of heat vulnerability, particularly in selecting indicators and establishing indices. This approach is not only applicable to Australian capital cities but also holds significant potential for broader application in other regions globally. The standardised indicator selection and rigorous validation process ensure the reliability and accuracy of the CHVI, making it adaptable for various urban and climatic contexts. Future research could extend this methodology to different geographical regions, considering local environmental, sociodemographic, and health conditions, to create tailored heat vulnerability assessments. For instance, by adjusting the selected indicators to reflect the specific characteristics and data availability of different regions, the CHVI can be customised for various urban and climatic contexts. Additionally, incorporating local climatic data and urban infrastructure details will enhance the index's accuracy and relevance, ensuring effective heat vulnerability assessments tailored to unique regional conditions. This adaptability underscores the CHVI's utility in guiding targeted heat mitigation and adaptation strategies worldwide, promoting thermal equity and enhancing climate resilience across diverse urban landscapes.

The fine-scale heat vulnerability datasets generated by this study can improve both public and governmental understanding of heat vulnerability in the Australian context, contributing to various aspects such as education, awareness, disaster risk reduction, and policy formulation. Specifically, this study aids in increasing public awareness of heat vulnerability and thermal inequity, thereby facilitating the development of targeted and effective policies and strategies for mitigating heat vulnerability and adapting to the increasing effects of global warming. For high vulnerability zones, enhancing urban infrastructure with more cooling centres and green spaces, alongside comprehensive heat action plans and improved healthcare access, is essential. Medium vulnerability areas would benefit from community-based outreach, residential retrofitting incentives, and increased green infrastructure. Low vulnerability zones should focus on continuous monitoring, integrating heat mitigation in urban planning, and public awareness campaigns to maintain resilience. These targeted recommendations aim to effectively address the specific needs of each classified vulnerable zone, ensuring both efficacy and equity in mitigation efforts. Finally, this study contributes to the achievement of several SDGs, including Sustainable Cities and Communities, Climate Action, and Reduced Inequalities, which are crucial for creating sustainable and resilient cities and communities.

Acknowledgements

This chapter, with permission from the copyright holder, is a reproduced version of the following journal article: Li, F., Yigitcanlar, T., Nepal, M., Nguyen, K., Dur, F., & Li, W. (2024). Assessing heat vulnerability and multidimensional inequity: Lessons from indexing the performance of Australian capital cities. *Sustainable Cities and Society, 115*(1), 105875.

References

Adams, R. M., Evans, C., Wolkin, A., Thomas, T., & Peek, L. (2022). Social vulnerability and disasters: Development and evaluation of a CONVERGE training module for researchers and practitioners. *Disaster Prevention and Management: An International Journal, 31*(6), 13–29. https://doi.org/10.1108/DPM-04-2021-0131.

Agathangelidis, I., Cartalis, C., & Santamouris, M. (2019). Integrating urban form, function, and energy fluxes in a heat exposure indicator in view of intra-urban heat island assessment and climate change adaptation. *Climate*, *7*(6), 75. https://doi.org/10.3390/cli7060075.

Alonso, L., & Renard, F. (2020). A comparative study of the physiological and socio-economic vulnerabilities to heat waves of the population of the metropolis of Lyon (France) in a climate change context. *International Journal of Environmental Research and Public Health*, *17*(3), 1004. https://doi.org/10.3390/ijerph17031004.

Australian Bureau of Statistics (ABS). (2021). *Digital boundary files*. Australian Statistical Geography Standard (ASGS) Edition 3. Retrieved June 6, 2023, from https://www.abs.gov.au/statistics/standards/australian-statistical-geography-standard-asgs-edition-3/jul2021-jun2026/access-and-downloads/digital-boundary-files.

Australian Bureau of Statistics (ABS). (2022). *ABS Census 2021*. Retrieved June 6, 2023, from https://www.abs.gov.au/census/find-census-data/datapacks.

Australian Bureau of Statistics (ABS). (2023). *National, state and territory population*. Retrieved June 6, 2023, from https://www.abs.gov.au/statistics/people/population/national-state-and-territory-population/sep-2023.

Australian Museum. (2023). *Impacts of climate change*. Retrieved from https://australian.museum/learn/climate-change/climate-change-impacts.

Barron, L., Ruggieri, D., & Branas, C. (2018). Assessing vulnerability to heat: A geospatial analysis for the City of Philadelphia. *Urban Science*, *2*(2), 38. https://doi.org/10.3390/urbansci2020038.

Basu, R. (2009). High ambient temperature and mortality: A review of epidemiologic studies from 2001 to 2008. *Environmental Health*, *8*, 1–13. https://doi.org/10.1186/1476-069X-8-40.

Bu, S., Smith, K. L., Masoud, F., & Sheinbaum, A. (2024). Spatial distribution of heat vulnerability in Toronto, Canada. *Urban Climate*, *54*, 101838. https://doi.org/10.1016/j.uclim.2024.101838.

Burbidge, M., Smith Larsen, T., Feder, S., & Yan, S. (2022). Don't blame it on the sunshine! An exploration of the spatial distribution of heat injustice across districts in Antwerp, Belgium. *Local Environment*, *27*(2), 160–176. https://doi.org/10.1080/13549839.2021.2005007.

Buscail, C., Upegui, E., & Viel, J. F. (2012). Mapping heatwave health risk at the community level for public health action. *International Journal of Health Geographics*, *11*, 1–9. https://doi.org/10.1186/1476-072X-11-38.

Byrne, J., Ambrey, C., Portanger, C., Lo, A., Matthews, T., Baker, D., & Davison, A. (2016). Could urban greening mitigate suburban thermal inequity?: The role of residents' dispositions and household practices. *Environmental Research Letters*, *11*(9), 095014. https://doi.org/10.1088/1748-9326/11/9/095014.

Cai, Z., Tang, Y., Chen, K., & Han, G. (2019). Assessing the heat vulnerability of different local climate zones in the old areas of a Chinese megacity. *Sustainability*, *11*(7), 2032. https://doi.org/10.3390/su11072032.

Campbell, S., Remenyi, T. A., White, C. J., & Johnston, F. H. (2018). Heatwave and health impact research: A global review. *Health & Place*, *53*, 210–218. https://doi.org/10.1016/j.healthplace.2018.08.017.

Central Intelligence Agency. (2024). *The World Factbook: Urbanization*. Retrieved June 6, 2023, from https://www.cia.gov/the-world-factbook/field/urbanization/

Chakraborty, J. (2024). Heatwave frequency and disability status: Thermal inequities in the US South. *Disability and Health Journal*, 101665. https://doi.org/10.1016/j.dhjo.2024.101665.

Chen, B., Xie, M., Feng, Q., Li, Z., Chu, L., & Liu, Q. (2021). Heat risk of residents in different types of communities from urban heat-exposed areas. *Science of the Total Environment, 768*, 145052. https://doi.org/10.1016/j.scitotenv.2021.145052.

Cheng, W., Li, D., Liu, Z., & Brown, R. D. (2021). Approaches for identifying heat-vulnerable populations and locations: A systematic review. *Science of The Total Environment, 799*, 149417. https://doi.org/10.1016/j.scitotenv.2021.149417.

Christenson, M., Geiger, S. D., Phillips, J., Anderson, B., Losurdo, G., & Anderson, H. A. (2017). Heat vulnerability index mapping for Milwaukee and Wisconsin. *Journal of Public Health Management and Practice, 23*(4), 396–403. https://doi.org/10.1097/PHH.0000000000000352.

Chuang, W. C., & Gober, P. (2015). Predicting hospitalization for heat-related illness at the census-tract level: Accuracy of a generic heat vulnerability index in Phoenix, Arizona (USA). *Environmental Health Perspectives, 123*(6), 606–612. https://doi.org/10.1289/ehp.1307868.

Climate Council. (2014). *Heatwaves: Hotter, longer and more often*. Retrieved June 6, 2023, from https://www.climatecouncil.org.au/uploads/9901f6614a2cac7b2b888f55b4dff9cc.pdf.

Climate Council. (2024). *How hot will your neighbourhood be by 2050*? Retrieved June 6, 2023, from https://www.climatecouncil.org.au/resources/how-hot-will-your-neighbourhood-be-by-2050/.

Colucci, A. R., Vecellio, D. J., & Allen, M. J. (2023). Thermal (In) equity and incarceration: A necessary nexus for geographers. *Environment and Planning E: Nature and Space, 6*(1), 638–657. https://doi.org/10.1177/25148486211063488

Conlon, K. C., Mallen, E., Gronlund, C. J., Berrocal, V. J., Larsen, L., & O'Neill, M. S. (2020). Mapping human vulnerability to extreme heat: A critical assessment of heat vulnerability indices created using principal components analysis. *Environmental Health Perspectives, 128*(9), 097001. https://doi.org/10.1289/EHP4030.

Cresswell, K. (2023). A Florida urban heat risk index: Assessing weighting and aggregation approaches. *Urban Climate, 51*, 101646. https://doi.org/10.1016/j.uclim.2023.101646.

CSIRO & Bureau of Meteorology. (2022). *State of the climate 2022*. Commonwealth of Australia. https://www.csiro.au/en/research/environmental-impacts/climate-change/state-of-the-climate.

Dai, G., Li, R., & Ma, S. (2022). Research on the equity of health resource allocation in TCM hospitals in China based on the Gini coefficient and agglomeration degree: 2009–2018. *International Journal for Equity in Health, 21*(1), 145. https://doi.org/10.1186/s12939-022-01749-7

Degirmenci, K., Desouza, K. C., Fieuw, W., Watson, R. T., & Yigitcanlar, T. (2021). Understanding policy and technology responses in mitigating urban heat islands: A literature review and directions for future research. *Sustainable Cities and Society, 70*, 102873. https://doi.org/10.1016/j.scs.2021.102873

Deininger, K., & Squire, L. (1996). A new data set measuring income inequality. *The World Bank Economic Review, 10*(3), 565–591. https://doi.org/10.1093/wber/10.3.565.

Delso, J., Martín, B., & Ortega, E. (2018). A new procedure using network analysis and kernel density estimations to evaluate the effect of urban configurations on pedestrian mobility. The case study of Vitoria–Gasteiz. *Journal of Transport Geography, 67,* 61–72. https://doi.org/10.1016/j.jtrangeo.2018.02.001.

Dialesandro, J., Brazil, N., Wheeler, S., & Abunnasr, Y. (2021). Dimensions of thermal inequity: Neighborhood social demographics and urban heat in the Southwestern US. *International Journal of Environmental Research and Public Health, 18*(3), 941. https://doi.org/10.3390/ijerph18030941.

Dizdaroglu, D., Yigitcanlar, T., & Dawes, L. (2012). A micro-level indexing model for assessing urban ecosystem sustainability. *Smart and Sustainable Built Environment, 1*(3), 291–315. https://doi.org/10.1108/20466091211287155

Dong, J., Peng, J., He, X., Corcoran, J., Qiu, S., & Wang, X. (2020). Heatwave-induced human health risk assessment in megacities based on heat stress-social vulnerability-human exposure framework. *Landscape and Urban Planning, 203,* 103907. https://doi.org/10.1016/j.landurbplan.2020.103907

Elizabeth Loughnan, M., J. Tapper, N., Phan, T., & A. McInnes, J. (2014). Can a spatial index of heat-related vulnerability predict emergency service demand in Australian capital cities? *International Journal of Emergency Services, 3*(1), 6–33. https://doi.org/10.1108/IJES-10-2012-0044

Ellena, M., Breil, M., & Soriani, S. (2020). The heat-health nexus in the urban context: A systematic literature review exploring the socioeconomic vulnerabilities and built environment characteristics. *Urban Climate, 34,* 100676. https://doi.org/10.1016/j.uclim.2020.100676

Ellena, M., Melis, G., Zengarini, N., Di Gangi, E., Ricciardi, G., Mercogliano, P., & Costa, G. (2023). Micro-scale UHI risk assessment on the heat-health nexus within cities by looking at socioeconomic factors and built environment characteristics: The Turin case study (Italy). *Urban Climate, 49,* 101514. https://doi.org/10.1016/j.uclim.2023.101514

El-Zein, A., & Tonmoy, F. N. (2015). Assessment of vulnerability to climate change using a multi-criteria outranking approach with application to heat stress in Sydney. *Ecological Indicators, 48,* 207–217. https://doi.org/10.1016/j.ecolind.2014.08.012.

Estoque, R. C., Ooba, M., Seposo, X. T., Togawa, T., Hijioka, Y., Takahashi, K., & Nakamura, S. (2020). Heat health risk assessment in Philippine cities using remotely sensed data and social-ecological indicators. *Nature Communications, 11*(1), 1581. https://doi.org/10.1038/s41467-020-15218-8.

Forceville, G., Lemonsu, A., Goria, S., Stempfelet, M., Host, S., Alessandrini, J. M., Cordeau, E., & Pascal, M. (2024). Spatial contrasts and temporal changes in fine-scale heat exposure and vulnerability in the Paris region. *Science of the Total Environment, 906,* 167476. https://doi.org/10.1016/j.scitotenv.2023.167476

Gabbe, C. J., Chang, J. S., Kamson, M., & Seo, E. (2023). Reducing heat risk for people experiencing unsheltered homelessness. *International Journal of Disaster Risk Reduction, 96,* 103904. https://doi.org/10.1016/j.ijdrr.2023.103904

Guo, X., Huang, G., Jia, P., & Wu, J. (2019). Estimating fine-scale heat vulnerability in Beijing through two approaches: Spatial patterns, similarities, and divergence. *Remote Sensing, 11*(20), 2358. https://doi.org/10.3390/rs11202358

Harlan, S. L., Brazel, A. J., Prashad, L., Stefanov, W. L., & Larsen, L. (2006). Neighborhood microclimates and vulnerability to heat stress. *Social Science & Medicine, 63*(11), 2847–2863. https://doi.org/10.1016/j.socscimed.2006.07.030.

He, C., Ma, L., Zhou, L., Kan, H., Zhang, Y., Ma, W., & Chen, B. (2019). Exploring the mechanisms of heat wave vulnerability at the urban scale based on the application of big data and artificial societies. *Environment International, 127,* 573–583. https://doi.org/10.1016/j.envint.2019.01.057.

Heaton, M. J., Sain, S. R., Greasby, T. A., Uejio, C. K., Hayden, M. H., Monaghan, A. J., Boehnert, J., Sampson, K., Banerjee, D., Nepal, V., & Wilhelmi, O. V. (2014). Characterizing urban vulnerability to heat stress using a spatially varying coefficient model. *Spatial and Spatio-Temporal Epidemiology, 8,* 23–33. https://doi.org/10.1016/j.sste.2014.01.002.

Ho, J. Y., Shi, Y., Lau, K. K., Ng, E. Y., Ren, C., & Goggins, W. B. (2023). Urban heat island effect-related mortality under extreme heat and non-extreme heat scenarios: A 2010–2019 case study in Hong Kong. *Science of the Total Environment, 858,* 159791. https://doi.org/10.1016/j.scitotenv.2022.159791.

Hulley, G., Shivers, S., Wetherley, E., & Cudd, R. (2019). New ECOSTRESS and MODIS land surface temperature data reveal fine-scale heat vulnerability in cities: A case study for Los Angeles County, California. *Remote Sensing, 11*(18), 2136. https://doi.org/10.3390/rs11182136.

Intergovernmental Panel on Climate Change (IPCC). (2021). *Climate change 2021: The physical science basis – Summary for policymakers (AR6).* IPCC: Switzerland.

Johnson, D. P. (2022). Population-based disparities in US urban heat exposure from 2003 to 2018. *International Journal of Environmental Research and Public Health, 19*(19), 12314. https://doi.org/10.3390/ijerph191912314.

Johnson, D. P., Stanforth, A., Lulla, V., & Luber, G. (2012). Developing an applied extreme heat vulnerability index utilizing socioeconomic and environmental data. *Applied Geography, 35*(1–2), 23–31. https://doi.org/10.1016/j.apgeog.2012.04.006.

Kamal, N. I. A., Ash'aari, Z. H., Abdullah, A. M., Kusin, F. M., Mohamat Yusuff, F., Sharaai, A. H., Muharam, F. M., & Mohd Ariffin, N. A. (2022). Extreme heat vulnerability assessment in tropical region: A case study in Malaysia. *Climate and Development, 14*(5), 472–486. https://doi.org/10.1080/17565529.2021.1937030

Kamruzzaman, M., Deilami, K., & Yigitcanlar, T. (2018). Investigating the urban heat island effect of transit oriented development in Brisbane. *Journal of Transport Geography, 66,* 116–124. https://doi.org/10.1016/j.jtrangeo.2017.11.016.

Kershaw, S. E., & Millward, A. A. (2012). A spatio-temporal index for heat vulnerability assessment. *Environmental Monitoring and Assessment, 184,* 7329–7342. https://doi.org/10.1007/s10661-011-2502-z.

Kim, D. W., Deo, R. C., Lee, J. S., & Yeom, J. M. (2017). Mapping heatwave vulnerability in Korea. *Natural Hazards, 89,* 35–55. https://doi.org/10.1007/s11069-017-2951-y.

Kim, Y., Li, D., Xu, Y., Zhang, Y., Li, X., Muhlenforth, L., Xue, S., & Brown, R. (2023). Heat vulnerability and street-level outdoor thermal comfort in the city of Houston: Application of google street view image derived SVFs. *Urban Climate, 51,* 101617. https://doi.org/10.1016/j.uclim.2023.101617.

Kovats, R. S., & Hajat, S. (2008). Heat stress and public health: A critical review. *Annual Review of Public Health, 29,* 41–55. https://doi.org/10.1146/annurev.publhealth.29.020907.090843

KPMG. (2024). *38% of Australians face high risk from extreme heat.* Retrieved from https://kpmg.com/au/en/home/media/press-releases/2024/01/38–of-australians-face-high-risk-from-extreme-heat.html.

Kumari, M. R., & Kitchley, J. L. (2024). A framework to assess the contextual composite heat vulnerability index for a heritage city in India – A case study of Madurai. *Sustainable Cities and Society, 101,* 105119. https://doi.org/10.1016/j.scs.2023.105119.

Kwon, Y. J., Lee, D. K., & Kwon, Y. H. (2020). Is sensible heat flux useful for the assessment of thermal vulnerability in Seoul (Korea)? *International Journal of Environmental Research and Public Health, 17*(3), 963. https://doi.org/10.3390/ijerph17030963.

Li, F., Yan, Q., Bian, Z., Liu, B., & Wu, Z. (2020). A POI and LST adjusted NTL urban index for urban built-up area extraction. *Sensors, 20*(10), 2918. https://doi.org/10.3390/s20102918.

Li, F., Yan, Q., Zou, Y., & Liu, B. (2021). Extraction accuracy of urban built-up area based on nighttime light data and POI: A case study of Luojia 1-01 and NPP/VIIRS nighttime light images. *Geomatics and Information Science of Wuhan University, 46*(6), 825–835. https://doi.org/10.13203/j.whugis20190266

Li, F., Yigitcanlar, T., Li, W., Nepal, M., Nguyen, K., & Dur, F. (2024a). Understanding urban heat vulnerability: Scientometric analysis of five decades of research. *Urban Climate, 56*, 102035. https://doi.org/10.1016/j.uclim.2024.102035.

Li, F., Yigitcanlar, T., Nepal, M., Thanh, K., & Dur, F. (2022). Understanding urban heat vulnerability assessment methods: A PRISMA review. *Energies, 15*(19), 6998. https://doi.org/10.3390/en15196998.

Li, F., Yigitcanlar, T., Nepal, M., Nguyen, K., & Dur, F. (2024b). A novel urban heat vulnerability analysis: Integrating machine learning and remote sensing for enhanced insights. *Remote Sensing, 16*(16), 3032. https://doi.org/10.3390/rs16163032.

Loughnan, M., Nicholls, N., & Tapper, N. J. (2012). Mapping heat health risks in urban areas. *International Journal of Population Research, 2012*. https://doi.org/10.1155/2012/518687

Macnee, R. G., & Tokai, A. (2016). Heat wave vulnerability and exposure mapping for Osaka City, Japan. *Environment Systems and Decisions, 36*, 368–376. https://doi.org/10.1007/s10669-016-9607-4.

Mallen, E., Stone, B., & Lanza, K. (2019). A methodological assessment of extreme heat mortality modeling and heat vulnerability mapping in Dallas, Texas. *Urban Climate, 30*, 100528. https://doi.org/10.1016/j.uclim.2019.100528.

Mitchell, B. C., & Chakraborty, J. (2015). Landscapes of thermal inequity: Disproportionate exposure to urban heat in the three largest US cities. *Environmental Research Letters, 10*(11), 115005. https://doi.org/10.1088/1748-9326/10/11/115005

Mortoja, M., & Yigitcanlar, T. (2020). Local drivers of anthropogenic climate change: Quantifying the impact through a remote sensing approach in Brisbane. *Remote Sensing, 12*(14), 2270. https://doi.org/10.3390/rs12142270.

Mortoja, M., Yigitcanlar, T., & Mayere, S. (2020). What is the most suitable methodological approach to demarcate peri-urban areas? A systematic review of the literature. *Land Use Policy, 95*, 104601. https://doi.org/10.1016/j.landusepol.2020.104601.

Naserikia, M., Hart, M. A., Nazarian, N., Bechtel, B., Lipson, M., & Nice, K. A. (2023). Land surface and air temperature dynamics: The role of urban form and seasonality. *Science of The Total Environment, 905*, 167306. https://doi.org/10.1016/j.scitotenv.2023.167306.

Nayak, S. G., Shrestha, S., Kinney, P. L., Ross, Z., Sheridan, S. C., Pantea, C. I., Hsu, W. H., Muscatiello, N., & Hwang, S. A. (2018). Development of a heat vulnerability index for New York State. *Public Health, 161*, 127–137. https://doi.org/10.1016/j.puhe.2017.09.006.

Niu, Y., Li, Z., Gao, Y., Liu, X., Xu, L., Vardoulakis, S., Yue, Y., Wang, J., & Liu, Q. (2021). A systematic review of the development and validation of the heat vulnerability index: Major factors, methods, and spatial units. *Current Climate Change Reports, 7*(3), 87–97. https://doi.org/10.1007/s40641-021-00173-3

Oh, K. Y., Lee, M. J., & Jeon, S. W. (2017). Development of the Korean climate change vulnerability assessment tool (VESTAP)—Centered on health vulnerability to heat waves. *Sustainability, 9*(7), 1103. https://doi.org/10.3390/su9071103.

Paranunzio, R., Dwyer, E., Fitton, J. M., Alexander, P. J., & O'Dwyer, B. (2021). Assessing current and future heat risk in Dublin city, Ireland. *Urban Climate, 40*, 100983. https://doi.org/10.1016/j.uclim.2021.100983.

Parry, M. L. (Ed.). (2007). *Climate change 2007-impacts, adaptation and vulnerability: working group II contribution to the fourth assessment report of the IPCC* (Vol. 4). Cambridge University Press.

Pitot, M., Yigitcanlar, T., Sipe, N., & Evans, R. (2006). Land Use and Public Transport Accessibility Index (LUPTAI) tool: The development and pilot application of LUPTAI for the Gold Coast. In *Proceedings of the 29th Australian Transport Research Forum (pp. 1-18). Planning and Transport Research Centre (PATREC).*

Reid, C. E., Mann, J. K., Alfasso, R., English, P. B., King, G. C., Lincoln, R. A., Margolis, H. G., Rubado, D. J., Sabato, J. E., West, N. L., & Woods, B. (2012). Evaluation of a heat vulnerability index on abnormally hot days: An environmental public health tracking study. *Environmental Health Perspectives, 120*(5), 715–720. https://doi.org/10.1289/ehp.1103766.

Reid, C. E., O'neill, M. S., Gronlund, C. J., Brines, S. J., Brown, D. G., Diez-Roux, A. V., & Schwartz, J. (2009). Mapping community determinants of heat vulnerability. *Environmental Health Perspectives, 117*(11), 1730–1736. https://doi.org/10.1289/ehp.0900683.

Rinner, C., Patychuk, D., Bassil, K., Nasr, S., Gower, S., & Campbell, M. (2010). The role of maps in neighborhood-level heat vulnerability assessment for the city of Toronto. *Cartography and Geographic Information Science, 37*(1), 31–44. https://doi.org/10.1559/152304010790588089.

Saverino, K. C., Routman, E., Lookingbill, T. R., Eanes, A. M., Hoffman, J. S., & Bao, R. (2021). Thermal inequity in Richmond, VA: The effect of an unjust evolution of the urban landscape on urban heat islands. *Sustainability, 13*(3), 1511. https://doi.org/10.3390/su13031511.

Sharma, A., Woodruff, S., Budhathoki, M., Hamlet, A. F., Chen, F., & Fernando, H. J. S. (2018). Role of green roofs in reducing heat stress in vulnerable urban communities—A multidisciplinary approach. *Environmental Research Letters, 13*(9), 094011. https://doi.org/10.1088/1748-9326/aad93c.

Song, J., Huang, B., Kim, J. S., Wen, J., & Li, R. (2020). Fine-scale mapping of an evidence-based heat health risk index for high-density cities: Hong Kong as a case study. *Science of the Total Environment, 718*, 137226. https://doi.org/10.1016/j.scitotenv.2020.137226

Spencer, J., & Angeles, G. (2007). Kernel density estimation as a technique for assessing availability of health services in Nicaragua. *Health Services and Outcomes Research Methodology, 7*, 145–157. https://doi.org/10.1007/s10742-007-0022-7.

Steffen, W. (2015). *Quantifying the impact of climate change on extreme heat in Australia*. Climate Council of Australia Limited.

Steffen, W., Hughes, L., & Pearce, A. (2015). *The heat is on: Climate change, extreme heat, and bushfires in Western Australia*. Climate Council of Australia.

Steffen, W., Mallon, K., Kompas, T., Dean, A., & Rice, M. (2019). *Compound costs: How climate change is damaging Australia's economy*. Climate Council.

Sun, Y., Zhang, X., Zwiers, F. W., Song, L., Wan, H., Hu, T., Yin, H., & Ren, G. (2014). Rapid increase in the risk of extreme summer heat in Eastern China. *Nature Climate Change, 4*(12), 1082–1085. https://doi.org/10.1038/NCLIMATE2410.

Sun, Y., Zhu, S., Wang, D., Duan, J., Lu, H., Yin, H., Tan, C., Zhang, L., Zhao, M., Cai, W., & Wang, Y. (2024). Global supply chains amplify economic costs of future extreme heat risk. *Nature*. https://doi.org/10.1038/s41586-024-07147-z

Szagri, D., Nagy, B., & Szalay, Z. (2023). How can we predict where heatwaves will have an impact? A literature review on heat vulnerability indexes. *Urban Climate, 52*, 101711. https://doi.org/10.1016/j.uclim.2023.101711.

Thomas, T. H., & Butters, C. P. (2018, September). Thermal equity, public health and district cooling in hot climate cities. In *Proceedings of the Institution of Civil Engineers-Municipal Engineer* (Vol. 171, No. 3, pp. 163–172). Thomas Telford Ltd. https://doi.org/10.1680/jmuen.17.00020.

Tomlinson, C. J., Chapman, L., Thornes, J. E., & Baker, C. J. (2011). Including the urban heat island in spatial heat health risk assessment strategies: A case study for Birmingham, UK. *International Journal of Health Geographics, 10*, 1–14. https://doi.org/10.1186/1476-072X-10-42.

Tuczek, M., Degirmenci, K., Desouza, K. C., Watson, R. T., Yigitcanlar, T., & Breitner, M. H. (2022). Mitigating urban heat with optimal distribution of vegetation and buildings. *Urban Climate, 44*, 101208. https://doi.org/10.1016/j.uclim.2022.101208.

Uejio, C. K., Wilhelmi, O. V., Golden, J. S., Mills, D. M., Gulino, S. P., & Samenow, J. P. (2011). Intra-urban societal vulnerability to extreme heat: The role of heat exposure and the built environment, socioeconomics, and neighborhood stability. *Health & Place, 17*(2), 498–507. https://doi.org/10.1016/j.healthplace.2010.12.005.

United Nations. (2023). *Global indicator framework for the Sustainable Development Goals and targets of the 2030 Agenda for Sustainable Development*. https://unstats.un.org/sdgs/indicators/indicators-list/

Wang, Q., Zhang, Y., Ban, J., Zhu, H., Xu, H., & Li, T. (2021). The relationship between population heat vulnerability and urbanization levels: A county-level modeling study across China. *Environment International, 156*, 106742. https://doi.org/10.1016/j.envint.2021.106742.

Wang, S., Sun, Q. C., Huang, X., Tao, Y., Dong, C., Das, S., & Liu, Y. (2023). Health-integrated heat risk assessment in Australian cities. *Environmental Impact Assessment Review, 102*, 107176. https://doi.org/10.1016/j.eiar.2023.107176.

Wei, H., Chen, B., Huang, K., Gao, M., Fan, B., Zhang, T., Tu, Y., & Xu, B. (2024). Moderating AC usage can reduce thermal disparity between indoor and outdoor environments. *Environmental Science & Technology*. https://doi.org/10.1021/acs.est.4c00424.

Wilson, B., & Chakraborty, A. (2019). Mapping vulnerability to extreme heat events: Lessons from metropolitan Chicago. *Journal of Environmental Planning and Management, 62*(6), 1065–1088. https://doi.org/10.1080/09640568.2018.1462475.

Wold, S., Esbensen, K., & Geladi, P. (1987). Principal component analysis. *Chemometrics and Intelligent Laboratory Systems, 2*(1–3), 37–52. https://doi.org/10.1016/0169-7439(87)80084-9.

Wolf, T., & McGregor, G. (2013). The development of a heat wave vulnerability index for London, United Kingdom. *Weather and Climate Extremes, 1*, 59–68. https://doi.org/10.1016/j.wace.2013.07.004.

Wolf, T., McGregor, G., & Analitis, A. (2014). Performance assessment of a heat wave vulnerability index for greater London, United Kingdom. *Weather, Climate, and Society, 6*(1), 32–46. https://doi.org/10.1175/WCAS-D-13-00014.1.

Wu, X., Liu, Q., Huang, C., & Li, H. (2022). Mapping heat-health vulnerability based on remote sensing: A case study in Karachi. *Remote Sensing, 14*(7), 1590. https://doi.org/10.3390/rs14071590.

Wüstemann, H., Kalisch, D., & Kolbe, J. (2017). Access to urban green space and environmental inequalities in Germany. *Landscape and Urban Planning, 164*, 124–131. https://doi.org/10.1016/j.landurbplan.2017.04.002.

Yang, J., Huang, X., & Liu, X. (2014). An analysis of education inequality in China. *International Journal of Educational Development, 37*, 2–10. https://doi.org/10.1016/j.ijedudev.2014.03.002.

Yigitcanlar, T., Dodson, J., Gleeson, B., & Sipe, N. (2007). Travel self-containment in master planned estates: Analysis of recent Australian trends. *Urban Policy and Research, 25*(1), 129–149. https://doi.org/10.1080/08111140701255823.

Yigitcanlar, T., Fabian, L., & Coiacetto, E. (2008). Challenges to urban transport sustainability and smart transport in a tourist city: The Gold Coast, Australia. *Open Transportation Journal, 2*, 29–46. http://dx.doi.org/10.2174/1874447800802010029.

Yu, Q., Yue, D., Wang, J., Zhang, Q., Li, Y., Yu, Y., Chen, J., & Li, N. (2017). The optimization of urban ecological infrastructure network based on the changes of county landscape patterns: A typical case study of ecological fragile zone located at Deng Kou (Inner Mongolia). *Journal of Cleaner Production, 163*, S54–S67. https://doi.org/10.1016/j.jclepro.2016.05.014.

Zeng, P., Sun, F., Shi, D., Liu, Y., Zhang, R., Tian, T., & Che, Y. (2022). Integrating anthropogenic heat emissions and cooling accessibility to explore environmental justice in heat-related health risks in Shanghai, China. *Landscape and Urban Planning, 226*, 104490. https://doi.org/10.1016/j.landurbplan.2022.104490.

Zha, F., Lu, L., Wang, R., Zhang, S., Cao, S., Baqa, M. F., Li, Q., & Chen, F. (2024). Understanding fine-scale heat health risks and the role of green infrastructure based on remote sensing and socioeconomic data in the megacity of Beijing, China. *Ecological Indicators, 160*, 111847. https://doi.org/10.1016/j.ecolind.2024.111847.

Zhang, R., Yang, J., Ma, X., Xiao, X., & Xia, J. C. (2023). Optimal allocation of local climate zones based on heat vulnerability perspective. *Sustainable Cities and Society, 99*, 104981. https://doi.org/10.1016/j.scs.2023.104981.

Zhang, W., McManus, P., & Duncan, E. (2018). A raster-based subdividing indicator to map urban heat vulnerability: A case study in Sydney, Australia. *International Journal of Environmental Research and Public Health, 15*(11), 2516. https://doi.org/10.3390/ijerph15112516.

Zhang, W., Zheng, C., & Chen, F. (2019). Mapping heat-related health risks of elderly citizens in mountainous area: A case study of Chongqing, China. *Science of the Total Environment, 663*, 852–866. https://doi.org/10.1016/j.scitotenv.2019.01.240.

10

AI-Based Mapping of Urban Vulnerability

10.1 Introduction and Background

The growing concern over heat vulnerability has intensified due to rapid urbanisation and climate change, exacerbating the urban heat island (UHI) effect (Bu et al., 2024; Chuang & Gober, 2015; Forceville et al., 2024; Reid et al., 2009). Cities worldwide face escalating heat-related risks, particularly in Europe, North America, Asia, and Oceania, where heat extremes have led to severe consequences. In Europe, the 2003 and 2022 heatwaves caused tens of thousands of deaths, while North America has faced increasingly severe heat events (Ballester et al., 2023; Macintyre et al., 2018; Mallen et al., 2019; Reid et al., 2009). South and Southeast Asia, including India and Pakistan, have endured deadly heat extremes with significant human and economic losses (Bakhsh et al., 2018; Kitchley, 2024). Oceania, particularly Australia, has experienced record-breaking heat and bushfires, straining urban centres and vulnerable populations (Adnan et al., 2022; Wang et al., 2023). Research shows urban populations, especially vulnerable groups like the elderly, children, and low-income residents, are disproportionately affected, exacerbating social equity and public health challenges (Aubrecht & Özceylan, 2013; Cheng et al., 2021).

Australian capital cities, characterised by high population densities and significant urban development, are particularly susceptible to heat-related risks (Elizabeth Loughnan et al., 2014; Kamruzzaman et al., 2018; Wang et al., 2023). Between 2001 and 2018, heatwaves in Australia caused about 350 deaths, with the elderly being the most affected (Coates et al., 2022). According to the Department of Human Services (2009), the 2009 heatwave in Victoria led to a 62% increase in all-cause mortality and a 12% increase in overall emergency department presentations. Furthermore, the economic impact of heatwaves is substantial. For instance, heatwave events impose an annual financial burden of approximately $87 million on the state of Victoria (Victorian Government Department of Environment, Land, Water and Planning, 2019). This substantial cost, which represents 0.025% of the state's Gross State Product (GSP), arises from increased healthcare expenditures, lost productivity, and infrastructure damage (Economics, 2018). Addressing heat vulnerability in these urban areas is crucial for enhancing public health, promoting social equity, and improving urban resilience against climate change impacts.

DOI: 10.1201/9781003737049-13

Heat vulnerability assessments have gained significant global attention due to the rising impacts of elevated temperatures, particularly frequent and intense heat events, which threaten human health and increase heat-related morbidity and mortality (Cheng et al., 2021; Degirmenci et al., 2021). Various methodologies have been developed to identify vulnerable populations and areas needing urgent intervention, typically involving three processes: Indicator selection, modelling, and validation (Li et al., 2022). Indicator selection, a critical foundation for modelling and validation, is guided by the population vulnerability framework and the risk triangle framework (Crichton, 1999; Parry, 2007). The former emphasises exposure, sensitivity, and adaptive capacity, while the latter focuses on hazard, exposure, and vulnerability. Selected indicators often include demographic and socioeconomic factors (e.g., age, income, social isolation, population density) and health-related factors (e.g., chronic illnesses, medical infrastructure) (Depietri et al., 2013; Knowlton et al., 2009; Mallen et al., 2019). Reid et al. (2009), for instance, identified age, social isolation, and health conditions as key indicators in US census tracts. Environmental factors, both natural (e.g., land surface temperature (LST), vegetation cover) and built (e.g., cooling space access, building density (BD)), are also crucial (Song et al., 2020; Hulley et al., 2019). Estoque et al. (2019) highlighted LST and vegetation cover in assessing vulnerability in Philippine cities. However, indicator selection varies regionally due to contextual and data constraints, limiting cross-study comparisons (Li et al., 2022).

The modelling stage of heat vulnerability assessment focuses on developing composite indices, such as the heat vulnerability index (HVI), which integrate various indicators to measure vulnerability across populations or regions. Reid et al. (2009) developed an HVI to identify vulnerable communities and target mitigation efforts. Since then, many HVIs have been introduced globally, including those by Elizabeth Loughnan et al. (2014) and Li et al. (2024c) for Australian cities, Chuang and Gober (2015) and Christenson et al. (2017) for US cities like Phoenix and Milwaukee, and Kitchley (2024) for Madurai, India. A key issue is indicator weighting, as indicators contribute unequally to vulnerability. Methods can be explicit, using equal weights or expert judgement (Guo et al., 2019; Oh et al., 2017), or statistical, employing techniques like principal component analysis (PCA) or the analytical hierarchy process (AHP) (Estoque et al., 2020; Tate, 2012). Equal weighting assumes all indicators are equally important, while PCA groups indicators statistically. Liu et al. (2020) compared these methods in Hangzhou, finding equal weighting better correlated with heat-related mortality. The choice of method significantly affects outputs and depends on the study's context.

Validation is the final and often underdeveloped stage of heat vulnerability assessment, testing model accuracy by comparing outputs to real-world health outcomes, such as heat-related morbidity and mortality (Conlon et al., 2020; Hu et al., 2017; McGeehin & Mirabelli, 2001). Regression methods, including linear and Poisson regression, are commonly used.

For instance, Maier et al. (2020) utilised multivariate Poisson regression to assess their HVI model against heat-related deaths in Georgia, while Reid et al. (2009) demonstrated significant correlations between HVI scores and hospital admissions across five US states. Correlation coefficients like Pearson's or Spearman's are also frequently applied; Zhang et al. (2019) validated Chongqing's HVI using high Pearson correlation coefficients, and Elizabeth Loughnan et al. (2014) used Spearman correlations to validate Australian city-level vulnerability maps. However, validation results often show low agreement between assessed and observed vulnerability (Conlon et al., 2020; Wang et al., 2023; Weber et al., 2015), highlighting the need for more robust methods.

Despite advancements in heat vulnerability research, current studies face several significant limitations. The main issue is the inconsistent use and interpretation of indicators across studies, complicating result comparison and universal application (Li et al., 2022). For instance, the research of Cai et al. (2019) uses only 3 indicators to assess heat vulnerability, while El-Zein and Tonmoy (2015) incorporate 22 indicators. This significant disparity in the number of indicators illustrates the lack of standardisation, leading to varied assessments of vulnerability. Additionally, Cai et al. (2019) interpret building age as sensitivity and housing price as adaptive capacity, based on the assumption that housing conditions and prices reflect the socioeconomic status of residents. However, this interpretation is weakly supported by existing studies, as it overlooks critical factors like health diversity and demographics, which play a key role in determining vulnerability (Christenson et al., 2017; Elizabeth Loughnan et al., 2014). These variations in indicator selection and interpretation highlight the subjective nature of current heat vulnerability models (HVMs) (Cheng et al., 2021; Li et al., 2022), making it difficult to generalise findings across different regions and contexts.

Additionally, both explicit and statistical weighting methods used to construct these models can introduce biases (Cheng et al., 2021). Explicit weighting, such as equal weights and expert judgement (Guo et al., 2019; Oh et al., 2017), can introduce subjective biases due to human decision-making. This subjectivity is based on experts' opinions or predetermined assumptions about the relative importance of each indicator, which may not accurately reflect their real-world impact. Statistical weighting methods are data-driven but still may fail to fully capture the intricate, nonlinear relationships between indicators and heat vulnerability. Validation is a frequently overlooked process, as highlighted by Li et al. (2022). Statistics indicate that two-thirds of current heat vulnerability studies lack validation of their vulnerability results. Among the remaining one-third, many studies show a lack of rigour and thoroughness in the validation of their HVMs, often reporting low correlations between the models' assessments and actual heat vulnerability outcomes (Conlon et al., 2020; Wang et al., 2023; Weber et al., 2015). This underscores the necessity for more rigorous validation processes (Hulley et al., 2019; Kitchley, 2024; Weber et al., 2015).

Geographical and demographic variability further complicates the application of HVMs. The effectiveness of these models can vary significantly across different regions and populations due to differences in climate, urban infrastructure, and social factors (Oh et al., 2017; Wu et al., 2022). As a result, models developed for one area may not be directly applicable to another. Data availability and accuracy also pose substantial challenges. Reliable data on critical indicators are scarce, particularly in developing regions, which can undermine the accuracy of HVMs (Cheng et al., 2021; Macintyre et al., 2018). While HVMs offer valuable insights for urban planning and public health interventions, their integration into final policymaking is limited because the theoretical potential often faces practical obstacles, such as the complexity of regional differences, data scarcity, and model errors (Cheng et al., 2021). There is a pressing need to bridge the gap between model findings and actionable strategies for heat mitigation and adaptation to ensure that the insights from these studies are effectively translated into real-world applications.

This chapter addresses limitations in traditional heat vulnerability studies by integrating machine learning algorithms with multi-source data to enhance model accuracy, applicability, and reliability. Machine learning efficiently processes large, complex datasets and identifies hidden patterns traditional models overlook (Li et al., 2023). For example, in traffic management, it predicts flow and optimises public transport routes, aiding urban planners in managing dynamic systems (Almukhalfi et al., 2024; Modi et al., 2022). Similarly, in UHI studies, machine learning enhances temperature prediction and analysis (Liu et al., 2024; Tanoori et al., 2024) and is widely used to predict air pollutants like $PM_{2.5}$ and NO_2, enabling targeted interventions (Kim et al., 2021; Meng et al., 2018; Tella et al., 2021). By providing accurate, data-driven insights, it supports evidence-based, regionally tailored strategies for urban challenges. Indicator selection inconsistency is mitigated through a systematic review, categorising them into demographic, socioeconomic, health, and environmental factors (Li et al., 2022). Bias in weighting is reduced by deriving relationships directly from data. Advanced algorithms like random forest (RF) and gradient boosting (GB) capture nonlinear relationships without relying on subjective expert judgement (Chen et al., 2021b; Li et al., 2023; Yang et al., 2022).

This study addresses validation inadequacies in traditional models by implementing rigorous training, cross-validation, and evaluation metrics to ensure generalisability across scenarios. To account for geographical and demographic variability, the models incorporate diverse indicators and are applied across urban contexts in Australian capital cities. High-quality remote sensing, detailed census data, and extensive points of interests (POIs) enhance spatial and temporal resolution, enabling precise identification of vulnerable areas often overlooked by traditional HVMs due to coarser data (Wang et al., 2023; Wu et al., 2022). These datasets, with their wide coverage and accessibility, are adaptable for large-scale applications. This study further improves traditional HVMs by refining weight allocation and validation processes, addressing issues like inconsistent indicators, weighting biases, and inadequate validation. By validating with real-world data, such as heat-related morbidity and mortality, the results align closely with actual vulnerability patterns (Niu et al., 2021). These enhancements yield robust, fine-scale assessments that provide actionable insights for urban planning and public health, bridging the gap between model outputs and practical applications.

To date, there is a lack of machine learning applications in assessing heat vulnerability. Therefore, the primary objective of this study is to fill this gap by evaluating the effectiveness of various machine learning algorithms in assessing heat vulnerability across Australian capital cities using heat-related indicators. By doing so, the research aims to identify key indicators of heat vulnerability and highlight regional disparities. The findings are expected to inform urban planners and policymakers in developing targeted mitigation and adaptation strategies, ultimately enhancing urban sustainability and climate resilience. The study's significance lies in its potential to advance the application of machine learning in environmental research and provide actionable insights for improving public health and social equity in urban areas. This ensures more efficient and equitable resource allocation in urban environments.

This chapter is structured as follows: Section 10.2 outlines the study area, methods for model development, evaluation, and heat vulnerability analysis. Section 10.3 compares machine learning performance, analyses heat vulnerability across cities, and conducts a sensitivity analysis of indicators. Section 10.4 discusses findings in the context of urban planning and policy interventions, while the conclusion summarises key insights and implications for mitigating heat vulnerability and promoting thermal equity.

10.2 Research Design

10.2.1 Study Area

This study focuses on Australia's eight major metropolitan regions: Greater Sydney (GSYD), Greater Melbourne (GMEL), Greater Brisbane (GBNE), Greater Perth (GPER), Greater Adelaide (GADL), Greater Hobart (GHBA), Greater Darwin (GDRW), and the Australian Capital Territory (ACT) (Figure 10.1). These regions, which are key economic, cultural, and population centres, face increasing heat vulnerability due to urban expansion and climate change (Australian Bureau of Meteorology, 2021). The Australian Census 2021 shows that the capital cities account for about 66.95% of the total population in all 39,862 Statistical Area Level 1 (SA1s) (Australian Bureau of Statistics (ABS), 2022). The rapid urbanisation in these areas has resulted in extensive hard surfaces, reducing green spaces and water bodies, thereby

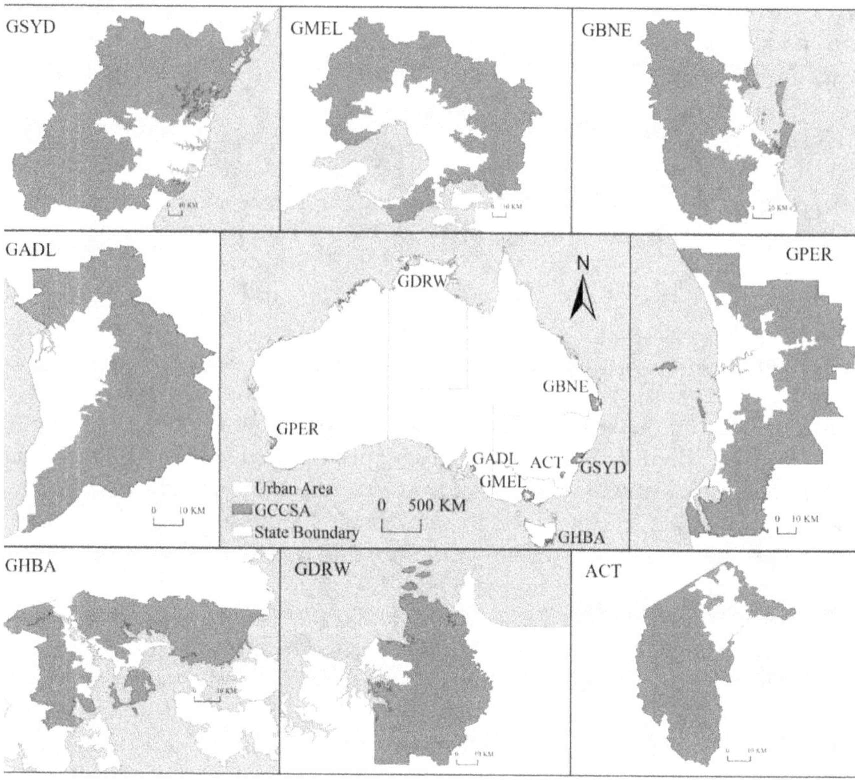

FIGURE 10.1
Study areas: Australian Capital cities.

exacerbating the UHI effect (Elizabeth Loughnan et al., 2014; Harmay et al., 2021; Iping et al., 2019). Cities like Sydney and Melbourne experience higher temperatures in their central areas compared to suburban regions (Elizabeth Loughnan et al., 2014; Zhang et al., 2018).

Climate change has intensified this issue, with more frequent and severe heatwaves posing serious health risks to residents, especially vulnerable groups such as the elderly, children, and low-income populations (Chambers, 2020; Green et al., 2016). Researching these metropolitan areas' heat vulnerability can help develop effective public health strategies, create fairer urban planning policies, and provide data-driven insights for informed decision-making in urban planning, public health, and environmental protection. Australia's metropolitan areas offer rich scenarios for studying heat vulnerability, improving residents' quality of life and health, and providing valuable insights for global regions facing climate change challenges.

10.2.2 Methodology

A comprehensive framework comprising four major steps—indicator selection, data collection and processing, machine learning selection and development, and heat vulnerability analysis—has been developed to systematically assess heat vulnerability (see Figure 10.2). The indicator selection step involves identifying and choosing relevant heat-related indicators based on established population vulnerability and heat risk frameworks. These indicators encompass demographic and socioeconomic characteristics such as age, economic status, social isolation, education level, and population density; health conditions like personal illness status, availability of medical resources, and disability status; and environmental factors, including LST, vegetation cover (normalised difference vegetation index, NDVI), availability of cooling spaces, land cover/use, and BD (normalised difference built-up index, NDBI). Labelled data, which include heat-related morbidity and mortality, are crucial in this context as they provide a concrete outcome measure against which the predictive machine learning models can be trained and validated. This labelled data distinguishes between direct heat-related illnesses and indirect heat-related conditions, allowing for a more nuanced understanding of how different factors contribute to heat vulnerability. By incorporating labelled data, the machine learning models can learn patterns and associations between the indicators and the heat-related health outcomes, thereby enhancing the accuracy and reliability of the assessments.

Data collection and processing utilises a variety of sources to ensure a comprehensive and multidimensional approach, including remote sensing data from Landsat 8, Sentinel 2, and Digital Earth Australia (DEA); statistical datasets from the Australia Census 2021; and POI gathered from OpenStreetMap. Data processing tools such as Google Earth Engine (GEE) and ArcGIS Pro are employed to handle operations, including data filtering and cleaning, LST, NDVI, and NDBI calculations, kernel density analysis, and data standardisation, conducted at the SA1 scale to facilitate detailed spatial analysis.

In the machine learning selection and development step, the study employs a selection of machine learning algorithms—RF, GB, ridge regression (RR), linear regression (LR), XGBoost (XGB), support vector machine (SVM), multi-layer perceptron (MLP), and k-nearest neighbours (KNN)—to develop assessment models for heat vulnerability. The process involves model training, validation, and testing using Google Colab, hyperparameter optimisation to enhance model performance, and evaluation using R-square, mean absolute error (MAE), and mean squared error (MSE) metrics.

The final step, heat vulnerability analysis, involves a comprehensive analysis of heat vulnerability across Australian capital cities through spatial analysis visualising the spatial distribution of vulnerability, statistical analysis quantifying heat vulnerability, and sensitivity analysis assessing the influence of different indicators on the heat vulnerability outcomes.

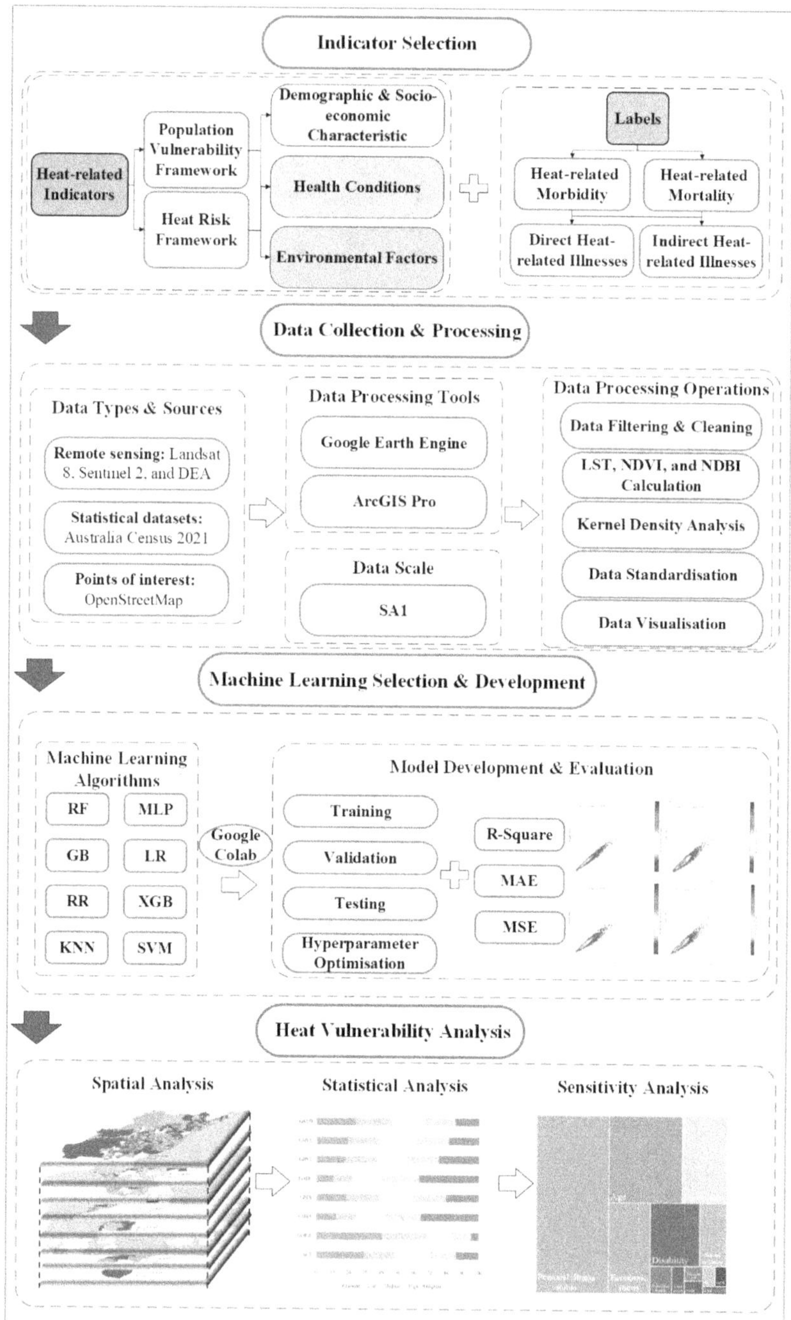

FIGURE 10.2
Research framework.

Heat vulnerability is categorised into five levels based on the ranking percentages of the values assessed by the machine learning model with the highest accuracy (RF): Lowest (0%–20%), low (20%–40%), medium (40%–60%), high (60%–80%), and highest (80%–100%) (Wang et al., 2023). The novelty of this framework lies in its integrative approach, combining diverse data sources and advanced machine learning techniques to comprehensively assess and analyse heat vulnerability. Unlike previous studies that may focus on limited indicators or traditional statistical methods, this study leverages the power of remote sensing, extensive demographic data, and machine learning to provide a more accurate and granular analysis of heat vulnerability. This framework addresses the research gap by offering a multidimensional assessment of vulnerability and utilising a robust methodological approach to ensure reliability and precision in the findings.

10.2.2.1 Indicator Selection

Referring to a rigorous and systematic literature review of measures assessing heat vulnerability (Li et al., 2022), this study meticulously selects indicators representing the vulnerability components (Table 10.1). The findings were further conceptualised and visualised as a conceptual framework in the study of Li et al. (2024a) (Appendix 10.A). The review employed the PRISMA (Preferred Reporting Items for Systematic Reviews and Meta-Analyses) approach to evaluate and synthesise findings from 76 relevant peer-reviewed articles. The goal of the review was to identify and synthesise methods used to assess urban heat vulnerability. To achieve this, the study established specific inclusion criteria, requiring that the articles be peer-reviewed, published in English, available in full text, and relevant to urban heat vulnerability assessments. The initial search yielded 368 records using keywords such as 'urban heat', vulnerab*, and method* from an academic search engine provided by the Queensland University of Technology Library, which offers access to all major databases, including Scopus and Web of Science. After reviewing abstracts and conducting two rounds of full-text screening, 76 peer-reviewed articles were included, all of which were directly relevant to the study's objectives and met the inclusion criteria. The review extracted indicators commonly used in heat vulnerability assessments from these articles, re-categorising them into three main groups: Demographic and socioeconomic characteristics, health conditions, and environmental factors (both natural and built) (Table 10.1). This thorough review ensures the representativeness and relevance of the indicators while enhancing general applicability and minimising subjective judgement, thus providing a standard for indicator selection in future heat vulnerability research. Additionally, the study (Li et al., 2024a), which focuses on the development of a standardised and advanced heat vulnerability assessment methodology, further conceptualised and visualised the findings, serving as a conceptual framework (Appendix 10.A).

TABLE 10.1

Information of Heat Vulnerability Indicators

Categories	Indicators	Descriptions	Units	Data Ranges	Vulnerability Elements	Data Source
Demographic and socioeconomic characteristics	Age	Population aged below 5 or over 65	Persons	[0,1535]	Sensitivity	Census 2021
	Economic status	Population with low incomes less than $800 weekly	Persons	[0,2196]	Sensitivity	Census 2021
	Social isolation	Population living alone	Persons	[0,644]	Sensitivity	Census 2021
	Education	Population with a low education level below Grade 12	Persons	[0,1585]	Sensitivity	Census 2021
	Population density	Population density of each SA1	Persons/km^2	[0,158534]	Sensitivity	WorldPop
Health conditions	Personal illness status	Population with heat-related long-term physical/mental illness	Persons	[0,2106]	Sensitivity	Census 2021
	Availability of medical resources	Availability of medical resources such as hospital/pharmacy/doctor/chemist	Points/km^2	[0,6.53]	Adaptive capacity	OpenStreetMap
	Disability	Population with disability	Persons	[0,525]	Sensitivity	Census 2021
Environmental factors (natural)	LST	Daytime LST	°C	[18.11,47.08]	Exposure	Landsat 8
	Vegetation cover	Mean value of NDVI	None	[−1,1]	Adaptive capacity	Sentinel 2
Environmental factors (built)	Availability of cooling spaces	Availability of public places with free cooling services	Points//km^2	[0,9.25]	Adaptive capacity	OpenStreetMap
	Land cover/use	Area of developed urban land cover	km^2	[0,13.97]	Sensitivity	Digital Earth Australia
	Building density	Mean value of NDBI	None	[−1,1]	Sensitivity	Sentinel 2

According to the Intergovernmental Panel on Climate Change, vulnerability is defined by three key elements: Exposure, sensitivity, and adaptive capacity (Parry, 2007). Exposure refers to the degree to which a system or population is subjected to the impacts of climate change, such as extreme heat. Sensitivity represents the extent to which a system or group responds to these impacts. Adaptive capacity refers to the ability of a system to cope with, adapt to, and mitigate these impacts. In the context of heat vulnerability, these components interact to determine the overall level of vulnerability to extreme heat, collectively contributing to increased heat-related mortality and morbidity (Li et al., 2024c). The re-categorisation reflects a balanced approach that encompasses various dimensions of heat vulnerability: Demographic and socioeconomic characteristics play a crucial role in reflecting the population's sensitivity to heat hazards. Age, which captures the population aged below 5 or over 65, is a critical indicator due to the increased physiological vulnerability of these age groups to heat-related morbidity and mortality. Economic status, measured by the number of individuals with incomes below $800 per week (the Australian minimum wage), highlights the financial limitations that impede access to heat mitigation resources. Social isolation, representing individuals living alone, reflects the lack of social support, which increases vulnerability during extreme heat. Education, measured by the population with educational attainment below Grade 12 (excluding higher or vocational education), affects preparedness and awareness of heat risks and appropriate responses. High population density increases anthropogenic heat and exacerbates heat retention, while also placing pressure on access to public resources and infrastructure needed to mitigate heat hazards. In areas with higher densities, residents tend to experience increased sensitivity to exacerbated thermal environments.

Health conditions focus on the physiological and medical factors that further contribute to sensitivity or adaptive capacity in the face of extreme heat. Personal illness status, which encompasses the population with long-term heat-related physical or mental illnesses (see Appendix 10.B), captures heightened sensitivity due to pre-existing conditions that impair the ability to withstand heat exposure. Disability similarly reflects an increased sensitivity to heat hazards, as individuals with disabilities often face mobility and healthcare access challenges. Conversely, the availability of medical resources, such as hospitals, pharmacies, doctors, and chemists, plays a vital role in enhancing adaptive capacity by facilitating timely medical interventions during heat events.

Environmental factors, comprising both natural and built environments, influence the population's exposure, sensitivity, and adaptive capacity to heat. In terms of exposure, LST directly measures heat intensity, with higher temperatures indicating greater exposure to heat. Vegetation cover, represented by the mean value of the NDVI, serves as an indicator of adaptive capacity, as areas with higher vegetation cover experience reduced heat exposure due to cooling effects such as shading and evapotranspiration. In the built environment, the availability of public cooling spaces significantly enhances adaptive capacity, providing vulnerable populations with access to cooling services. Land cover/use, which measures the area of developed urban land, is associated with sensitivity, as is BD, represented by the mean value of the NDBI. Both tend to increase heat retention, reduce ventilation, and exacerbate the UHI effect, thereby intensifying the overall sensitivity of urban residents to heat exposure.

Unlike prior research, Li et al.'s (2022) review introduces a novel re-categorisation of indicators into three balanced categories, enhancing clarity and balance for a more comprehensive assessment of heat vulnerability. This approach addresses two critical aspects: The representativeness and comprehensiveness of the indicators, and the mitigation of inconsistencies and biases in their selection. By systematically grouping widely accepted and recently validated indicators (Cheng et al., 2021; Li et al., 2022), the framework ensures up-to-date and methodologically rigorous inputs. This structured categorisation avoids over-reliance on any single type of indicator, supporting a holistic view of vulnerability and enabling more accurate urban planning and public health outcomes. The balance and clarity within the framework not only enhance its robustness but also provide a valuable reference for standardising indicator selection and categorisation in heat vulnerability studies.

10.2.2.2 Data Collection and Processing

The demographic, socioeconomic, and health data are mainly retrieved from the Australian Census 2021. Only the population density data are obtained from WorldPop with a 100 m resolution. The environmental factors are divided into natural and built indicators, reflecting different dimensions of exposure, sensitivity, and adaptive capacity. Among the natural factors, LST is obtained from Landsat 8 satellite data with a spatial resolution of 30 m, providing an analysis of daytime surface temperature distribution to assess heat exposure in urban areas. Vegetation cover (NDVI), based on Sentinel 2 satellite data with a resolution of 10 m, evaluates the amount of green vegetation, serving as an adaptive capacity indicator by reflecting the role of green spaces in mitigating heat exposure. For the built factors, the availability of cooling spaces is derived from OpenStreetMap POI data, identifying public places offering cooling services (e.g., cooling centres, libraries), which are critical for assessing the community's adaptive capacity during heatwaves. Land cover/use, sourced from DEA with a resolution of 25 m, represents the extent of urban development and serves as a sensitivity indicator to measure the impact of urban construction on heat exposure. Building density (NDBI) is obtained from Sentinel 2 data with a 10-m resolution, indicating the density and layout of urban buildings. It is used to assess the contribution of buildings to the UHI effect, serving as a key measure of heat sensitivity.

In this study, the labelled data encompass heat-related morbidity and mortality attributable to heat-related illnesses from Australian Bureau of Statistics (ABS), either directly or indirectly (Appendix 10.B). The research aggregates heat-related morbidity data at the SA1 level (ABS, 2022) and all-cause mortality data at the Statistical Area Level 2 (SA2) level (ABS, 2023b) from Australian capital cities. Subsequently, it estimates SA2 heat-related mortality data based on state-level death rates from heat-related diseases (ABS, 2023a) and resamples it to the SA1 scale according to the population density (Estoque et al., 2020; Wang et al., 2023) (see Eqs. (10.1) and (10.2)). Ultimately, the mean values of heat-related morbidity and mortality at the SA1 scale are computed to serve as the labelled data for analysis, which ranges from 0 to 0.4 after screening outliers.

$$D_{h,\mathrm{SA1}} = D_{a,\mathrm{SA2}} \times R_{h,\mathrm{State}} \times \frac{P_{\mathrm{SA1}}}{P_{\mathrm{SA2}}} \tag{10.1}$$

$$V_{\mathrm{SA1}} = \frac{1}{2}(M_{h,SA1} + D_{\mathrm{h,SA1}}) \tag{10.2}$$

where $D_{h,\mathrm{SA1}}$ is the heat-related mortality in SA1s, $D_{a,\mathrm{SA2}}$ is the all-cause mortality in SA2s, $R_{h,\mathrm{State}}$ is the heat-related mortality rate in states, P_{SA1} and P_{SA2} are populations in SA1s and SA2s, respectively, V_{SA1} is the heat vulnerability in SA1s, and $M_{h,\mathrm{SA1}}$ is heat-related morbidity in SA1s.

In Australia, SA1 and SA2 are geographic units used by the ABS for census and data collection under the Australian Statistical Geography Standard (ASGS). SA1 is the smallest unit, typically covering areas with a small population and defined by visible or socially relevant boundaries, such as roads or natural features. It allows for detailed analysis of localised demographic, social, and economic characteristics. In contrast, SA2 is a larger geographic area that combines multiple SA1s, representing broader communities like suburbs or small towns. SA2 regions are designed to reflect areas of social and economic interaction and are used for more aggregated analysis, such as healthcare, education, or employment trends. The relationship between SA1 and SA2 ensures that data can be analysed both at a fine-grained local level and a broader community level, enabling comprehensive insights into Australia's population and geography.

To quantify availabilities of medical resources and cooling spaces, this chapter utilises kernel density analysis to explore the spatial distribution patterns of medical resources and cooling services from POIs (Eq. (10.3)). Kernel density analysis smooths the data's discreteness, generating a continuous probability density estimate that aids in the intuitive understanding of the spatial distribution of resources and services (Yu et al., 2017). It also produces specific density values for each location, thereby enabling further quantitative analysis. By comparing the density values across different areas, researchers can quantitatively evaluate the accessibility and availability of medical resources and cooling services, thus providing a foundation for policy formulation and resource optimisation.

$$\hat{f}(x) = \frac{1}{nh} \sum_{i=1}^{n} K\left(\frac{x - x_i}{h}\right) \tag{10.3}$$

where $\hat{f}(x)$ is the estimated density at point x, n is the number of data points, h is the bandwidth (smoothing parameter), x_i are the data points, K is the kernel function, which is a symmetric, non-negative function that integrates to one.

All the data are pre-processed, calculated, standardised, and visualised at the SA1 scale as continuous values using GEE and ArcGIS Pro 2.9.1, making them suitable for machine learning operations. The dataset comprises a total of 39,862 data points; please see the data units, ranges, and sources in Table 10.1. Before training machine learning models, collinearity diagnostics have been conducted by using SPSS 29. The results in Appendix 10.C show that the variance inflation factor (VIF) values for all indicators are below 10, with most being under 5, indicating that the degree of collinearity among the input variables is within an acceptable range (Liu et al., 2022; Zhao et al., 2018). Additionally, nonlinear models such as RF, GB, and KNN are generally robust to the effects of multicollinearity, as they do not depend on the estimation of linear coefficients. Instead, these models prioritise the relative importance of variables, rather than their linear interrelationships. Consequently, the presence of multicollinearity typically exerts minimal influence on their performance and accuracy. An exploratory data analysis using scatter plots was conducted to provide a preliminary understanding of the relationships between various input indicators and the labelled data (see Appendix 10.D). The analysis reveals that personal physiological and health characteristics, such as age, illness status, and disability, are positively associated with heat vulnerability, indicating that individuals with these conditions may face greater risks during extreme heat events. Socioeconomic factors, including economic status, education, and social isolation, show more scattered or weaker positive relationships with heat vulnerability. In contrast, there are relatively weak negative correlations between the availability of cooling spaces and medical resources and heat vulnerability, indicating a certain mitigation effect. Environmental characteristics, such as LST, vegetation cover, and BD, appear to moderately correlate with heat vulnerability, while the role of factors like population density and land cover/use remains less clear, potentially indicating nonlinear effects. Overall, these initial insights highlight the multifaceted nature of heat vulnerability and suggest the need for more advanced modelling to fully understand these dynamics.

10.2.2.3 Machine Learning Algorithm Selection and Development

This study employs eight machine learning regression algorithms to assess heat vulnerability in Australian capital cities. All the algorithms are running on the Google Collaboratory (Colab) cloud-based platform using Python 3.10.12 and the machine learning library Scikit-learn 1.2.2. These algorithms were chosen because they represent a diverse set of typical and robust methods that have demonstrated high performance across various datasets and tasks in urban studies (Li et al., 2023).

RF and GB, for instance, are known for their ability to handle complex, nonlinear relationships and have been broadly used in urban studies to predict environmental phenomena, such as air quality and UHIs (Jato-Espino et al., 2022; Yang et al., 2022). XGB, with its efficiency and accuracy, has also been widely applied in urban studies for tasks like disaster management and pollution monitoring (Chi et al., 2022; Ghaffarian et al., 2021). RR and LR are traditional methods that have been extensively utilised in urban health studies to identify key socio-demographic and environmental predictors (Aheto et al., 2021; Kan et al., 2019). SVM is recognised for its robustness in high-dimensional spaces and has been used in urban planning to model urban sprawl and building information identification (Lu et al., 2014; Shafizadeh-Moghadam et al., 2017). MLP, as a type of neural network, excels in capturing intricate patterns in large datasets and has been applied in urban studies for predicting energy consumption and urban land cover classification (Afzal et al., 2023; Zhang et al., 2018). KNN, though simpler, is effective for spatial analysis and has been used in studies mapping urban land cover and building footprints (Karalas et al., 2016; Schlosser et al., 2020).

The selection of these algorithms also allows for a comprehensive comparison with existing literature, where traditional statistical models and simpler machine learning approaches have been predominantly used. However, current heat vulnerability research lacks exploration and application of these machine learning algorithms. This study addresses this gap by applying these advanced algorithms to heat vulnerability research, aiming to explore their effectiveness and provide deeper insights into the underlying factors influencing heat-related health risks. By integrating these advanced algorithms, this study aims to enhance the accuracy and reliability of heat vulnerability assessments, thereby contributing to the growing body of research in urban environmental studies.

The selected algorithms are briefly introduced below.

RF: RF builds multiple decision trees and combines their predictions to improve accuracy and stability. It is robust against overfitting, handles high-dimensional data, and works well with missing values, capturing nonlinear relationships effectively. The ensemble nature of RF makes it suitable for identifying key factors influencing heat vulnerability (Jato-Espino et al., 2022; Wang et al., 2022a). The formula is

$$\hat{y} = \frac{1}{T}\sum_{t=1}^{T} h_t(x) \tag{10.4}$$

where T is the number of decision trees, and h_t is the prediction of the tth tree.

MLP: MLP is a type of neural network with one or more hidden layers, capable of learning complex function mappings. It has a strong nonlinear modelling capability, making it suitable for handling complex patterns and high-dimensional data in heat vulnerability. However, it requires substantial data and computational resources (Afzal et al., 2023; Zhang et al., 2018). The formula is

$$\hat{y} = f\left(\sum_{i=1}^{n} w_i x_i + b\right) \tag{10.5}$$

where f is the activation function, and w_i and b are weights and biases.

GB: GB sequentially adds weak learners, usually decision trees, to improve model performance. It is known for high accuracy and the ability to handle nonlinear relationships effectively. GB's robust performance in high-dimensional and complex data makes it ideal for heat vulnerability regression (Xiao et al., 2021; Yang et al., 2022). The formula is

$$\hat{y} = \sum_{t=1}^{T} \alpha_t h_t(x) \tag{10.6}$$

where α_t is the weight, and h_t is the tth weak learner.

LR: LR assumes a linear relationship between dependent and independent variables. It is simple, easy to interpret, and computationally efficient, making it suitable for situations with clear linear relationships. However, its ability to handle nonlinear relationships is limited (Kan et al., 2019; Zhao et al., 2022). The formula is

$$\hat{y} = \beta_0 + \sum_{i=1}^{n} \beta_i x_i \tag{10.7}$$

where β_i are the regression coefficients.

RR: RR is a LR with a regularisation term that helps mitigate multicollinearity issues. It prevents overfitting and improves model generalisability, performing well in handling high-dimensional data, making it a good choice for heat vulnerability analysis (Aheto et al., 2021; Okumus & Terzi, 2021). The formula is

$$\hat{y} = \beta_0 + \sum_{i=1}^{n} \beta_i x_i \quad with \quad \sum_{i=1}^{n} \beta_i^2 < \lambda \tag{10.8}$$

where λ is the regularisation parameter.

XGB: XGB is an efficient implementation of GB that enhances performance and accuracy. It handles nonlinear relationships well and is suitable for large-scale datasets. XGB's efficiency and robustness make it ideal for complex heat vulnerability data (Chi et al., 2022; Ghaffarian et al., 2021). The formula is

$$\hat{y} = \sum_{k=1}^{K} f_k(x), where\ f_k \in F \tag{10.9}$$

where F is the space of regression trees.

KNN: KNN makes predictions based on the distance between data points. It is simple and non-parametric, suitable for small datasets. However, it relies heavily on data distribution, making it more applicable to cases with clear local patterns in heat vulnerability (Karalas et al., 2016; Schlosser et al., 2020). The formula is

$$\hat{y} = \frac{1}{k} \sum_{i=1}^{k} y_i \tag{10.10}$$

where y_i are the labels of the KNN.

SVM: SVM finds the optimal hyperplane in high-dimensional space for classification and regression. It handles high-dimensional data well and has good generalisation ability, though it is computationally intensive. SVM is suitable for small samples and complex data structures in heat vulnerability analysis (Lu et al., 2014; Shafizadeh-Moghadam et al., 2017). The formula is

$$\hat{y} = w^T \phi(x) + b \tag{10.11}$$

where $\phi(x)$ is a nonlinear mapping function.

10.2.2.4 Hyperparameter Optimisation

During the hyperparameter optimisation process, this study partitioned the data into 70% for training and 30% for testing. This study employed a manual grid search method for hyperparameter tuning, utilising five-fold cross-validation for validation (Fan et al., 2024). Specifically, this study defined a grid of potential hyperparameters and applied five-fold cross-validation, dividing the training data into five subsets. In each iteration, four subsets were used for training and one for validation. This process was repeated for each combination of hyperparameters, with performance metrics recorded at each step. The optimal hyperparameter combination was selected based on cross-validation results, ensuring the model's best performance and generalisation capability. The detailed hyperparameter combinations, learning curves, and optimal results are presented in Appendices 10.E and 10.F.

10.2.2.5 Performance Evaluation Metrics of Machine Learning Algorithms

This study uses the coefficient of determination (R^2), MSE, and MAE as metrics to evaluate the performance of machine learning algorithms in assessing heat vulnerability. These metrics are selected because they provide a balance between measuring model fit and prediction error, ensuring a comprehensive evaluation of both model accuracy and precision (Wang et al., 2022b; Zhou et al., 2024). Additionally, they are widely recognised and commonly used performance evaluation metrics in urban studies that utilise machine learning, making them appropriate for this context (Gao et al., 2023; Kokkinos et al., 2021; Palaniyappan & Vinopraba, 2024).

R^2 represents the proportion of variance in the dependent variable that is predictable from the independent variables. The closer the value is to 1, the better the model's explanatory power. The formula is as follows:

$$R^2 = 1 - \frac{\sum_{i=1}^{n}\left(y_i - \hat{y}_i\right)^2}{\sum_{i=1}^{n}\left(y_i - \bar{y}\right)^2} \tag{10.12}$$

where y_i are the actual values, $\hat{y}_i$ are the predicted values, and $\bar{y}$ is the mean of the actual values.

MSE measures the average of the squares of the errors between predicted and actual values. A lower MSE indicates higher prediction accuracy. The formula is

$$MSE = \frac{1}{n}\sum_{i=1}^{n}\left(y_i - \hat{y}_i\right)^2 \tag{10.13}$$

MAE measures the average of the absolute differences between predicted and actual values. A lower MAE indicates lower prediction error. The formula is

$$MAE = \frac{1}{n}\sum_{i=1}^{n}\left|y_i - \hat{y}_i\right| \tag{10.14}$$

By calculating and comparing these metrics, we can comprehensively evaluate the performance of different machine learning algorithms in assessing heat vulnerability. Their combination allows us to evaluate model performance from multiple perspectives, ensuring that both high accuracy and minimal errors are achieved, which is crucial for assessing the spatial distribution of urban heat vulnerability. This, in turn, provides data support and scientific basis for urban planning and policy formulation.

Figure 10.3 illustrates how machine learning models are developed to assess heat vulnerability, enhancing the reader's understanding. It begins with data pre-preparation, where various data types, including remote sensing, statistical, and POI data, are processed into continuous values suitable

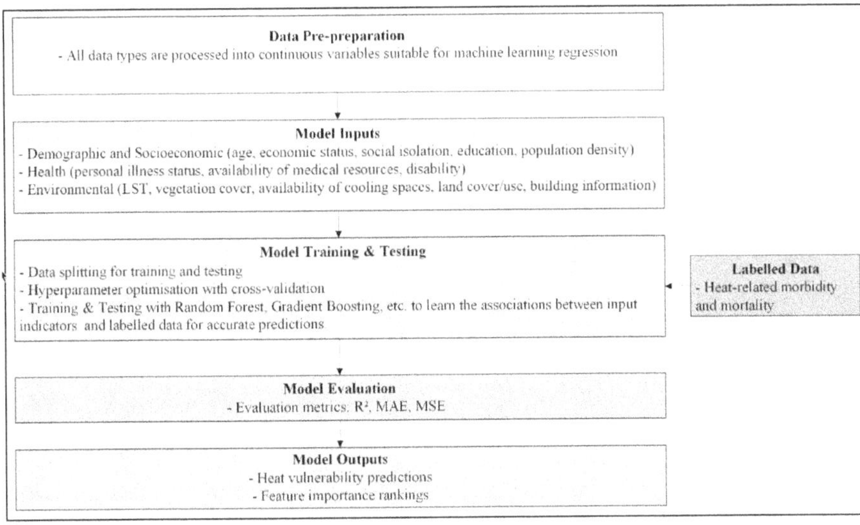

FIGURE 10.3
Workflow of machine learning development.

for machine learning regression. This ensures that all inputs are standardised and ready for use in model training. Next, the processed data are fed into the model as inputs, with indicators such as age, personal illness status, and LST forming the basis for assessing heat vulnerability. These inputs represent different perspectives influencing heat vulnerability. Machine learning algorithms, such as RF and GB, are employed to learn patterns and associations between these input indicators and the labelled data, which, in this case, are heat-related morbidity and mortality. The labelled data serve as the target variable that guides the model's learning process, allowing it to understand the relationships between the inputs and the health outcomes. During model training, the input data are split into training and testing sets, and techniques such as hyperparameter optimisation and cross-validation are applied to refine the model's performance. This step ensures that the models can generalise well to new data and accurately predict outcomes. Once the models are trained, their performance is evaluated using metrics such as R^2, MAE, and MSE to assess how well the models assess heat vulnerability. The final output of the machine learning process includes heat vulnerability predictions and feature importance rankings. These outputs not only map areas most vulnerable to heat but also provide insights into which variables are the most significant contributors to vulnerability.

10.2.2.6 Heat Vulnerability Analysis

The heat vulnerability analysis step involves a comprehensive evaluation of heat vulnerability across Australian capital cities using a multidimensional approach. This process integrates spatial, statistical, and sensitivity analyses to assess heat vulnerability in a systematic and robust manner.

Spatial Analysis: Spatial analysis is conducted using ArcGIS Pro 2.9 to map and visualise the geographical distribution of heat vulnerability across different SA1 units. This step helps to capture the spatial variability of vulnerability and identify potential hotspots. The visual representation of heat vulnerability supports further analysis and provides a clear geographic context for understanding the spatial distribution of heat vulnerability in both urban and suburban areas.

Statistical Analysis: To quantify heat vulnerability, the heat vulnerability values assessed by the machine learning model with the highest accuracy (RF) are categorised into five distinct levels based on the ranking percentages: Lowest (0%–20%), low (20%–40%), medium (40%–60%), high (60%–80%), and highest (80%–100%) (Wang et al., 2023). The statistical analysis is used to compute the distribution of SA1 units across these vulnerability levels in eight cities. This allows for a comparative assessment of regional disparities, helping to understand the distribution of vulnerability across various geographic areas and providing a structured way to assess which regions may require targeted interventions.

Sensitivity Analysis: Sensitivity analysis is conducted using feature importance rankings derived from the RF algorithm, partial dependence plots (PDP), and SHapley Additive exPlanations (SHAP). This approach identifies the demographic, socioeconomic, health, and environmental factors that contribute most significantly to heat vulnerability. Feature importance offers a global evaluation of each feature's contribution to the model. PDPs aid in interpreting feature importance by visualising the average effect of individual features, while SHAP values provide a more granular insight into how specific features influence individual assessments, ensuring both local and global interpretability (Liao et al., 2024). By combining these techniques, the sensitivity analysis yields insights into the interaction of various variables and their impact on vulnerability, informing the refinement of vulnerability models and policy strategies.

By combining spatial, statistical, and sensitivity analyses, this approach ensures a comprehensive evaluation of heat vulnerability and supports the development of targeted strategies for urban planning and public health.

10.3 Analysis and Results

10.3.1 Comparative Performance of Machine Learning Algorithms

This study evaluated eight machine learning algorithms to assess heat vulnerability in Australian capital cities, focusing on assessment accuracy and

TABLE 10.2

Training and Testing Accuracy Results of Machine Learning Algorithms

	Training Accuracy			Testing Accuracy		
ML Algorithms	R^2	MAE	MSE	R^2	MAE	MSE
RF	**0.9179**	**0.0051**	**0.0001**	**0.9089**	**0.0076**	**0.0001**
MLP	0.9115	0.0075	0.0001	0.9014	0.0078	0.0001
GB	0.9057	0.0080	0.0001	0.8962	0.0082	0.0001
LR	0.9018	0.0078	0.0001	0.9050	0.0078	0.0001
RR	0.9017	0.0078	0.0001	0.9049	0.0078	0.0001
XGBoost	0.8842	0.0084	0.0002	0.8782	0.0087	0.0002
KNN	0.8613	0.0082	0.0002	0.8544	0.0094	0.0002
SVM	0.8381	0.0111	0.0002	0.8343	0.0108	0.0002

error metrics (Table 10.2, Figure 10.4, and Appendix 10.G). The RF algorithm emerged as the best performer, achieving R^2 values of 0.9179 (training) and 0.9089 (testing), with an MAE of 0.0051 (training) and 0.0076 (testing), and an MSE of 0.0001 for both sets. These metrics highlight RF's ability to effectively model complex relationships in the data, making it suitable for visualising the spatial distribution of heat vulnerability in Section 3.2. The MLP also performed well, with R^2 values of 0.9115 (training) and 0.9014 (testing), and similar MAE and MSE values to RF. GB showed slightly lower but reliable performance, with R^2 of 0.9057 and 0.8962. LR and RR had almost identical results, both achieving R^2 values around 0.902, with MAE and MSE values of 0.0078 and 0.0001. XGB had slightly lower accuracy, with R^2 values of 0.8842 and 0.8782, but still demonstrated adequate assessment power. KNN showed moderate performance with R^2 values of 0.8613 (training) and 0.8544 (testing) and a slightly higher error. The SVM algorithm performed the worst, with R^2 values of 0.8381 and 0.8343, indicating limited effectiveness in this context. Overall, RF proved to be the most effective algorithm for heat vulnerability assessment, providing key insights for future research and urban heat management.

10.3.2 Spatial and Statistical Analysis of Heat Vulnerability

Heat vulnerability in Australian capital cities is categorised into five levels based on assessments from the RF algorithm, selected for its high accuracy. Figure 10.5 shows the spatial distribution of heat vulnerability across eight major Australian capital cities, revealing significant regional disparities and distinct urban characteristics.

GSYD: Heat vulnerability in GSYD displays a clear gradient, with the highest levels in densely populated inner-city areas (red and orange zones). Middle zones show moderate vulnerability (yellow and light green), with

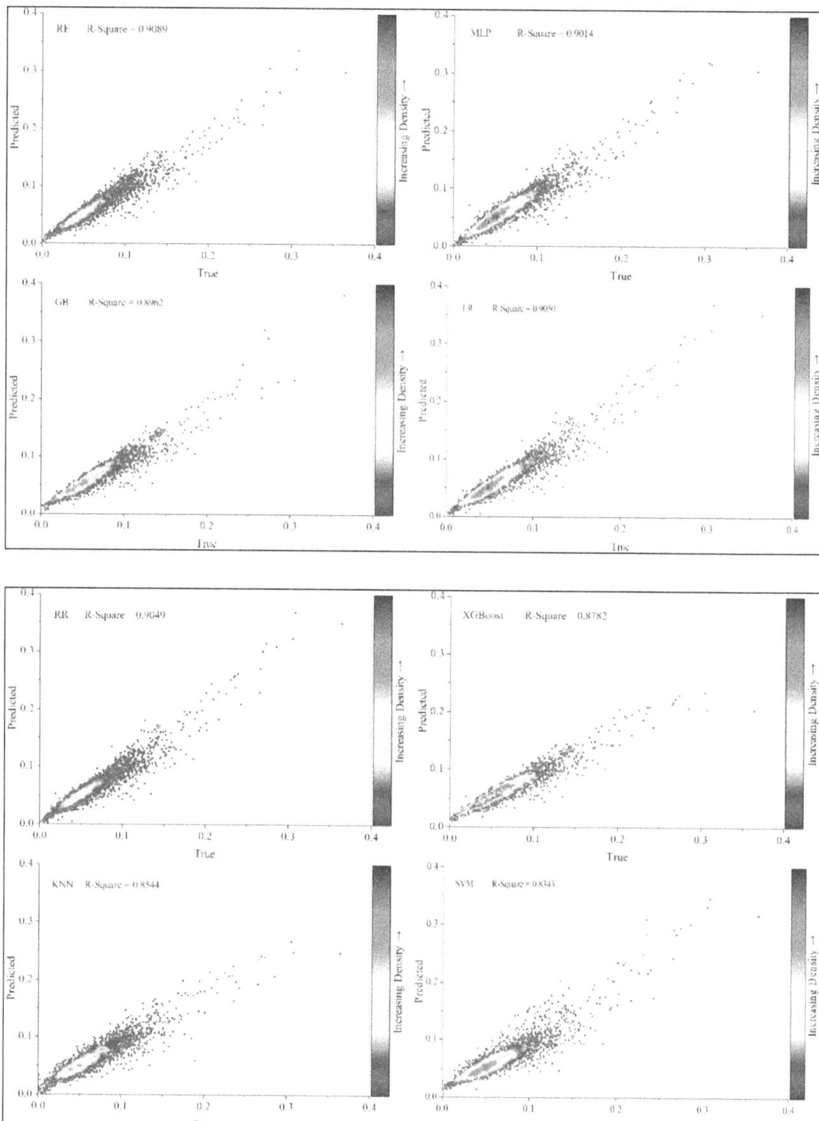

FIGURE 10.4
Testing results of the selected machine learning algorithms.

some pockets of high vulnerability. The outer suburbs and rural regions (dark green) have the lowest vulnerability, though isolated areas of higher vulnerability can still be found in the middle and northeastern coastal suburbs.

GMEL: In GMEL, heat vulnerability decreases outwards from the city centre. The highest vulnerability is found centrally, with gradual decreases in

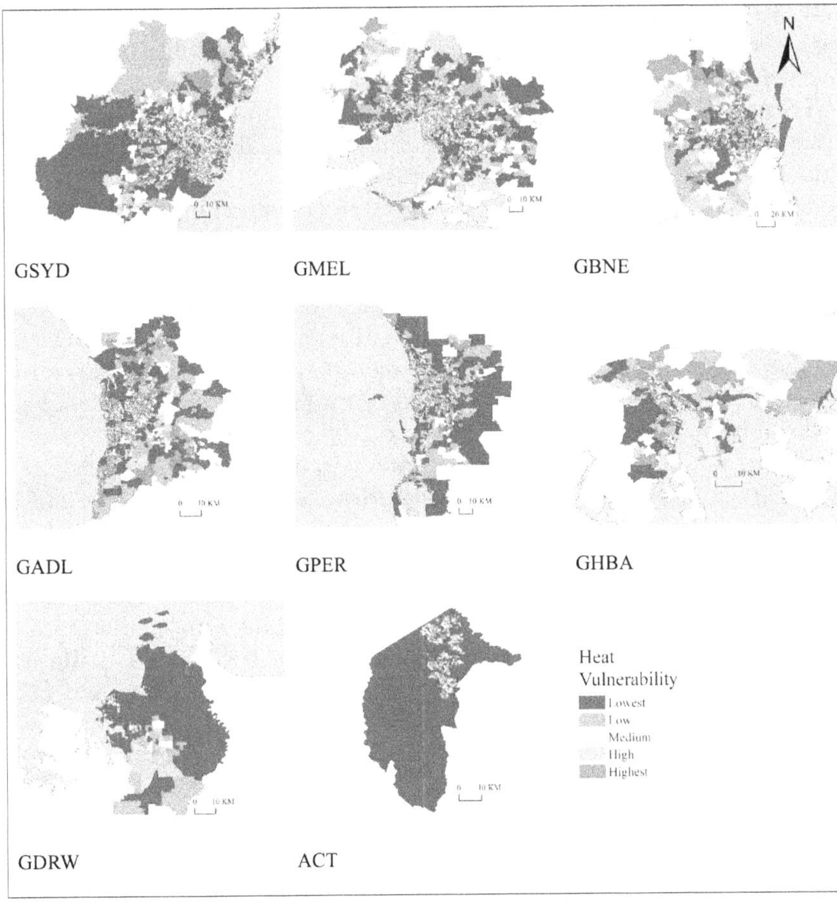

FIGURE 10.5
Spatial distribution of assessed heat vulnerability from RF at the SA1 scale in Australian capital cities.

the inner suburbs. The northwestern and southwestern regions have moderate to high vulnerability, while the outer suburban areas have the lowest.

GBNE: GBNE exhibits considerable variation in heat vulnerability. The northern areas are predominantly highly vulnerable, while the central region has lower vulnerability. The eastern coastal region shows elevated vulnerability, particularly in the southern part, while the southern areas generally have lower vulnerability.

GADL: GADL shows an urban–rural gradient, with high vulnerability concentrated in central and southwestern urban areas. Moderate vulnerability appears in central and southeastern regions, while low vulnerability is prevalent in the northern and northeastern suburbs and rural areas.

GPER: In GPER, lower heat vulnerability is seen in coastal and suburban areas, while the urban core experiences higher vulnerability. Inland areas have heterogeneous vulnerability.

GHBA: GHBA shows significant variation, with high vulnerability in central urban areas, lower vulnerability in peripheral suburban and coastal regions, and mixed levels in inland areas.

GDRW: Heat vulnerability in GDRW forms a distinct gradient, with high vulnerability in southern and central regions, surrounded by moderately vulnerable buffer zones. The northern areas have low vulnerability.

ACT: In ACT, high vulnerability is concentrated in the northern and central regions, surrounded by moderately vulnerable areas. The southern and peripheral areas have the lowest vulnerability, illustrating a gradient from high in the densely populated north and central regions to low in less populated southern areas.

In sum, high-heat vulnerability is common in the central business districts and inner suburbs of six major cities, including GSYD, GMEL, GBNE, GADL, GPER, and GHBA. These areas are characterised by high population density, extensive impervious surfaces, and limited green space, contributing to the UHI effect. In contrast, peripheral and suburban areas have lower vulnerability, benefiting from greater vegetation cover, lower population density, and natural cooling effects like parks and coastal breezes, particularly in Melbourne's eastern suburbs and the coastal areas of Brisbane and Perth. Rural and outlying regions, such as in GDRW and ACT, show the lowest vulnerability, with abundant green zones and minimal urban development. Unique regional patterns also emerge, such as higher vulnerability in northern GHBA's middle part and a balanced vulnerability distribution in Perth's western coastal and central areas. These findings underscore the complex interplay of socioeconomic factors, urban density, land use, and natural features in shaping heat vulnerability. Tailored mitigation and adaptation strategies are essential to address local needs and enhance urban resilience to rising heat risks.

The comparative analysis of heat vulnerability across eight major Australian capital cities reveals significant regional differences in the distribution of SA1 units across vulnerability categories. Figure 10.6 shows that ACT and GDRW have the highest proportions of SA1 units in the lowest vulnerability category, at 29.83% and 40.28%, respectively, indicating generally lower vulnerability in these areas. In contrast, GHBA and GADL have the largest proportions of SA1 units in the highest vulnerability category, at 35.64% and 36.91%, highlighting significant heat vulnerability concerns. GPER, GMEL, and GSYD show more balanced distributions across all categories, indicating diverse vulnerability profiles. GBNE also has a relatively even distribution, with a slight bias towards higher vulnerability. These results underscore the differing levels of heat vulnerability across Australian capital cities, highlighting the importance of region-specific mitigation and adaptation strategies.

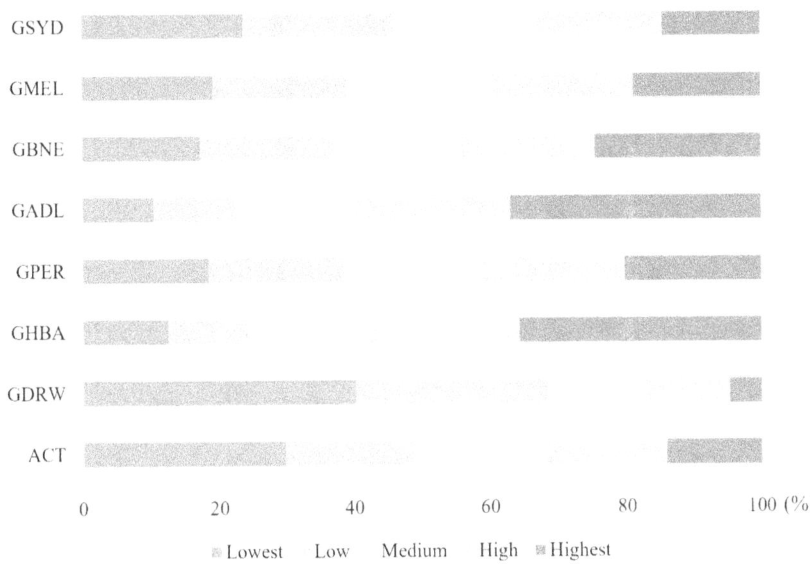

FIGURE 10.6
Proportions of SA1s in heat vulnerability levels across Australian capital cities.

10.3.3 Sensitivity Analysis

The sensitivity analysis results, using three interpretability techniques (Figure 10.7 and Appendix 10.H)—feature importance treemap, PDP, and mean SHAP values—collectively underscore the role of each heat-related indicator in determining heat vulnerability. As highlighted by both visualisations and SHAP value assessments, personal illness consistently emerges as the most influential factor. The treemap identifies it as the largest segment, while the SHAP values further confirm this, showing the highest mean value (0.0136) and a positive trend in the PDP. This suggests that individuals with chronic health conditions are particularly susceptible to heat stress, a finding of significant concern for public health interventions. In addition to illness status, age and education also rank highly in both analyses. The treemap demonstrates their notable contributions, with the elderly and very young populations being especially vulnerable, likely due to physical frailty and reduced ability to regulate body temperature. Similarly, the PDP for age displays a strong positive trend, with a SHAP value of 0.0045, confirming its substantial impact. Education, while slightly less influential than age and illness, still plays a significant role. Lower educational attainment is linked to higher vulnerability, likely due to a lack of heat risk awareness and fewer resources available for mitigation. Economic status and social isolation also feature prominently, albeit with somewhat lower SHAP values

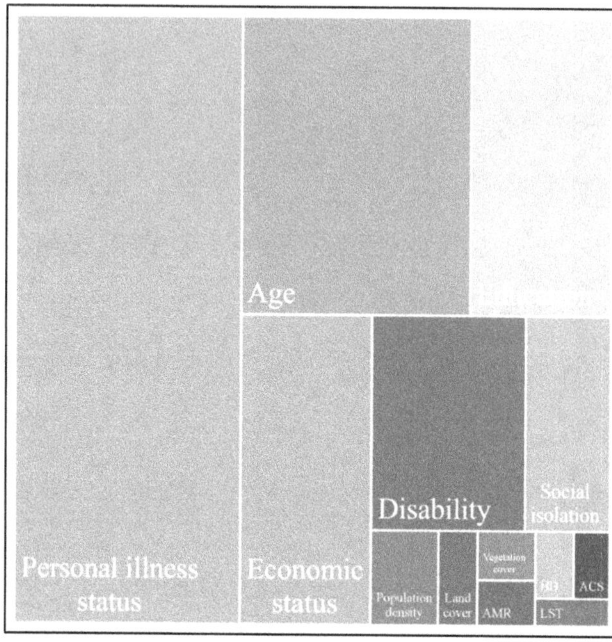

FIGURE 10.7
Treemap of feature importance for heat-related indicators in Australian capital cities. Notes: ACS for availability for cooling spaces; AMR for availability for medical resources; BD for building density.

(ranging from 0.0019 to 0.0026). Both the treemap and SHAP analyses suggest that these variables contribute to vulnerability by limiting access to cooling resources, healthcare, and social support. The influence of disability follows a similar pattern, with moderate but meaningful contributions to the model's assessments, further highlighting the importance of socioeconomic and demographic factors. Other indicators, primarily environmental and infrastructural, exhibit smaller SHAP values and appear as minor segments in the treemap. Population density and the availability of medical resources, for instance, have smaller SHAP values, indicating limited contribution to the model. Although their PDPs show some level of influence, particularly at lower values, their overall effect diminishes as their values increase. LST, vegetation cover, availability of cooling spaces, land cover, and BD also display minimal SHAP values, suggesting their marginal role in the model's outcomes. The PDP trends further support this, with some factors showing negative or fluctuating effects, reinforcing their limited relevance to the assessments. While these environmental and infrastructural factors play a role, their contribution to heat vulnerability is significantly overshadowed by health and socio-demographic variables. These findings underscore the complex interplay between individual health, social conditions, and the physical environment, with personal health and demographic characteristics emerging as the most critical determinants of heat vulnerability.

10.4 Findings and Discussion

The study successfully utilised eight machine learning algorithms to assess heat vulnerability in Australian capital cities, revealing several critical insights. The RF algorithm emerged as the most robust model, with R^2 values of 0.9179 for the training set and 0.9089 for the testing set, underscoring its superior ability to capture complex relationships within the data. Spatial analysis revealed significant regional disparities, with higher heat vulnerability in densely populated urban centres and areas with limited green spaces, while suburban and rural regions with higher vegetation cover exhibited lower vulnerability. This highlights a substantial imbalance in heat vulnerability between urban and suburban areas. Notably, the ACT and GDRW have the lowest proportions of highly vulnerable SA1 units, whereas GHBA and GADL show the highest, indicating significant regional disparities that necessitate tailored heat mitigation strategies. Sensitivity analysis revealed that personal health conditions and socio-demographic characteristics, such as personal illness status, age, and education level, play dominant roles in determining heat vulnerability, overshadowing the impact of environmental and infrastructural factors.

The study's findings can be attributed to several key factors. The superior performance of the RF algorithm is likely due to its ability to handle complex, nonlinear relationships and interactions among the various heat vulnerability indicators. This algorithm's ensemble approach combines multiple decision trees, effectively capturing intricate patterns in the data that simpler models might miss (Belgiu & Drˇagut, 2016; Ye et al., 2019). From the most evident environmental differences, the significant regional disparities in heat vulnerability can be explained by the varying urbanisation levels, population densities, and green space availability across different cities (Chen et al., 2018). Densely populated urban centres with extensive impervious surfaces and limited vegetation exacerbate the UHI effect, leading to higher heat exposure (Wang et al., 2023). Conversely, suburban and rural areas benefit from greater vegetation cover, which provides natural cooling and reduces heat exposure (Li et al., 2024b). However, the sensitivity analysis reveals that personal health conditions and socio-demographic characteristics—such as illness status, age, and education—play a more dominant role in determining individual vulnerability. The dominance of personal health conditions and socio-demographic characteristics in determining heat vulnerability aligns with the theoretical framework of social determinants of health, which posits that an individual's social and economic conditions significantly influence their health outcomes (Marmot, 2005; Solar & Irwin, 2010). Populations with chronic health conditions, the elderly, and those with lower education levels are less capable of coping with heat stress, making them more vulnerable. This finding supports existing models that emphasise the importance of addressing social inequalities to mitigate health risks.

When comparing this study's results with existing literature, several key points of convergence and divergence emerge. Previous research has consistently shown that urban centres with high population densities and limited green spaces tend to exhibit higher heat vulnerability due to the UHI effect (Aubrecht & Özceylan, 2013; Li et al., 2024c; Wang et al., 2023). This study's findings align with these observations, reinforcing the notion that urbanisation significantly impacts heat exposure and vulnerability. Moreover, the emphasis on socio-demographic factors such as age, health conditions, and education level as primary determinants of heat vulnerability is well-supported by existing models of social determinants of health (Marmot, 2005). This study advances the understanding of heat vulnerability by integrating machine learning algorithms, thereby enabling a more precise and refined identification of key vulnerability indicators. Comparative studies on Australian capital cities provide valuable context for evaluating the performance of different approaches. Elizabeth Loughnan et al. (2014), for instance, developed a HVI for Australian capital cities, employing stepwise LR for weighting and validating their index using Spearman correlation coefficients against emergency service demand during extreme heat events. Their findings indicated accuracy levels ranging from 0.186 to 0.807, with the majority of cities achieving accuracy below 0.60. Similarly, Wang et al. (2023) proposed a heat health risk index for Australian capital cities, utilising equal weighting and PCA, and validated the index using hospital admissions, deaths, and death rates during summer periods. However, their Pearson correlation coefficients between index values and validation variables were all below 0.1, suggesting weak linear relationships. More recently, Li et al. (2024c) introduced a comprehensive HVI, incorporating more representative and balanced indicators same as this study, and adopting a weighting approach referring to Wang et al. (2023). Their index was validated against heat-related deaths, yielding Pearson correlation coefficients of 0.92 and 0.68 for deaths estimated by population and death causes at the SA2 scale, corresponding to R^2 values of approximately 0.85 and 0.46 in linear models. In contrast, machine learning–based HVMs in this study demonstrated superior assessment performance, with R^2 values consistently exceeding 0.83 in both training and testing phases. Notably, the RF algorithm achieved R^2 values above 0.90, highlighting its robustness and accuracy. This comparison underscores the enhanced precision offered by machine learning approaches, particularly RF, in assessing heat vulnerability, thereby reinforcing the potential of these advanced analytical techniques in improving vulnerability assessments.

Traditional methods such as equal weighting and PCA, the two most frequently used methods, are limited in accurately demonstrating how indicators influence heat vulnerability. Equal weighting assumes that all indicators contribute equally, which overlooks the fact that different indicators may have varying levels of impact. It also fails to capture interactions between indicators, which can significantly affect outcomes. For example, Christenson et al. (2017) assigned equal weights to a total of 23 vulnerability indicators, such as air surface temperature, air quality, and poverty rate. While air quality can exacerbate heat vulnerability by contributing to higher temperatures through pollutants like ozone, air surface temperature has a more immediate and direct effect on thermal stress and heat-related health risks. Similarly, poverty exerts a more pervasive influence than the proportion of households without a vehicle, as it not only limits access to air conditioning and healthcare but also affects housing quality and insulation, further increasing heat exposure. Furthermore, although diabetes prevalence is recognised as a major risk factor for heat-related morbidity, it is less directly associated with heat-induced outcomes than heat stress, which represents a more immediate and universal health threat during extreme heat events. PCA, while useful for reducing dimensionality, compromises interpretability by transforming original indicators into principal components, which are difficult to relate back to the specific factors driving vulnerability. For instance, in the research of Hulley et al. (2019), the green vegetation fraction is grouped into a component with income and education, which is interpreted as representing adaptability through socioeconomic status, despite vegetation being more closely related to environmental factors than to socioeconomic adaptability. Similarly, the elderly population is included in a component with population density and building height, labelled representing sensitivity through urban congestion, even though the elderly population has no direct contribution to urban congestion compared to the other two indicators. Moreover, PCA does not account for the relationship between indicators and the target variable, making it less suited for understanding direct influences on heat vulnerability. This study adopts an integrated approach combining feature importance, PDP, and SHAP values to explore the influence of heat-related indicators on heat vulnerability and to enhance result interpretability (Liao et al., 2024). Feature importance provides an overall evaluation of each feature's contribution, while PDP reveals the marginal effect of individual features. SHAP values explain the contributions of features at the individual level. Together, these methods offer a comprehensive understanding of feature influence from global, marginal, and individual perspectives. This methodological innovation challenges traditional approaches that rely on simpler statistical models, suggesting that machine learning can provide deeper insights into the complex interplay of factors influencing heat vulnerability.

The methodological framework utilised in this study advances existing heat vulnerability assessment practices in terms of indicator selection, weight allocation, and result validation. This study adopts the theoretical framework for indicator selection and categorisation proposed by Li et al. (2020), which incorporates widely accepted and representative indicators re-categorised into balanced groups. The high accuracy of the results demonstrates that this study provides a solid empirical validation of the indicator framework, helping to address limitations in current studies by providing standardised and comprehensive indicator references. Existing studies lack a standard for indicator selection, leading to inconsistent indicator use and interpretation. For instance, the number of indicators used varies significantly, such as three indicators selected in Cai et al.'s (2019) study compared to 22 indicators utilised in El-Zein and Tonmoy's (2015) work. Subjective indicator selection also exists; for example, Prosdocimi and Klima (2020) used access to sewage and garbage services to assess heat vulnerability, which is rarely used and lacks supporting references. Traditional HVMs typically rely on explicit methods (such as equal weights and expert judgement) (Guo et al., 2019; Oh et al., 2017) or statistical techniques (such as PCA and AHP) (Estoque et al., 2020; Tate, 2012) to determine the weight of each indicator, which can introduce subjective biases and make it difficult to capture the nonlinear relationships between indicators and heat vulnerability. In this study, machine learning algorithms are used to automatically learn the relationships directly from multidimensional data, reducing biases introduced by human subjectivity and statistical limitations, thereby making the weight assignment more objective and accurate. A lack of validation processes and low validation accuracy are other key limitations in existing studies. Reviews reveal that only 25% of reviewed studies involved validation processes, and validation results varied significantly. For example, the R^2 value between assessed heat vulnerability and observed deaths was less than 0.01 in Conlon et al. (2020), and only 0.32 in Kim et al. (2017). This study addresses this by involving heat-related morbidity and mortality as labelled data and implementing strict validation methods, such as cross-validation and test set evaluation, thereby improving the accuracy and reliability of the assessment results. Moreover, this framework comprehensively evaluates heat vulnerability by integrating remote sensing data, detailed census data, and POI information. These multi-source data not only enhance the spatial and temporal resolution of vulnerability assessments, enabling more accurate identification of vulnerable areas that traditional HVMs may overlook due to coarse data, but also improve the applicability of models across different geographic and demographic contexts.

The unique contribution of this research lies in its comprehensive application of machine learning to assess heat vulnerability, coupled with the use of diverse data sources, including remote sensing and detailed census data. By providing a robust analytical framework and highlighting the critical role of social factors, this study offers valuable insights that can inform more targeted and effective urban planning and public health policies. Machine learning offers more accurate and fine-scale heat vulnerability identification than traditional HVMs, enabling urban planners to understand the spatial distribution and differences in heat vulnerability more efficiently and precisely. This highlights the importance of a more integrated approach that considers both social and environmental determinants in developing effective heat mitigation strategies. Furthermore, machine learning identifies key factors related to heat vulnerability in this study. The study's findings that personal health conditions and socio-demographic characteristics play a more dominant role than environmental and infrastructural factors challenge some conventional perspectives that prioritise physical and environmental interventions (Gill et al., 2007; Liou et al., 2021; Stone et al., 2010). These allow policymakers to tailor interventions directly to the most vulnerable populations, such as by providing cooling services, enhancing medical support, and increasing public awareness by organising training modules about heat risks. Machine learning can also incorporate real-time data from remote sensing and detailed census data to formulate adaptive policies that adjust with changing data (Sohani et al., 2022), allowing for the development of dynamic intervention strategies that evolve with urban changes. With machine learning, more adaptable models can be developed for different geographic contexts through adaptive learning. Expanding geographic scope and using higher-resolution datasets can enhance the generalisability of findings, allowing for better-suited local policies. Finally, machine learning enhances the understanding of the interaction between social factors and environmental exposure through its ability to handle complex non-linear relationships, integrate multi-source data, and extract features from high-dimensional datasets, leading to urban planning decisions that simultaneously consider both the physical environment and social determinants. This integration of advanced data analytics and multidisciplinary perspectives represents a significant advancement in the field, paving the way for future research to further explore and address the multifaceted nature of heat vulnerability.

The findings of this study have substantial practical implications, particularly for policymaking and urban planning. By identifying key factors that contribute to heat vulnerability, this research provides critical insights for developing targeted heat mitigation and adaptation strategies. Policymakers can leverage these insights to prioritise interventions aimed at improving the health and social conditions of vulnerable populations, such as the elderly and those with chronic health conditions. For example, improving heat-related health infrastructure is crucial for protecting vulnerable populations (Chen et al., 2021a). The Australian government can increase the number of air-conditioned spaces in communities, provide cooling centres for the elderly and people with chronic illnesses, and enhance the supply of emergency medical resources during extreme heat periods, thereby reducing health risks associated with high temperatures. Moreover, raising public awareness of heat risks helps prevent heat-related health issues (Wang et al., 2023). Through educational campaigns, the government can increase public understanding of heat risks and encourage people, especially those with lower levels of education, to take preventive measures during high temperatures, such as staying hydrated, avoiding outdoor activities during the hottest parts of the day, and recognising symptoms of heat-related illnesses. Enhancing access to cooling and medical resources and increasing public awareness about heat risks can significantly reduce heat-related health impacts (He et al., 2019; Lee et al., 2015). Furthermore, the study underscores the importance of integrating green infrastructure into urban planning. Cities with higher vegetation cover and green spaces exhibit lower heat vulnerability, suggesting that expanding urban green spaces can mitigate the UHI effect and improve the overall thermal comfort of urban environments (Gascon et al., 2016; Xiang et al., 2024).

This insight supports the development of policies that promote the creation and maintenance of green spaces, which not only provide cooling benefits but also enhance urban biodiversity and residents' wellbeing. This study also aligns with the Sustainable Development Goals (SDGs), particularly goals 3, 11, 13, and 15 (United Nations, 2023). This approach aids in identifying and protecting high-risk populations, thereby improving public health and wellbeing (Goal 3). It optimises urban planning and green space distribution, enhancing the resilience and sustainability of communities (Goal 11). Extraordinary assessment capabilities support climate action by providing scientific evidence for effective mitigation and adaptation strategies (Goal 13). Overall, this method contributes to a more sustainable and climate-resilient future.

Despite this study's efforts to enhance model accuracy, applicability, and reliability by integrating machine learning algorithms and multi-source data, several limitations may still affect the interpretation and application of its findings. Firstly, census data may not be capable of fully capturing the characteristics of population mobility and unrecorded residents. Australian census data are collected based on places of usual residence and cannot fully reflect people's movement across study units, such as commuting and working. This limitation may affect the assessment of heat vulnerability due to potential inaccuracies in capturing socio-demographic features. Additionally, there are limited numbers of undocumented populations due to factors like homelessness, language barriers, and residence in remote areas, which may result in findings that do not accurately reflect their heat vulnerability. To mitigate this limitation, future research could integrate dynamic data sources, such as mobile phone data, social media data, and municipal data, to capture dynamic information on population distribution and enrich the information about unrecorded individuals. Collaborations with non-governmental organisations could also be considered, as they can help organise survey events to collect data, enhancing the comprehensiveness of heat vulnerability assessments. Incorporating more advanced machine learning algorithms to train these dynamic multi-source datasets could also be highly beneficial. For instance, graph neural networks can capture complex spatial relationships between different regions, identifying the influence of neighbourhood interactions on heat vulnerability. Moreover, recurrent neural networks can be used to capture the impact of dynamic population movement patterns on heat vulnerability at different time points.

Secondly, although Landsat 8 has a high spatial resolution of 30 m, its temporal resolution is lower, with a 16-day interval. Short-term temperature variations may not be fully captured, particularly during extreme heat events. Additionally, the imaging time of Landsat 8 is during the daytime, without nighttime imaging, which may not reflect heat exposure throughout the entire day. Future studies could incorporate MODIS data and air temperature data, which have higher temporal resolutions of one day and 1 hour, respectively, to capture rapid changes in the thermal environment. Furthermore, MODIS data also provide information on variations in both daytime and nighttime temperatures. Air temperatures complement LST because the sensations of humans to these temperatures are different, enabling a more comprehensive assessment of human exposure to extreme heat. Thirdly, although Australian capital cities exhibit a variety of geographic and climatic conditions, all of them are highly urbanised, with advanced levels of economic development, education, and infrastructure. This means that the results may not be directly applicable to less urbanised and developed regions, suggesting the need for validation of the findings across a diverse range of settings, including less developed regions or cities with lower urbanisation rates. Expanding the geographic scope could enhance the study's adaptability, applicability, and generalisability, aiding in the development of tailored and effective policies and strategies. Additionally, indicators such as land cover/use and BD can capture the broad impacts of urban development on thermal comfort characteristics, including increased heat retention and reduced ventilation. However, these indicators do not fully depict insulation and ventilation within buildings, which also play critical roles in mitigating heat vulnerability. It is essential to consider these factors in future studies to develop a more comprehensive framework for assessing thermal environments, thereby enhancing the precision and reliability of heat vulnerability results. By addressing these limitations, future research can build on this study's foundation, providing more comprehensive and universally applicable insights into heat vulnerability and informing more effective and inclusive urban planning and public health policies.

10.5 Conclusion

This study set out to evaluate heat vulnerability in Australian capital cities by integrating machine learning algorithms and multi-source data. The primary objective was to identify key heat vulnerability indicators and highlight regional disparities to inform urban planners and policymakers in developing targeted mitigation and adaptation strategies.

The findings reveal that the RF algorithm is the most effective, achieving a training R^2 of 0.9179 and a testing R^2 of 0.9089, underscoring its ability to capture complex relationships within the data. Spatial analysis highlighted significant regional disparities, with higher heat vulnerability in densely populated urban areas and lower vulnerability in greener suburban and rural regions. This imbalance necessitates tailored heat mitigation strategies. Sensitivity analysis further revealed that personal health conditions and socio-demographic characteristics, such as illness status, age, and education level, play dominant roles in determining heat vulnerability, overshadowing environmental and infrastructural factors.

The findings of this study have substantial practical implications for policymakers and urban planners in mitigating heat vulnerability in Australian capital cities. Targeted healthcare interventions are essential, particularly in areas identified as highly vulnerable, where establishing cooling centres and enhancing community healthcare services can significantly reduce heat-related health risks, especially for the elderly and those with chronic conditions. Public awareness campaigns should also be prioritised, educating vulnerable populations about the risks of heat exposure and promoting preventive measures. Expanding urban green infrastructure is another crucial step; increasing green cover, such as parks and green roofs, can help reduce UHI effects and improve overall urban resilience (Goonetilleke et al., 2014; Vaeztavakoli et al., 2018; Yigitcanlar, 2010). Furthermore, integrating heat mitigation strategies into urban planning, such as mandating green spaces and reflective building materials in vulnerable areas, will help in reducing heat retention and mitigating its impact. Lastly, enhancing accessibility to cooling and medical resources, particularly for the most vulnerable populations, should be part of a broader urban cooling strategy, ensuring that these groups have the support needed during extreme heat events.

The significance of this study lies in its comprehensive approach of combining machine learning with diverse data sources to provide valuable insights for urban planning and public health policies. The research confirms that improving the health and social conditions of vulnerable populations and integrating green infrastructure are crucial for mitigating heat risks. Nonetheless, the study acknowledges limitations, such as potential inaccuracies in data sources and the need for broader geographic applicability. Future research should integrate higher-resolution datasets, expand geographic scope, and employ a wider range of machine learning techniques to improve the robustness and applicability of vulnerability assessments.

In conclusion, this study advances the application of machine learning in environmental research, offering a nuanced understanding of heat vulnerability. By highlighting the critical role of social factors, it provides actionable insights for enhancing urban resilience and sustainability in the face of climate change. Future research should continue to explore the complex interplay of social and environmental determinants, further enhancing our ability to address heat vulnerability effectively.

Acknowledgements

This chapter, with permission from the copyright holder, is a reproduced version of the following journal article: Li, F., Yigitcanlar, T., Nepal, M., Nguyen, K., Dur, F., & Li, W. (2025). Mapping heat vulnerability in Australian capital cities: A machine learning and multi-source data analysis. *Sustainable Cities and Society, 119*(1), 106079.

References

Adnan, M., Dewan, A., Botje, D., Shahid, S., & Hassan, Q. K. (2022). Vulnerability of Australia to heatwaves: A systematic review on influencing factors, impacts, and mitigation options. *Environmental Research, 213*, 113703.

Afzal, S., Ziapour, B. M., Shokri, A., Shakibi, H., & Sobhani, B. (2023). Building energy consumption prediction using multilayer perceptron neural network-assisted models; Comparison of different optimization algorithms. *Energy, 282*, 128446.

Aheto, J. M. K., Duah, H. O., Agbadi, P., & Nakua, E. K. (2021). A predictive model, and predictors of under-five child malaria prevalence in Ghana: How do LASSO, Ridge and Elastic net regression approaches compare. *Preventive Medicine Reports, 23*, 101475.

Almukhalfi, H., Noor, A., & Noor, T. H. (2024). Traffic management approaches using machine learning and deep learning techniques: A survey. *Engineering Applications of Artificial Intelligence, 133*, 108147.

Aubrecht, C., & Özceylan, D. (2013). Identification of heat risk patterns in the US National Capital Region by integrating heat stress and related vulnerability. *Environment International, 56*, 65–77.

Australian Bureau of Meteorology. (2021). *National climate statement*. Climate Change in Australia. Retrieved July 5, 2024, from https://www.climatechangeinaustralia.gov.au/en/changing-climate/national-climate-statement/

Australian Bureau of Statistics (ABS). (2022). *ABS census 2021*. Retrieved June 6, 2023, from https://www.abs.gov.au/census/find-census-data/datapacks

Australian Bureau of Statistics (ABS). (2023a, September 27). *Causes of death, Australia*. 2022. https://www.abs.gov.au/statistics/health/causes-death/causes-death-australia/2022

Australian Bureau of Statistics (ABS). (2023b, October 18). *Deaths, Australia*. 2022. https://www.abs.gov.au/statistics/people/population/deaths-australia/latest-release

Bakhsh, K., Rauf, S., & Zulfiqar, F. (2018). Adaptation strategies for minimizing heat wave induced morbidity and its determinants. *Sustainable Cities and Society, 41*, 95–103.

Ballester, J., Quijal-Zamorano, M., Méndez Turrubiates, R. F., Pegenaute, F., Herrmann, F. R., Robine, J. M., Basagaña, X., Tonne, C., Antó, J. M., & Achebak, H. (2023). Heat-related mortality in Europe during the summer of 2022. *Nature Medicine, 29*(7), 1857–1866.

Belgiu, M., & Drˇagut, L. (2016). Random forest in remote sensing: A review of applications and future directions. *ISPRS Journal of Photogrammetry and Remote Sensing, 114*, 24–31.

Bu, S., Smith, K. L., Masoud, F., & Sheinbaum, A. (2024). Spatial distribution of heat vulnerability in Toronto, Canada. *Urban Climate, 54*, 101838.

Cai, Z., Tang, Y., Chen, K., & Han, G. (2019). Assessing the heat vulnerability of different local climate zones in the old areas of a Chinese megacity. *Sustainability, 11*(7), 2032.

Chambers, J. (2020). Global and cross-country analysis of exposure of vulnerable populations to heatwaves from 1980 to 2018. *Climatic Change, 163*(1), 539–558.

Chen, B., Xie, M., Feng, Q., Li, Z., Chu, L., & Liu, Q. (2021a). Heat risk of residents in different types of communities from urban heat-exposed areas. *Science of the Total Environment, 768*, 145052.

Chen, C.-C., Wang, Y.-R., Yeh, H.-Y., Lin, T.-H., Huang, C.-S., & Wu, C.-F. (2021b). Estimating monthly $PM_{2.5}$ concentrations from satellite remote sensing data, meteorological variables, and land use data using ensemble statistical modeling and a random forest approach. *Environmental Pollution, 291*.

Chen, Q., Ding, M., Yang, X., Hu, K., & Qi, J. (2018). Spatially explicit assessment of heat health risk by using multi-sensor remote sensing images and socioeconomic data in Yangtze River Delta, China. *International Journal of Health Geographics, 17*, 1–15.

Cheng, W., Li, D., Liu, Z., & Brown, R. D. (2021). Approaches for identifying heat-vulnerable populations and locations: A systematic review. *Science of The Total Environment, 799*, 149417.

Chi, Y., Fan, M., Zhao, C., Yang, Y., Fan, H., Yang, X., Yang, J., & Tao, J. (2022). Machine learning-based estimation of ground-level NO_2 concentrations over China. *Science of The Total Environment, 807*, 150721.

Christenson, M., Geiger, S. D., Phillips, J., Anderson, B., Losurdo, G., & Anderson, H. A. (2017). Heat vulnerability index mapping for Milwaukee and Wisconsin. *Journal of Public Health Management and Practice, 23*(4), 396–403.

Chuang, W. C., & Gober, P. (2015). Predicting hospitalization for heat-related illness at the census-tract level: Accuracy of a generic heat vulnerability index in Phoenix, Arizona (USA). *Environmental Health Perspectives, 123*(6), 606–612.

Coates, L., van Leeuwen, J., Browning, S., Gissing, A., Bratchell, J., & Avci, A. (2022). Heatwave fatalities in Australia, 2001–2018: An analysis of coronial records. *International Journal of Disaster Risk Reduction, 67*, 102671. https://doi.org/10.1016/j.ijdrr.2021.102671

Conlon, K. C., Mallen, E., Gronlund, C. J., Berrocal, V. J., Larsen, L., & O'Neill, M. S. (2020). Mapping human vulnerability to extreme heat: A critical assessment of heat vulnerability indices created using principal components analysis. *Environmental Health Perspectives, 128*(9), 097001.

Crichton, D. (1999). The risk triangle. *Natural Disaster Management, 102*(3), 102–103.

Degirmenci, K., Desouza, K., Fieuw, W., Watson, R., & Yigitcanlar, T. (2021). Understanding policy and technology responses in mitigating urban heat islands: A literature review and directions for future research. *Sustainable Cities and Society, 70*, 102873.

Department of Human Services. (2009). *January 2009 heatwave in Victoria: An assessment of health impacts*. Victorian Government Department of Human Services.

Depietri, Y., Welle, T., & Renaud, F. G. (2013). Social vulnerability assessment of the Cologne urban area (Germany) to heat waves: Links to ecosystem services. *International Journal of Disaster Risk Reduction*, *6*, 98–117.

Economics, N. C. (2018). *Heatwaves in Victoria: A vulnerability assessment*. Department of Environment, Land, Water and Planning, Victoria.

Elizabeth Loughnan, M., J. Tapper, N., Phan, T., & A. McInnes, J. (2014). Can a spatial index of heat-related vulnerability predict emergency service demand in Australian capital cities? *International Journal of Emergency Services*, *3*(1), 6–33.

El-Zein, A., & Tonmoy, F. N. (2015). Assessment of vulnerability to climate change using a multi-criteria outranking approach with application to heat stress in Sydney. *Ecological Indicators*, *48*, 207–217.

Estoque, R. C., Ooba, M., Seposo, X. T., Togawa, T., Hijioka, Y., Takahashi, K., & Nakamura, S. (2020). Heat health risk assessment in Philippine cities using remotely sensed data and social-ecological indicators. *Nature Communications*, *11*(1), 1581.

Estoque, R. C., Togawa, T., Ooba, M., Gomi, K., Nakamura, S., Hijioka, Y., & Kameyama, Y. (2019). A review of quality of life (QOL) assessments and indicators: Towards a "QOL-Climate" assessment framework. *Ambio*, *48*(6), 619–638.

Fan, D., Xue, K., Zhang, R., Zhu, W., Zhang, H., Qi, J., Zhu, Z., Wang, Y., & Cui, P. (2024). Application of interpretable machine learning models to improve the prediction performance of ionic liquids toxicity. *Science of The Total Environment*, *908*, 168168.

Forceville, G., Lemonsu, A., Goria, S., Stempfelet, M., Host, S., Alessandrini, J. M., Cordeau, E. & Pascal, M. (2024). Spatial contrasts and temporal changes in fine-scale heat exposure and vulnerability in the Paris region. *Science of the Total Environment*, *906*, 167476.

Gao, Y., Zhao, J., & Han, L. (2023). Quantifying the nonlinear relationship between block morphology and the surrounding thermal environment using random forest method. *Sustainable Cities and Society*, *91*, 104443.

Gascon, M., Triguero-Mas, M., Martínez, D., Dadvand, P., Rojas-Rueda, D., Plasència, A., & Nieuwenhuijsen, M. J. (2016). Residential green spaces and mortality: A systematic review. *Environment International*, *86*, 60–67.

Ghaffarian, S., Roy, D., Filatova, T., & Kerle, N. (2021). Agent-based modelling of post-disaster recovery with remote sensing data. *International Journal of Disaster Risk Reduction*, *60*, 102285.

Gill, S. E., Handley, J. F., Ennos, A. R., & Pauleit, S. (2007). Adapting cities for climate change: The role of the green infrastructure. *Built Environment*, *33*(1), 115–133.

Goonetilleke, A., Yigitcanlar, T., Ayoko, G. A., & Egodawatta, P. (2014). *Sustainable urban water environment: Climate, pollution and adaptation* (pp. 245–346). Edward Elgar.

Green, H. K., Andrews, N., Armstrong, B., Bickler, G., & Pebody, R. (2016). Mortality during the 2013 heatwave in England – How did it compare to previous heatwaves? A retrospective observational study. *Environmental Research*, *147*, 343–349.

Guo, X., Huang, G., Jia, P., & Wu, J. (2019). Estimating fine-scale heat vulnerability in Beijing through two approaches: Spatial patterns, similarities, and divergence. *Remote Sensing*, *11*(20), 2358.

Harmay, N. S. M., Kim, D., & Choi, M. (2021). Urban heat island associated with land use/land cover and climate variations in Melbourne, Australia. *Sustainable Cities and Society*, *69*, 102861.

He, C., Ma, L., Zhou, L., Kan, H., Zhang, Y., Ma, W., & Chen, B. (2019). Exploring the mechanisms of heat wave vulnerability at the urban scale based on the application of big data and artificial societies. *Environment International, 127*, 573–583.

Hu, K., Yang, X., Zhong, J., Fei, F., & Qi, J. (2017). Spatially explicit mapping of heat health risk utilizing environmental and socioeconomic data. *Environmental Science & Technology, 51*(3), 1498–1507.

Hulley, G., Shivers, S., Wetherley, E., & Cudd, R. (2019). New ECOSTRESS and MODIS land surface temperature data reveal fine-scale heat vulnerability in cities: A case study for Los Angeles County, California. *Remote Sensing, 11*(18), 2136.

Iping, A., Kidston-Lattari, J., Simpson-Young, A., Duncan, E., & McManus, P. (2019). (Re)presenting urban heat islands in Australian cities: A study of media reporting and implications for urban heat and climate change debates. *Urban Climate, 27*, 420–429.

Jato-Espino, D., Manchado, C., Roldán-Valcarce, A., & Moscard´o, V. (2022). ArcUHI: A GIS add-in for automated modelling of the urban heat island effect through machine learning. *Urban Climate, 44*, Article 101203.

Kamruzzaman, M., Deilami, K., & Yigitcanlar, T. (2018). Investigating the urban heat island effect of transit oriented development in Brisbane. *Journal of Transport Geography, 66*, 116–124.

Kan, H. J., Kharrazi, H., Chang, H. Y., Bodycombe, D., Lemke, K., & Weiner, J. P. (2019). Exploring the use of machine learning for risk adjustment: A comparison of standard and penalized linear regression models in predicting health care costs in older adults. *PLoS ONE, 14*(3), e0213258.

Karalas, K., Tsagkatakis, G., Zervakis, M., & Tsakalides, P. (2016). Land classification using remotely sensed data: Going multilabel. *IEEE Transactions on Geoscience and Remote Sensing, 54*(6), 3548–3563.

Kim, M., Brunner, D., & Kuhlmann, G. (2021). Importance of satellite observations for high-resolution mapping of near-surface NO2 by machine learning. *Remote Sensing of Environment, 264*, 112573.

Kim, Y. H., Min, S. K., Zhang, X., Sillmann, J., & Sandstad, M. (2020). Evaluation of the CMIP6 multi-model ensemble for climate extreme indices. *Weather and Climate Extremes, 29*, 100269.

Kitchley, J. L. (2024). A framework to assess the contextual composite heat vulnerability index for a heritage city in India – A case study of Madurai. *Sustainable Cities and Society, 101*, 105119.

Knowlton, K., Rotkin-Ellman, M., King, G., Margolis, H. G., Smith, D., Solomon, G., Trent, R., & English, P. (2009). The 2006 California heat wave: Impacts on hospitalizations and emergency department visits. *Environmental Health Perspectives, 117*(1), 61–67.

Kokkinos, K., Karayannis, V., Nathanail, E., & Moustakas, K. (2021). A comparative analysis of statistical and computational intelligence methodologies for the prediction of traffic-induced fine particulate matter and NO_2. *Journal of Cleaner Production, 328*, 129500.

Lee, T. M., Markowitz, E. M., Howe, P. D., Ko, C. Y., & Leiserowitz, A. A. (2015). Predictors of public climate change awareness and risk perception around the world. *Nature Climate Change, 5*(11), 1014–1020.

Li, F., Yigitcanlar, T., Li, W., Nepal, M., Nguyen, K., & Dur, F. (2024b). Understanding urban heat vulnerability: Scientometric analysis of five decades of research. *Urban Climate, 56*, 102035.

Li, F., Yigitcanlar, T., Nepal, M., Thanh, K. N., & Dur, F. (2022). Understanding urban heat vulnerability assessment methods: A PRISMA review. *Energies, 15*(19), 6998.

Li, F., Yigitcanlar, T., Nepal, M., Nguyen, K., & Dur, F. (2023). Machine learning and remote sensing integration for leveraging urban sustainability: A review and framework. *Sustainable Cities and Society, 96*, 104653.

Li, F., Yigitcanlar, T., Nepal, M., Thanh, K. N., & Dur, F. (2024a). A novel urban heat vulnerability analysis: Integrating machine learning and remote sensing for enhanced insights. *Remote Sensing, 16*(16), 3032.

Li, F., Yigitcanlar, T., Nepal, M., Nguyen, K., Dur, F., & Li, W. (2024c). Assessing heat vulnerability and multidimensional inequity: Lessons from indexing the performance of Australian capital cities. *Sustainable Cities and Society, 105875.*

Li, Q., Shan, H., Tang, Y., & Yao, V. (2024). Corporate climate risk: Measurements and responses. *The Review of Financial Studies, 37*(6), 1778–1830.

Liao, W., Fang, J., Ye, L., Bak-Jensen, B., Yang, Z., & Porte-Agel, F. (2024). Can we trust explainable artificial intelligence in wind power forecasting? *Applied Energy, 376*, 124273.

Liou, Y. A., Nguyen, K. A., & Ho, L. T. (2021). Altering urban greenspace patterns and heat stress risk in Hanoi city during master plan 2030 implementation. *Land Use Policy, 105*, 105405.

Liu, J., Liu, F., Zheng, C., Zhou, D., & Wang, L. (2022). Optimizing asphalt mix design through predicting the rut depth of asphalt pavement using machine learning. *Construction and Building Materials, 356*, 129211.

Liu, X., Yue, W., Yang, X., Hu, K., Zhang, W., & Huang, M. (2020). Mapping urban heat vulnerability of extreme heat in Hangzhou via comparing two approaches. *Complexity, 2020*(1), 9717658.

Liu, Y., An, Z., & Ming, Y. (2024). Simulating influences of land use/land cover composition and configuration on urban heat island using machine learning. *Sustainable Cities and Society, 108*, 105482.

Lu, Z., Im, J., Rhee, J., & Hodgson, M. (2014). Building type classification using spatial and landscape attributes derived from LiDAR remote sensing data. *Landscape and Urban Planning, 130*, 134–148.

Macintyre, H. L., Heaviside, C., Taylor, J., Picetti, R., Symonds, P., Cai, X. M., & Vardoulakis, S. (2018). Assessing urban population vulnerability and environmental risks across an urban area during heatwaves – Implications for health protection. *Science of the Total Environment, 610*, 678–690.

Maier, P., Liebmann, L., Hasel, K., Berg, R., Lehner, F., Bügelmayer-Blaschek, M., Formayer, H., & Suna, D. (2025). Climatological hazard indicators for a robust and integrated energy infrastructure in Austria. *Scientific Data, 12*(1), 1862.

Mallen, E., Stone, B., & Lanza, K. (2019). A methodological assessment of extreme heat mortality modeling and heat vulnerability mapping in Dallas, Texas. *Urban Climate, 30*, 100528.

Marmot, M. (2005). Social determinants of health inequalities. *The Lancet, 365*(9464), 1099–1104.

McGeehin, M. A., & Mirabelli, M. (2001). The potential impacts of climate variability and change on temperature-related morbidity and mortality in the United States. *Environmental Health Perspectives, 109*(suppl 2), 185–189.

Meng, X., Hand, J. L., Schichtel, B. A., & Liu, Y. (2018). Space-time trends of $PM_{2.5}$ constituents in the conterminous United States estimated by a machine learning approach, 2005–2015. *Environment International, 121*, 1137–1147.

Modi, Y., Teli, R., Mehta, A., Shah, K., & Shah, M. (2022). A comprehensive review on intelligent traffic management using machine learning algorithms. *Innovative Infrastructure Solutions*, *7*(1), 128.

Niu, Y., Li, Z., Gao, Y., Liu, X., Xu, L., Vardoulakis, S., Yue, Y., Wang, J., & Liu, Q. (2021). A systematic review of the development and validation of the heat vulnerability index: Major factors, methods, and spatial units. *Current Climate Change Reports*, *7*(3), 87–97.

Oh, K. Y., Lee, M. J., & Jeon, S. W. (2017). Development of the Korean climate change vulnerability assessment tool (VESTAP)—Centered on health vulnerability to heat waves. *Sustainability*, *9*(7), 1103.

Okumus, D. E., & Terzi, F. (2021). Evaluating the role of urban fabric on surface urban heat island: The case of Istanbul. *Sustainable Cities and Society*, *73*, 103128.

Palaniyappan, B., & Vinopraba, T. (2024). Dynamic pricing for load shifting: Reducing electric vehicle charging impacts on the grid through machine learning-based demand response. *Sustainable Cities and Society*, *103*, 105256.

Parry, M. L. (Ed.). (2007). *Climate change 2007-impacts, adaptation and vulnerability: Working group II contribution to the fourth assessment report of the IPCC* (Vol. 4). Cambridge University Press.

Prosdocimi, D., & Klima, K. (2020). Health effects of heat vulnerability in Rio de Janeiro: A validation model for policy applications. *SN Applied Sciences*, *2*, 1–11.

Reid, C. E., O'neill, M. S., Gronlund, C. J., Brines, S. J., Brown, D. G., Diez-Roux, A. V., & Schwartz, J. (2009). Mapping community determinants of heat vulnerability. *Environmental Health Perspectives*, *117*(11), 1730–1736.

Schlosser, A. D., Szabó, G., Bertalan, L., Varga, Z., Enyedi, P., & Szabó, S. (2020). Building extraction using orthophotos and dense point cloud derived from visual band aerial imagery based on machine learning and segmentation. *Remote Sensing*, *12*(15), 2397.

Shafizadeh-Moghadam, H., Asghari, A., Tayyebi, A., & Taleai, M. (2017). Coupling machine learning, tree-based and statistical models with cellular automata to simulate urban growth. *Computers, Environment and Urban Systems*, *64*, 297–308.

Sohani, A., Sayyaadi, H., Miremadi, S. R., Samiezadeh, S., & Doranehgard, M. H. (2022). Thermo-electro-environmental analysis of a photovoltaic solar panel using machine learning and real-time data for smart and sustainable energy generation. *Journal of Cleaner Production*, *353*, 131611.

Solar, O., & Irwin, A. (2010). *A conceptual framework for action on the social determinants of health*. WHO Document Production Services. https://www.who.int/publications/i/item/9789241500852

Song, J., Huang, B., Kim, J. S., Wen, J., & Li, R. (2020). Fine-scale mapping of an evidence-based heat health risk index for high-density cities: Hong Kong as a case study. *Science of the Total Environment*, *718*, 137226.

Stone, B., Hess, J. J., & Frumkin, H. (2010). Urban form and extreme heat events: Are sprawling cities more vulnerable to climate change than compact cities? *Environmental Health Perspectives*, *118*(10), 1425–1428.

Tanoori, G., Soltani, A., & Modiri, A. (2024). Machine learning for urban heat island (UHI) analysis: Predicting land surface temperature (LST) in urban environments. *Urban Climate*, *55*, 101962.

Tate, E. (2012). Social vulnerability indices: A comparative assessment using uncertainty and sensitivity analysis. *Natural Hazards*, *63*, 325–347.

Tella, A., Balogun, A. L., Adebisi, N., & Abdullah, S. (2021). Spatial assessment of PM10 hotspots using random forest, K-nearest neighbour and Naïve Bayes. *Atmospheric Pollution Research, 12*(10), 101202.

United Nations. (2023). *Global indicator framework for the Sustainable Development Goals and targets of the 2030 Agenda for Sustainable Development*. https://unstats.un.org/sdgs/indicators/Global%20Indicator%20Framework%20after%202023%20refinement_Eng.pdf

Vaeztavakoli, A., Lak, A., & Yigitcanlar, T. (2018). Blue and green spaces as therapeutic landscapes: Health effects of urban water canal areas of Isfahan. *Sustainability, 10*(11), 4010.

Victorian Government Department of Environment, Land, Water and Planning. (2019). *The economic impact of heatwaves on Victoria*. Victorian Government. https://www.climatechange.vic.gov.au/__data/assets/pdf_file/0011/413030/The-economic-impact-of-heatwaves-on-Victoria.pdf

Wang, E., Xia, J., Li, J., Sun, X., & Li, H. (2022b). Parameters exploration of SOFC for dynamic simulation using adaptive chaotic grey wolf optimization algorithm. *Energy, 261*, 125146.

Wang, Q., Wang, X., Zhou, Y., Liu, D., & Wang, H. (2022a). The dominant factors and influence of urban characteristics on land surface temperature using random forest algorithm. *Sustainable Cities and Society, 79*, 103722.

Wang, S., Sun, Q. C., Huang, X., Tao, Y., Dong, C., Das, S., & Liu, Y. (2023). Health-integrated heat risk assessment in Australian cities. *Environmental Impact Assessment Review, 102*, 107176.

Weber, S., Sadoff, N., Zell, E., & de Sherbinin, A. (2015). Policy-relevant indicators for mapping the vulnerability of urban populations to extreme heat events: A case study of Philadelphia. *Applied Geography, 63*, 231–243.

Wu, X., Liu, Q., Huang, C., & Li, H. (2022). Mapping heat-health vulnerability based on remote sensing: A case study in Karachi. *Remote Sensing, 14*(7), 1590.

Xiang, Y., Yuan, C., Cen, Q., Huang, C., Wu, C., Teng, M., & Zhou, Z. (2024). Heat risk assessment and response to green infrastructure based on local climate zones. *Building and Environment, 248*, 111040.

Xiao, L., Lo, S., Liu, J., Zhou, J., & Li, Q. (2021). Nonlinear and synergistic effects of TOD on urban vibrancy: Applying local explanations for gradient boosting decision tree. *Sustainable Cities and Society, 72*, 103063.

Yang, N., Shi, H., Tang, H., & Yang, X. (2022). Geographical and temporal encoding for improving the estimation of $PM_{2.5}$ concentrations in China using end-to-end gradient boosting. *Remote Sensing of Environment, 269*, 112828.

Ye, T., Zhao, N., Yang, X., Ouyang, Z., Liu, X., Chen, Q., Hu, K., Yue, W., Qi, J., Li, Z., & Jia, P. (2019). Improved population mapping for China using remotely sensed and points-of-interest data within a random forests model. *The Science of the Total Environment, 658*, 936–946.

Yigitcanlar, T. (2010). *Rethinking sustainable development: Urban management, engineering, and design*. IGI Global.

Yu, Q., Yue, D., Wang, J., Zhang, Q., Li, Y., Yu, Y., Chen, J., & Li, N. (2017). The optimization of urban ecological infrastructure network based on the changes of county landscape patterns: A typical case study of ecological fragile zone located at Deng Kou (Inner Mongolia). *Journal of Cleaner Production, 163*, S54–S67.

Zhang, C., Pan, X., Li, H., Gardiner, A., Sargent, I., Hare, J., & Atkinson, P. M. (2018). A hybrid MLP-CNN classifier for very fine resolution remotely sensed image classification. *ISPRS Journal of Photogrammetry and Remote Sensing, 140*, 133–144.

Zhang, H., Zhang, M., Jin, J., Fei, K., Ji, D., Wu, C., Zhu, J., He, J., Chai, Z., Xie, J., & Dong, X. (2020). Description and climate simulation performance of CAS-ESM version 2. *Journal of Advances in Modeling Earth Systems, 12*(12), p.e2020MS002210.

Zhao, J., Ji, G., Tian, Y., Chen, Y., & Wang, Z. (2018). Environmental vulnerability assessment for mainland China based on entropy method. *Ecological Indicators, 91*, 410–422.

Zhao, P., Zhu, N., Chong, D., & Hou, Y. (2022). Developing a new heat strain evaluation index to classify and predict human thermal risk in hot and humid environments. *Sustainable Cities and Society, 76*, 103440.

Zhou, Z., Yu, L., Wang, Y., Tian, Y., & Li, X. (2024). Innovative approach to daily carbon dioxide emission forecast based on ensemble of quantile regression and attention BILSTM. *Journal of Cleaner Production, 460*, 142605.

Appendix

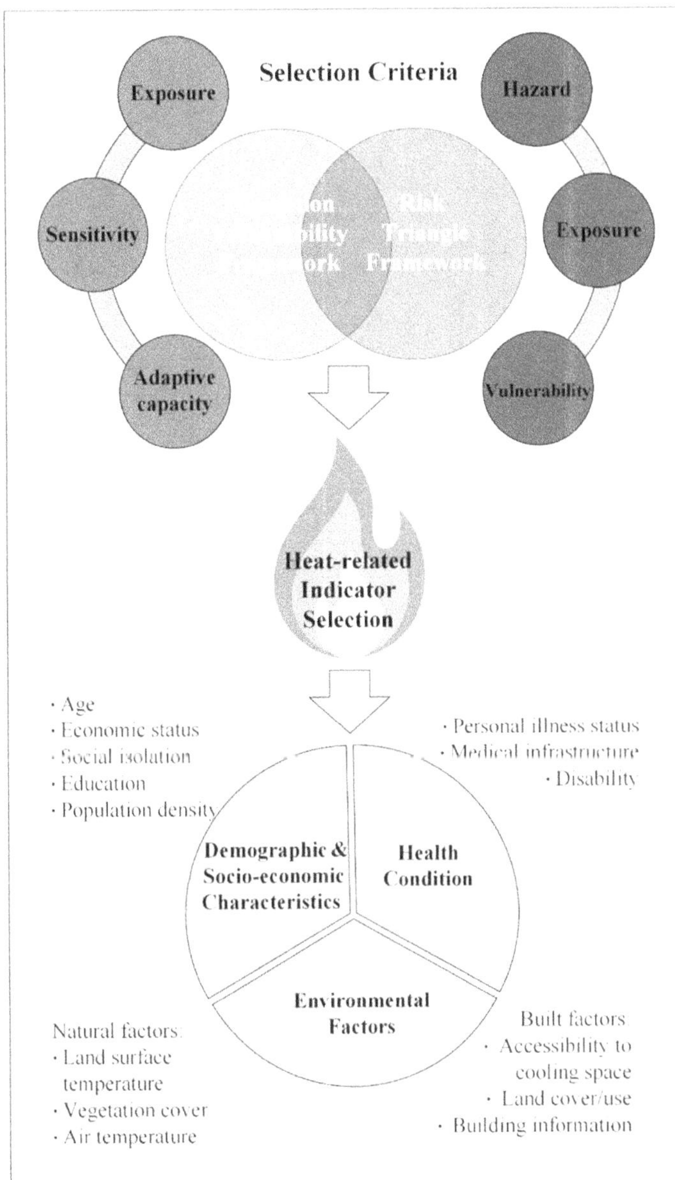

APPENDIX 10.A
Heat-related Indicator Selection Conceptual Framework (Li et al., 2022; Li et al., 2024a).

APPENDIX 10.B

Heat-Related Death Data of Australia's Capital Cities for Validation

q		NSW	VIC	QLD	SA	WA	TAS	NT	ACT
Direct heat-related illnesses	Heat stroke (X30)	2	6	6	6	6	0	6	0
	Dehydration (E86)	27	34	18	11	5	1	1	5
	Hyperpyrexia (R50.9)	0	0	2	0	0	0	0	0
Indirect heat-related illnesses	Cardiovascular diseases (R50.9)	13659	10198	7773	3399	3628	1200	217	548
	Respiratory diseases (J00-J99)	4287	3103	2356	1049	1091	381	101	154
	Diabetes (E10-E14)	1735	1290	989	418	493	162	88	51
	Renal disease (N00-N29)	973	946	421	247	240	57	38	30
	Nervous disorders (G00-G99)	3074	2737	1815	988	1073	208	45	162
	Mental health conditions (F00-F99)	3661	2368	2119	995	868	278	53	165
Total		27418	20682	15499	7113	7404	2287	549	1115
Total deaths		52485	41093	31367	13607	14993	4435	1141	2162
Death rate of total deaths (%)		52.24	50.33	49.41	52.27	49.38	51.57	48.12	51.57

APPENDIX 10.C

Collinearity Diagnostics Results

Indicators	Tolerance	VIF
Age	0.288	3.474
Economic status	0.394	2.535
Social isolation	0.526	1.903
Education	0.304	3.293
Population density	0.732	1.367
Personal illness status	0.162	6.177
Availability of medical resources	0.148	6.737
Disability	0.394	2.535
LST	0.569	1.756
Vegetation cover	0.284	3.525
Availability of cooling spaces	0.146	6.862
Land cover/use	0.937	1.068
Building density	0.316	3.160

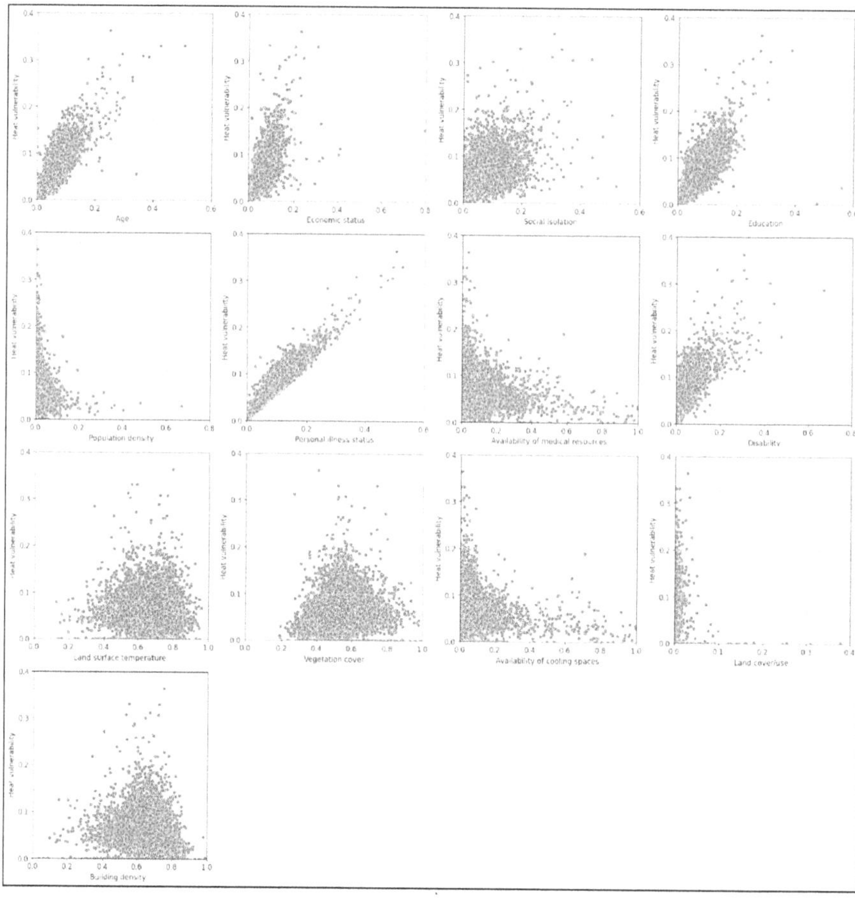

APPENDIX 10.D
Exploratory data analysis results – Scatter plots of heat-related indicators versus labelled data.

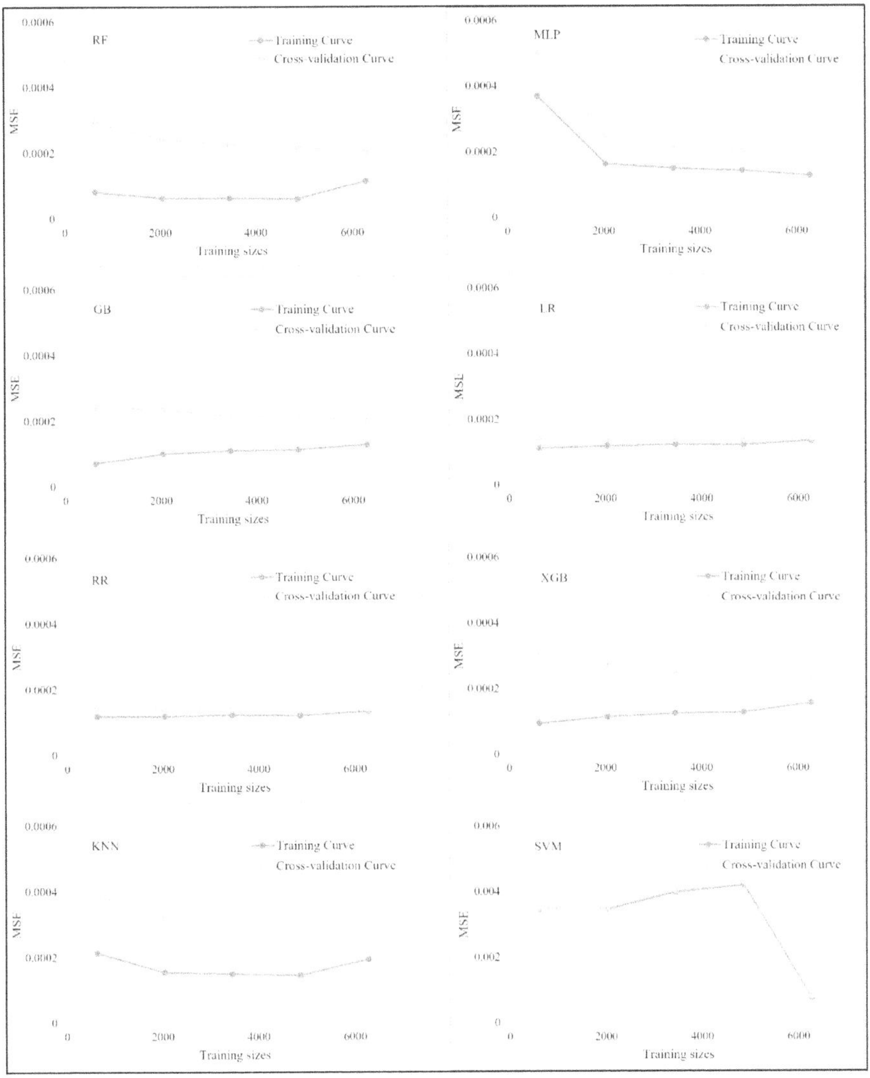

APPENDIX 10.E
Learning curves of machine learning algorithms.

APPENDIX 10.F

Hyperparameter Optimisation Results of Machine Learning Algorithms

ML Algorithms	Hyperparameters	Possible Values	Optimal Values
RF	n_estimators	[100, 150, 200]	150
	max_depth	[10, 15, 20]	20
	min_samples_split	[2, 5, 10]	2
	min_samples_leaf	[1, 2, 4]	4
	max_features	['sqrt', 'log2']	log2
MLP	hidden_layer_sizes	[(50, 50, 50), (50, 100, 50), (100,)]	(50, 100, 50)
	activation	['tanh', 'relu']	relu
	solver	['sgd', 'adam']	adam
	alpha	[0.0001, 0.05]	0.0001
	learning_rate	['constant', 'adaptive']	adaptive
GB	n_estimators	[100, 150, 200, 250, 300]	250
	learning_rate	[0.01, 0.05, 0.1]	0.01
	max_depth	[3, 4, 5]	4
	min_samples_split	[2, 5, 10]	10
	min_samples_leaf	[1, 2, 4]	2
	subsample	[0.8, 0.9, 1.0]	0.9
LR	fit_intercept	[True, False]	True
RR	alpha	[0.1, 1, 10, 100]	0.1
	fit_intercept	[True, False]	True
XGB	n_estimators	[50, 100, 150, 200, 250]	200
	learning_rate	[0.01, 0.05, 0.1]	0.01
	max_depth	[3, 4, 5]	5
	subsample	[0.8, 0.9, 1.0]	0.8
	colsample_bytree	[0.8, 0.9, 1.0]	0.8
	gamma	[0, 0.1, 0.2]	0
	reg_alpha	[0, 0.01, 0.1]	0.01
	reg_lambda	[1, 0.1, 0.01]	1
KNN	n_neighbors	[6, 7, 8, 9, 10, 11, 12, 13, 14, 15]	7
	weights	['uniform', 'distance']	uniform
	metric	['euclidean', 'manhattan', 'minkowski']	manhattan
SVM	C	[0.1, 1, 10, 100, 1000]	1000
	epsilon	[0.05, 0.1, 0.15, 0.2, 0.25, 0.3, 0.4, 0.5, 0.6, 0.7]	0.1
	kernel	['linear', 'poly', 'rbf', 'sigmoid']	poly
	gamma	['scale', 'auto', 0.001, 0.01, 0.1, 1]	0.1
	degree	[2, 3, 4, 5] For 'poly' kernel	2
	coef0	[0.0, 0.5, 1, 2, 3] For 'poly' and 'sigmoid' kernels	2

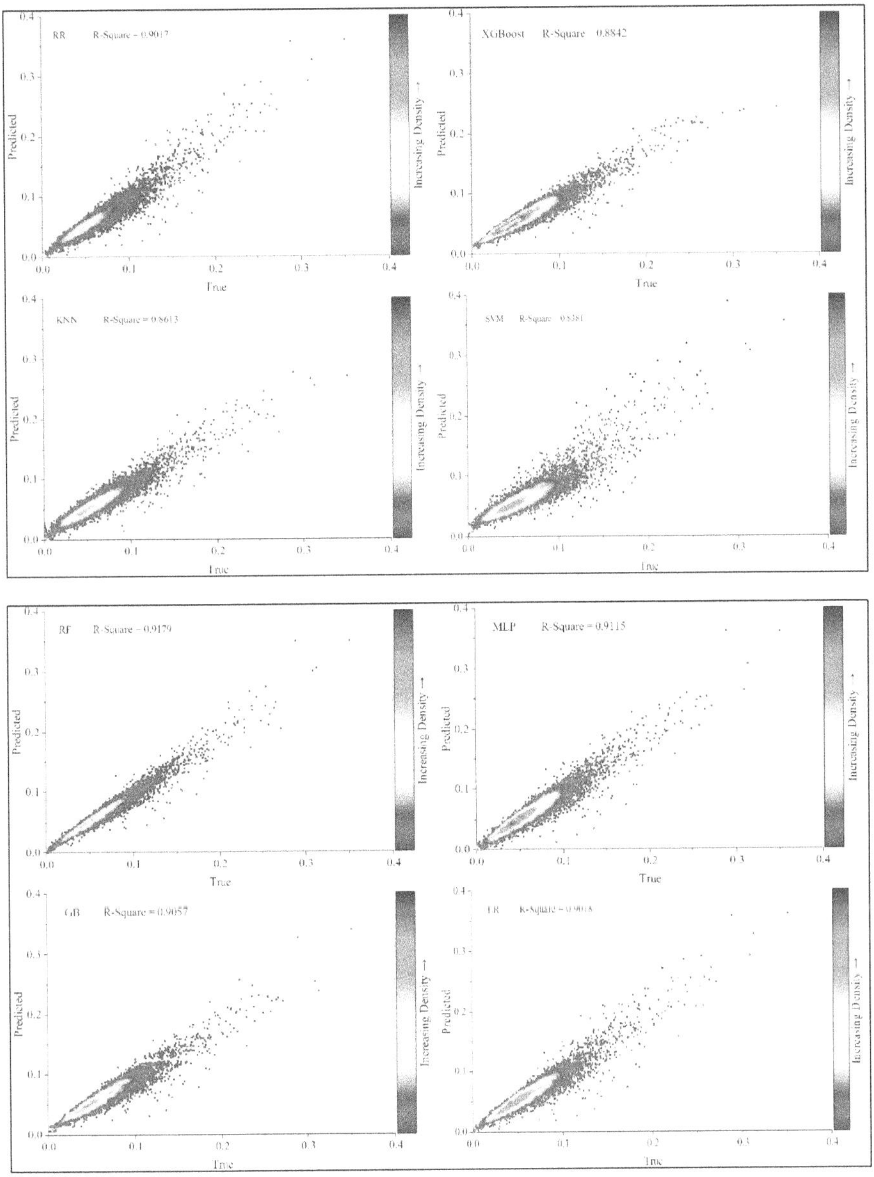

APPENDIX 10.G
Training results of machine learning algorithms.

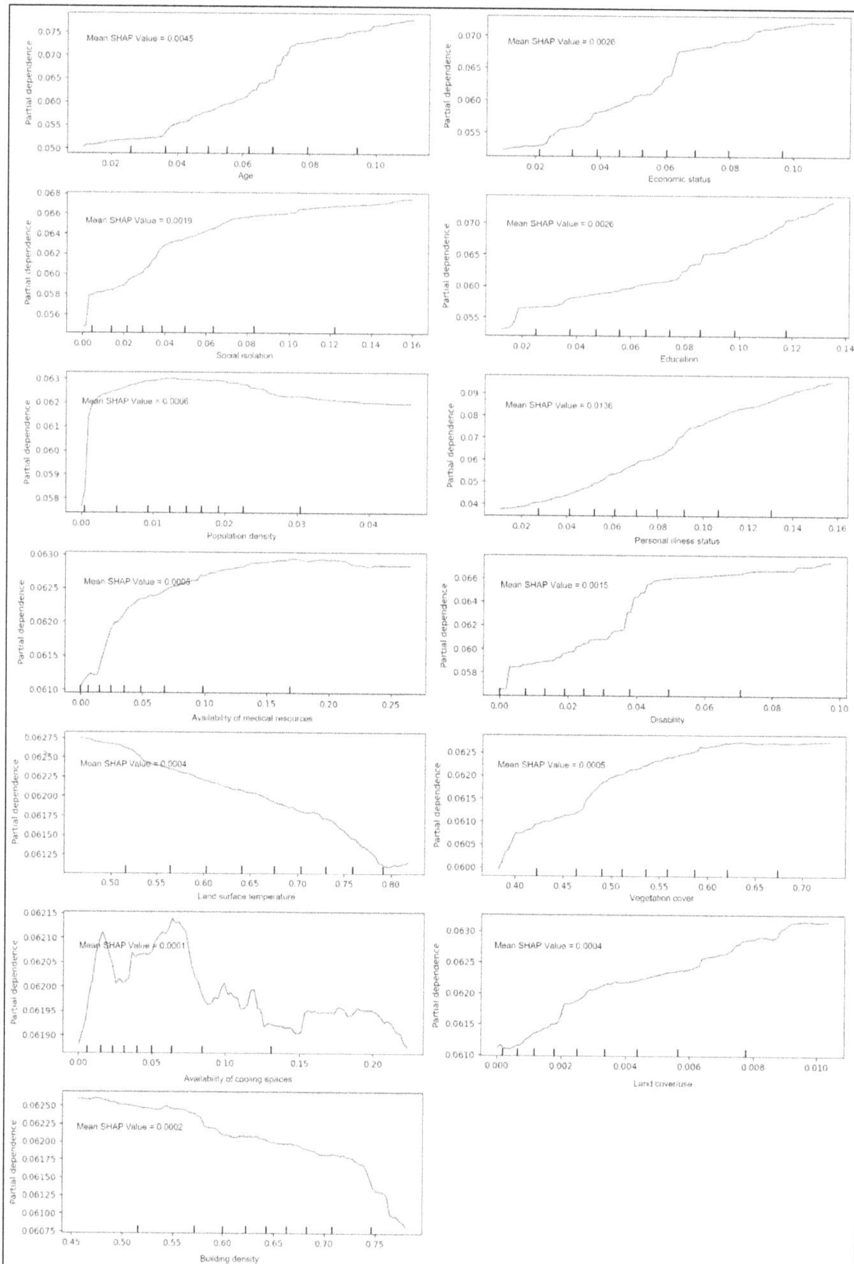

APPENDIX 10.H
Patial Dependence Plots and Mean SHAP Values of Each Heat-related Indicator.

Index

Note: Page numbers in **bold** and *italics* refer to tables and figures, respectively.

For Product Safety Concerns and Information please contact our EU representative GPSR@taylorandfrancis.com
Taylor & Francis Verlag GmbH, Kaufingerstraße 24, 80331 München, Germany

www.ingramcontent.com/pod-product-compliance
Lightning Source LLC
Chambersburg PA
CBHW070246230726
48635CB00030B/331
* 9 7 8 1 0 4 1 2 3 2 3 4 6 *